T0155901

Graduate Texts in Mathematics **123**

*Readings in Mathematics*

**Springer**
*New York*
*Berlin*
*Heidelberg*
*Barcelona*
*Budapest*
*Hong Kong*
*London*
*Milan*
*Paris*
*Santa Clara*
*Singapore*
*Tokyo*

# Graduate Texts in Mathematics
*Readings in Mathematics*

Ebbinghaus/Hermes/Hirzebruch/Koecher/Mainzer/Neukirch/Prestel/Remmert: *Numbers*
Fulton/Harris: *Representation Theory: A First Course*
Remmert: *Theory of Complex Functions*


# Undergraduate Texts in Mathematics
*Readings in Mathematics*

Anglin: *Mathematics: A Concise History and Philosophy*
Anglin/Lambek: *The Heritage of Thales*
Bressoud: *Second Year Calculus*
Hairer/Wanner: *Analysis by Its History*
Hämmerlin/Hoffmann: *Numerical Mathematics*
Isaac: *The Pleasures of Probability*
Samuel: *Projective Geometry*

H.-D. Ebbinghaus   H. Hermes
F. Hirzebruch   M. Koecher   K. Mainzer
J. Neukirch   A. Prestel   R. Remmert

# Numbers

With an Introduction by K. Lamotke
Translated by H.L.S. Orde
Edited by J.H. Ewing

With 24 Illustrations

 Springer

Heinz-Dieter Ebbinghaus
Hans Hermes
Mathematisches Institut
  Universität Freiburg
Albertstraße 23b, D-79104
  Freiburg, Germany

Friedrich Hirzebruch
Man-Planck-Institut für
  Mathematik
Gottfried-Claren-Straße 26
D-53225 Bonn, Germany

Klaus Lamotke (*Editor of*
  *German Edition*)
Mathematisches Institut
  der Universität zu Köln
Weyertal 86–90, D50931
  Köln, Germany

Max Koecher (1924–1990)
Reinhold Remmert
Mathematisches Institut
  Universität Münster
Einsteinstraße 62
D-48149 Münster, Germany

Klaus Mainzer
Lehrstuhl für Philosophie und
  Wissenschaftstheorie
Universität Augsburg
Universitätsstraße 10
D-86195 Augsburg, Germany

H.L.S. Orde (*Translator*)
Bressenden
Biddenden near Ashford
Kent TN27 8DU, UK

Jürgen Neukirch
Fachbereich Mathematik
Universitätsstraße 31
D-93053 Regensburg, Germany

Alexander Prestel
Fakultät für Mathematik
  Universität Konstanz
Postfach 5560, D-78434
  Konstanz, Germany

John H. Ewing (*Editor of*
  *English Edition*)
Department of Mathematics
Indiana University
Bloomington, IN 47405, USA

*Editorial Board*

Mathematics Subject Classification (1991): 11-XX, 11-03

Library of Congress Cataloging-in Publication Data
Zahlen, Grundwissen Mathematik 1. English
    Numbers / Heinz-Dieter Ebbinghaus . . . [et al.]: with an
  introduction by Klaus Lamotke; translated by H.L.S. Orde; edited
  by John H. Ewing.
       p.   cm.—(Readings in mathematics)
    Includes bibliographical references.
    ISBN 0-387-97497-0
    1. Number theory.   I. Ebbinghaus, Heinz-Dieter.   II. Ewing,
John H.   III. Series: Graduate texts in mathematics. Readings
in mathematics.
QA241.Z3413 1991
512'.7—dc20                                    89-48588

Printed on acid-free paper.

This book is a translation of the second edition of *Zahlen*, Grundwissen Mathematik 1, Springer-Verlag, 1988. The present volume is the first softcover edition of the previously published hardcover version (ISBN 0-387-97202-1).

Camera-ready copy prepared using LaT$_E$X.
Printed and bound by R.R. Donnelley & Sons, Harrisonburg, Virginia.
Printed in the United States of America.

9 8 7 6 5 4 3 (Corrected third printing, 1995)

ISBN 0-387-97497-0 Springer-Verlag New York Berlin Heidelberg
ISBN 3-540-97497-0 Springer-Verlag Berlin Heidelberg New York

# Preface to the English Edition

A book about numbers sounds rather dull. This one is not. Instead it is a lively story about one thread of mathematics—the concept of "number"—told by eight authors and organized into a historical narrative that leads the reader from ancient Egypt to the late twentieth century. It is a story that begins with some of the simplest ideas of mathematics and ends with some of the most complex. It is a story that mathematicians, both amateur and professional, ought to know.

Why write about numbers? Mathematicians have always found it difficult to develop broad perspective about their subject. While we each view our specialty as having roots in the past, and sometimes having connections to other specialties in the present, we seldom see the panorama of mathematical development over thousands of years. *Numbers* attempts to give that broad perspective, from hieroglyphs to $K$-theory, from Dedekind cuts to nonstandard analysis. Who first used the standard notation for $\pi$ (and who made it standard)? Who were the "quaternionists" (and can their zeal for quaternions tell us anything about the recent controversy concerning Chaos)? What happened to the endless supply of "hypercomplex numbers" or to quaternionic function theory? How can the study of maps from projective space to itself give information about algebras? How did mathematicians resurrect the "ghosts of departed quantities" by reintroducing infinitesimals after 200 years? How can games be numbers and numbers be games? This is mathematical culture, but it's not the sort of culture one finds in scholarly tomes; it's lively culture, meant to entertain as well as to inform.

This is not a book for the faint-hearted, however. While it starts with material that every undergraduate could (and should) learn, the reader is progressively challenged as the chapters progress into the twentieth century. The chapters often tell about people and events, but they primarily tell about mathematics. Undergraduates can certainly read large parts of this book, but mastering the material in late chapters requires work, even for mature mathematicians. This is a book that can be read on several levels, by amateurs and professionals alike.

The German edition of this book, *Zahlen*, has been quite successful. There was a temptation to abbreviate the English language translation by making it less complete and more compact. We have instead tried to produce a faithful translation of the entire original, which can serve as a scholarly reference as well as casual reading. For this reason, quotations

are included along with translations and references to source material in
foreign languages are included along with additional references (usually
more recent) in English.

Translations seldom come into the world without some labor pains. Authors and translators never agree completely, especially when there are
eight authors and one translator, all of whom speak both languages. My
job was to act as referee in questions of language and style, and I did so in
a way that likely made neither side happy. I apologize to all.

Finally, I would like to thank my colleague, Max Zorn, for his helpful
advice about terminology, especially his insistence on the word "octonions"
rather than "octaves."

March 1990                                                         John Ewing

# Preface to Second Edition

The welcome which has been given to this book on numbers has pleasantly surprised the authors and the editor. The scepticism which some of us had felt about its concept has been dispelled by the reactions of students, colleagues and reviewers. We are therefore very glad to bring out a second edition—much sooner than had been expected. We have willingly taken up the suggestion of readers to include an additional chapter by J. NEUKIRCH on $p$-adic numbers. The chapter containing the theorems of FROBENIUS and HOPF has been enlarged to include the GELFAND–MAZUR theorem. We have also carefully revised all the other chapters and made some improvements in many places. In doing so we have been able to take account of many helpful comments made by readers for which we take this opportunity of thanking them. P. ULLRICH of Münster who had already prepared the name and subject indexes for the first edition has again helped us with the preparation of the second edition and deserves our thanks.

Oberwolfach, March 1988                    Authors and Publisher

# Preface to First Edition

The *basic mathematical knowledge* acquired by every mathematician in the course of his studies develops into a unified whole only through an awareness of the multiplicity of relationships between the individual mathematical theories. Interrelationships between the different mathematical disciplines often reveal themselves by studying historical development. One of the main underlying aims of this series is to make the reader aware that mathematics does not consist of isolated theories, developed side by side, but should be looked upon as an organic whole.

The present book on numbers represents a departure from the other volumes of the series inasmuch as seven authors and an editor have together contributed thirteen chapters. In conversations with one another the authors agreed on their contributions, and the editor endeavored to bring them into harmony by reading the contributions with a critical eye and holding subsequent discussions with the authors. The other volumes of the series can be studied independently of this one.

While it is impossible to name here all those who have helped us by their comments, we should nevertheless like to mention particularly Herr Gericke (of Freiburg) who helped us on many occasions to present the historical development in its true perspective.

K. Peters (at that time with Springer-Verlag) played a vital part in arranging the first meeting between the publisher and the authors. The meetings were made possible by the financial support of the Volkswagen Foundation and Springer-Verlag, as well as by the hospitality of the Mathematical Research Institute in Oberwolfach.

To all of these we extend our gratitude.

Oberwolfach, July 1983                              Authors and Editor

# Contents

*Preface to the English Edition*     v

*Preface to Second Edition*     vii

*Preface to First Edition*     ix

*Introduction*, K. Lamotke     1

**Part A. From the Natural Numbers, to the Complex Numbers, to the *p*-adics**     7

*Chapter 1. Natural Numbers, Integers, and Rational Numbers.* K. Mainzer     9

§1.   Historical     9
    1. Egyptians and Babylonians. 2. Greece. 3. Indo-Arabic Arithmetical Pratice. 4. Modern Times
§2.   Natural Numbers     14
    1. Definition of the Natural Numbers. 2. The Recursion Theorem and the Uniqueness of ℕ. 3. Addition, Multiplication and Ordering of the Natural Numbers. 4. PEANO's Axioms
§3.   The Integers     19
    1. The Additive Group ℤ. 2. The Integral Domain ℤ. 3. The Order Relation in ℤ
§4.   The Rational Numbers     22
    1. Historical. 2. The Field ℚ. 3. The Ordering of ℚ
References     23

*Chapter 2. Real Numbers.* K. Mainzer     27

§1.   Historical     27
    1. HIPPASUS and the Pentagon. 2. EUDOXUS and the Theory of Proportion. 3. Irrational Numbers in Modern Mathematics. 4. The Formulation of More Precise Definitions in the Nineteenth Century
§2.   DEDEKIND Cuts     36
    1. The Set ℝ of Cuts. 2. The Order Relation in ℝ.

3. Addition in $\mathbb{R}$. 4. Multiplication in $\mathbb{R}$

§3.  Fundamental Sequences                                                                   39
     1. Historical Remarks. 2. CAUCHY's Criterion for
     Convergence. 3. The Ring of Fundamental Sequences.
     4. The Residue Class Field $F/N$ of Fundamental Sequences
     Modulo the Null Sequence. 5. The Completely Ordered Residue
     Class Field $F/N$

§4.  Nesting of Intervals                                                                     43
     1. Historical Remarks. 2. Nested Intervals and Completeness

§5.  Axiomatic Definition of Real Numbers                                                     46
     1. The Natural Numbers, the Integers, and the Rational
     Numbers in the Real Number Field. 2. Completeness Theorem.
     3. Existence and Uniqueness of the Real Numbers

References                                                                                    51


Chapter 3. Complex Numbers. R. Remmert                                                        55


§1.  Genesis of the Complex Numbers                                                           56
     1. CARDANO (1501–1576). 2. BOMBELLI (1526–1572).
     3. DESCARTES (1596–1650), NEWTON (1643–1727)
     and LEIBNIZ (1646–1716). 4. EULER (1707–1783).
     5. WALLIS (1616–1703), WESSEL (1745–1818) and
     ARGAND (1768–1822). 6. GAUSS (1777–1855).
     7. CAUCHY (1789–1857). 8. HAMILTON (1805–1865).
     9. Later Developments

§2.  The Field $\mathbb{C}$                                                                   65
     1. Definition by Pairs of Real Numbers. 2. The Imaginary
     Unit $i$. 3. Geometric Representation. 4. Impossibility of
     Ordering the Field $\mathbb{C}$. 5. Representation by Means
     of $2 \times 2$ Real Matrices

§3.  Algebraic Properties of the Field $\mathbb{C}$                                           71
     1. The Conjugation $\mathbb{C} \to \mathbb{C}$, $z \mapsto \bar{z}$. 2. The Field
     Automorphisms of $\mathbb{C}$. 3. The Natural Scalar Product
     $\mathrm{Re}(w\bar{z})$ and Euclidean Length $|z|$. 4. Product Rule and
     the "Two Squares" Theorem. 5. Quadratic Roots and
     Quadratic Equations. 6. Square Roots and $n$th Roots

§4.  Geometric Properties of the Field $\mathbb{C}$                                           78
     1. The Identity $\langle w, z \rangle^2 + \langle iw, z \rangle^2 = |w|^2 |z|^2$. 2. Cosine Theorem
     and the Triangle Inequality. 3. Numbers on Straight Lines
     and Circles. Cross-Ratio. 4. Cyclic Quadrilaterals and Cross-
     Ratio. 5. PTOLEMY's Theorem. 6. WALLACE's Line.

§5.  The Groups $O(\mathbb{C})$ and $SO(2)$                                                   85
     1. Distance Preserving Mappings of $\mathbb{C}$. 2. The Group $O(\mathbb{C})$.
     3. The Group $SO(2)$ and the Isomorphism $S^1 \to SO(2)$.

4. Rational Parametrization of Properly Orthogonal 2 × 2
Matrices.

§6.     Polar Coordinates and *n*th Roots                                       89
1. Polar Coordinates. 2. Multiplication of Complex Numbers
in Polar Coordinates. 3. DE MOIVRE's Formula. 4. Roots
in Unity.

*Chapter 4. The Fundamental Theorem of Algebra.*
R. Remmert                                                                      97

§1.     On the History of the Fundamental Theorem                                98
1. GIRARD (1595–1632) and DESCARTES (1596–1650).
2. LEIBNIZ (1646–1716). 3. EULER (1707–1783). 4. D'ALEMBERT
(1717–1783). 5. LAGRANGE (1736–1813) and LAPLACE
(1749–1827). 6. GAUSS's Critique. 7. GAUSS's Four Proofs.
8. ARGAND (1768–1822) and CAUCHY (1798–1857). 9. The
Fundamental Theorem of Algebra: Then and Now.
10. Brief Biographical Notes on Carl Friedrich GAUSS
§2.     Proof of the Fundamental Theorem Based on ARGAND                        111
1. CAUCHY's Minimum Theorem. 2. Proof of the Fundamental
Theorem. 3. Proof of ARGAND's Inequality. 4. Variant of the
Proof. 5. Constructive Proofs of the Fundamental Theorem.
§3.     Application of the Fundamental Theorem                                  115
1. Factorization Lemma. 2. Factorization of Complex
Polynomials. 3. Factorization of Real Polynomials. 4. Existence
of Eigenvalues. 5. Prime Polynomials in $\mathbb{C}[Z]$ and $\mathbb{R}[X]$.
6. Uniqueness of $\mathbb{C}$. 7. The Prospects for "Hypercomplex
Numbers."
Appendix. Proof of the Fundamental Theorem, after LAPLACE                       120
1. Results Used. 2. Proof. 3. Historical Note

*Chapter 5. What is $\pi$?* R. Remmert                                          123

§1.     On the History of $\pi$                                                 124
1. Definition by Measuring a Circle. 2. Practical Approxi-
mations. 3. Systematic Approximation. 4. Analytical Formulae.
5. BALTZER's Definition. 6. LANDAU and His Contemporary
Critics
§2.     The Exponential Homomorphism exp: $\mathbb{C} \to \mathbb{C}^{\times}$ 131
1. The Addition Theorem. 2. Elementary Consequences.
3. Epimorphism Theorem. 4. The Kernel of the Exponential
Homomorphism. Definition of $\pi$. Appendix. Elementary Proof
of Lemma 3.
§3.     Classical Characterizations of $\pi$                                    137
1. Definitions of $\cos z$ and $\sin z$. 2. Addition Theorem.

3. The Number $\pi$ and the Zeros of $\cos z$ and $\sin z$. 4. The
Number $\pi$ and the Periods of $\exp z$, $\cos z$ and $\sin z$. 5. The
Inequality $\sin y > 0$ for $0 < y < \pi$ and the Equation $e^{i\frac{\pi}{2}} = i$.
6. The Polar Coordinate Epimorphism $p: \mathbb{R} \to S^1$. 7. The
Number $\pi$ and the Circumference and Area of a Circle.

§4.    Classical Formulae for $\pi$                                    142
       1. LEIBNIZ's Series for $\pi$. 2. VIETA's Product Formula for $\pi$.
       3. EULER's Product for the Sine and WALLIS's Product for $\pi$.
       4. EULER's Series for $\pi^2, \pi^4, \ldots$. 5. The WEIERSTRASS
       Definition of $\pi$. 6. The Irrationality of $\pi$ and Its Continued
       Fraction Expansion. 7. Transcendence of $\pi$.

*Chapter 6. The p-Adic Numbers.* J. Neukirch                           155

§1.    Numbers as Functions                                            155
§2.    The Arithmetic Significance of the $p$-Adic Numbers             162
§3.    The Analytical Nature of $p$-Adic Numbers                       166
§4.    The p-Adic Numbers                                              173
References                                                             177

**Part B. Real Division Algebras**                                     179

*Introduction*, M. Koecher, R. Remmert                                 181

*Repertory. Basic Concepts from the Theory of Algebras*,
       M. Koecher, R. Remmert                                          183

       1. Real Algebras. 2. Examples of Real Algebras. 3. Subalgebras
       and Algebra Homomorphisms. 4. Determination of All One-
       Dimensional Algebras. 5. Division Algebras. 6. Construction
       of Algebras by Means of Bases

*Chapter 7. Hamilton's Quaternions.* M. Koecher, R. Remmert            189

Introduction                                                           189
§1.    The Quaternion Algebra $\mathbb{H}$                             194
       1. The Algebra $\mathbb{H}$ of the Quaternions. 2. The Matrix
       Algebra $\mathcal{H}$ and the Isomorphism $F: \mathbb{H} \to \mathcal{H}$. 3. The
       Imaginary Space of $\mathbb{H}$. 4. Quaternion Product, Vector
       Product and Scalar Product. 5. Noncommutativity of $\mathbb{H}$. The
       Center. 6. The Endomorphisms of the $\mathbb{R}$-Vector Space $\mathbb{H}$.
       7. Quaternion Multiplication and Vector Analysis. 8. The
       Fundamental Theorem of Algebra for Quaternions.
§2.    The Algebra $\mathbb{H}$ as a Euclidean Vector Space            206
       1. Conjugation and the Linear Form $\mathbb{R}e$. 2. Properties of

the Scalar Product. 3. The "Four Squares Theorem".
4. Preservation of Length, and of the Conjugacy Relation Under
Automorphisms. 5. The Group $S^3$ of Quaternions of
Length 1. 6. The Special Unitary Group $SU(2)$ and the
Isomorphism $S^3 \to SU(2)$.

§3.　The Orthogonal Groups $O(3)$, $O(4)$ and Quaternions　　　　213
1. Orthogonal Groups. 2. The Group $O(\mathbb{H})$. CAYLEY's
Theorem. 3. The Group $O(\operatorname{Im}\mathbb{H})$. HAMILTON's Theorem.
4. The Epimorphisms $S^3 \to SO(3)$ and $S^3 \times S^3 \to SO(4)$.
5. Axis of Rotation and Angle of Rotation. 6. EULER's
Parametric Representation of $SO(3)$.

*Chapter 8. The Isomorphism Theorems of FROBENIUS, HOPF
and GELFAND–MAZUR. M. Koecher, R. Remmert*　　　　221

Introduction　　　　　　　　　　　　　　　　　　　　　　　221
§1.　Hamiltonian Triples in Alternative Algebras　　　　　　223
1. The Purely Imaginary Elements of an Algebra.
2. Hamiltonian Triple. 3. Existence of Hamiltonian Triples in
Alternative Algebras. 4. Alternative Algebras.
§2.　FROBENIUS's Theorem　　　　　　　　　　　　　　　227
1. FROBENIUS's Lemma. 2. Examples of Quadratic Algebras.
3. Quaternions Lemma. 4. Theorem of FROBENIUS (1877)
§3.　HOPF's Theorem　　　　　　　　　　　　　　　　　230
1. Topologization of Real Algebras. 2. The Quadratic Mapping
$\mathcal{A} \to \mathcal{A}$, $x \mapsto x^2$. HOPF's Lemma. 3. HOPF's Theorem.
4. The Original Proof by HOPF. 5. Description of All
2-Dimensional Algebras with Unit Element
§4.　The GELFAND–MAZUR Theorem　　　　　　　　　　238
1. BANACH Algebras. 2. The Binomial Series. 3. Local
Inversion Theorem. 4. The Multiplicative Group $\mathcal{A}^\times$. 5. The
GELFAND–MAZUR Theorem. 6. Structure of Normed Associative
Division Algebras. 7. The Spectrum. 8. Historical Remarks
on the GELFAND–MAZUR Theorem. 9. Further Developments

*Chapter 9. CAYLEY Numbers or Alternative Division Algebras.
M. Koecher, R. Remmert*　　　　　　　　　　　　　249

§1.　Alternative Quadratic Algebras　　　　　　　　　　250
1. Quadratic Algebras. 2. Theorem on the Bilinear Form.
3. Theorem on the Conjugation Mapping. 4. The Triple
Product Identity. 5. The Euclidean Vector Space $\mathcal{A}$ and
the Orthogonal Group $O(\mathcal{A})$
§2.　Existence and Properties of Octonions　　　　　　　256
1. Construction of the Quadratic Algebra $\mathbb{O}$ of Octonions.

2. The Imaginary Space, Linear Form, Bilinear Form, and
Conjugation of $\mathbb{O}$. 3. $\mathbb{O}$ as an Alternative Division
Algebra. 4. The "Eight-Squares" Theorem. 5. The Equation
$\mathbb{O} = \mathbb{H} \oplus \mathbb{H}p$. 6. Multiplication Table for $\mathbb{O}$

§3.   Uniqueness of the CAYLEY Algebra                              261
      1. Duplication Theorem. 2. Uniqueness of the CAYLEY Algebra
      (Zorn 1933). 3. Description of $\mathbb{O}$ by ZORN's Vector Matrices


*Chapter 10. Composition Algebras. HURWITZ's Theorem—*
*Vector-Product Algebras. M. Koecher, R. Remmert*            265


§1.   Composition Algebras                                         267
      1. Historical Remarks on the Theory of Composition.
      2. Examples. 3. Composition Algebras with Unit Element.
      4. Structure Theorem for Composition Algebras with Unit
      Element
§2.   Mutation of Composition Algebras                             272
      1. Mutation of Algebras. 2. Mutation Theorem for Finite-
      Dimensional Composition Algebras. 3. HURWITZ's Theorem
      (1898)
§3.   Vector-Product Algebras                                      275
      1. The Concept of a Vector-Product Algebra. 2. Construction
      of Vector-Product Algebras. 3. Specification of all Vector-
      Product Algebras. 4. MALCEV-Algebras. 5. Historical Remarks


*Chapter 11. Division Algebras and Topology.*
*F. Hirzebruch*                                               281


§1.   The Dimension of a Division Algebra Is a Power of 2          281
      1. Odd Mappings and HOPF's Theorem. 2. Homology and
      Cohomology with Coefficients in $F_2$. 3. Proof of HOPF's
      Theorem. 4. Historical Remarks on Homology and Cohomology
      Theory. 5. STIEFEL's Characteristic Homology Classes
§2.   The Dimension of a Division Algebra Is 1, 2, 4 or 8          290
      1. The mod 2 Invariants $\alpha(f)$. 2. Parallelizability of
      Spheres and Division Algebras. 3. Vector Bundles.
      4. WHITNEY's Characteristic Cohomology Classes. 5. The Ring
      of Vector Bundles. 6. Bott Periodicity. 7. Characteristic
      Classes of Direct Sums and Tensor Products. 8. End of
      the Proof. 9. Historical Remarks
§3.   Additional Remarks                                           299
      1. Definition of the HOPF Invariant. 2. The HOPF
      Construction. 3. ADAMS's Theorem on the HOPF Invariants.
      4. Summary. 5. ADAMS's Theorem About Vector Fields
      on Spheres

References                                                                   301

**Part C. Infinitesimals, Games, and Sets**                                  303

*Chapter 12. Nonstandard Analysis.* A. Prestel                               305

§1.  Introduction                                                            305
§2.  The Nonstandard Number Domain $^*\mathbb{R}$                            309
     1. Construction of $^*\mathbb{R}$. 2. Properties of $^*\mathbb{R}$
§3.  Features Common to $\mathbb{R}$ and $^*\mathbb{R}$                      316
§4.  Differential and Integral Calculus                                      321
     1. Differentiation. 2. Integration
Epilogue                                                                     326
References                                                                   327

*Chapter 13. Numbers and Games.* H. Hermes                                   329

§1.  Introduction                                                            329
     1. The Traditional Construction of the Real Numbers.
     2. The CONWAY Method. 3. Synopsis
§2.  CONWAY Games                                                            331
     1. Discussion of the DEDEKIND Postulates. 2. CONWAY's
     Modification of the DEDEKIND Postulates. 3. CONWAY Games
§3.  Games                                                                   334
     1. The Concept of a Game. 2. Examples of Games. 3. An
     Induction Principle for Games
§4.  On the Theory of Games                                                  336
     1. Winning Strategies. 2. Positive and Negative Games.
     3. A Classification of Games
§5.  A Partially Ordered Group of Equivalent Games                          339
     1. The Negative of a Game. 2. The Sum of Two Games.
     3. Isomorphic Games. 4. A Partial Ordering of Games.
     5. Equality of Games
§6.  Games and CONWAY Games                                                  343
     1. The Fundamental Mappings. 2. Extending to CONWAY
     Games the Definitions of the Relations and Operations Defined
     for Games. 3. Examples
§7.  CONWAY Numbers                                                          346
     1. The CONWAY Postulates (C1) and (C2). 2. Elementary
     Properties of the Order Relation. 3. Examples
§8.  The Field of CONWAY Numbers                                            349
     1. The Arithmetic Operations for Numbers. 2. Examples.
     3. Properties of the Field of Numbers
References                                                                   353

*Chapter 14. Set Theory and Mathematics.*
        H.-D. Ebbinghaus                                                355

Introduction                                                           355
§1.    Sets and Mathematical Objects                                   358
       1. Individuals and More Complex Objects. 2. Set
       Theoretical Definitions of More Complex Objects.
       3. Urelements as Sets
§2.    Axiom Systems of Set Theory                                     363
       1. The RUSSELL Antinomy. 2. ZERMELO's and the ZERMELO–
       FRAENKEL Set Theory. 3. Some Consequences. 4. Set
       Theory with Classes
§3.    Some Metamathematical Aspects                                   372
       1. The VON NEUMANN Hierarchy. 2. The Axiom of Choice.
       3. Independence Proofs
Epilogue                                                               378
References                                                             378

*Name Index*                                                           381

*Subject Index*                                                        387

*Portraits of Famous Mathematicians*                                   393

# Introduction

*K. Lamotke*

Mathematics, according to traditional opinion, deals with numbers and figures. In this book we do not begin, as EUCLID began, with figures but with numbers.

Mathematical research over the last hundred years has created abstract theories, such as set theory, general algebra, and topology, whose ideas have now penetrated into the teaching of mathematics at the elementary level. This development has not been ignored by the authors of this book; indeed, they have willingly taken advantage of it in that the authors assume the reader to be familiar with the basic concepts of (naive) set theory and algebra. On the other hand, a first volume on numbers should emphasize the fact that modern research in mathematics and its applications is, to a considerable extent, linked to what was created in the past. In particular, the traditional number system is the most important foundation of all mathematics.

The book that we now present is divided into three parts, of which the first, which may be regarded as the heart, describes the structure of the number-system, from the natural numbers to the complex and *p*-adic numbers. The second part deals with its further development to 'hypercomplex numbers,' while in the third part two relatively new extensions of the real number system are presented. The six chapters of the first part cover those parts of the subject of 'numbers' that every mathematician ought to have heard or read about at some time. The other two parts are intended to satisfy the appetite of a reader who is curious to learn something beyond the basic facts. On the whole, "the structure of number systems" would be a more accurate description of the content of this book.

We should now like to say a few words in more detail about the various contributions, the aims that the authors have set out to achieve, and the reasons that have induced us to bring them together in the form in which they are presented here.

## PART A

Since the end of the last century it has been customary to construct the number system by beginning with the natural numbers and then extending the structure step-by-step to include the integers, the rational numbers, the real numbers, and finally the complex numbers. That is not, however, the way in which the concept of number developed historically. Even in ancient times, the rational numbers (fractions and ratios) and certain irrational numbers (such as $\pi$, the ratio of the circumference to the radius of a circle, and square-roots) were known in addition to the natural numbers. The system of (positive) rational and irrational numbers was also described theoretically by Greek philosophers and mathematicians, but it was done within the framework of an autonomous theory of commensurable and incommensurable proportions, and it was not thought of as an extension of the natural numbers. It was not until after many centuries of working numerically with proportions that the realization dawned in the 17th century that a number is something that bears the same relationship to (the unit) one as a line segment bears to another given segment (of unit length). Negative numbers, which can be shown to have been in use in India in the 6th century, and complex numbers, which CARDAN took into consideration in 1545 as a solution of a quadratic equation, were still looked upon as questionable for a long time afterwards. In the course of the 19th century the construction that we use today began to emerge.

Each chapter contains a contribution that includes a description of the historical development of the fundamental concepts. These contributions are not intended to replace a history of the number concept, but are aimed at contributing towards a better understanding of the modern presentation by explaining the historical motivation.

In this sense, Chapter 1, §1 begins with the oldest of the representations of numbers that have been handed down to us by tradition, and leads into §2 in which the ideas involved in counting are given axiomatically following the methods introduced by DEDEKIND, by using the concepts of set-theory.

In the ensuing step-by-step construction of the number-system certain themes constantly recur. (1) The step from one stage to the next is prompted each time by the desire to solve problems that can be formulated but not solved in terms of numbers defined so far. (2) The number system of the next stage is constructed, with the help of the operations of set-theory, as an extension of the existing system designed to make the initial problem solvable. For this the following items are necessary. (3) The existing computational operations and relations must be carried over to the new system. (4) The validity of all the computational rules in the new context has to be checked. The processes (1) to (3) are always carried out, in the chapters that follow, but item (4) usually involves tedious verifications, which soon become a matter of routine. Here the authors allow themselves to carry out only a few of them by way of example, and to leave the rest as a routine exercise for the reader.

By the end of Chapter 1 the rational numbers have thus been reached. In Chapter 2, §2 they are extended to the real number system, by means of Dedekind cuts. The preceding §1 begins with the discovery of the irrational numbers by the Pythagoreans and describes the philosophical and mathematical attempts in earlier times that finally led to DEDEKIND's construction. CANTOR's method of completing the rational number system, through the use of fundamental sequences, is described in §3. Here the historical roots stretched back only a few decades, but the procedure turned out later to be fruitful, because valuation rings, metric spaces, topological vector spaces, and general uniform structures can all be completed in exactly the same way. The third approach to the real numbers, described in §4, follows WEIERSTRASS. It is based on the idea, going back to ancient times, of enclosing a number whose exact value is not easily determined, within small intervals bounded by rational numbers. This idea still finds application today in the estimation of errors in numerical computation.

By §2 of Chapter 2, a system of axioms for the real numbers has been formulated. In §5 it is shown that they characterize these numbers to within isomorphism. In that section the structure of the number system is reconstituted from these axioms, and numerous different formulations of the concept of the "completeness" of the real numbers are compared with one another.

Chapters 3 to 5 are devoted to the complex numbers. Using linear algebra as a tool, it is easy for us today to describe them as pairs of real numbers, which can be added like vectors and multiplied according to an explicitly specified rule. This definition, in §2 of Chapter 3, is preceded by a summary of the historical development that shows how it took 300 years from the discovery of the complex numbers until, with the advent of GAUSS, they became generally understood and accepted. One basic thought runs through the history until GAUSS: The complex numbers make possible the impossible. Above all, they make it possible to solve all equations of the second or higher degree. Chapter 4 is devoted to demonstrating this result, known as the fundamental theorem of algebra. Two proofs, going back to ARGAND and LAPLACE respectively, are presented which require no complex function theory.

As far as complex numbers are concerned, the reader may be surprised to find that the whole of Chapter 5 is devoted to the special number $\pi$. Now as explained in Chapter 3, and used in Chapter 4, the representation by polar co-ordinates is an essential feature of the complex number system. To provide a deeper understanding of this representation, the complex exponential function exp is treated in Chapter 5. This function is closely connected with $\pi$, because $\exp(z) = 1$ if and only if $z$ is an integral multiple of $2\pi i$. Indeed this relation serves as a definition of $\pi$, and all the other commonly used descriptions of $\pi$ (that is, as a number associated with the circle, as the value of an integral, as the limit of an infinite series or infinite product) may be deduced from it.

The complex numbers formed the point of departure for one of the greatest creations of 19th century mathematics, complex function theory.

In modern number theory, the $p$-adic numbers have equal importance with the reals. Chapter 6 contains two approaches to the $p$-adic numbers. At the beginning of the twentieth century, HENSEL created the $p$-adic numbers by modeling them on the power series and Laurent series of complex function theory. One can also view them, however, in a different way as a natural completion of the field of rationals. Just as the reals are the completion of the rationals using the usual absolute value, the $p$-adic numbers can be thought of as the completion when the absolute value is replaced by a $p$-adic valuation. We only hint at the importance of the $p$-adic numbers for number theory in this chapter.

## PART B

With the complex numbers the construction of the number system is in a sense completed. If, following the model provided by the complex numbers, which form a two-dimensional real vector-space, one tries to make higher-dimensional real vector spaces into hypercomplex number systems (nowadays usually called algebras), then either infinite dimension must be allowed or else familiar field axioms must be given up such as the commutativity or associativity of multiplication, or the possibility of performing division. If too many of such axioms are given up, then there is an overwhelming flood of new number systems. To act as a kind of flood barrier, in Part B of this book, we shall confine ourselves mostly to finite-dimensional systems in which division is possible.

The four-dimensional division algebra of quaternions, and the eight dimensional one of octonions, which were discovered shortly after one another in the year 1843, are discussed in detail in Chapters 7 and 9 respectively. Just as the complex numbers allow the Euclidean geometry of the plane to be described in an often amazingly simple way (§4, Chapter 3 contains a few samples), so the quaternions are suited to description of three- and four-dimensional geometry. All this is gone into in Chapter 7 as well.

The other chapters in Part B deal, from various points of view, with the uniqueness of the four algebras of the real numbers, the complex numbers, the quaternions and the octonions. If commutativity alone is abandoned, then the quaternion algebra is the only possibility (FROBENIUS 1877; proof in the second part of Chapter 8). If one retains commutativity but is prepared to give up associativity, real and complex numbers are the only possibilities (H. HOPF 1940; proof in the third part of Chapter 8). The proof uses non-trivial topological methods. By the same methods the theorem of GELFOND and MAZUR can be proved (1938; fourth part of Chapter 8): The real numbers, the complex numbers, and the quaternions are the only possible normal associative real division algebras, even when infinite-dimensional algebras are admitted. If both commutativity and associativity are abandoned but still a weaker form of associativity represented by the law $x(xy) = x^2y$ and $(xy)y = xy^2$ is retained, then the octonions represent the only possibility (ZORN 1933; proof at the end of Chapter 9).

Another characterization of the four algebras was found by HURWITZ in 1898; they are the only possible division algebras with unit element, which are at the same time Euclidean vector-spaces with a norm-preserving multiplication ($\|x\| \cdot \|y\| = \|x \cdot y\|$). This is closely connected with the fact that the product of two natural numbers, each of which is the sum of 2, 4 or 8 squares, is itself a sum of a like number of squares, and that the corresponding statement for $n$ squares is true only when $n = 2$, 4 or 8. Chapter 10 deals with these things.

So far all the results are given with proofs that assume some linear algebra, differential calculus of several variables, and the rudiments of algebra and topology. Chapter 11 deals with the most far-reaching result; namely, that finite-dimensional division algebras are possible only when the number of dimensions is 1, 2, 4 or 8. Here the conclusion can be drawn without any other assumption. This theorem was proved, to the great surprise of algebraists, in 1958 by BOTT, KERVAIRE and MILNOR, and moreover, as with HOPF's results, by topological methods. This time however the whole extensive apparatus of algebraic topology has to be employed, and in Chapter 11 only an outline of the proof can be sketched.

HAMILTON regarded his discovery of quaternions in the year 1843 as one of the most important events in the history of mathematics. However, it turned out, that quaternions (and even more so octonions) come far behind complex numbers in importance. Non-commutativity has proved to be an insurmountable obstacle to the creation of a quaternionic analysis.

## PART C

The real number system has appeared for some time to be a completed edifice from the standpoint of mathematical research, but some new ideas have emerged fairly recently.

In the year 1960 ROBINSON discovered how an infinitesimal calculus modelled on that of the 17th and 18th century, and operating with infinitesimal quantities, could be precisely defined and operated on a secure foundation. To do this, he extended the field of real numbers to an ordered field of non-standard numbers incorporating infinitely small as well as infinitely large numbers. The construction of this extension is described in Chapter 12. It requires no greater effort than, for example, CANTOR's construction of the real numbers (cf. §3 of Chapter 2); and the differential and integral calculus based on infinitesimal quantities will seem to some readers to be simpler and more intuitive than the customary methods. Unfortunately there is a price to be paid. All statements needing 'translation' from real numbers to non-standard numbers, have first to be expressed in a formal language; and this means that mathematicians need to delve rather more deeply into formal logic than most of them are accustomed to do.

CONWAY's ingenious idea is still more recent, about ten years later. He hit upon a way of defining a large ordered number field *ab initio* without any intermediate steps by a process of iterated Dedekind-cut operations, and to interpret the elements of this field as "games" that could be ordered by making use of the concept of a winning strategy. All this is defined and explained in Chapter 13.

In the two Chapters, 12 and 13, it is ideas in the main that are presented and we do not go into all the details. For Conway's construction, naive set theory does not entirely suffice. Chapter 14 therefore contains an account of the fundamental principles of the axiomatic set theory developed by ZERMELO and FRAENKEL. This chapter is also intended for a reader of the first two chapters of this book who, when the natural numbers and their extensions to this system are introduced, does not wish to rely on a naively understood set theory. From a strictly logical standpoint this chapter should be at the beginning, but we have taken heed of SCHILLER's advice (in a letter to GOETHE, dated the 5th February 1796): "Wo es die Sache leidet, halte ich es immer für besser, nicht mit dem Anfang anzufangen, der immer das Schwerste ist." which could be roughly translated as "I always think it better, whenever possible, not to begin at the beginning, as it is always the most difficult part."

# Part A

# From the Natural Numbers, to the Complex Numbers, to the $p$-adics

# 1

# Natural Numbers, Integers, and Rational Numbers

*K. Mainzer*

Die ganzen Zahlen hat der liebe Gott gemacht, alles andere ist Menschenwerk (KRONECKER, Jahresber. DMV 2, S. 19).

[God made the whole numbers, all the rest is the work of Man.]

Die Zahlen sind freie Schöpfungen des menschlichen Geistes, sie dienen als ein Mittel, um die Verschiedenheit der Dinge leichter und schärfer aufzufassen (DEDEKIND, Was sind und was sollen die Zahlen? Braunschweig 1887, S. III).

[Numbers are free creations of the human intellect, they serve as a means of grasping more easily and more sharply the diversity of things.]

## §1. HISTORICAL

**1. Egyptians and Babylonians.** Symbols for numbers are found in the earliest remains of human writing. Even in the *early stone age* we find them in the form of notches in bones or as marks on the walls of caves. It was the age when man lived as a hunter and today we can only speculate as to whether |||| for example was intended to represent the size of the kill. Number systems mark the beginning of arithmetic. The first documents go back to the earliest civilizations in the *valley of the Nile, Euphrates and Tigris.* Hieroglyphs for the numbers 10 000, 100 000 and 1 000 000 are to be found on a mace of King Narmer, of the first *Egyptian dynasty* (*circa* 3000 BC). The numbers are reproduced schematically below:

| 1 | 10 | 100 | 1000 |
|---|----|-----|------|
| I | ∩ | ℮ | ᛘ |

| 10 000 | 100 000 | 1 000 000 |
|--------|---------|-----------|

The pictures used may refer to practical occurrences connected with the relevant numbers; for example $\mathcal{C}$ may be a symbol for a measuring tape with 100 units. On the other hand it is also possible that the symbols represent objects whose initial letter is the same as that for the word for the corresponding number. New numbers are formed by an additive notation based on juxtaposition, for example, ⌇⌇⌇ = 221 000 or ❘∩ = 10 010. Thus addition and subtraction present no problem. For example, ∩ ‖ = 12 added to ∩❘ = 11 gives ∩∩‖‖ = 23. Multiplication and division are reduced to a succession of doubling and halving operations. The resulting fractions are expressed as sums of unit fractions (fractions whose numerator is 1), the sign ⌒ being used to indicate that the number symbol above which it is placed represents the denominator of a unit fraction. Thus for example the fraction 1/12 is written as ∩‖ (with ⌒ above). To represent the fraction 3/12, the calculation three times one-twelfth is performed as follows:

$$1 \quad \tfrac{1}{12} \quad (\text{that is once times } \tfrac{1}{12} = \tfrac{1}{12})$$

$$2 \quad \tfrac{1}{6} \quad (\text{doubling})$$

so that the fraction 3/12 is written as $\tfrac{1}{6}\tfrac{1}{12}$, that is, ‖‖ ∩‖.

To perform calculations of this kind with *general fractions,* one needs to be able to express the halves and doubles of unit fractions as sums of unit fractions with odd denominators. The Rhind papyrus (about 1650 BC) contains tables giving such decompositions of the fraction $2/n$ for odd integers $n$. (For details of Egyptian calculation, see the *Moscow papyrus* [28] and the *Rhind papyrus* [23].)

The *Babylonians* used cuneiform symbols on clay tablets. These were based on a mixed decimal and sexagesimal position notation: ▼ stood for 1, $60^1$, $60^2$, ...; while < stood for 10, $10 \cdot 60^1$, $10 \cdot 60^2$, ... and so on. A zero symbol was not always used by the Babylonians, and they never used a mark like our decimal point. In a positional notation the role of the zero is that of a sign marking a "gap." A sign of this kind, two small wedge marks ⦚ , is already to be found in an old Babylonian text from Susa (*Text* 12, p. 4), but only in isolated instances (TROPFKE [29], p. 28).

In the absence of such a sign, the positional value has to be deduced in each case from the context. Thus, for example, << ▼ < could mean any of the numbers $21 \cdot 60 + 10$ or $21 \cdot 60^2 + 10 \cdot 60^1$ or $21 \cdot 60^2 + 10$ and so on. Examples of sexagesimal fractions are <<< for $0.30 = 30/60 = 1/2$ or ▼▼▼<< for $0.64 = 6\tfrac{1}{60^1} + 40 \cdot \tfrac{1}{60^2} = \tfrac{1}{9}$. (For details of Babylonian calculation see NEUGEBAUER [20], BRUINS–RUTTEN [7].)

The Babylonians show themselves to have been highly talented arithmeticians and algebraists. They developed sophisticated tables for use in calculations involving multiplication and division, and for solving quadratic and cubic equations. They gave rules for solving mixed quadratic equations by the process of "completing the square" and even for solving mixed cubic equations with the help of tables of $x^2(x+1)$. We shall also be mentioning their methods of approximating the roots of equations in Chapter 2. At all events it is safe to assert that the Babylonians, with their skillful and ingenious methods of calculation exercised a considerable influence on the subsequent development of arithmetic and algebra.

**2. Greece.** The number system of the Greeks was decadic, though not positional. The *earlier system* used individual symbols for the decadic steps, which were the initial letters of the corresponding words for the numbers concerned. By combining the symbol for 5 with the other symbols, the intermediate steps of 50, 500, ... could be represented, so that the set of symbols ran as follows:

| I | Γ | Δ | Γ̄ | Η | Γ̄ | Χ | Γ̄ | Μ | Γ̄ |
|---|---|---|---|---|---|---|---|---|---|
| 1 | 5 | 10 | 50 | 100 | 500 | 1000 | 5000 | 10 000 | 50 000 |

The *later system* of representing numbers by letters (about 450 BC) was used in mathematical texts. It comprised the 24 letters of the standard Greek alphabet with three further symbols from oriental tradition:

$$
\begin{array}{lll}
1-9 & \alpha,\beta,\gamma,\delta,\varepsilon,\varsigma,\zeta,\eta,\theta & (\ \varsigma\ =6) \\
10-90 & \iota,\kappa,\lambda,\mu,\nu,\xi,o,\pi,\varsigma & (\ \varsigma\ =90) \\
100-900 & \rho,\sigma,\tau,\upsilon,\varphi,\chi,\psi,\omega,\text{ꙇ} & (\ \text{ꙇ}\ =900) \\
1000-9000 & {}_{\prime}\alpha, {}_{\prime}\beta,\ldots & \text{(written with a subscript accent} \\
& & \text{on the left)} \\
10\,000 & M & (M=Mυριάς)
\end{array}
$$

Addition of numbers was indicated by the juxtaposition of the corresponding symbols, so that for example $\iota\beta=10+2=12$, $\sigma\kappa\beta=200+20+2=222$, ${}_{\prime}\alpha\tau\varepsilon=1000+300+5=1305$. The number of tens of thousands (myriads) was written above the symbol $M$, so that, for example

$$
\overset{\beta}{M}\ {}_{\prime}\varepsilon\mu\gamma=25\,000+40+3=25\,043.
$$

Unit fractions were usually indicated by a superscript accent to the right of the letter denoting the denominator of the fraction. More general fractions were written in various different ways (for example, by writing the letter for the numerator underneath the letter for the denominator). The Greek system, unlike our decimal notation, was therefore not purely positional and calculation was rather tedious.

Alongside an arithmetic with numbers represented by symbols, one can find from an early stage a representation of numbers by *counters* (such as the beads of an abacus, pebbles and so on), which was a means by which arithmetical theorems were discovered. Thus ARISTOTLE mentions the Pythagorean EURYTOS who is said "to have determined what is the number (ἀριθμός) of what object and imitated the shapes of living things by pebbles (ψῆφοι) after the manner of those who bring numbers into the forms of triangle or square" (ARISTOTLE [1], 1092b, 10.12). For example, the odd numbers can be arranged in succession in the manner illustrated below to form the squares

$$
\begin{array}{cccc}
 & & \circ\ \circ\ \circ & \\
 & \circ\ \circ & \bullet\ \bullet\ \circ & \\
\circ & \bullet\ \circ & \bullet\ \bullet\ \circ & \cdots \\
1 & 1+3 & 1+3+5 &
\end{array}
$$

By dividing the squares into sections parallel to one of the diagonals and counting the number of pebbles in each line we can read off

$$2^2 = 1 + 2 + 1, \qquad 3^2 = 1 + 2 + 3 + 2 + 1,$$

and, in general,

$$n^2 = 1 + 2 + \cdots + n + \cdots + 2 + 1,$$

so that $1 + 2 + \cdots + (n-1) = \frac{1}{2}(n^2 - n)$ (ARISTOTLE [2], III 4, 203a, 13–15, BECKER [3], p. 34ff).

While the Egyptians and Babylonians contented themselves with developing highly sophisticated numerical techniques, the Pythagoreans became primarily interested in the philosophical significance of numbers. In their philosophy the entire universe was characterized by numbers and their relationships, and thus the problem arose of defining generally what a number was. EUCLID defines in the *Elements*, VII, 2, a number as "the multitude made up of units" having previously (*Elements*, VII, 1) said that a unit is "that by virtue of which each of existing things is called one." As a unit is not composed of units, neither EUCLID nor ARISTOTLE regard a unit as a number, but rather as "the basis of counting, or as the origin of number." There is an echo of this Euclidean definition in CANTOR's definition of the cardinal number as a set composed of nothing but units (CANTOR [8], p. 283).

Apart from this *definition of number*, which is oriented towards the idea of counting, one can also find in ARISTOTLE the following statement: that which is divisible into discrete parts is called πλῆθος (multitude), and the bounded (finite) multiplicity is called the number (ARISTOTLE [1], 1020a, 7.14).

The Greeks thus regarded as numbers, only the natural numbers excluding unity; fractions were treated as ratios of numbers, and irrational numbers as relationships between incommensurable magnitudes in geometry (cf. Chapter 2).

**3. Indo-Arabic Arithmetical Practice.** Between 300 BC and 600 AD the present-day positional decimal notation with 0 and its own particular symbols $1, \ldots, 9$, came into existence in India, presumably under Babylonian influence. Thus, for example, from the primitive forms $-$, $=$, there arose at first the symbols ¬, ⊅, which eventually developed into 1, 2. The *Indian notation* was taken over by the *Arabs,* not least by their astronomers. The Indians had signs for *positive* and *negative* numbers; namely, "dhana" or "sva" (denoting ownership) and "rina" or "kśaya" (diminution, debit). Arithmetic rules for handling positive and negative numbers are found in the works of BRAHMAGUPTA (born 598) (JUSHKEWITSCH [15], p. 126). However, there is nothing to indicate that negative numbers were generally recognized as solutions of equations. Thus negative solutions to such problems as those where it was a question of finding the number of monkeys in a horde were regarded as meaningless. On the other hand, a negative solution to a problem involving distances was on at least one occasion interpreted as a distance measured in the opposite direction.

The Indian mathematician SRIDHARA (about 850–950) laid down arithmetical rules for operations with *zero*, symbols for which had already appeared among the Egyptians (the symbol ⌣ is to be found in an inscription of the second century BC in a temple of Edfu), the Greeks (the symbol $o$, which is possibly the initial letter of the word 'ουδέν = nothing), and the Indians (who from the 5th century AD used the word "sunya" for the void). The Arabs used the word "al-sifr" for zero, from which was derived the word "cifra,"[1] which was still used by GAUSS with the meaning zero (JUSCHKEWITZ [15], p. 107, LEPSIUS [19] and GAUSS [12], p. 8). A dot or a circle was used as a symbol for zero in India, from the seventh century AD onwards.

**4. Modern Times.** Indo-arabic arithmetical practices were disseminated throughout the Western world by arithmetical textbooks in the 13th to the 16th centuries (for example, those of LEONARDO of PISA, RIESE, STIFEL) and made possible the subsequent successes of the Italian mathematicians of the Renaissance (such as DEL FERRO, CARDAN, and FERRARI) in the solution of algebraic equations. STIFEL says, in talking about negative numbers, that they are not just "meaningless twaddle" but on the contrary that it is "not without usefulness" to *feign* numbers below zero, that is to fabricate fictitious numbers that are less than nothing (STIFEL [27], p. 248 *et seq.*).

In the new *algebra* of the Renaissance, zero and the negative numbers acquired a new function as they made it possible to assimilate several types of equations under one category. From the time of DESCARTES equations

---

[1] See the English word 'cypher' one of whose meanings is zero.

have been written in the form

$$a_n x^n + a_{n-1} x^{n-1} + \cdots + a_0 = 0$$

(though without coefficient suffixes in the case of DESCARTES) where the coefficients $a_i$ may be positive, negative or zero.

Although mathematicians have, from the very beginning of their science, operated with numbers and discovered theorems about numbers, it was not until the *19th century* that they gave mathematically serviceable definitions of the concept of number. Their foremost consideration was initially to provide secure foundations for analysis by defining more precisely the real numbers. It was not until after DEDEKIND and CANTOR (and others) had defined real numbers by means of sets of rational numbers (see Chapter 2) that the classical definitions of the natural numbers in terms of logic and set theory then followed. The realization that the extensions of the natural numbers to the integers and the rationals could still essentially be regarded as a topic of algebra was closely bound up with the introduction of the fundamental algebraic ideas of ring theory and field theory.

## §2. NATURAL NUMBERS

Counting with the help of number symbols marks the beginning of arithmetic. Computation presupposes counting. Until well into the nineteenth century, efforts were made to trace the idea of number back to its origins in the psychological process of counting. The psychological and philosophical terminology used for this purpose met with criticism, however, after FREGE's *logic* and CANTOR's *set theory* had provided the logico-mathematical foundations for a critical assessment of the number concept. DEDEKIND, who had been in correspondence with CANTOR since the early 1870's, proposed in his book *Was sind und was sollen die Zahlen?* [9] (published in 1888, but for the most part written in the years 1872–1878) a "set-theoretical" definition of the natural numbers, which other proposed definitions by FREGE and CANTOR and finally PEANO's axiomatization were to follow. That the numbers, axiomatized in this way, are uniquely defined, (up to isomorphism) follows from DEDEKIND's recursion theorem.

From now on we shall take as known the basic concepts of set theory (although the reader may consult the last chapter of this book).

**1. Definition of the Natural Numbers.** The *natural numbers* form a set $\mathbb{N}$, containing a distinguished element 0, called zero, together with a successor function $S: \mathbb{N} \rightarrow \mathbb{N}$, of $\mathbb{N}$ into itself, which satisfies the following axioms:

(S1) *S is injective,*

(S2) $0 \notin S(\mathbb{N})$,

(S3) *If a subset $M \subset \mathbb{N}$ contains zero and is mapped into itself by $S$, then $M = \mathbb{N}$.*

The successor function $S$ describes, in the language of set theory, the process of counting. The idea is that $S$ assigns to every natural number $n$ its successor $S(n)$. Thus $1 := S(0)$, $2 := S(1)$, $3 := S(2)$ and so on. The first axiom asserts that in counting one never encounters the same number more than once. The second axiom expresses the fact that 0 is the starting point of the counting process, or, alternatively that 0 is never encountered as a successor during the process. Many mathematicians prefer, as did DEDEKIND, to begin the counting process with 1. The third axiom is the set theoretic formulation of the

*Principle of complete induction.* If a certain property $E$ is possessed by the number 0 (the commencement of the induction) and if, for every number $n$ which has the property $E$, its successor $S(n)$ also has the property $E$ (the induction step), then this property is possessed by all the natural numbers.

The equivalence of this principle to the third axiom is seen when the property $E$ is replaced by the subset $M$ of numbers possessing the property. Instead of saying "$n$ has the property $E$" we can also say "the proposition $E$ applies to $n$" or "$E(n)$ holds." The principle of induction is not some new kind of syllogism of mathematicians set apart from the ordinary rules of inference in logic; it is merely the use of axiom S3 to prove that certain statements are valid for all natural numbers.

A set $M$ is said to be *infinite* if there exists an injective mapping $f: M \rightarrow M$, of $M$ into itself, such that $f(M) \neq M$. This definition expresses the fact that only infinite sets can be mapped injectively onto one of their proper subsets. Historically this was the definition given by DEDEKIND in *Was sind und was sollen die Zahlen*? Instead of speaking of injective mappings, DEDEKIND used the term (§5, No. 64) "ähnliche Abbildungen" [similarity mappings].

**Theorem.** *There exists an infinite set, if and only if there is a set $\mathbb{N}$ satisfying the axioms* (S1)–(S3).

**Proof.** If there is such a set $\mathbb{N}$, then by axioms (S1) and (S2), there must also exist an infinite set (putting $f = S$).

Let $A$ be an infinite set. Then by definition there is an injective mapping $f: A \rightarrow A$ with $f(A) \neq A$. Consequently there must also be an element $0 \in A$ with $0 \notin f(A)$. Let $I$ be the class of all sets $M \subset A$ with $0 \in M$ and $f(M) \subset M$. By hypothesis $I \neq \emptyset$. Thus we can define the intersection $\bigcap_{M \in I} M$. This set satisfies the axioms (S1)–(S3), if one takes $f \mid M$ as the successor function $S$.                                                                               □

*Remark.* DEDEKIND also gave a proof of the existence of an infinite set, but it was based on the inconsistent concept of the set of all sets (5, No. 66). A similar unsuccessful attempt is to be found in BOLZANO's *Paradoxien des Unendlichen* [4, §13]. We assume, under the *axiom of infinity* (see Chapter 13), that there are infinite sets. In our proof $\mathbb{N}$ is a "smallest" infinite set contained in an infinite set. DEDEKIND therefore speaks of "simple infinite systems" (§6, No. 71). The construction of $\mathbb{N}$ given in the proof depends on the choice of $A$, $f$ and $0$. The fact that $\mathbb{N}$, the successor function $S$, and $0$, are all uniquely defined to within isomorphism, will be shown in paragraph 2 (uniqueness theorem). According to VON NEUMANN, there is a canonically defined set-theoretic model for $\mathbb{N}$, on the basis of the Zermelo–Fraenkel set theory (VON NEUMANN [21], see also Chapter 13).

FREGE and CANTOR defined the natural numbers as "finite potencies" and "finite cardinal numbers" respectively (FREGE [11], p. 73 et seq., CANTOR [8], p. 119, see also Chapter 13). This formulation is also found in RUSSELL [25], p. 116 and BOURBAKI [6], I, Chap. III, §4, Def. 1.

**2. The Recursion Theorem and the Uniqueness of $\mathbb{N}$.** New concepts for natural numbers are for the most part introduced recursively. One also talks of inductive definitions. For example, addition may be defined inductively by successively stipulating that $m + 0 := m$, $m + 1 := S(m)$, $m + 2 := S(m + 1)$, and generally $m + S(n) := S(m + n)$. The justification establishing that this recursive procedure gives a meaningful definition, is provided by the following result.

**Recursion Theorem** (DEDEKIND 1888). *Let $A$ be an arbitrary set containing an element $a \in A$, and $g$ a given mapping $g: A \to A$ of $A$ into itself. Then there is one and only one mapping $\varphi: \mathbb{N} \to A$ with the two properties $\varphi(0) = a$ and $\varphi \circ S = g \circ f$.*

The mapping $\varphi$ is said to be defined recursively starting from $\varphi(0) = a$, by the recursion formula $\varphi(n + 1) = g(\varphi(n))$.

**Proof.** To show the *uniqueness* of the mapping $\varphi$, we consider two mappings $\varphi_1$, $\varphi_2$ from $\mathbb{N}$ to $A$ with the stated properties. We show, by induction on $n$, that $\varphi_1(n) = \varphi_2(n)$ for all $n$. The induction begins with $\varphi_1(0) = a = \varphi_2(0)$. Since, by the inductive hypothesis, $\varphi_1(n) = \varphi_2(n)$ it follows that

$$\varphi_1(S(n)) = g(\varphi_1(n)) = g(\varphi_2(n)) = \varphi_2(S(n)).$$

To prove the *existence* of $\varphi$, we consider all subsets $H \subset \mathbb{N} \times A$ having the two properties (1) $(0, a) \in H$ and (2) for all $n$, $b$, if $(n, b) \in H$, then $(S(n), g(b)) \in H$. Since the whole set $\mathbb{N} \times A$ is such a set $H$, and all sets $H$ contain the element $(0, a)$, the intersection $D$ of all the $H$ is the smallest subset of $\mathbb{N} \times A$ satisfying (1) and (2). We now assert that $D$ is the graph of a mapping $\varphi: \mathbb{N} \to A$, and prove this assertion by complete induction:

(*) To every $n \in \mathbb{N}$, there is just one $b$, such that $(n, b) \in D$.

To begin the induction we note that, by (1), $(0, a) \in D$. If $(0, c) \in D$ were possible with $c \neq a$, then one could remove $(0, c)$ from $D$, and the remaining set $D \setminus \{(0, c)\}$ would still have the properties (1) and (2), in contradiction to the fact that $D$ is the smallest set of this kind.

We now complete the inductive argument as follows. By the inductive hypothesis there is just one $b$, such that $(n, b) \in D$. By (2) we then have $(S(n), g(b)) \in D$. If $(S(n), c) \in D$ and $c \neq g(b)$ were possible, then one could remove $(S(n), c)$ from $D$ and by the same argument as was used at the start of the induction, we should arrive at a contradiction. Now that the proposition (*) has been proved, $D$ can be written, as the graph of a mapping $\varphi: N \to A$, namely $D = \{(n, \varphi(n)) \mid n \in \mathbb{N}\}$. The property (1) of $D$ means that $\varphi(0) = a$, and the property (2) that $(S(n), g(\varphi(n))) \in D$, and hence $\varphi \circ S(n) = g \circ \varphi(n)$ for all $n$.                                    □

*Example.* The $n$th power $c^n$ of a real number $c$ is defined by the recursion formula $c^{n+1} = c^n \cdot c$ starting from $c^0 = 1$. Here we apply the Recursion theorem with $A = \mathbb{R}$ (the set of real numbers), $a = 1$ and $g(b) = b \cdot c$.

As a first application of the Recursion theorem we shall now prove the uniqueness of $\mathbb{N}$.

**Uniqueness Theorem.** *Let $\mathbb{N}'$ be a set with a successor function $S'$, a distinguished element $0'$ and satisfying the axioms (S1)–(S3). Then $\mathbb{N}$ and $\mathbb{N}'$ are canonically isomorphic, that is, there exists just one bijective mapping $\varphi: \mathbb{N} \to \mathbb{N}'$ with $\varphi(0) = 0'$ and $S' \circ \varphi = \varphi \circ S$.*

**Proof.** By the Recursion theorem, applied to $A = \mathbb{N}'$, $a = 0'$ and $\varphi = S'$, there is just one mapping $\varphi: \mathbb{N} \to \mathbb{N}'$ with $\varphi(0) = 0'$ and $\varphi \circ S = S' \circ \varphi$. By interchanging the roles of $\mathbb{N}$ and $\mathbb{N}'$ one obtains a corresponding mapping $\psi \circ \mathbb{N}' \to \mathbb{N}$ with $\psi(0') = 0$ and $\psi \circ S' = S \circ \psi$. To prove that $\psi \circ \varphi = \mathrm{id}$ (the identity mapping), we use the uniqueness assertion of the Recursion theorem for $A = \mathbb{N}$, $a = 0$, and $g = S$. Both $\psi \circ \varphi$ and $\mathrm{id}$ are mappings $\Phi: \mathbb{N} \to \mathbb{N}$, for which $\Phi(\theta) = 0$ and $\Phi \circ S = S \circ \Phi$ and therefore $\psi \circ \varphi$ must be the same as $\mathrm{id}$. Similarly $\varphi \circ \psi = \mathrm{id}$.                    □

**3. Addition, Multiplication and Ordering of the Natural Numbers.** For every fixed natural number $m$, the *addition* $m + n$ is defined, starting from $m + 0 = m$, by the recursion formula $m + S(n) = S(m + n)$. Here again the Recursion theorem is being applied for $A = \mathbb{N}$, $a = m$, $g = S$ and $\varphi(n) = m + n$. In particular, it follows for $1 := S(0)$ that $m + 1 = S(m)$ is the successor of $m$.

All the well-known *rules of addition* now must be proved. We shall confine ourselves to the proof of the *associative law* and refer the reader to the classical work by LANDAU [18], Chapter 1, §2.

**Theorem.** *For all $k, m, n \in \mathbb{N}$, $(k + m) + n = k + (m + n)$.*

**Proof.** The induction begins with $n = 0$, for which $n = 0$: $(k + m) + 0 = k + m = k + (m + 0)$. The inductive argument from $n$ to $n + 1$ runs as follows:

$$(k + m) + (n + 1) \overset{*}{=} ((k + m) + n) + 1 \overset{**}{=} (k + (m + n)) + 1$$
$$\overset{*}{=} k + ((m + n) + 1) \overset{*}{=} k + (m + (n + 1)).$$

The steps marked with $*$ use the recursive formula for addition. Those marked with $**$ use the inductive hypothesis.                    □

One can easily convince one's self in this way that $\mathbb{N}$ *is a commutative semigroup with cancellation law, in respect of addition.* The *cancellation law* asserts that $n + k = m + k$ implies $n = m$, for all $k, m, n \in \mathbb{N}$.

Analogously to addition, the operation of *multiplication* $m \cdot n$, by a fixed number $m$, can be defined, starting from $m \cdot 0 = 0$, recursively by the formula $m \cdot (n + 1) = m \cdot n + m$. All the well-known arithmetical rules of multiplication again require proofs, for which we refer the reader to LANDAU [18], Chapter 1, §4.

An order relation $\leq$ may be defined on $\mathbb{N}$ as follows: the relation $n \leq m$ holds if and only if there is a $t \in \mathbb{N}$ such that $n + t = m$. The usual properties of an order relation, namely 1) reflexivity, 2) antisymmetry and 3) transitivity hold good, that is to say for all $m, n, l \in \mathbb{N}$:

1) $n \leq n$.

2) if $n \leq m$ and $m \leq n$, then $m = n$.

3) If $n \leq m$ and $m \leq l$, then $n \leq l$.

We write $m < n$ if and only if $m \leq n$ and $m \neq n$. The ordering is *linear* (or *total*, as opposed to a *partial* order), that is to say for all $l, m, n \in \mathbb{N}$ it follows from $m \leq n$ that $m + l \leq n + l$ (and the corresponding statements are true with $<$ in place of $\leq$). Analogous statements also hold for multiplication, that is, $m \leq n$ implies $m \cdot l \leq n \cdot l$ with the corresponding statements with $<$ instead of $\leq$ being true (provided $l \neq 0$).

**4. PEANO's Axioms.** Following the Italian mathematician PEANO (1858–1932) the natural numbers can also be described in terms of the following axioms for the basic concepts $\mathbb{N}$, 0 and $S$:

(P1) $0 \in \mathbb{N}$.

(P2) if $n \in \mathbb{N}$ then $S(n) \in \mathbb{N}$.

(P3) if $n \in \mathbb{N}$ then $S(n) \neq 0$.

(P4) if $0 \in \mathbb{N}$ and if it always follows from $n \in E$ that $S(n) \in E$, then $\mathbb{N} \subset E$.

(P5) if $m, n \in \mathbb{N}$, then $S(m) = S(n)$ implies that $m = n$.

If (P1)–(P5) are interpreted set theoretically, then they are equivalent to the definition in §1.1. In contrast to DEDEKIND, however, PEANO was not primarily interested in a set theoretical construction of the natural numbers, but in their axiomatization in a formal language. In this sense, (P4) should be read as meaning: if zero has the property $E$ and if, from the fact that $n$ has the property $E$, it always follows, that the successor $S(n)$ has the property $E$, then the property $E$ follows from the property $\mathbb{N}$ of being a natural number. We shall not pursue this particular aspect, namely that of a formal language, any further here, but it will be of importance later in the transition from standard to non-standard numbers discussed in Chapter 12.

Historically PEANO in 1889 laid down a set of nine axioms (with 1 as the distinguished element) in his *Arithmetices principia nova methodo exposita* [24]. On the relationship between his system and DEDEKIND's definition he writes "Utilius quoque mihi fuit recens scriptum: R. DEDEKIND, Was sind und was sollen die Zahlen, Braunschweig 1888, in quo quaestiones, quae ad numerorum fundamenta pertinent, acute examinantur." ([24], p. 22). [The recent work by DEDEKIND *Was sind und was sollen die Zahlen*, Brunswick 1888, in which questions relating to the foundations of numbers are acutely analyzed, was also particularly useful to me.]

## §3. THE INTEGERS

Subtraction cannot be done without restriction in the domain of the natural numbers. While the negative numbers ("false" numbers as they were called by DESCARTES) had at first been treated warily, like roots and imaginary numbers, as fictitious expressions, KRONECKER in the 19th century described integers as the "natural starting point for the development of the concept of number" (see TROPFKE [29], p. 126; KRONECKER [16]). KRONECKER's famous quip that the Good Lord made the integers and that all the rest is the work of man is well known. However, according to DEDEKIND even the positive integers were not simply "given by nature" but rather "free creations of the human mind," namely, set-theoretic concepts. Algebraically, it is a question of extending the additive semigroup of the natural numbers to the group of integers, and central to this topic is the algebraic concept of an integral domain, which was introduced by

KRONECKER [17] in his "Grundzüge einer arithmetischen Theorie der algebraischen Grössen" (§5) [Foundations of an arithmetical theory of algebraic magnitudes] as the so-called "Integritätsbereich."

**1. The Additive Group $\mathbb{Z}$.** The systematic introduction of the integers is motivated by the following considerations. Every integer can be expressed as a difference $a - b$ between two natural numbers $a$ and $b$. This suggests that the integer $a - b$ should be described by the pair $(a, b)$, but of course one must be careful to remember that other pairs $(c, d)$ can describe the same number $a - b = c - d$, in fact whenever $a + d = b + c$. We therefore proceed as follows.

We consider the *relation*, defined on $\mathbb{N} \times \mathbb{N}$, by

$$(a, b) \sim (c, d) \quad \text{if and only if} \quad a + d = b + c.$$

We then establish that this is an equivalence relation. For example, transitivity may be proved as follows: if $(a, b) \sim (c, d)$ and $(c, d) \sim (e, f)$ then by definition, $a + d = b + c$ and $c + f = d + e$. By addition we obtain $a + d + c + f = b + c + d + e$ and by cancellation of $c + d$ we obtain $a + f = b + e$, that is $(a, b) \sim (e, f)$. (We have also made use of the commutativity and associativity of addition.)

The *integers* may now be defined as equivalence classes of the relation $\sim$. The class represented by $(a, b)$, is denoted by $[a, b]$. The set of all integers (a set of equivalence classes) is denoted by $\mathbb{Z}$.

We can define on $\mathbb{N} \times \mathbb{N}$ a componentwise addition, $(a, b) + (c, d) := (a + c, b + d)$. The commutative and associative laws hold, and the zero element is $(0, 0)$. This addition is compatible with the relation $\sim$, that is to say, if $(a', b') \sim (a, b)$ and $(c', d') \sim (c, d)$ then $(a' + c', b' + d') \sim (a + c, b + d)$. It is therefore meaningful to introduce in $\mathbb{Z}$, an *addition* $\mathbb{Z} \times \mathbb{Z} \rightarrow \mathbb{Z}$, $[a, b] + [c, d] := [a + c, b + d]$, which is likewise commutative and associative and which has $[0, 0]$ as zero element. By passing to equivalence classes (integers) we have gained more. Every integer $[a, b]$ has an inverse, namely, the integer $[b, a]$. We have established the following.

**Theorem.** *The integers form a commutative group with respect to addition.*

The element inverse to $\alpha \in \mathbb{Z}$ is uniquely determined, and is denoted by $-\alpha$. *Subtraction* in $\mathbb{Z}$ is defined by $\alpha - \beta := \alpha + (-\beta)$.

The mapping $\iota: \mathbb{N} \rightarrow \mathbb{Z}$, $a \rightarrow [a, 0]$ is injective and compatible with addition. It is usual to identify $\mathbb{N}$ with the subset of $\mathbb{Z}$, $\iota(\mathbb{N}) \subset \mathbb{Z}$, isomorphic to it. The integer $[a, b]$ is then written as $a - b$, and we have thus justified the notation, which provided the motivation. If one uses $\mathbb{N}^+ = \mathbb{N} \setminus \{0\}$, one can represent $\mathbb{Z}$ as a union of three disjoint sets $\mathbb{Z} = -N^+ \cup \{0\} \cup N^+$. Depending on whether $a > b$, $a = b$ or $a < b$ the integer $[a, b] = a - b$ lies in $\mathbb{N}^+$, in $\{0\}$ or in $-\mathbb{N}^+$.

The construction of the integers is an algebraic one. Instead of starting from $\mathbb{N}$, one could have begun with any commutative semigroup $H$ and constructed from it as above a commutative group $G$. If the cancellation law does not hold in $H$ some modifications are required: we define $(a, b) \sim (c, d)$ if and only if there is an $e$ such that $a + d + e = b + c + e$. However in this case $\iota : H \to G$ is no longer injective.

**2. The Integral Domain $\mathbb{Z}$.** The representation of integers as differences provides a motivation for the definition of their multiplication. We should like $(a - b) \cdot (c - d)$ to be equal to $(ac + bd) - (ad + bc)$ and accordingly this leads to the following definition:

$$[a, b] \cdot [c, d] = [ac + bd, ad + bc] \quad \text{for} \quad a, b, c, d \in \mathbb{N}.$$

This definition is independent of the particular choice of the representative pairs.

**Theorem.** *The integers form an integral domain with respect to addition and multiplication (that is, a commutative ring without zero divisors and with identity element).*

Incidentally, $\mathbb{Z}$ is the smallest integral domain containing $\mathbb{N}$ as a subset: to every domain of integrity $R \supset \mathbb{N}$ there is just one monomorphism (that is, injective mapping, compatible with $+$ and $\cdot$) $\varphi : \mathbb{Z} \to R$ with $\varphi \mid \mathbb{N} = $ inclusion of $\mathbb{N}$ in $\mathbb{R}$.

**3. The Order Relation in $\mathbb{Z}$** is defined by

$$a \leq b \quad \text{if and only if} \quad b - a \in \mathbb{N}.$$

**Theorem.** *The ring $\mathbb{Z}$ of integers is linearly (completely) ordered by the relation $\leq$. For all $a, b, c \in \mathbb{Z}$ the relation $a \leq b$ implies $a + c \leq b + c$ and, when $c > 0$, $a \cdot c \leq b \cdot c$ as well.*

The natural numbers other than zero are thus the integers $> 0$, the so-called positive integers. A number $a$ is said to be *negative* whenever $-a$ is positive.

*Remarks.* Every commutative ring $R$ expressible as a disjoint union $R = -P \cup \{0\} \cup P$ where $P$ is additively and multiplicatively closed, can be totally ordered by the relation $a \leq b$ if $b - a \in P \cup \{0\}$.

Historically, it was also DEDEKIND who introduced the idea of defining integers by pairs from $\mathbb{N} \times \mathbb{N}$. In a letter from the 82-year-old mathematician written in 1913 to a former student, DEDEKIND ([10], p. 490) describes an extension of the domain $N$ of natural numbers to the domain $G$ of the integers. LANDAU [18] first constructs the rational numbers $\geq 0$ from $\mathbb{N}$, and then extends this set by means of the negative rational numbers, to the field $\mathbb{Q}$ (see §4) obtaining $\mathbb{Z}$ as a subring of $\mathbb{Q}$.

## §4. THE RATIONAL NUMBERS

**1. Historical.** Division, as the inverse of multiplication, cannot be done without restriction in the domain of integers. Fractions, which make division always possible, were already considered in early times. They were never surrounded by such mystery as were the negative numbers, which were thought of as being in some never-never land below "nothing," or the irrational and imaginary numbers, which we still have to discuss. The first systematic treatment of rationals is found in Book VII of EUCLID's *Elements,* which deals with the ratios of natural numbers. The idea, which is so familiar to us, of interpreting ratios as fractions and of extending in this way the domain of whole numbers first arises in comparatively modern times. The first theoretical investigations stem from the nineteenth century.

BOLZANO [5] in a posthumously published paper entitled "Reine Zahlenlehre" developed a theory of rational numbers, and in fact a theory of those sets of numbers that are *closed with respect to the four elementary arithmetic operations.* One also finds, in a paper by OHM [22] (the brother of the famous physicist) an intention to define the rational numbers "solely through the basic truths relating to addition, substraction, multiplication and division."

Their foremost consideration was therefore the investigation of certain arithmetical relationships, and not a philosophical question about the nature of number. Finally, with HANKEL ([13], p. 2), in his *Theorie der complexen Zahlensysteme* of 1867, it comes down to this: The laws of these operations determine "the system of conditions ... which are necessary and sufficient to define the operation formally." Apart from the rational numbers, the notion of a field (as a concept, even if not yet under this name) had also been discussed in the writings of ABEL and GALOIS, where, for example, a root of an equation is adjoined to the rationals and an investigation is made of all possible expressions that can be formed from it by means of the four operations, addition, subtraction, multiplication and division. KRONECKER in 1853 speaks in his theory of algebraic quantities of "domains of rationality" (KRONECKER [17], §1), and DEDEKIND, at first of "rational domains" and finally of "fields" in the case of real and complex numbers (DEDEKIND [12], p. 224). Number fields were also investigated by WEBER [30] and HILBERT [14]. In 1910 STEINITZ [26] gave an abstract definition of this fundamental algebraic concept. STEINITZ also brought out clearly the fact that behind this extension of the integers to the rational numbers there lies a general algebraic construction, namely, that of the embedding of an integral domain in a field by the formation of fractions.

**2. The Field $\mathbb{Q}$.** Following the example of WEBER in his *Lehrbuch der Algebra* of 1895, we shall introduce fractions as equivalence classes of integers, and guided by the relation

$$\frac{a}{b} = \frac{c}{d} \quad \text{if and only if} \quad ad = bc,$$

we start from the equivalence relation $\sim$ defined on $\mathbb{Z} \times (\mathbb{Z} \setminus \{0\})$ by

$$(a, b) \sim (c, d) \quad \text{if and only if} \quad ad = bc.$$

These definitions are independent of the particular choice of representatives. In LANDAU [18], Chapter 2, §§3–4, is given a detailed proof of the

**Theorem.** *The set $\mathbb{Q}$ of rational numbers, with the addition and multiplication defined above, constitutes a field.*

$\mathbb{Z}$ is mapped isomorphically on the subring $\iota(\mathbb{Z}) \subset \mathbb{Q}$ by the mapping $\iota \colon \mathbb{Z} \to \mathbb{Q}$, $a \mapsto \frac{a}{1}$. $\mathbb{Z}$ is usually identified with $\iota(\mathbb{Z})$. The field $\mathbb{Q}$ is the smallest field containing $\mathbb{Z}$ as a subring.

**3. The Ordering of $\mathbb{Q}$.** A fraction $a/b$ is said to be positive if $a, b$ are both positive or both negative. The set $P$ of positive fractions is closed with respect to the operations $+$ and $\cdot$. $\mathbb{Q}$ is expressible as a union of disjoint sets $-P \cup \{0\} \cup P$. As in the remark in 3.3 a total order relation on $\mathbb{Q}$ can be defined by $r \leq s$ if and only if $s - r \in P \cup \{0\}$ which coincides with the order on $\mathbb{Z}$ defined in 3.3.

The order relation in $\mathbb{Q}$ is *Archimedean*, that is, for all positive rational numbers $r, s \in \mathbb{Q}$ there exists a natural number $n$ with $s < n \cdot r$. To prove this, we write $s = p/h$ and $r = q/h$ as fractions whose numerators and denominators are natural numbers and with a common denominator $h$. The truth of the statement then follows as soon as it has been proved that $p < n \cdot q$ for natural numbers $> 0$. The latter can be demonstrated for a fixed $q \geq 1$ by induction over $p = 1, 2, \ldots$ A noteworthy property, which distinguishes the field $\mathbb{Q}$ from the ring of integers $\mathbb{Z}$ is its *density:* for all $r, s \in \mathbb{Q}$ with $r < s$, a $t \in \mathbb{Q}$ can always be found such that $r < t < s$. One can, for example, choose the arithmetic mean $t := \frac{1}{2}(r + s)$.

REFERENCES

[1] ARISTOTELES: Metaphysik, Aristotelis Opera, ed. I. Bekker, Berlin 1831, repr. Darmstadt 1960

[2] ARISTOTELES: Physik, Aristotelis Opera, ed. I. Bekker, Berlin 1831, repr. Darmstadt 1960

[3] BECKER, O.: Grundlagen der Mathem. in geschichtlicher Entwicklung, Freiburg/München 1954, ²1964, Frankfurt 1975

[4] BOLZANO, B.: Paradoxien des Unendlichen, ed. F. Přihonský, Leipzig 1851, Berlin ²1889, ed. A. Höfler, Leipzig 1920, mit Einl., Anm., Reg. u. Bibliographie ed. B. van Rootselaar, Hamburg 1955, ²1975

[5] BOLZANO, B.: Reine Zahlenlehre, in: B. Bolzano — Gesamtausgabe (eds. E. Winter, J. Berg, F. Kambartel, J. Loužil, B. v. Rootselaar), Reihe II Nachlaß, A. Nachgelassene Schriften, Bd. 8 Größenlehre II, Reine Zahlenlehre, Stuttgart/Bad Cannstatt 1976

[6] BOURBAKI, N.: Eléments de mathématique, Paris, 1939 and later

[7] BRUINS, E.M., RUTTEN, M.: Textes mathématiques de Suse, Mémoires de la Mission Archéologique en Iran, Tome 34, Paris 1961

[8] CANTOR, G.: Gesam. Abh. mathem. u. philos. Inhalts, Berlin 1932, repr. Berlin 1980

[9] DEDEKIND, R.: Was sind und was sollen die Zahlen? Braunschweig 1888, ¹⁰1965, repr. 1969

[10] DEDEKIND, R.: Mathem. Werke Bd. 3, Braunschweig 1932, repr. New York 1969

[11] FREGE, G.: Die Grundlagen der Arithmetik. Eine logisch mathematische Untersuchung über den Begriff der Zahl, Breslau 1884, repr. Darmstadt/Hildesheim 1961

[12] GAUSS, C.F.: Werke Bd. 3, Göttingen 1876

[13] HANKEL, H.: Theorie der complexen Zahlensysteme, Leipzig 1867

[14] HILBERT, D.: Über den Zahlbegriff, in: Jahresber. d. Deutschen Math. Verein. 1900, 180–184

[15] JUSCHKEWITSCH, A.P.: Geschichte der Mathematik im Mittelalter, dt. Leipzig 1964

[16] KRONECKER, L.: Über den Zahlbegriff, in: Journ. f. d. reine u. angew. Mathem. 101 1887, 339, in: Math. Werke Bd. 3, Leipzig 1899/1931, repr. New York 1968, 249–274

[17] KRONECKER, L.: Grundzüge einer arithmetischen Theorie der algebraischen Größen, in: J. f. d. reine u. angew. Mathem. 1882, 1–122, in: Math. Werke Bd. 2, Leipzig 1897, repr. New York 1968, 237–287

[18] LANDAU, E.: Grundlagen der Analysis, Leipzig 1930, repr. Darmstadt 1963

[19] LEPSIUS, R.: Über eine Hieroglyphische Inschrift am Tempel in Edfu, in: Abh. d. Kgl. Akad. d. Wiss., Berlin 1855, 69–111

[20] NEUGEBAUER, O.: Mathem. Keilschrifttexte, Quellen u. Studien A3, Berlin I, II 1935, III 1937

[21] NEUMANN, J.v.: Zur Einführung der transfiniten Zahlen, in: Acta Szeged 1 1923, 199–202, repr. in: *A.H. Taub* (ed.), Collected Works, Oxford/London/Paris Bd. 1 1961, 24–33

[22] OHM, M.: Die reine Elementarmathematik Bd. 1, Berlin $^2$1834

[23] PAPYRUS RHIND, (Hrsg. A. Eisenlohr) Leipzig 1877; A.B. Chace, The Rhind Mathem. Papyrus, Oberlin I 1927, II 1929

[24] PEANO, G.: Arithmetices principia nova exposita, in: Opere scelte Bd. II, Rom 1958, 20–55

[25] RUSSELL, B.: The principles of mathematics, London 1903, $^7$1956

[26] STEINITZ, E.: Algebraische Theorie der Körper, in: J. f. d. reine u. angew. Math. 137 1910, 167–309

[27] STIFEL, M.: Arithmetica integra, Nürnberg 1544

[28] STRUWE, W.W.: Papyrus des staatl. Museums der schönen Künste in Moskau, Quellen u. Studien A1 1930

[29] TROPFKE, J.: Geschichte der Elementarmathematik, Bd. 1 Arithmetik und Algebra, vollst. neu bearb. von H. Gericke, K. Reich u. K. Vogel, Berlin $^4$1980

[30] WEBER, H.: Lehrbuch der Algebra Bd. 1 1895, repr. der 3. Aufl. New York 1961

FURTHER READINGS

[31] AABOE, ASGER: Episodes from the Early History of Mathematics (New Mathematical Library, No. 13) New York: Random House and L.W. Singer, 1964

[32] CAJORI, FLORIEN: A History of Mathematical Notations, 2 vols. Chicago: Open Court Publishing, 1928–1929

[33] KRAMER, EDNA: The Nature and Growth of Modern Mathematics, Princeton: Princeton University Press, 1970

[34] MENNINGER, KARL: Number Words and Number Symbols: A Cultural History of Numbers, Cambridge, Mass.: The M.I.T. Press, 1969

[35] NEUGEBAUER, OTTO: The Exact Sciences in Antiquity, 2nd ed. New York: Harper and Row, 1962

[36] ORE, OYSTEIN: Number Theory and Its History, New York: McGraw-Hill, 1948

[37] RESNIKOFF, H.L. AND WELLS, R.O. JR.: Mathematics in Civilization, New York: Dover, 1984

[38] SONDHEIMER, ERNST AND ROGERSON, ALAN: Numbers and Infinity: A Historical Account of Mathematical Concepts, Cambridge: Cambridge University Press, 1981

[39] VAN DER WAERDEN, B.L.: Science Awakening, tr. by Arnold Dresden, New York: Oxford University Press, 1961; New York: John Wiley, 1963 (paperback ed.)

# 2

# Real Numbers

*K. Mainzer*

λέγω δ' εἶναι συνεχὲς ὅταν ταὐτὸ γένηται καὶ ἓν τὸ
ἑκατέρου πέρας οἷς ἅπτονται, καὶ ὥσπερ σημαίνει τοὔνομα,
συνέχηται
(ARISTOTLE, Physics 227a, 11–12).

[I call it holding together if it is the same and a single thing
that becomes the boundary for each of the parts to which
they cling and, as the word signifies, it is kept together.]

Continuum est totum cuius duae quaevis partes cointegrantes
(seu quae simul sumtae toti coincidunt) habent aliquid com-
mune, ... saltem habent communem terminum)
(G.W. LEIBNIZ, Mathem. Schr. VII, 284).

[A continuum is a whole when any two component parts
thereof (or more precisely any two parts which together make
up the whole) have something in common, ... at the very
least they have a common boundary.]

Zerfallen alle Punkte der Geraden in zwei Klassen von der Art,
daß jeder Punkt der ersten Klasse links von jedem Punkt der
zweiten Klasse liegt, so existiert ein und nur ein Punkt, welcher
diese Einteilung aller Punkte in zwei Klassen, diese Zerschnei-
dung der Geraden in zwei Stücke, hervorbringt (R. DEDEKIND,
Stetigkeit und irrationale Zahlen, Braunschweig 1872, 10).

[If the points of a line are divided into two classes, in such a
way that each point of the first class lies to the left of every
point of the second class, then there exists one and only
one point of division which produces this particular subdivision
into two classes, this cutting of the line into two parts.]

## §1. HISTORICAL

**1. HIPPASUS and the Pentagon.** When today we define the real numbers as elements of a completely ordered field, we tend to forget the magnitude of the intellectual and philosophical crisis brought about by the
discovery that there were things outside the grasp of the rational numbers.
Indeed, if we can trust later legends, the discoverer incurred the wrath of
the Gods. We mean of course the discovery ascribed to the 5th century

B.C. Pythagorean, HIPPASUS of METAPONT, that there are line segments whose ratios are *incommensurable*. The discovery is said to have caused a great shock in Pythagorean circles because it finally called into question one of the basic tenets of their philosophy, that everything was expressible in terms of whole numbers.

To understand the effects of this crisis, one has to remember that the Pythagoreans were not only active as a highly influential mathematical school, who were the first to raise the requirement for exact mathematical science and who insisted on a strict education in arithmetic, geometry, astronomy and music for their members, but that in addition to all this they pledged themselves to an orderly way of life. Until the uprising of 445 BC, they had been a dominant force throughout Southern Italy. In this political turmoil, HIPPASUS is presumed to have played an important role (see IAMBLICHUS [14], p. 77, 6f; also FRITZ [10], HELLER [11]).

The treatment of ratios of line-segments had come out of traditionally employed practices in measurement. A segment $a$ of a line had traditionally been measured by laying along the line unit measures $e$, one after the other, along the line, as many times as were necessary:

$$a = \underbrace{e + \cdots + e}_{m \text{ times}} = m \cdot e.$$

Two segments $a_0$ and $a_1$ are said to be commensurable if they can both be measured, in this sense, with the same unit of measurement $e$, so that $a_0 = m \cdot e$ and $a_1 = n \cdot e$ with $m$, $n$ being two natural numbers. In this case the ratio $a_0 : a_1$ of the line segments is equal to the ratio $m : n$ of two natural numbers.

The method of finding a common measure of two line segments $a_0$, $a_1$ had already been practiced, before the days of Greek philosophy and science, by craftsmen, by a process of alternate "taking away." EUCLID described the process in his *Elements* which now goes by the name of the Euclidean algorithm. The smaller segment $a_1$ is taken away from the larger segment $a_0$ as many times as possible, until the residue left is smaller than $a_1$, so that, if $a_2$ is this residue, then

$$a_0 = n_1 a_1 + a_2 \quad \text{with} \quad a_2 < a_1.$$

One then continues in the same way:

$$a_1 = n_2 a_2 + a_3 \quad \text{with} \quad a_3 < a_2,$$
$$a_2 = n_3 a_3 + a_4 \quad \text{with } a_4 < a_3,$$
$$\vdots$$

If $a_0$ and $a_1$ have a common measure, the process comes to an end after a finite number of steps, so that there is a $k$ with $a_{k-1} = n_k a_k$, and $a_k$ is a common measure of $a_0$ and $a_1$.

At first, it was probably felt intuitively that this process would always terminate, and that therefore there would always be a common measure. In modern language, however, all that this procedure shows is that every ratio of line segments can be developed as a *continued fraction*

$$a_0 : a_1 = n_1 + a_2 : a_1$$
$$= n_1 + \cfrac{1}{a_1 : a_2} = n_1 + \cfrac{1}{n_2 + a_3 : a_2}$$
$$= n_1 + \cfrac{1}{n_2 + \frac{1}{a_2 : a_3}} = \cdots = n_1 + \cfrac{1}{n_2 + \frac{1}{n_3 + \cdots}}$$

which is finite when $a_0$ and $a_1$ is commensurable.

The badge or symbol of their order used by the Pythagoreans was the *Pentagram*, which still retained its magical potency in mediaeval astrology and according to legend was used by Faust to exorcize Mephistopheles. There is good reason to believe that HIPPASUS by working from this symbol found that two of the lines therein were incommensurable (see IAMBLICHUS [15], p. 132, 11–12; for references to the sources see FRITZ [10], HELLER [11], TROPFKE [23]).

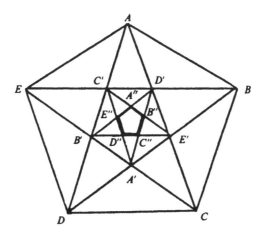

To see this, we begin with the regular pentagon $ABCDE$ in which all five diagonals have been drawn. The diagonals intersect to form a smaller regular pentagon $A'B'C'D'E'$ in the middle. Because of symmetry, each side of a regular pentagon is parallel to one of the diagonals. Thus, the triangle $AED$ and $BE'C$ have their corresponding sides parallel and are therefore similar, so that $AD : AE = BC : BE'$. Now $BE' = BD - BC$, since $BC = AE = DE'$, as $EA$ is parallel to $DB$, and $DE$ is parallel to $AC$. Consequently for any regular pentagon

diagonal: side = side:(diagonal − side).

If we denote the diagonal by $a_0$, the side by $a_1$ and their difference by $a_2 = a_0 - a_1$, then $a_0 : a_1 = a_1 : a_2$ and in particular $a_2 < a_1$. If we now form the difference $a_3 = a_1 - a_2$, we obtain the same equation between the ratios $a_1 : a_2 = a_2 : a_3$, and in particular $a_3 < a_2$. The process can clearly be continued indefinitely:

$$a_2 = a_0 - a_1, \quad a_3 = a_1 - a_2, \quad a_4 = a_2 - a_3 \cdots$$

$$a_0 : a_1 = a_1 : a_2 = a_2 : a_3 = a_3 : a_4 = \cdots .$$

The Euclidean algorithm for $a_0$ and $a_1$, namely

$$a_0 = 1 \cdot a_1 + a_2,$$
$$a_1 = 1 \cdot a_2 + a_3,$$
$$a_2 = 1 \cdot a_3 + a_4$$
$$\ldots \ldots$$

never terminates, thus the side $a_1$ and diagonal $a_0$ of the pentagon are not commensurable.

We obtain for the ratio, the continued fraction

$$a_0 : a_1 = 1 + \cfrac{1}{1 + \cfrac{1}{1 + \cfrac{1}{1 + \cfrac{1}{1 + \cdots}}}}$$

It follows from $a_0 : a_1 = a_1 : (a_0 - a_1)$ that $a_0 : a_1 = \frac{1}{2}(1 + \sqrt{5})$. This ratio is known as the *golden section*. The fact that the Euclidean algorithm never terminates can be seen at once from the diagram, which shows that each pentagon always has a smaller one within it so that there is an infinity of pentagons, whose sides are of length $a_1, a_3, a_5, \ldots$ and diagonals of length $a_2, a_4, a_6, \ldots$ respectively.

**2. EUXODUS and the Theory of Proportion.** The Babylonians worked with rational approximations to *irrational* (incommensurable) ratios. For example, they used the sexagesimal fractions 1; 25 and 1; 24, 51, 10 as approximations to $\sqrt{2}$. But we owe to Greek mathematics the fundamental discovery that $\sqrt{2}$, the ratio of the diagonal to the side of a square, is incommensurable. In EUCLID's *Elements* X, §115a, we find the following proof. Let $a$ be the side and $d$ the diagonal of a square. If they were commensurable then the same number would have to be both odd and even, which is absurd. For, clearly $d^2 = 2a^2$, and since $d$ and $a$ have been assumed to be commensurable, $d : a = m : n$ where $m, n$ are natural numbers which may be taken to be the smallest possible. Then $d^2 : a^2 = m^2 : n^2$, but since $d^2 = 2a^2$ it follows that $m^2 = 2n^2$. Thus $m^2$ is even, and hence $m$ is even, say $m = 2l$. Now since $m$, $n$ are by hypothesis the smallest numbers satisfying $d : a = m : n$, they must be relatively prime and this implies that $n$ must be odd. Since $m = 2l$, it follows that $m^2 = 4l^2$ and thus since $m^2 = 2n^2$, we have $n^2 = 2l^2$ which implies that $n^2$ and hence $n$ are both even.

However, the irrationality of $\sqrt{2}$ was certainly known before EUCLID. According to PLATO (*Theaetetus* 147d) the irrationality of certain square roots such as $\sqrt{3}, \sqrt{5}, \ldots, \sqrt{17}$ had been demonstrated earlier by THEODORUS of CYRENE. In PLATO's *Laws* (819d–820c) there is a passage where the Athenian stranger speaks of the shameful ignorance of the generality of Greeks who are unaware that not all geometrical quantities are commensurable with one another and adds that it was only late (in life, or possibly late in the day) that he himself learned the truth. (See HEATH's *History of Greek Mathematics*, p. 156.)

A decisive factor in the rapid progress of Greek mathematics was the distinctive logic. The form of inference known as *reductio ad absurdum* (proving the truth of a proposition by showing that the assumption of its falsity leads to a contradiction) allows them to give the first "impossibility" proofs and the first precise statements about the "infinite." As HERMANN WEYL wrote, Mathematics became for the first time, in the hands of the Greeks, the "science of the infinite."

It was the brilliant stroke of a genius, EUDOXUS of KNIDOS, the contemporary and acquaintance of PLATO, that created a geometrical theory of proportion capable of dealing with incommensurable as well as commensurable magnitudes. This theory has come down to us in Book V of EUCLID's *Elements*. EUDOXUS starts off from (positive) geometrical magnitudes of a like kind; for example, line segments $a, b, \ldots$ or areas $A, B, \ldots$. He postulates that magnitudes of the same kind can be added, and tacitly assumes that the addition obeys the commutative and associative law. Magnitudes of the same kind are ordered: $a < b$ if and only if there exists a $c$ such that $a + c = b$. It is assumed that when $a \neq b$, one of the two relations $a < b$ or $b < a$ must hold. Integral multiples are defined by repeated addition, so that $m \cdot a = a + \cdots + a$ with $m$ summands on the right. The axiom now usually called the axiom of ARCHIMEDES is assumed. This states that for any given $a, b$ there exists a natural number $n$ for which $a < n \cdot b$. Thus infinitely small quantities are excluded. (It was reserved for a later age to allow these, see Chapter 12 in this connection.)

The *ratios* between geometrical magnitudes of the same kind, which do not necessarily have to be commensurable with one another (ratios of line segments, of areas, and so on) form the subject of the theory. To enable such ratios to be compared with one another, the following is given (Definition 5 in Book V of EUCLID's *Elements* in Heath's translation): "Magnitudes are said to be in the same ratio, the first to the second and the third to the fourth when, if any equimultiples whatever be taken of the first and third, and any equimultiples whatever of the second and fourth, the former equimultiples alike exceed, are alike equal to, or alike fall short of, the latter equimultiples respectively taken in corresponding order." Expressed in modern mathematical language this means: we define $a : b = A : B$ as being equivalent to the statement "$n \cdot a > m \cdot b$ if and only if $nA > mB$, $n \cdot a = m \cdot b$ if and only if $n \cdot A = m \cdot B$, and $n \cdot a < n \cdot b$ if and only if $nA < nB$," where $m, n$ are any two natural numbers.

Many of the theorems in the theory of proportion can nowadays be interpreted simply as arithmetical laws governing calculations with real numbers. It should always be remembered, however, that the Greeks never at any time regarded rational ratios, let alone irrational ratios, as extensions of the domain of natural numbers. They saw them as a concept *sui generis*. The objectives of the theory of proportion were geometrical results such as, for instance, the accurate substantiation of numerous formulae relating to areas and volumes. The geometrical proofs of these, which for the most part use *reductio ad absurdum* arguments, may seem to us long-winded and involved. But it was not until the 19th century, that more elegant methods, developed mainly since the Renaissance, could be provided with a justification as rigorous as that which had been customary in Greek mathematics.

**3. Irrational Numbers in Modern (that is, post-mediaeval) Mathematics.** After the geometrical theory of proportion of the Greeks, we now turn to the arithmetic aspect which becomes important for the development of mathematics in the modern era. Its history can be traced back to the practical calculation of approximate values, which had been practiced since very early times by mathematicians interested in astronomy and civil engineering. After the Babylonians, we need especially to remember ARCHIMEDES who, in his determination of the circumference of a circle, succeeded in showing that $\pi$ lay between $3\frac{1}{7}$ and $3\frac{10}{71}$ and PTOLEMY (*circa* 150 AD) the great astronomer of the Ancient and Mediaeval world, who chose the sexagesimal fraction $3;8,30$ as a mean between $3\frac{1}{7} = 3;8,34$ and $3\frac{10}{71} = 3;8,27$. The process of *nesting of intervals* is applied here.

While Greek mathematics was showing little interest in arithmetical calculations, which were kept very much in the background compared with geometrical constructions and proofs of propositions by logical inference, the development of the number concept gained a decisive impetus from the influence of Indo-Arabic algebra. Thus, for example, the Arab mathematician ABŪ KĀMIL (*circa* 850–930) was able to work with expressions involving square roots, using such rules as, among others:

$$\sqrt{p} + \sqrt{q} = \sqrt{p + q + 2 \cdot \sqrt{pq}}$$

(TROPFKE [23], p. 135). One begins to operate with new expressions without realizing that they are a new type of number. This process received a further impetus through the discovery, in the 16th century, of the formulae for the solution of cubic and biquadratic equations. The reader will find more on this subject in Chapter 3, §1.

M. STIFEL [22] still wrote, in his *Arithmetica integra* of 1544 "So wie eine unendliche Zahl keine Zahl ist, so ist eine irrationale Zahl keine wahre Zahl, weil sie sozusagen unter einem Nebel der Unendlichkeit verborgen ist." [Just as an infinite number is no number, so an irrational number is not a true number, because it is so to speak concealed under a fog of infinity.]

This "fog of infinity" is already defined rather more precisely by STEVIN (1548–1620) as an infinite sequence of decimal fractions, representing a sequence of nested intervals, which he develops, for example, in finding successive approximations to the solution of the equation $x^3 = 300x + 33\,900\,000$. He writes: "... et procédant ainsi infiniment, l'on approche infiniment plus près au requis" [and proceeding in this way unendingly, one approaches infinitely closer to the required value] (S. STEVIN [21], p. 353).

In the *Geometrie* of 1637 by DESCARTES, the operations of addition, subtraction, multiplication, division and root extraction of line segments are defined in such a way that the result of the operation is again a line segment in each case. Whereas the product of two line segments had hitherto always been interpreted as a rectangle, DESCARTES obtains the product as the fourth proportional in the Intercept theorem, when the first intercept is taken to be of unit length, so that 1 is to $b$ as $a$ is to $a \cdot b$.

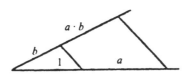

The development of the number concept received a new boost through the *infinitesimal calculus* in the 17th and 18th century. Here the theory of series, especially from the time of LEIBNIZ and the brothers BERNOULLI, opened up new possibilities for the representation of numbers. In the *Arithmetica infinitorum* of 1655, by WALLIS (1616–1703), we find, for example, the infinite product $\frac{\pi}{2} = \frac{2}{1} \cdot \frac{2}{3} \cdot \frac{4}{3} \cdot \frac{4}{5} \cdot \frac{6}{5} \cdot \frac{6}{7} \cdots$.

Representations of numbers by infinite sums and infinite products were not defined however—as has usually been the case since CAUCHY and WEIERSTRASS—as convergent sequences, using the concept of a limit. Instead, a sum such as

$$\sum_{k=1}^{\infty} \frac{1}{k(k+1)}$$

was said to differ from 1 by an "infinitesimal" or "infinitely small" quantity. EULER [9] formulated in 1734 a convergence criterion for series in the language of infinitesimals. Apart from the "finite" and "actual" (real) numbers, which found their application as values in measurement, there ap-

peared to be also "infinitesimal" numbers and "ideal" numbers. In the 19th
century such terms were banned from mathematics as being too imprecise
and "psychologizing" a form of expression, and were felt to be superfluous
after the clarification which had been brought about by the introduction
of the concept of a limit. It is only with the comparatively recent non-
standard analysis (see Chapter 12) that infinitely small numbers have once
more come into fashion and achieved full respectability.

**4. The Formulation of More Precise Definitions in the Nineteenth
Century.** CAUCHY in his *Cours d'analyse* of 1821, formulated the convergence criterion called after him and considered it as self-evident as the
laws of arithmetic. The completeness of the system of real numbers, the
property which CAUCHY is here expressing, had however already been assumed before him. Thus, for example, LEIBNIZ assumed that a continuous
line drawn on a surface, and lying partly within and partly without some
portion of that surface, must intersect the boundary of that portion.

In 1817, BOLZANO [4] proved the *Intermediate value theorem* under the
assumption of the CAUCHY criterion. However, it should be pointed out
that he already had this criterion at his disposal before CAUCHY. Recently,
a BOLZANO manuscript was discovered containing an unpublished draft of
a book entitled *Grössenlehre* (Theory of Quantities) in which he attempted
to base the theory of real numbers on firmer foundations by using sequences
of intervals.

With WEIERSTRASS consideration of the foundations of the real number
system entered into the basic mathematical curriculum. All that has come
down to us of this, however, are some notes written by his pupils and which
were in part criticized by WEIERSTRASS. The central idea of the concept
of a real number as visualized by WEIERSTRASS [24] is expressed in terms
of the principle of nesting of intervals. He also uses this to prove his well-
known Limit-point Theorem (see, also DUGAC [8]). A systematic definition
of real numbers in terms of nested intervals was given by BACHMANN [1]
in 1892.

Another method of defining real numbers was introduced by CANTOR in
his theory of *fundamental sequences* (see 2). Shortly before, MERAY (1835–
1911) had used (though CANTOR was not aware of this) this approach to
the definition of irrational numbers by regarding them as "fictive" limits of
convergent sequences and, harking back to the discovery in classical times,
calling them "nombres incommensurables."

Finally, DEDEKIND (1831–1916) in his famous book *Stetigkeit und Irrationalzahlen* [7] published in 1872 took up the theory of proportion of
EUDOXUS and presented it in a modernized form with exemplary clarity.
DEDEKIND's definition expresses our geometrical intuition of the continuum, which has been so deeply rooted since the days of classical antiquity. This intuition tells us that the points of a straight line are defined
by "the bisection of a line into two parts" (DEDEKIND) by "the common

boundary between two parts, which together constitute the whole" (LEIB-
NIZ) or by the "extremities of two parts which touch" (ARISTOTLE) see
§1). The question of whether EUDOXUS and EUCLID with their theory of
proportion had satisfactorily settled the matter of defining the irrational
numbers led to some controversy in connection with the work published by
DEDEKIND in 1872. Thus LIPSCHITZ wrote to DEDEKIND in 1876: "... Ich
kann nur sagen, daß (ich) die von Euclid V, 5 aufgestellte Definition ... für
genauso befriedigend halte, als Ihre Definition. Aus diesem Grunde würde
ich wünschen, daß namentlich die Behauptung wegfiele, daß solche Sätze
wie $\sqrt{2} \cdot \sqrt{3} = \sqrt{6}$ bisher nicht wirklich bewiesen seien." [I can only say
that I personally find the definition in Euclid V, 5 just as satisfactory as
yours. For this reason I would have liked to have seen omitted, in particular, the statement that such propositions as $\sqrt{2} \cdot \sqrt{3} = \sqrt{6}$ have never yet
really been proved.] Characteristic is LIPSCHITZ's remark: "Was Sie an der
Vollständigkeit des Gebietes erwähnen, die aus Ihren Principien abgeleitet
wird, so fällt dieselbe in der Sache mit der Grundeigenschaft einer Linie
zusammen, ohne die kein Mensch sich eine Linie vorstellen kann." [What
you say in regard to the completeness of the domain, deduced from your
principles, in point of fact merely coincides with the basic property of a
line, without which no one can possibly imagine a line.]
     While LIPSCHITZ thus expresses an attitude recalling that of the math-
ematicians of the previous century who were frequently content to rely on
an intuitive understanding of the foundations of their science, DEDEKIND
stands at the start of an era heralding a new methodical approach. He
is concerned—as were CANTOR, FREGE, PEANO and others—to formulate
explicitly and precisely the concepts on which mathematics is founded. And
so DEDEKIND writes to LIPSCHITZ with particular reference to the concept
of completeness: "...Aber Euklid schweigt vollständig über diesen, für die
Arithmetik wichtigsten Punkt, und deshalb kann ich Ihrer Ansicht nicht
zustimmen, daß bei Euklid die vollständigen Grundlagen für die Theorie
der irrationalen Zahlen zu finden seien." ["...But Euclid is completely silent
on this, the most important point for arithmetic, and therefore I cannot
share your opinion that a complete theory of irrational numbers is to be
found in Euclid."]
     The real number concept became a problem area once more in the dis-
cussions of the nineteen twenties between HILBERT and BROUWER on the
foundations of mathematics, after RUSSELL had derived contradictions from
the so-called "naive" set theory of CANTOR and FREGE, and after it was
found that even the new axiomatized versions of set theory could not be
proved to be consistent, and, as GÖDEL showed, were inherently incapable
of being proved consistent by finite methods. Within mathematical logic
these considerations led to an interesting discussion, which continues up to
the present day, of more limited concepts such as, for example, computable
numbers, and constructive real numbers (see, BISHOP [3], HERMES [12],
LORENZEN [18]).

## §2. DEDEKIND CUTS

The incompleteness of the field $\mathbb{Q}$ of rational numbers can be repaired by making "cuts" in $\mathbb{Q}$, which in an entirely natural way can be completely and totally (= linearly) ordered. Addition and multiplication are defined for these new objects in such a way that they form a field. Altogether these cuts possess the following properties (R1)–(R3), which are nowadays usually taken as a set of axioms for the real numbers.

A set $(K, +, \cdot, \leq)$ with the two (internal) compositions $+$ and $\cdot$, and the binary relation $\leq$ is said to be the *set of real numbers* if and only if the following axioms are satisfied:

(R1) $(K, +, \cdot)$ *is a field.*

(R2) $\leq$ *is a linear order relation on $K$, compatible with addition and multiplication.*

(R3) *Completeness: any non-empty subset $M$ of $K$, bounded below, has an infimum in $K$.*

A lower bound $s$ of an ordered set $M$ is said to be an *infimum* of $M$ (the standard abbreviation is inf $M$) if all lower bounds of $M$ are $\leq s$. Thus inf $M$ is clearly the greatest lower bound of $M$.

**1. The Set $\mathbb{R}$ of Cuts.** A Dedekind cut is an ordered pair $(\alpha, \beta)$ of two sets, $\alpha$ (the "left" or "lower" set) and $\beta$ (the "right" or "upper" set) with $\alpha, \beta \subset \mathbb{Q}$, satisfying the following conditions:

(D1) *Every rational number belongs to one of the two sets $\alpha$, $\beta$.*

(D2) *Neither $\alpha$ nor $\beta$ are empty.*

(D3) *Every element of $\alpha$ is less than every element of $\beta$.*

(D4) *$\beta$ has no least element ($\beta$ has no minimum).*

Every cut is uniquely determined by its left and right set each of which determines the other. We may therefore from now on identify it with its right-hand set $\beta$, which has the following properties:

(D'1) *$\beta$ and its complementary set $\bar{\beta} = \mathbb{Q} \setminus \beta$ are non-empty.*

(D'2) *If $r \in \beta$, $s \in \mathbb{Q}$ and $r < s$ then $s \in \beta$.*

(D'3) *$\beta$ has no least element (minimum).*

In the following treatment we shall use Greek letters $\alpha, \beta, \ldots$ to denote right-hand sets and call a Dedekind cut a real number. The set of all Dedekind cuts is denoted by $\mathbb{R}$.

Every rational number $s$ defines the cut: $\underline{s} := \{r : r \in \mathbb{Q}, s < r\}$, which is described as *rational*. A cut $\alpha$ is rational, if and only if $\bar{\alpha}$ has a largest element (maximum). $\mathbb{Q}$ is embedded in $\mathbb{R}$ by the mapping $\mathbb{Q} \to \mathbb{R}$, $s \mapsto \underline{s}$.

Not all cuts are rational. For example $\sqrt{2}$, that is the cut defined by $\alpha := \{r : r \in \mathbb{Q}, r > 0, r^2 > 2\}$, is not rational. It is easily verified that $\alpha$ satisfies the first two axioms for a cut. To verify the third we need to show that for every $r \in \alpha$, there is an $s \in \alpha$ satisfying $s < r$. To this end we choose $s := \frac{2r+2}{r+2} \geq 0$. Since $r - s = \frac{r^2-2}{r+2}$ and $r^2 > 2$, the inequality $r \geq 0$ entails $s < r$. Since $s^2 - 2 = \frac{2(r^2-2)}{(r+2)^2}$ and $r^2 > 2$ we have $s^2 > 2$. The cut $\alpha$ is *irrational* because the complementary set $\bar{\alpha}$ has no maximum element. For $r \in \bar{\alpha}$ with $r \geq 0$ (and thus $r^2 < 2$) we again choose $s$ as above. It then follows, since $s^2 < 2$, that $s \in \bar{\alpha}$ and $r < s$.

**2. The Order Relation in $\mathbb{R}$.** For any two cuts (right-hand sets) the order relation $\alpha < \beta$ is defined by the set-theoretic inclusion relation $\beta \subset \alpha$. The reflexivity, transitivity and antisymmetry of this relation is easily proved. The ordering is total (linear). For, suppose $\alpha \neq \beta$, and say $r \in \alpha$, with $r \notin \beta$. Then $r \in \bar{\beta}$, and for every $s \in \beta$ it follows that $r < s$, and hence $s \in \alpha$, or in other words, $\beta \subset \alpha$. The ordering is complete in the sense of the axiom (R3). To see this, let $A$ be a set of cuts bounded from below. Then $\beta = \bigcup_{\alpha \in A} \alpha$ is a cut. (Since $A$ is bounded below there is a $c \in \mathbb{Q}$ with $c \notin \beta$.) The second and third cut axioms for $\beta$ are easily checked as is the fact that $\beta$ is an infimum of $A$.

If we carry out the Dedekind cut construction once again on $\mathbb{R}$, we obtain nothing new. To every cut $a$ in $\mathbb{R}$ there corresponds a $\gamma \in \mathbb{R}$ such that $a = \{\alpha \in \mathbb{R} : \gamma < \alpha\}$. In fact, we simply take the infimum $\gamma = \bigcup_{\alpha \in a} \alpha$ of $a$.

This fact is expressed by the third of the quotations which stand at the head of this chapter. The other two quotations (from ARISTOTLE and LEIBNIZ) show that the basic underlying idea of the connected continuum is very old.

The embedding of $\mathbb{Q}$ in $\mathbb{R}$ (see 1) is compatible with the order relation. The rational numbers are dense in $\mathbb{R}$: given any two cuts (real numbers) $\alpha$ and $\beta$, there exists an $r \in \mathbb{Q}$ such that $\alpha < \underline{r} < \beta$.

**3. Addition in $\mathbb{R}$.** For any two cuts $\alpha$ and $\beta$ in $\mathbb{R}$, the *sum* $\alpha + \beta$ is defined as the set $\{r + s : r \in \alpha, s \in \beta\}$. The three characteristic properties of a cut follow for $\alpha + \beta$, from the corresponding properties of $\alpha$ and $\beta$, and so $\alpha + \beta \in \mathbb{R}$. On the subset $\mathbb{Q}$ of $\mathbb{R}$ the sum coincides with the one defined by the usual addition of rational numbers. As far as the order relation is concerned it is immediately clear that if $\alpha$, $\beta$ are any two cuts such that

$\alpha < \beta$, then $\alpha + \gamma < \beta + \gamma$ for every $\gamma$ belonging to $\mathbb{R}$.

**Theorem.** *The set $\mathbb{R}$ is an ordered commutative group with respect to addition, with (the cut) zero as its neutral element.*

**Proof.** Associativity, commutativity and $\alpha + \underline{0} = \alpha$ follow immediately from the definition of addition. The inverse of a cut $\alpha \in \mathbb{R}$ is defined as $-\alpha := \{-r : r \in \bar{\alpha}, r \neq \max \bar{\alpha}\}$. ($-\max \bar{\alpha}$ has to be excluded to ensure that the condition (D'3) is satisfied.) For the proof that $\alpha + (-\alpha) = \underline{0}$, the inclusion $\subset$ is easily checked. Conversely, suppose $r \in \underline{0}$ and thus $r > 0$; we have to show that $r \in \alpha + (-\alpha)$. Since $\bar{\alpha}$ and $\alpha$ come arbitrarily close to each other, there is an $s \in \bar{\alpha}$ and a $t \in \alpha$ such that $0 < t - s < r$. Without loss of generality we may suppose $s \neq \max \bar{\alpha}$, $-s \in -\alpha$, and therefore $t - s \in \alpha + (-\alpha)$, and because $r > t - s$, we must also have $r \in \alpha + (-\alpha)$. □

**4. Multiplication in $\mathbb{R}$.** In the case where the cuts $\alpha$, $\beta$ are both $\geq 0$, the product is defined in the way that obviously suggests itself, namely, by $\alpha \cdot \beta = \{r \cdot s : r \in \alpha, s \in \beta\}$. One can then check in routine fashion that $\alpha \cdot \beta$ satisfies the axioms (D'1) to (D'3) for a cut; that this multiplication is associative and commutative; that $\underline{1}$ is a unit element; that the distributive law holds; and that multiplication is order preserving.

The difficulties begin with the existence of multiplicatively inverse elements. If $\alpha > 0$ is a cut, we define

$$\alpha^{-1} := \{r^{-1} : r \in \bar{\alpha}, r > 0, r \neq \max \bar{\alpha}\}.$$

We leave it for the reader to check that $\alpha^{-1}$ is in fact a cut and that $\alpha \cdot \alpha^{-1} \subset \underline{1}$. To prove that $\alpha \cdot \alpha^{-1} = \underline{1}$ it only remains to show that $\underline{1} \subset \alpha \cdot \alpha^{-1}$, which can be seen as follows. Suppose $r \in \underline{1}$, and thus $r - 1 > 0$. Suppose $q \in \alpha^{-1}$. By the principle of Archimedes (see Chapter I, §4.2) for rational numbers, there is a natural number $n$ for which $q < n \cdot (r - 1)$. We now follow the same procedure as that used in the proof that $\alpha + (-\alpha) = \underline{0}$ (see 3 above). Since $\alpha$ and $\bar{\alpha}$ come arbitrarily close to one another, an $s \in \bar{\alpha}$ and a $t \in \alpha$ can be found such that $0 < t - s < n^{-1}$, where, without loss of generality, it may be assumed that $s \neq \max \bar{\alpha}$ and $q^{-1} < s$. Then $s^{-1} \in \alpha^{-1}$, and hence $t \cdot s^{-1} \in \alpha \cdot \alpha^{-1}$. Now $t \cdot s^{-1} < (s + n^{-1})s^{-1} = 1 + n^{-1}s^{-1} < 1 + n^{-1}q < r$ and therefore $r \in \alpha \cdot \alpha^{-1}$.

A further difficulty lies in the fact that the definition given above, namely, $\alpha \cdot \beta = \{r \cdot s : r \in \alpha, s \in \beta\}$, makes sense only when $\alpha \geq 0$, $\beta \geq 0$ because otherwise it does not define a cut. In order to multiply with negative cuts as well, we adopt the procedure already used in defining the multiplication of integers (see Chapter I, §3.2). We first show that every cut $\gamma$ can be written as the difference of two non-negative cuts $\alpha \geq 0$ and $\beta \geq 0$, so that $\gamma = \alpha - \beta$. The product of $\gamma = \alpha - \beta$ and $\gamma' = \alpha' - \beta'$ where $\alpha'$, $\beta'$ are also

$\geq 0$, can then be defined by the expression obtained by multiplying out

$$\gamma \cdot \gamma' = (\alpha - \beta) \cdot (\alpha' - \beta') := \alpha \cdot \alpha' + \beta \cdot \beta' - \alpha \cdot \beta' - \beta \cdot \alpha'.$$

It is easily checked that the cut so defined depends only on $\gamma$ and $\gamma'$ and not on the particular difference representations chosen. When $\gamma$ and $\gamma'$ are both $\geq 0$ the new definition agrees with the old. This latter point is easily seen by considering the representation $\gamma = \gamma - 0$, $\gamma' = \gamma' - 0$. However, it is a tedious if routine business to verify that all the axioms for a field are verified. E. LANDAU who carries out this task in detail in [16], writes in his "Vorwort für den Kenner" [Foreword for the expert]: "Ein anderer hat sich meine zum Teil langweilige Mühe nicht gemacht." [... but no one else has undertaken this task which is in part rather boring.] In his "Vorwort für den Lernenden" [Foreword for the student] on the other hand, he says: "Bitte vergiß alles, was Du auf der Schule gelernt hast; denn Du hast es nicht gelernt." [Please forget all you learnt at school because you never learnt it.]

It is undoubtedly true that when we set out to justify all the operations with numbers which have been so familiar to us from our school days, we have to take great care to use only what has already been proved, and not to assume things to be true merely because they are so familiar to us.

## §3. FUNDAMENTAL SEQUENCES

**1. Historical Remarks.** The definition of real numbers by means of fundamental sequences, which goes back to CANTOR and MÉRAY [19], uses the idea that every real number is the limit of a sequence of rational numbers, in which the differences between the successive terms become arbitrarily small. Such a sequence is known as a "fundamental sequence," and is illustrated below, the successive terms $r_1, r_2, \ldots$ being indicated by subscripts.

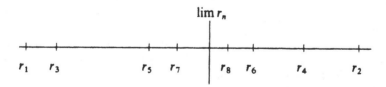

CANTOR's contribution to the theory of irrational numbers forms part (§9) of a larger work, *Grundlagen einer allgemeinen Mannigfaltigkeitslehre* [Foundations of a general theory of manifolds (that is, sets in present-day terminology)] published in 1883, in which he develops his new theory of sets. In addition to his own definition, CANTOR also mentions the approach taken by WEIERSTRASS and the work of DEDEKIND. In CANTOR's view the logical clarity of DEDEKIND's definition has to be set against the "great disadvantage" that "numbers in analysis never present themselves in the form of "cuts," and therefore have first of all to be brought into this form by elaborate artifices." On the other hand, CANTOR leaves no doubt that he regards his form of definition as the "simplest and most natural of all." He mentions as contributing to the historical development of this approach a paper of his own published in 1871 (*Math. Ann:* 5, p. 123) and a book by LIPSCHITZ [17].

Quite apart from its use in the definition of real numbers, the CANTOR construction with fundamental sequences has turned out to be the most fruitful, inasmuch as it can also be used for the completion of metric spaces. In this sense one has to agree with CANTOR when he asserts, in speaking of his construction: "Man hat an ihr den Vorteil, daß sie sich dem analytischen Kalkül am unmittelbarsten anpaßt." [It has the advantage of being the one most immediately suited to analytical calculations.] In the following section the basic facts about sequences will be assumed.

**2. CAUCHY's Criterion for Convergence.** In accordance with CANTOR's basic idea, real numbers can be described by convergent rational sequences. Two rational sequences $(r_n)$ and $(s_n)$ have the same (real) limit, if and only if the sequence of their differences $(r_n - s_n)$ converges to zero. It is natural therefore to define the real numbers as equivalence classes of convergent rational sequences; two sequences being equivalent when their difference sequence converges to zero. For this definition to be meaningful, the convergence of a sequence has to be characterized without making use of its limit. This can be done with the help of *Cauchy's criterion*, which will be used to define the sequences concerned.

A sequence $(r_n)$ of rational numbers is said to be a *fundamental sequence* or *Cauchy sequence*, if, for every rational $\varepsilon > 0$, there is an index $k$, such that $|r_m - r_n| < \varepsilon$ for all $m, n \geq k$.

The rational sequence $(r_n)$ is said to be *rationally convergent* if there is a rational number $r$, such that for every $\varepsilon > 0$, there exists an index $k$, with $|r_n - r| < \varepsilon$ for all $n \geq k$. In that case $r$ is defined uniquely, and one writes $r = \lim r_n$. Every rationally convergent sequence is a fundamental sequence.

On the other hand there are fundamental sequences which do not converge rationally. Every non-periodic decimal fraction provides an example, for example, the one for $\sqrt{2}$, where

$$r_0 = 1; \quad r_1 = 1.4; \quad r_2 = 1.41; \quad r_3 = 1.414; \quad r_4 = 1.4142; \ldots .$$

To give another example, where the law for formation of the terms of the sequence is shown explicitly, we consider the continued fraction for the ratio $\frac{1}{2}(1 + \sqrt{5})$ corresponding to the golden section (see, 1.1). This continued fraction is defined recursively by the sequence $r_0 = 1$, $r_{n+1} = 1 + \frac{1}{1+r_n}$. To prove that this is a fundamental sequence, we shall show that $|r_{n+1} - r_n| <$

$\frac{1}{2}|r_n - r_{n-1}|$. For

$$r_{n+1} - r_n = 1 + \frac{1}{1 + r_n} - \left(1 + \frac{1}{1 + r_{n-1}}\right) = \frac{r_{n-1} - r_n}{(1 + r_n)(1 + r_{n-1})}$$

and $r_{n-1}, r_n \geq 1$. It follows therefore, by complete induction, that

$$|r_{n+1} - r_n| < 2^{-n}|r_1 - r_0| = 2^{-n-1},$$

and hence

$$|r_{n+k} - r_n| \leq |r_{n+k} - r_{n+k-1}| + |r_{n+k-1} - r_{n+k-2}| + \cdots + |r_{n+1} - r_n|$$
$$< 2^{-n-k} + 2^{-n-k-1} + \cdots + 2^{-n-1} < 2^{-n}.$$

For any given $\varepsilon > 0$, we can choose $l$ so that $2^{-l} \leq \varepsilon$. Consequently $|r_{n+k} - r_n| < \varepsilon$ for all $n \geq 1$ and all $k$.

**3. The Ring of Fundamental Sequences.** The set $F$ of all fundamental sequences becomes a ring when addition and multiplication are defined termwise:

$$(r_n) + (s_n) := (r_n + s_n) \quad \text{and} \quad (r_n) \cdot (s_n) = (r_n \cdot s_n).$$

It is verified as follows, that the sum and product are likewise fundamental sequences. For any given $\varepsilon > 0$, $k$ may be chosen large enough to ensure that $|r_m - r_n| < \frac{1}{2}\varepsilon$ and $|s_m - s_n| < \frac{1}{2}\varepsilon$ for all $m, n \geq k$. Then $|r_m + s_m - r_n - s_n| \leq |r_m - r_n| + |s_m - s_n| < \varepsilon$. In the case of the product we first use the fact that fundamental sequences are bounded so that there is a $c \geq 1$, such that $|r_n|, |s_n| \leq c$. For any given $\varepsilon > 0$, we choose $k$ large enough to ensure that $|r_m - r_n|, |s_m - s_n| < \frac{1}{2}\frac{\varepsilon}{c}$ for all $m, n \geq k$. Then

$$|r_m s_m - r_n s_n| = |r_m(s_m - s_n) + s_n(r_m - r_n)|$$
$$\leq |r_m||s_m - s_n| + |s_n||r_m - r_n| < c\frac{1}{2}\frac{\varepsilon}{c} + c\frac{1}{2}\frac{\varepsilon}{c} = \varepsilon.$$

$\mathbb{Q}$ can be embedded as a subring in $F$ by associating with each $r \in \mathbb{Q}$ the constant sequence $(r, r, r, \ldots)$.

**4. The Residue Class Field $F/N$ of Fundamental Sequences Modulo the Null Sequence.** A rational sequence $(r_n)$ is said to be a *null sequence* when $\lim r_n = 0$. The set $N$ of null sequences is an *ideal* in $F$, or in other words, (1) if $(r_n)$ and $(s_n)$ are null sequences, then so is $(r_n + s_n)$ and (2) if $(r_n)$ is a null sequence and $(s_n)$ any fundamental sequence, then $(r_n \cdot s_n)$ is a null sequence.

Two fundamental sequences are said to be equivalent if their difference is a null sequence. (The reader should check that this does in fact define an equivalence relation.) The equivalence class represented by $(r_n)$ is $(r_n) +$

$N := \{(r_n + h_n): (h_n) \in N\}$. It is called the residue class of $r_n$ modulo $N$. As $N$ is an ideal, the residue classes can be added and multiplied: $((r_n) + N) + ((s_n) + N) = (r_n + s_n) + N$ and $((r_n) + N) \cdot ((s_n) + N) = (r_n \cdot s_n) + N$. The set $F/N$ of residue classes in this way constitutes a commutative ring with unit element. It contains $\mathbb{Q}$ as a subset, where we identify each rational $r$ with its associated class of constant sequences modulo $N$.

**Theorem.** *The residue classes of the fundamental sequences modulo the null sequences form a field $F/N$.*

**Proof.** For every $(r_n) + N$ with $(r_n) \notin N$ we have to be able to define a class which is its multiplicative inverse. The obvious candidate is $(1/r_n) + N$. However, for this we need to have $r_n \neq 0$. In point of fact we are entitled to assume this. Since $(r_n) \notin N$, only a finite number of terms of the sequence are equal to zero. We replace these by 1. This does not alter the class of $(r_n) + N$. We now have to show that $(1/r_n)$ is a fundamental sequence: since $(r_n) \notin N$ and all $r_n$ are nonzero, there is $\delta > 0$ such that $|r_n| > \delta$ for all $n$. For any given $\varepsilon > 0$ we choose the index $k$ large enough to ensure that $|r_m - r_n| < \delta^2 \varepsilon$ for all $m, n \geq k$. Then

$$\left| \frac{1}{r_m} - \frac{1}{r_n} \right| = \frac{|r_m - r_n|}{|r_m r_n|} < \frac{\delta^2 \varepsilon}{\delta\delta} = \varepsilon.$$

Following CANTOR we now define the *field of real numbers* as $\mathbb{R} := F/N$.                                                                     □

## 5. The Completely Ordered Residue Class Field $F/N$.

A rational fundamental sequence $(r_n)$ is said to be *positive* if there is a rational $\varepsilon > 0$ such that $r_n > \varepsilon$ for almost all (that is, for all but a finite number of) indices $n$. Let $P$ be the set of positive fundamental sequences. Clearly $P + N \subset P$, $P + P \subset P$ and $P \cdot P \subset P$. The set $F$ of all fundamental sequences can be expressed as a union of disjoint subsets $F = -P \cup N \cup P$. We can therefore obtain a well defined total ordering on $F/N$ by defining

$$(r_n) + N \geq (s_n) + N \quad \text{if and only if} \quad (r_n - s_n) \in P \cup N.$$

The sum and product of positive elements in $F/N$ are themselves positive. On the subset $\mathbb{Q} \subset F/N$, the ordering coincides with the usual ordering of the rational numbers.

It follows from the definition of positive rational fundamental sequences that for every $\rho \in F/N$ with $\rho > 0$, there is an $r \in \mathbb{Q}$, with $0 < r < \rho$. It makes no difference, therefore, to the definition of convergence in $F/N$, whether one allows all positive $\varepsilon \in F/N$, or only those that belong to $\mathbb{Q}$. It is equally true that for every $\sigma \in F/N$ there is an $s \in \mathbb{Q}$, with $s \geq \sigma$. (This is trivial for $\sigma < 0$, and if not one can choose an $r \in \mathbb{Q}$, such that $0 < r \leq \sigma^{-1}$ and take $s = r^{-1}$.)

The ordering of $F/N$ is *Archimedean*, for if $\alpha$, $\beta$ are both positive and belong to $F/N$, a natural number $n$ such that $n\alpha > \beta$ can be found in the following manner. We choose $a, b \in \mathbb{Q}$, such that $0 < a < \alpha$ and $\beta < b$. Since $\mathbb{Q}$ is Archimedically ordered there is an $n$ such that $na > b$, and thus $n\alpha \geq na > b \geq \beta$.

The field $F/N$ was so constructed that (1) every $\rho \in F/N$ is the limit of a rational sequence $(r_n)$ and (2) every rational fundamental sequence in $F/N$ converges. We can improve (2) to the following.

**Theorem.** *Cauchy's criterion for convergence is valid in $F/N$.* A sequence $(\rho_n)$ in $F/N$ converges if and only if the following condition is satisfied: for every $\varepsilon > 0$ there is an index $k$, such that

$$|\rho_m - \rho_n| < \varepsilon \quad \text{for all} \quad m, n \geq k.$$

**Proof.** By (1) there is, for every $\rho_n$, an $r_n \in \mathbb{Q}$, such that $|\rho_n - r_n| < \frac{1}{n}$. Then $(r_n)$ is a fundamental sequence: for any given $\varepsilon > 0$ we choose the index $k$ so that $\frac{1}{k} < \frac{1}{3}\varepsilon$ and $|\rho_m - \rho_n| < \frac{1}{3}\varepsilon$ for all $m, n \geq k$. Then

$$|r_m - r_n| \leq |r_m - \rho_m| + |\rho_m - \rho_n| + |\rho_n - r_n| < \frac{1}{m} + \frac{1}{3}\varepsilon + \frac{1}{n} < \varepsilon.$$

By (2) the sequence $(r_n)$ converges to a $\rho \in F/N$, and hence $(\rho_n)$ also converges to $\rho$, because to any given $\varepsilon > 0$ one can choose the index $l$ sufficiently large to ensure that $\frac{1}{l} < \frac{1}{2}\varepsilon$ and $|\rho - r_n| < \frac{1}{2}\varepsilon$ for all $n \geq 1$ and thus $|\rho - \rho_n| \leq |\rho - r_n| + |r_n - \rho_n| < \frac{1}{2}\varepsilon + \frac{1}{n} \leq \varepsilon$ for all $n \geq l$. $\qquad\square$

Numerous different formulations for the completeness of totally ordered fields will be given in 5.2 and compared with one another. In particular it will emerge among other things that the completeness axiom (R3) is equivalent to the assertion that the ordering is Archimedean and that the Cauchy criterion for convergence holds. Thus the *Cantor field $F/N$ satisfies all the axioms for the real numbers.* Any two fields satisfying these axioms will be shown in 5.3 to be canonically isomorphic. In particular therefore $F/N$ is isomorphic to the field of Dedekind cuts.

## §4. NESTING OF INTERVALS

**1. Historical Remarks.** The idea of fitting intervals, one within another, to form a so-called nest of intervals is an old one and is found above all in applied mathematics in connection with the calculation of approximate values. In *Babylonian* times, we already find the sexagesimal fractions $1; 25 = 1 + \frac{25}{60}$ and $1; 24, 51, 10 = 1 + \frac{24}{60} + \frac{51}{60^2} + \frac{10}{60^3}$ as approximations for $\sqrt{2}$ (see, NEUGEBAUER AND SACHS [20], p. 42). These can be obtained by the following general process for enclosing $\sqrt{a}$ within smaller and smaller intervals, which is applicable to any $a > 1$:

$$a > \sqrt{a} > 1,$$

$$x_0 = \frac{1}{2}(a+1) > \sqrt{a} > \frac{a}{x_0},$$

$$x_1 = \frac{1}{2}\left(x_0 + \frac{a}{x_0}\right) > \sqrt{a} > \frac{a}{x_1},$$

$$x_2 = \frac{1}{2}\left(x_1 + \frac{a}{x_1}\right) > \sqrt{a} > \frac{a}{x_2}.$$

In fact, when $a = 2$, we obtain the values $x_0 = \frac{3}{2} = 1;30$, $x_1 = \frac{1}{2}\left(\frac{3}{2} + \frac{4}{3}\right) = \frac{17}{12} = 1;25$ and $x_2 = \frac{1}{2}\left(\frac{17}{12} + \frac{24}{17}\right) = \frac{577}{408} = 1;24,51,10$. However, the general process is not explicitly given as such in the Babylonian texts, so that we are relying on a plausible assumption. This process can be regarded as an application of the proposition that the geometric mean lies between the harmonic mean and the arithmetic mean: $\frac{2ab}{a+b} < \sqrt{a \cdot b} < \frac{a+b}{2}$, to the particular case $b = 1$. This was already known to the Pythagoreans, as a fragment from ARCHYTAS OF TARENTUM shows (see BECKER [2], p. 78 *et seq.*).

The determination of the area of a circle as lying between those of inscribed and circumscribed polygons is another example of the nesting of intervals. It was STEVIN who around the year 1594 used the technique of calculating with decimals and defined a real number by the nesting of intervals (see, 1.3). In the 19th century nested intervals were used in proving some of the central theorems of analysis. An attempt to define real numbers by certain sequences of intervals in order to prove CAUCHY's criterion for convergence goes back to BOLZANO [4]. WEIERSTRASS [24] uses the nesting of intervals to prove his theorem on limit points (the theorem that a bounded infinite set has a limit point). Finally, BACHMANN in his *Vorlesungen über die Theorie der Irrationalzahlen* (Leipzig, 1892) introduces real numbers by systematically making use of nested intervals.

**2. Nested Intervals and Completeness.** The introduction of real numbers by means of nested intervals is motivated by the following situation. We consider a sequence of intervals $I_1, I_2, \ldots, I_n, \ldots$, on the arithmetical line continuum (or real axis) each of which is contained within the one which precedes it, and such that the length of $I_n$, the $n$th interval, tends to zero as $n$ increases. (In the particular case of decimal intervals the length of $I_n$ is $10^{-n}$, and the endpoints of $I_n$ are integral multiples of $10^{-n}$.) We require that corresponding to every such sequence of nested intervals there should exist one and only one point on the real axis which is contained in all the intervals of the sequence:

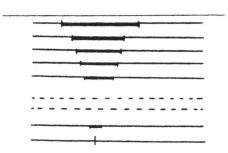

A rational sequence of nested intervals, or more shortly a rational *net*, is a sequence of closed intervals $[r_n, s_n]$ with $r_n, s_n \in \mathbb{Q}$, such that $I_n \supset I_{n+1}$ for all $n$, and $\lim(s_n - r_n) = 0$. A net $(J_n)$ is said to be *finer* than $(I_n)$, if $J_n \subseteq I_n$ for all $n$. We say that $(I_n)$ and $(I'_n)$ are *equivalent* if there is a net $(J_n)$ which is finer than each, and we say that $(J_n)$ is a *refinement* of $(I_n)$ and of $(I'_n)$. This is so, if and only if $r''_n = \max(r_n, r'_n) \leq s''_n = \min(s_n, s'_n)$ because $I''_n = [r''_n, s''_n]$ is then a common refinement. We can now define *real numbers* as equivalence classes of nets. The rational numbers are embedded in these real numbers inasmuch as to every $r \in \mathbb{Q}$, corresponds the equivalence class containing the (constant) net $(I_n)$ defined by $I_n := [r, r]$ for all $n$.

An example of a net of nested intervals is $([e_n, e'_n])$ where $e_n := \left(1 + \frac{1}{n}\right)^n$ and $e'_n := \left(1 + \frac{1}{n}\right)^{n+1}$. This net defines the real number $e = 2.71828\ldots$, introduced by EULER, which is of fundamental importance in analysis in the theory of the logarithmic and exponential functions (see also Chapter 5).

At this point addition, multiplication and an ordering for these equivalence classes of nets ought to be defined and the axioms (R1)–(R3) stated at the beginning of §2 ought to be verified. We shall not adopt this course, however, but instead set up a direct correspondence between nets and Dedekind cuts (§2) on the one hand, and between nets and fundamental sequences (§3) on the other.

Corresponding to a given net $([r_n, s_n])$ we form the sets $\alpha := \{x : x \in \mathbb{Q},$ and $x \leq s_n$ for all $n\}$ and $\beta' := \{y : y \in \mathbb{Q}$ and $y > r_n$ for all $n\}$. If $\beta'$ contains a least element, we remove it and form the set $\beta := \beta' - \{\min \beta'\}$. Then $(\alpha, \beta)$ has the properties (D1)–(D4) of the Dedekind cut (see 2.1). If we refine the net, the cut remains unchanged. Conversely, to every Dedekind cut $(\alpha, \beta)$ there corresponds a net $([r_n, s_n])$ with $r_n \in \alpha$ and $s_n \in \beta$. We begin with any $r_0 \in \alpha$, $s_0 \in \beta$ and proceed recursively: having obtained $r_n, s_n$ we form the arithmetic mean $d_n = \frac{1}{2}(r_n + s_n)$ and define

$$[r_{n+1}, s_{n+1}] = \begin{cases} [d_n, s_n], & \text{if} \quad d_n \in \alpha, \\ [r_n, d_n], & \text{if} \quad d_n \in \beta. \end{cases}$$

All nets $[r_n, s_n]$ with $r_n \in \alpha$ and $s_n \in \beta$ are equivalent. We associate $(\alpha, \beta)$

with the equivalence class. The two correspondences that have thus been defined are mappings inverse to one another. If the rational numbers are regarded firstly as equivalence classes of constant nets, and secondly as rational cuts, then the former is the image of the latter and vice versa in the correspondence which has just been described.

The direct relationship between nets and fundamental sequences rests on the following facts: (1) every bounded, monotone sequence is a fundamental sequence. (2) to every rational fundamental sequence $(a_n)$ corresponds a monotonically increasing rational sequence $(r_n)$ and a monotonically decreasing rational sequence $(s_n)$, such that $(r_n - a_n)$ and $(s_n - a_n)$ are null sequences. Now if $([r_n, s_n])$ is a given net of nested intervals, $(r_n)$ and $(s_n)$ are fundamental sequences, and $(s_n - r_n)$ is a null sequence. If the net is refined to $([r'_n, s'_n])$, $(r'_n - r_n)$ is a null sequence. The correspondence $([r_n, s_n]) \mapsto (r_n)$ therefore induces a well defined mapping of equivalence classes of rational nets of nested intervals into the Cantor field $F/N$ of fundamental sequences modulo the null sequences. Conversely, corresponding to any given fundamental sequence $(a_n)$ one can choose a monotonically increasing sequence $(r_n)$ and a monotonically descreasing sequence $(s_n)$ by the rule (2), and then $([r_n, s_n])$ will be a net. If one had started from another fundamental sequence $(a'_n)$ instead of from $(a_n)$ so that $(a'_n - a_n)$ were a null sequence, and had then chosen $(r'_n)$ and $(s'_n)$ by the rule (2), then clearly $([r_n, s_n])$ would be equivalent to $([r'_n, s'_n])$. We therefore have a well defined mapping of the fundamental sequences modulo the null sequence into the set of equivalence classes of nets of nested intervals. This mapping is inverse to the one described above.

The *practical advantages* of nested intervals over cuts or fundamental sequences are as follows. If the real number $x$ is described by $(I_n)$ the position of $x$ on the number axis is fixed within defined bounds by each $I_n$. On the other hand with a fundamental sequence $(r_n)$, the knowledge of one $r_n$ still tells us nothing about the position of $x$. Again, the description of $x$ as a cut $(\bar{\alpha}, \beta)$ can result from a definition of the set $\alpha$ by means of statements which say nothing directly about the position of $x$.

The theoretical disadvantage of using the nested interval approach is that introducing the $\leq$ relation between equivalence classes of nets of nested intervals and verifying the field properties for addition and multiplication is somewhat troublesome.

## §5. AXIOMATIC DEFINITION OF REAL NUMBERS

While axiomatic methods were at first used only in geometry (see, EUCLID's *Elements*), it was not until comparatively recently with the publication of HILBERT's *Grundlagen der Geometrie* [13] [Foundations of geometry] that they were also used for real numbers. The axiomatic treatment that follows will however be based not on the system of axioms proposed by HILBERT (in §13 of [13], where it is called "the theory of ratios," following the tradition set by EUCLID in his *Elements*), but on the axioms (R1)–(R3) of §2.

**1. The Natural Numbers, the Integers, and the Rational Numbers in the Real Number Field** should all be recoverable once the latter has been defined axiomatically by (R1)–(R3). For this purpose only (R1) and (R2) are needed. Thus let $K$ be a totally ordered field, or in other words let $K$ satisfy the axioms (R1) and (R2) of §2. We shall say that a subset $M \subset K$ is inductive, if $0 \in M$ and $x+1 \in M$ whenever $x \in M$. For example, $K$ itself and the subset $K^+ = \{x : x \in K, \ x \geq 0\}$ are both inductive. The intersection $N$ of all inductive subsets of $K$, is the smallest inductive subset of $K$. It fulfils, with the successor function $S(x) := x+1$ the axioms (S1)–(S3) for the natural numbers, formulated in 2.1 of Chapter 1. By the Uniqueness theorem (2.2 of Chapter 1) the set $N \subset K$ can therefore be identified unambiguously with $\mathbb{N}$.

Let $Z \subset K$ be the smallest subring containing 1. By complete induction, it follows that $\mathbb{N} \subset Z$. Thus $Z$, as the smallest ring that contains $\mathbb{N}$, is in a unique way isomorphic to $\mathbb{Z}$ (see 3.2 of Chapter 1).

Let $Q \subset K$ be the smallest subfield. It contains the smallest subring $\mathbb{Z}$, and hence $Q$ is in a unique way isomorphic to $\mathbb{Q}$ (see Chapter 1, §4.2).

*The ordered field $K$ has the Archimedean property (that is, given any two elements $a, b > 0$ in $K$, an $n \in \mathbb{N}$ can always be found such that $na > b$) if and only if $\mathbb{Q}$ is dense in $K$, that is to say, between any two elements $x < y$ in $K$, there is an $r \in \mathbb{Q}$, such that $x < r < y$.*

This proposition has already been proved in one direction (when $\mathbb{Q}$ is dense in $K$) in §3.5 (with $K = F/N$). Conversely, if $a = 1$ and $b = (y-x)^{-1}$ there is an $n \in \mathbb{N}$ with $(y-x)^{-1} < n$. Moreover, we can now find an $m \in \mathbb{Z}$, such that $\frac{m}{n} \leq x < \frac{m+1}{n}$ and then $x < \frac{m+1}{n} \leq x + \frac{1}{n} < y$, the last inequality being a consequence of $(y-x)^{-1} < n$.

**2. Completeness Theorem.** Each of the three different methods of constructing the real numbers, by cuts, by fundamental sequences, and by nested intervals, is based on a different formulation of the idea of completeness. We shall now show that each is equivalent to the completeness axiom (R3) of §2.

*Let $K$ be a totally ordered field, that is, suppose the axioms (R1) and (R2) of §2 to be satisfied for $K$. Then the following statements are equivalent.*

  (a) *Every subset of $K$ that is bounded below possesses an infimum (greatest lower bound).*

  (a′) *Every subset of $K$ that is bounded above possesses a supremum (least upper bound).*

  (b) *If $(\alpha, \beta)$ is a cut in $K$ (that is, the axioms (D1)–(D4) of §2 are* *satisfied when elements of $K$ instead of rational numbers are taken) then $\alpha$ contains a maximum element.*

(c) *Every monotonically decreasing sequence, bounded below, converges in $K$.*

(c') *Every monotonically increasing sequence, bounded above, converges in $K$.*

(d) *The field $K$ has an Archimedean ordering and every fundamental sequence (Cauchy sequence) of elements of $K$ converges in $K$.*

(e) *The field $K$ has an Archimedean ordering and for every sequence of nested intervals $I_0 \supset I_1 \supset \cdots \supset I_n \cdots$ in $K$, for which the lengths of $I_n$ converge to zero with increasing n, there exists one and only one s lying in all the intervals $I_n$.*

(a) and (a') are obviously equivalent: if and only if $M$ is bounded below is $-M = \{-x : x \in M\}$ bounded above, and $-\inf M = \sup(-M)$. Similarly (c) and (c') are equivalent. The complete equivalence of all the assertions will follow from the implications (a) $\to$ (b) $\to$ (c) $\to$ (d) $\to$ (e) $\to$ (a) which we shall prove in turn.

(a) $\to$ (b): The set $\beta$ is bounded below, every $a \in \alpha$ is a lower bound. By (a) $\beta$ has an infimum. Since $\beta$ has no minimum, $\inf \beta \in \alpha$. Since $a < b$ holds for all $a \in \alpha$ and $b \in \beta$, we have $a \leq \inf \beta$ for all $a \in \alpha$, that is $\inf \beta$ is the maximum of $\alpha$.

(b) $\to$ (c): Let $(b_n)$ be a monotonically decreasing sequence, bounded below. We can define a cut $(\alpha, \beta)$ by $\alpha = \{x : x \leq b_n \text{ for all } n\}$ and $\beta = \{y : \text{there is an } n \text{ such that } b_n < y\}$. By (b) the set $\alpha$ has a maximum $s$. We can now show that $(b_n)$ converges to $s$. To prove this suppose $\varepsilon > 0$ be given, then there is an index $k$ such that $b_k < s + \varepsilon$ because if $s + \varepsilon$ were $\leq b_k$ for all $k$, we should have $s + \varepsilon \in \alpha$, in contradiction to $s = \max \alpha$. As $(b_n)$ is monotonically decreasing, $b_m \leq b_k$ for all $m \geq k$, and since $s \leq b_m$ for all $m$, we therefore have $s \leq b_m \leq b_k < s + \varepsilon$ for all $m \geq k$.

(c) $\to$ (d): The Archimedean property of the ordering of $K$ can be proved as follows. Let $a, b$ be $> 0$, and suppose that $na \leq b$ for all $n \in \mathbb{N}$. Then $(na)$ would be a monotonically increasing sequence bounded above, which by (c) would converge to some $s$. There would therefore also be an index $k$ such that $s - a < na < s$ for all $n \geq k$. Between $s - a$ and $s$ there is however room for only one term $na$ of the sequence $(na)$.

To prove that every fundamental sequence converges, we need two lemmas.

(1) Every sequence $(a_n)$ has a monotonic subsequence.

(2) Every fundamental sequence is bounded.

We shall postpone the proof of (1) and (2) for a moment and first show that every fundamental sequence $(a_n)$ converges. Let $(a_n)$ be a monotonic subsequence. It is bounded and hence $s = \lim_{j \to \infty} a_{n_j}$, exists. We assert that $s = \lim_{n \to \infty} a_n$; for given any $\varepsilon > 0$, one can choose the index $k$ so that $|a_m - a_n| < \frac{1}{2}\varepsilon$ for all $m, n \geq k$, and there is then a $j$ such that $n_j \geq k$ and $|a_{n_j} - s| < \frac{1}{2}\varepsilon$. It now follows that $|a_n - s| \leq |a_n - a_{n_j}| + |a_{n_j} - s| < \varepsilon$ for all $n \geq k$.

**Proof of Lemma (1).** We shall say that the sequence $(a_n)$ has a *peak* $a_k$ for the index $k$, if $a_k \geq a_n$ for all $n \geq k$. If there is an infinity of peaks then they form a monotonic non-increasing sequence. If there are no peaks or only a finite number of peaks, there is a last index $k$ beyond which there are no peaks. We begin our subsequence with $n_0 = k + 1$. Since $a_{n_0}$ is not a peak there is an $n_1 > n_0$, for which $a_{n_1} > a_{n_0}$. Since $a_{n_1}$ is not a peak, there is an $n_2 > n_1$, such that $a_{n_2} > a_{n_1}$ and so on. We have thus found by recursion a monotonically increasing subsequence $(a_{n_j})$.

**Proof of Lemma (2).** Let $(a_n)$ be a fundamental sequence. There is an index $k$ such that $|a_m - a_n| < 1$ for all $m, n \geq k$. In particular therefore all subsequent terms $a_n$ for $n \geq k$ lie within the bounded interval $(a_k - 1, a_k + 1)$. The finitely many initial terms $a_0, \ldots, a_{k-1}$ of the sequence obviously also form a bounded set, and consequently the set of all terms $a_n$ with $n \in \mathbb{N}$ is also bounded.

(d) $\rightarrow$ (e): Let $([a_n, b_n])$ be a sequence of nested intervals. Then $(a_n)$ is a fundamental sequence, because, for every $k$ and all $m, n \geq k$, $a_m, a_n$ lie in $[a_k, b_k]$, and hence $|a_m - a_n| < b_k - a_k$. Since $\lim(b_n - a_n) = 0$ we can therefore ensure that $|a_m - a_n| < \varepsilon$ by choosing $k$ large enough. By (d), $s = \lim a_n$ exists. Since $(a_n)$ increases monotonely, $a_n \leq s$ for all $n$. As $a_k \leq b_n$ for all $k$ and $n$, we also have $s \leq b_n$ for all $n$, and thus $s \in [a_n, b_n]$ for every $n$. Since $b_n - a_n$ becomes arbitrarily small as $n$ increases, $s$ is defined unambiguously.

(e) $\rightarrow$ (a): Let $M$ be a non-empty subset of $K$, bounded below. We can construct a sequence of nested intervals $([a_n, b_n])$, in which all the $a_n$ are lower bounds of $M$, while none of the $b_n$ are lower bounds of $M$. We begin with any lower bound $a_0$ and a $b_0$ which is not. We then proceed recursively: having already defined $[a_n, b_n]$ we form the arithmetic mean $d_n = \frac{1}{2}(a_n + b_n)$ and define

$$[a_{n+1}, b_{n+1}] = \begin{cases} [d_n, b_n], & \text{if } d_n \text{ is a lower bound} \\ [a_n, d_n], & \text{if } d_n \text{ is not a lower bound.} \end{cases}$$

Then $b_{n+1} - a_{n+1} = \frac{1}{2}(b_n - a_n)$, so that $b_n - a_n = 2^{-n}(b_0 - a_0)$. As the ordering is Archimedean, $\lim(b_n - a_n) = 0$. By (e) there is just one $s$ which lies in all the intervals $[a_n, b_n]$. Now $c$ is a lower bound of $M$, for otherwise there would be an $x \in M$ with $x < c$, and since every $a_n \leq x$ we should have $b_n - a_n \geq c - a_n \geq c - x$ which would contradict $\lim(b_n - a_n) = 0$. This $c$ is the greatest of the lower bounds, because if $b > c$ were a lower bound, we should have to have $b_n > b$ and $b_n - a_n > b - a_n > b - c$ in contradiction to $\lim(b_n - a_n) = 0$.                                              □

The list (a)–(e) of equivalent statements by no means exhausts all the possible formulations. One could for example also mention the HEINE–BOREL covering property or the fact that every bounded infinite subset contains a limit point. The student learns about these and other results, as consequences of the property of completeness, in every introductory course on analysis.

There are totally ordered fields in which every fundamental sequence converges, but in which the ordering is not Archimedean. An example of this will be given in Chapter 12 where the real numbers will be extended to the field $^*\mathbb{R}$ of non-standard numbers. In this extended field there are infinitely small and infinitely large numbers, and for this reason $^*\mathbb{R}$ is not Archimedean, while every fundamental sequence is constant and therefore convergent. Just how much the Archimedean axiom restricts the possibilities is shown clearly by the following result due to HÖLDER [13a], see also CARTAN [6]: An ordered group is Archimedean if and only if it is isomorphic to a subgroup of the additive group of real numbers. One does not even have to assume that the group is commutative; it follows from the other hypotheses.

**3. Existence and Uniqueness of the Real Numbers.** We now show that the axiom system (R1)–(R3) for the real numbers characterizes them unambiguously. Let $F/N$ be the Cantor field of fundamental sequences modulo the null sequences.

**Theorem.** *Every ordered field $K$ satisfying the axioms (R1)–(R3) is isomorphic to $F/N$ in one and only one way.*

**Proof.** The mapping $\varphi \colon K \to F/N$ is defined as follows. Let $x$ be an element of $K$; since $\mathbb{Q}$ is dense in $K$, there is a rational fundamental sequence $(x_n)$ with $\lim x_n = x$. We set $\varphi(x) = (x_n) \bmod N$. This definition does not depend on the choice of $(x_n)$ because, for any other choice, say $(x_n')$ the differences $x_n' - x_n$ form a null sequence. As the limit is compatible with the sum and product, $\varphi$ is a homomorphism. Clearly $\varphi$ maps the rationals on to themselves, and in particular $\varphi$ is not the null homomorphism, while its kernel must be the null ideal, or in other words $\varphi$ is injective. So far we have used only the fact that $K$ is Archimedean. From the hypothesis that every (rational) fundamental sequence in $K$ converges, it follows that $\varphi$ is also surjective, and hence an isomorphism.

The *uniqueness* of $\varphi$ is a consequence of the following result, which is also of interest in itself.

*The field of real numbers has no automorphisms apart from the identity mapping.*

By the "field of real numbers" is here meant any field $K$ which satisfies the axioms (R1)–(R3). To prove this we start from the fact that $K$ must contain the field $\mathbb{Q}$ of the rationals. Every automorphism $\sigma$ of $K$ maps $\mathbb{Q}$ identically on to itself, since $\sigma(0) = 0$ and $\sigma(1) = 1$ and it follows therefore by complete induction that $\sigma \mid \mathbb{N} = \mathrm{id}_{\mathbb{N}}$. As every element of $\mathbb{Q}$ can be expressed in the form $(a - b)/c$ with $a, b, c \in \mathbb{N}$, it then follows that $\sigma \mid \mathbb{Q} = \mathrm{id}_{\mathbb{Q}}$.

The ordering relation in $K$ can be defined on the basis of the field structure alone. We have $x \geq y$ if and only if there exists a $z \in K$ such that $z^2 = x - y$. It follows that every automorphism $\sigma$ is order preserving. If now a sequence $(x_\nu)$ converges to $x$ in $K$, the image sequence $(\sigma(x_\nu))$ must converge to $\sigma(x)$, or in other words $\sigma$ is continuous. As $\mathbb{Q}$ is dense in $K$, there is, for every $x \in K$, a sequence in $\mathbb{Q}$ which converges to $x$. This sequence is mapped identically on to itself by $\sigma$. Regarded as an image sequence it converges to $\sigma(x)$. Since a limit is uniquely defined, $\sigma(x) = x$. $\qquad\square$

In Chapters 1 and 2, we have created $\mathbb{R}$, starting from an infinite set, and using the methods of set theory to construct in succession the sets $\mathbb{N}$, $\mathbb{Z}$ and $\mathbb{Q}$ on the way. The existence of $\mathbb{R}$ is therefore assured, provided that we accept the validity of this set theory. Expressed in other words we may say that the axioms (R1)–(R3) are consistent (that is, free from contradiction), provided that the set theory we have used is consistent. The problem of the consistency of set theory is dealt with in the last chapter.

## REFERENCES

[1] BACHMANN, P.: Vorlesungen über die Theorie der Irrationalzahlen, Leipzig 1892

[2] BECKER, O.: Grundlagen der Mathematik in geschichtlicher Entwicklung, Freiburg/München 1954

[3] BISHOP, E.: Foundations of Constructive Analysis, New York 1967

[4] BOLZANO, B.: Rein analytischer Beweis des Lehrsatzes, daß zwischen je zwei Werthen, die ein entgegengesetztes Resultat gewähren, wenigstens eine reele Wurzel der Gleichung liege, Prag 1817, Ostwalds Klassiker Nr. 153, Leipzig 1905

[5] BOLZANO, B.: Reine Zahlenlehre, 7. Abschnitt, in: Bernhard Bolzano – Gesamtausgabe, eds. E. Winter/J. Berg/F. Kambartel/I. Loužil/B.v. Rootselaar, Reihe II. A. Nachgelassene Schriften Bd. 8, Stuttgart/Bad Cannstatt 1976

[6] CARTAN, H.: Un théorème sur les groupes ordonnés, in: Bull. Sci. Math. 63 1939, 201–205

[7] DEDEKIND, R.: Stetigkeit und Irrationalzahlen, Braunschweig 1872, [7]1965

[8] DUGAC, P.: Eléments d'analyse de Karl Weierstraß, in: Arch. hist. ex. Sciences 10, 1973, 41–176

[9] EULER, L.: De progressionibus harmonicis observationes (1734/35), in: Op. omn. I, 14, 73–86

[10] FRITZ, K.v.: The Discovery of Incommensurability by Hippasus of Metapontum, in: Annals of Mathematics 46, 1945, 242–264

[11] HELLER, S.: Die Entdeckung der stetigen Teilung, Abh. d. Dt. Ak. Wiss. Berlin, Klasse für Mathematik, Physik u. Technik 1958, Nr. 6, Berlin 1958

[12] HERMES, H.: Aufzählbarkeit, Entscheidbarkeit, Berechenbarkeit, Einführung in die Theorie der rekursiven Funktionen, Berlin/ Heidelberg/New York 1971

[13] HILBERT, D.: Grundlagen der Geometrie, Leipzig 1899, ed. With supplements by P. Bernays, Stuttgard [11]1972

[13a] HÖLDER, O.: Die Axiome der Quantität und die Lehre vom Maß. Berichte Verh. Kgl. Sächs. Ges. Wiss., Leipzig, Math. Phys. Kl. 53, 1901, 1–64.

[14] IAMBLICHI: de communi mathematica scientia liber (ed. N. Festa), Leipzig 1891

[15] IAMBLICHI: de vita Pythagorica liber (ed. L. Deubner), Leipzig 1937

[16] LANDAU, E.: Grundlagen der Analysis (Das Rechnen mit ganzen, rationalen, irrationalen, komplexen Zahlen). Ergänzung zu den Lehrbüchern der Differential- und Integralrechnung, Leipzig 1930 (repr. Frankfurt 1970)

[17] LIPSCHITZ, R.: Grundlagen der Analysis, Bonn 1877

[18] LORENZEN, P.: Differential und Integral. Eine konstruktive Einführung in die klassische Analysis, Frankfurt 1965

[19] MERAY, C.: Remarques sur la nature des quantités définies par la condition de servir de limites à des variables données, in: Revue des Sociétés savantes. Sciences mathém., phys. et naturelles, 2$^e$ séries, t. IV, 1869

[20] NEUGEBAUER, O. and A. SACHS: Mathematical Cuneiform Texts, New Haven 1945

[21] STEVIN, S.: La practique d'arithmetique, Leiden 1685

[22] STIFEL, M.: Arithmetica integra, Nürnberg 1544, Buch II, Kap. 1

[23] TROPFKE, J.: Geschichte der Elementarmathematik. Vol. 1 Arithmetik und Algebra, Berlin/New York 1980

[24] WEIERSTRASS, K.: Einleitung in die Theorie der analytischen Funktionen. Vorlesung 1880/81. Nachschrift von A. Kneser

[25] VAN DER WAERDEN, B.L.: Die Pythagoreer: religiöse Bruderschaft und Schule d. Wiss., Zürich 1979

[26] VAN DER WAERDEN, B.L.: Algebra II, Berlin/Heidelberg/New York 1967

## FURTHER READING

[27] BURRILL, C.W.: Foundations of Real Numbers. New York: McGraw-Hill, 1967

[28] COHEN, L.W. and GERTRUDE EHRLICH: The Structure of the Real Number System. Princeton: D. Van Nostrand, 1963

[29] DEDEKIND, R.: Essays on the Theory of Numbers, tr. by W.W. Beman. Chicago: Open Court, 1901; New York: Dover, 1963

[30] DODGE, C.W.: Numbers and Mathematics. Boston: Prindle, Weber & Schmidt, 1969

[31] LANDAU, EDMUND: Foundations of Analysis, tr. by F. Steinhardt, New York: Chelsea, 1951

[32] NIVEN, IVAN: Irrational Numbers, Carus Mathematical Monograph No. 11. New York: John Wiley, 1956

# 3

# Complex Numbers

*R. Remmert*[1]

> Ex irrationalibus oriuntur quantitates impossibiles seu
> imaginariae, quarum mira est natura, et tamen non
> contemnenda utilitas (LEIBNIZ).
>
> [From the irrationals are born the impossible or imaginary
> quantities whose nature is very strange but whose use-
> fulness is not to be despised.]

The quadratic equation $x^2 + 1 = 0$ has no solutions in the field $\mathbb{R}$ of real
numbers, because every sum of squares $r^2 + 1$ with $r \in \mathbb{R}$ is positive.
A new epoch in the mathematics of modern times was inaugurated by
the recognition that this incompleteness of the real number system could
be obviated by yet another simple extension of the number domain, the
extension of $\mathbb{R}$ to the field $\mathbb{C}$ of complex numbers.

The development of the theory of complex numbers makes an impressive
chapter in the history of mathematical concepts. When they first made
their appearance at the time of the Renaissance these new numbers were
called *impossible quantities* (quantitates impossibiles), just as had happened earlier with the negative numbers. Mathematicians began to use
complex numbers in their calculations but at first warily and without really
accepting them. Until the end of the eighteenth century there was no precise foundation for the theory of imaginary numbers. A quantity $i = \sqrt{-1}$,
whose square $i^2 = -1$ was negative, remained unimaginable. Nevertheless,
despite this awkward fact, from the days of BOMBELLI, and certainly from
EULER onwards, imaginary numbers were used ever more successfully and
with greater assurance. Their applicability, exceeding all expectations; the
unassailability of the results achieved by their use; and above all the validity of the Fundamental Theorem of Algebra (see Chapter 4), eventually
helped to ensure their full recognition, especially after their representation
as points on a plane had enabled everyone to visualize them.

The genesis of the complex numbers is described in §1 of this chapter.
In §§2 to 5 we develop the *elementary theory* of these numbers as far as

---

[1] I am indebted to the Volkswagen Foundation for the award of a research grant
during the academic year 1980/81, as a result of which the work on Chapters 3,
4 and 5 of this book was very considerably facilitated.

can be done *without using the methods of analysis*. In §6 we deal with the
polar coordinate representation of complex numbers

$$z = |z|e^{i\varphi} = |z|(\cos\varphi + i\sin\varphi).$$

Here we have to draw upon properties of the exponential and of trigonometrical functions, whose proofs lie deeper. In particular we shall need $\pi$,
the ratio of the circumference to the diameter of a circle. This number $\pi$
forms the subject of Chapter 5 where it is discussed in detail.

Complex numbers provide the basis for the theory of holomorphic functions. This theory is dealt with in R. Remmert, *Theory of Complex Functions*, GTM/RIM 122, Springer-Verlag, 1990.

## §1. GENESIS OF THE COMPLEX NUMBERS

It is almost impossible for anyone today who already hears at school about
$i = \sqrt{-1}$ being a solution of $x^2 + 1 = 0$ to understand what difficulties the
complex (that is, imaginary) numbers presented to mathematicians and
physicists in former times.

We summarize below the important historical dates. As secondary source
material we have made use of the following books:

ARNOLD, W. UND WUSSING, H. (Herausgeber): Biographien bedeutender
   Mathematiker, Aulis Verlag Deubner u. Co KG, Köln 1978
BOYER, C.B.: A History of Mathematics, John Wiley and Sons, Inc., New
   York, London, Sidney 1968
CARTAN, E.: Nombres complexes. Exposé, d'après l'article allemand de E.
   Study (Bonn), Encyclop. Sci. Math. édition française 15, 1908; see also
   E. Cartan OEuvres II, 1, 107–247
COOLIDGE, J.L.: The Geometry of the Complex Domain. Oxford Univer-
   sity Press 1924; especially Chapter I
HANKEL, H.: Theorie der complexen Zahlensysteme, Leipzig 1867
KLINE, M.: Mathematical Thought from Ancient to Modern Times, Oxford
   University Press, New York 1972
MARKUSCHEWITSCH, A.I.: Skizzen zur Geschichte der Analytischen Funk-
   tionen, VEB Deutscher Verlag der Wissenschaften, Berlin 1955
STUDY, E.: Theorie der gemeinen und höheren complexen Grössen, Encykl.
   Math. Wiss. I.1, 147–183, Teubner Verlag Leipzig, 1898–1904
TROPFKE, J.: Geschichte der Elementarmathematik, 4. Aufl., Bd. 1: Arith-
   metik und Algebra, Vollständig neu bearbeitet von Kurt Vogel, Karin
   Reich und Helmuth Gericke; Walter de Gruyter, Berlin, New York 1980

The article by CARTAN essentially complements the one by STUDY and
goes into greater depth.

**1. CARDANO (1501–1576).** Imaginary quantities make their first appearance during the Renaissance. In 1539, Girolamo CARDANO, a mathematician and renowned physician in Milan, learned from TARTAGLIA a process for solving cubic equations; in 1545 he broke his promise never to divulge the secret to anyone. In 1570 he was imprisoned on a charge of having cast the horoscope of Christ. In 1571 he became a protégé of Pope Pius V who granted him an annuity for life. (See *Dictionary of Scientific Biography*, vol. 3.) In his book entitled *Artis magnae sive de regulis algebraicis liber unus* he tries to work with imaginary roots in dealing with quadratic equations: in Chapter 37 he boldly ascribes the solution $5 + \sqrt{-15}$ and $5 - \sqrt{-15}$ to the equation $x(10 - x) = 40$, saying: "Manifestum est, quod casus seu quaestio est impossibilis, sic tamen operabimur...". As the symbols written down appear to be meaningless, he calls $\sqrt{-15}$ a "quantitas sophistica" which should perhaps be translated as a "formal number."[2]

It is not clear whether CARDAN (to use the name by which he is usually known in English) was led to complex numbers through cubic or quadratic equations. While quadratic equations $x^2 + b = ax$, where the solution is given by the formula $x = \frac{1}{2}a \pm \sqrt{\frac{1}{4}a^2 - b}$ have no real roots (and are therefore *impossible* equations) when $a^2 < 4b$, cubic equations $x^3 = px + q$ *have real roots which are given as sums of imaginary cube roots.*[3]

Cardan points out in Chapter 12 that his formula

$$x = \sqrt[3]{q/2 + \sqrt{d}} + \sqrt[3]{q/2 - \sqrt{d}} \quad \text{with} \quad d := (q/2)^2 - (p/3)^3$$

*fails* in the case $(p/3)^3 > (q/2)^2$. He gives examples such as the equations $x^3 = 20x + 25$ and $x^3 = 30x + 36$ (which can be derived from the identity $x^3 = (x^2 - x)x + x^2$ by substituting 5 and 6 respectively): his formula leads to roots of negative numbers, but the equations are not *impossible* because the solutions $x = 5$ and $x = 6$ are obvious. Whether Cardan had seen this clearly is questionable.

**2. BOMBELLI (1526–1572).** CARDAN's algebra was further developed by Rafael BOMBELLI, whose "L'algebra," published in Bologna in 1572 probably originated between 1557 and 1560. BOMBELLI, without having thought too much about the nature of complex numbers, laid down eight

---

[2] In discussing the product $(5 + \sqrt{-15})(5 - \sqrt{-15})$ Cardan writes "dismissis incruciationibus," meaning no doubt that the (imaginary) cross product terms cancel each other. It is tempting to read in these words the additional meaning given by the translation "setting aside any intellectual scruples" (from friciatus— torture, mental anguish, etc.) and to assume that Cardan was indulging in a play on words—but this interpretation is probably unjustified.

[3] Nowadays it is well known that it is impossible to solve, by real radicals, an irreducible cubic equation over $\mathbb{Q}$ whose three roots are all real (the so-called *casus irreducibilis*). For further details on this see Van Der Waerden, *Algebra*, Part I, Springer-Verlag, Berlin-Heidelberg-New York, 7th ed. 1966, p. 194.

fundamental rules of computation. The last (in modern notation) is
$(-i)(-i) = -1$. BOMBELLI carries out correctly a few calculations and
knows for example that

$$(2 \pm i)^3 = 2 \pm 11i, \quad \text{so that} \quad \sqrt[3]{2 \pm \sqrt{-121}} = 2 \pm \sqrt{-1}.$$

He applies this identity to the equation $x^3 = 15x + 4$, where Cardan's
formula yields the solution

$$x = \sqrt[3]{2 + \sqrt{-121}} + \sqrt[3]{2 - \sqrt{-121}}.$$

The obvious solution 4 is given by $(2 + \sqrt{-1}) + (2 - \sqrt{-1})$ so that he arrives
with the help of complex numbers at real solutions. BOMBELLI was the first
to teach the art of correct formal computation with complex numbers.

**3.    DESCARTES (1596–1650), NEWTON (1642–1727) and
LEIBNIZ (1646– 1716).** René DESCARTES in his "La géométrie" (Leyden 1637) brings out the antithesis between real and imaginary. He says,
in essence, that one can imagine, for every equation, as many roots as are
indicated by the degree of the equation, but these imagined roots do not
always correspond to any real quantity. Incidentally, DESCARTES candidly
confesses that one is quite unable to visualize imaginary quantities.

Isaac NEWTON regarded complex roots as an indication of the insolu-
bility of a problem, expressing himself as follows: "But it is just that the
Roots of Equations should be impossible, lest they should exhibit the cases
of Problems that are impossible as if they were possible" (*Universal arithmetic,* 2nd ed., 1728, p. 193). In Newtonian times complex numbers had
not yet arisen anywhere in physics. Gottfried Wilhelm LEIBNIZ in a letter
to HUYGENS written in 1674 or 1675 (see LEIBNIZ *Math. Schriften,* ed.
GERHARDT, vol. 1, II, p. 12) enriched the theory of imaginaries by noting
the surprising relation

$$\sqrt{1 + \sqrt{-3}} + \sqrt{1 - \sqrt{-3}} = \sqrt{6}.$$

In 1702, in an article appearing in the Leipzig *Acta Eruditorum,* a journal
which he had founded, and the first scientific periodical to be published in
Germany[4] (see also *Math. Schriften,* ed. GERHARDT, vol. 5, p. 357) he calls
imaginary roots ... a subtle and wonderful resort of the divine spirit, a kind
of hermaphrodite between existence and non-existence (inter Ens et non
Ens Amphibio). LEIBNIZ had already, by 1712, claimed that $\log(-1)$ is an
imaginary number.

---

[4]The true founder of this periodical, modeled on the *Journal des Savants* was
Mencke. The *Acta Eruditorum* ceased publication in 1782.

**4. EULER (1707–1783).** This great Swiss mathematician had no scruples about making use of complex numbers in his calculations but intuitively used them correctly and in a masterly fashion. He was already aware, by 1728, of the transcendental relationship

$$i \log i = -\frac{1}{2}\pi \quad \text{or, what amounts to the same thing} \quad i^i = e^{-\frac{1}{2}\pi},$$

but he made no attempt to give a rigorous proof. In his famous textbook, the "Introductio in Analysin infinitorum" imaginary numbers first appear in §30, quite suddenly and completely unmotivated. They play a decisive role in §138 in the derivation of the "Euler formulae"

$$\cos x = \frac{1}{2}(e^{ix} + e^{-ix}) \quad \text{and} \quad \sin x = \frac{1}{2i}(e^{ix} - e^{-ix}).$$

Leonhard EULER's elementary textbook on algebra[5] was first published in 1768 in Russian in St. Petersburg and then later in a German edition in 1770 as the "Vollständige Anleitung zur Algebra" (*Opera Omnia* 1, 1–498, ed. WEBER, also reprinted in English translation as "Elements of Algebra" by Springer-Verlag, 1983). Euler had great difficulty in explaining and defining just what the imaginary numbers, which he had been handling so masterfully during the past forty years and more, really were. He points out that the square root of a negative number can be neither greater than zero, nor smaller than zero, nor yet equal to zero, and writes in Chapter 13, Article 143: "it is clear therefore that the square roots of negative numbers cannot be reckoned among the possible numbers: consequently we have to say that they are numbers which are impossible. This circumstance leads us to the concept of numbers, which by their very nature are impossible, and which are commonly called *imaginary numbers* or *fancied numbers* because they exist only in our fancy or imagination." One would smile nowadays at such a sentence if it had not been written by the great EULER. In his book on algebra, EULER occasionally makes some mistakes, for example, he argues that $\sqrt{-1}\sqrt{-4} = \sqrt{4} = 2$, because $\sqrt{a}\sqrt{b} = \sqrt{ab}$.

**5. WALLIS (1616–1703), WESSEL (1745–1818) and ARGAND (1768–1822).** The first vague notions on a correspondence between complex numbers and points on a plane were put forward by the English mathematician John WALLIS is his "De algebra tractatus," a work published in 1685. However his ideas remained muddled and exercised no influence on his contemporaries. The first representation of the points of a plane by complex numbers which has to be taken seriously was proposed by the Norwegian

---

[5]Euler, who had by then already become blind, dictated the book to an amanuensis who had formerly been a tailor by profession. It is said that Euler let the text stand only when he had satisfied himself that the writer had fully understood it (the ultimate aim of all applied didactics).

surveyor Caspar WESSEL. WESSEL, who was self-taught, wrote a memoir
"On the analytical representation of direction—an essay" which is to be
found in the Transactions of the Danish Academy for 1798. WESSEL's primary object was to be able to operate with directed line segments and he
thus hit upon the idea of representing them as complex numbers—not the
other way around. WESSEL introduced an imaginary axis, perpendicular to
the axis of real numbers (he wrote $\varepsilon$ for $\sqrt{-1}$) and interpreted vectors in the
plane as complex numbers. He defined the usual operations for vectors and
thus for complex numbers *geometrically* in a perfectly satisfactory manner.
Despite its considerable merit WESSEL's work remained unnoticed until a
French translation appeared in 1897.

A somewhat different geometrical interpretation of complex numbers was
given by the Swiss accountant Jean Robert ARGAND in his "Essai sur
une manière de représenter les quantités imaginaires dans les constructions
géometriques." ARGAND, who like WESSEL was also an amateur, interprets
$\sqrt{-1}$ as a *rotation* through a right angle in the plane and justifies this on
the grounds that two such rotations, that is, the product $\sqrt{-1}\sqrt{-1} = -1$,
are equivalent to a rotation through two right angles or in other words,
a reflection. (We shall describe this interpretation more fully in 6.2.) AR-
GAND's work also remained largely without influence, although in the older
literature there are frequently references to the ARGAND plane (or ARGAND
diagram).

There are good grounds for believing that, as early as 1749, EULER had
visualized complex numbers as points of a plane. In his paper "De la con-
troverse entre Mrs. LEIBNIZ et BERNOULLI sur les logarithmes des nombres
négatifs et imaginaires" (Mémoires de l'Académie des Sciences de Berlin
[5], (1749), 1751, 139–179; Opera Omnia, 1, Ser. XVII, 195–232) he says
(in French p. 230): ..."In every other case the number $x$ is imaginary: to
find it one has only to take an arc $g$ of the unit circle and determine its
sine and cosine. The number sought is then

$$x = \cos g + \sqrt{-1} \cdot \sin g."$$

**6. GAUSS (1777–1855).** Views on complex numbers first began to change
through the influence of Carl Friedrich GAUSS. He was aware of the interpretation of complex numbers as points of the complex plane from about
1796 and made use of it in 1799 in his dissertation where he proves the
fundamental theorem of algebra (see on this point, Chapter 4), though in a
carefully disguised form. In the year 1811 GAUSS wrote to BESSEL (*Werke*
8, p. 90): "...Just as one can think of the whole domain of real magnitudes
as being represented by an infinite straight line, so the *complete* domain of
all magnitudes, real and imaginary numbers alike, can be visualized as an
infinite plane, in which the point defined by the ordinate $a$ and the abscissa
$b$, likewise represents the magnitude $a + bi$." This is the representation by
real number pairs expressed in geometric language.

By 1815, at the latest, GAUSS was in full possession of the geometrical theory. But true dissemination of the idea of the complex number plane did not occur until 1831 with the publication of GAUSS's *Theoria Residuorum Biquadraticorum. Commentatio Secunda* (*Werke* 2, 93–148). In the now classical introductory review which he wrote summarizing this second memoir (*Werke* 2, 169–178) he sets out clearly his views in a manner which overcomes all logical objections. He coins the expression "complex number" and describes the attitude of his contemporaries to these numbers as follows: "but these imaginary numbers, as opposed to real quantities— formerly, and even now occasionally, though improperly called *impossible*— have been merely tolerated rather than given full citizenship and appear therefore more like a game played with symbols devoid of content in itself, to which one refrains absolutely from ascribing any visualizable substratum. In saying this one has no wish to belittle the rich tribute which this play with symbols has contributed to the treasury of relations between real numbers." As regards the aura of mystery which still clung to complex numbers, he writes (pp. 177–178): "If this subject has hitherto been considered from the wrong viewpoint and thus enveloped in mystery and surrounded by darkness, it is largely an unsuitable terminology which should be blamed. Had $+1$, $-1$ and $\sqrt{-1}$, instead of being called positive, negative and imaginary (or worse still impossible) unity, been given the names, say, of direct, inverse and lateral unity, there would hardly have been any scope for such obscurity." And later (after 1831, *Werke* 10, 1, p. 404) he says, looking back:

Bei allem dem sind die imaginären Grössen, so lange ihre Grundlage immer nur in einer Fiction bestand, in der Mathematik nicht sowohl wie eingebürgert, als viel mehr nur wie geduldet betrachtet, und weit davon entfernt geblieben, mit den reellen Grössen auf gleiche Linie gestellt zu werden. Zu einer solchen Zurücksetzung ist aber jetzt kein Grund mehr, nachdem die Metaphysik der imaginären Grössen in ihr wahres Licht gesetzt, und nachgewiesen ist, daß diese, eben so gut wie die negativen, ihre reale gegenständliche Bedeutung haben.

[It could be said in all this that so long as imaginary quantities were still based on a fiction, they were not, so to say, fully accepted in mathematics but were regarded rather as something to be tolerated; they remained far from being given the same status as real quantities. There is no longer any justification for such discrimination now that the metaphysics of imaginary numbers has been put in a true light and that it has been shown that they have just as good a real objective meaning as the negative numbers.]

It was the authority of GAUSS that first removed from complex numbers all aura of mysticism: his simple interpretation of complex numbers as points in the plane freed these fictive magnitudes from all mysterious and speculative associations and gave them the same full citizenship rights in mathematics as those enjoyed by the real numbers. "You have made possible the impossible" is a phrase used in a congratulatory address made to GAUSS in 1849 by the Collegium Carolinum in Brunswick (now the Technical University) on the occasion of the 50-year jubilee of his doctorate. The German Post Office issued a stamp in 1977 illustrating the Gaussian number plane to celebrate the bicentenary of his birth.

**7. CAUCHY (1789–1857).** The French mathematician Augustin-Louis CAUCHY did not regard the geometric interpretation of complex numbers as the last word on the subject. He wrote in 1821, in his "Cours d'Analyse de l'Ecole Royale Polytechnique": "On appelle expression imaginaire toute expression symbolique[6] de la forme $a + b\sqrt{-1}$, $a, b$ désignant deux quantités réelles ... toute equation imaginaire n'est que la représentation symbolique de deux équations entre quantités réelles." [We call an imaginary expression, any symbolic expression of the form $a + b\sqrt{-1}$, where $a, b$ denote two real quantities ... Every imaginary equation is only just the symbolic representation of two equations between real quantities.] (*Oeuvres* 3, 2 Ser., 17–331, p. 155). This conception of imaginary expressions as symbolic

---

[6]Cauchy also tries to explain what a *symbolic expression* is. He says (p. 153): "En analyse, on appelle expression symbolique ou symbole toute combinaison de signes algébriques qui ne signifie rien par elle-même ou à laquelle on attribue une valeur différente de celle qu'elle doit naturellement avoir." Hankel, who in 1867, in his book "Theorie des complexen Zahlensysteme" was wrestling with the metaphysics of the foundations of mathematics, called this amazing definition a *Gaukelspiel* (conjuring trick or illusion) and (p. 73) a *galimatias* (a meaningless jumble of words, nonsense). Incidentally the origin of this word is unknown, but according to *Meyers Enz. Lexik.* 1973, it is probably compounded from the low Latin *galli* a term used for certain disputants at the Sorbonne, and the Greek $\mu\alpha\theta\epsilon\iota\alpha$ (learning). He writes, somewhat aggressively (p. 14): "Ich glaube nicht zu viel zu sagen, wenn ich dies ein unerhörtes Spiel mit Worten nenne, das der Mathematik, die auf die Klarheit und Evidenz ihrer Begriffe stolz ist und stolz sein soll, schlecht ansteht." [I do not think I am exaggerating in calling this an outrageous play on words, ill becoming Mathematics, which is proud and rightly proud of the clarity and convincingness of its concepts.]

representations of two real numbers is, in contrast to GAUSS's geometric interpretation, purely algebraic.

CAUCHY was still, in 1847, and thus long after HAMILTON (see next paragraph) unsatisfied with the interpretation of the symbol $i$. In a note in the *Comptes rendus* entitled "Mémoire sur une nouvelle théorie des imaginaires, et sur les racines symboliques des équations et des équivalences" (*Oeuvres* 10, 1 Ser., 312–323) he gives a definition which makes it possible "...à réduire les expressions imaginaires, et la lettre $i$ elle même, à n'être plus que des quantités réelles." Using the concept of equivalence (with an explicit reference to the work of GAUSS on classes of quadratic forms) he now interprets computations involving complex numbers as computations with real polynomials modulo the polynomial $X^2 + 1$. In modern terminology this is equivalent to interpreting the field $\mathbb{C}$ of complex numbers as the splitting field of $X^2 + 1$ that is, $\mathbb{C} = \mathbb{R}[X]/(X^2 + 1)$. CAUCHY thus proves here a special case of what is now known as KRONECKER's theorem, the theorem that for every (abstract) field $K$ and every irreducible polynomial $f \in K[X]$ the residue class ring $L = K[X]/(f)$ is a finite extension field of $K$, in which $f$ has at least one zero.

**8. HAMILTON (1805–1865).** However helpful the geometric interpretation of complex numbers as points, or vectors on a plane may be ("seeing is believing"), a geometrical foundation for computation with such numbers is not entirely satisfactory ("On ne cherche pas à voir, mais à comprendre"). The important (if now seemingly trivial) step to the formal definition as an *ordered pair of real numbers* still remained to be taken. This first occurred in 1835 through Sir William Rowan HAMILTON, probably in the course of the preliminary studies preceding his discovery of quaternions. In his work with the strange title[7] "Theory of Conjugate Functions, or Algebraic Couples, with a Preliminary and Elementary Essay on Algebra as the Science of Pure Time" (*Math. Papers* 3, 3–96) is to be found (p. 81) for the first time the definition of complex numbers as ordered pairs of real numbers. HAMILTON defines addition and multiplication in such a way that the well-known arithmetical laws (the distributive, associative and commutative laws) remain valid. We shall be following HAMILTON's example when we introduce complex numbers in 2.1. GAUSS, in a letter of 1837 to Wolfgang BOLYAI, says that the representation by ordered pairs had already been familiar to

---

[7]This remarkable title owes its origin to Kant. Real numbers at that time were usually defined as the ratio of the length of a line segment to that of a given unit line segment. Now Kant had said that geometry belongs to space, and arithmetic—and therefore numbers—to time. Accordingly Hamilton, with Kant's perception of numbers in mind, defined numbers as ratios of time intervals. Naturally, from a purely mathematical standpoint, nothing was gained by this, but it is interesting to note that, long before Weierstrass and in ignorance of Bolzano, he sought to give a new definition of real numbers.

him since 1831.

**9. Later Developments.** Complex numbers during the last century began their tempestuous and triumphant march through every field of mathematics. For Bernhard RIEMANN (1826–1866) they are already a matter of course. In his 1851 Göttingen inaugural dissertation "Grundlagen für eine allgemeine Theorie der Funktionen einer veränderlichen complexen Grösse" (*Werke* 5–43) he philosophizes (pp. 37,38) "Die Einführung der complexen Größen in die Mathematik hat ihren Ursprung und nächsten Zweck in der Theorie einfacher durch Größenoperationen ausgedrückter Abhängigkeitsgesetze zwischen veränderlichen Größen. Wendet man diese Abhängigkeitsgesetze in einem erweiterten Umfang an, indem man den veränderlichen Größen, auf welche sie sich beziehen, complexe Werte gibt, so tritt eine sonst versteckt bleibende Harmonie und Regelmäßigkeit hervor." [The original purpose and immediate objective in introducing complex numbers into mathematics is to express laws of dependence between variables by simpler operations on the quantities involved. If one applies these laws of dependence in an extended context, by giving the variables to which they relate complex values, there emerges a regularity and harmony which would otherwise have remained concealed.] On the other hand, in 1854, the 23-year-old mathematician Richard DEDEKIND (1831–1916), who was a friend of RIEMANN's and who, in the words of BELL (*Men of mathematics*, p. 518) "... occupied a relatively obscure position for fifty years while men who were not fit to lace his shoes filled important and influential university chairs," judged the position differently. In his habilitation presentation[8] at Göttingen at which GAUSS was present (*Math. Werke* 3, p. 434), DEDEKIND said "Bis jetzt ist bekanntlich eine vorwurfsfreie Theorie der imaginären... Zahlen entweder nicht vorhanden, oder doch wenigstens noch nicht publiziert." [Until now we have had available no theory of complex numbers entirely free from reproach... or at least none has so far been published.]

Complex numbers soon begin to be used in Physics as well. Already in 1823 FRESNEL used complex numbers in his theory of total reflection (published in 1831). Nowadays physicists think nothing of talking about complex-valued physical objects: the basic equations of quantum mechanics are written, without any compunction, in the form:

$$pq - qp = \frac{h}{2\pi i}, \qquad \frac{h}{2\pi i}\frac{\partial \Psi}{\partial t} = -H\Psi.$$

Complex numbers have also long been used in electrical engineering; the electrical engineer writes $j$ instead of $i$ (as $i$ is reserved as the symbol for current intensity). It is a little known fact that one of the first computers ever built was a "complex number computer" to multiply and divide

---

[8] The oral thesis presented at German universities to qualify as a lecturer.

complex numbers. It was developed during the years 1938 to 1940 by the engineer STIBITZ in the Bell Telephone laboratories, and thus before ZUSE's programmable computer, and before the ENIAC in Princeton. Admittedly STIBITZ's machine, which worked with relays, was not a program-controlled machine. It was used successfully from 1940 to 1949 on network analysis computations, particularly on telephone switching problems.

The *numeri impossibiles* have thus during the course of the last few centuries taken a firm place in science and engineering; they are used confidently and consistently in calculations, without fear of encountering any contradictions, and mathematicians no longer worry about such philosophical questions as the ens or non-ens of $i = \sqrt{-1}$.

## §2. THE FIELD $\mathbb{C}$

We shall introduce complex numbers[9] following HAMILTON (see 1.8), as ordered pairs of real numbers. They form a commutative, 2-dimensional extension field $\mathbb{C}$ of the field $\mathbb{R}$. There is an element $i \in \mathbb{C}$ with $i^2 + 1 = 0$, and every complex number $z$ can be written uniquely in the form $x + iy$, with $x, y \in \mathbb{R}$. Complex numbers can also be described elegantly as *real $2 \times 2$ matrices*.

**1. Definition by Pairs of Real Numbers.** The set $\mathbb{R} \times \mathbb{R}$ of all ordered pairs of real numbers $z := (x, y)$ is an *Abelian group* with respect to the natural *addition* defined by

$$(1) \qquad (x_1, y_1) + (x_2, y_2) := (x_1 + x_2, y_1 + y_2).$$

We introduce a *multiplication* in $\mathbb{R} \times \mathbb{R}$ by the definition

$$(2) \qquad (x_1, y_1) \cdot (x_2, y_2) := (x_1 x_2 - y_1 y_2, x_1 y_2 + y_1 x_2)$$

which may at first sight appear to be rather artificial. It can then be easily verified that the commutative, associative and distributive laws hold. The element $e := (1, 0)$ is the unit element. Direct calculation shows that *if $z = (x, y) \neq 0$, then*

$$z^{-1} := \left( \frac{x}{x^2 + y^2}, \frac{-y}{x^2 + y^2} \right)$$

*is the inverse of $z$, that is $zz^{-1} = e$.*

---

[9] The adjective "complex" was first used in its present technical sense by Gauss in 1831. Until then he had also used the word "imaginary." Bézout had earlier used the expression complex number in an entirely different sense in his "Cours de mathématiques à l'usage des gardes du pavillon et de la marine. I partie. Eléments d'arithmétique" published in Paris in 1773 where, on page 105 *et seq.* he uses it to denote a number involving several different units of measure, e.g. days, hours and minutes.

The set $\mathbb{R} \times \mathbb{R}$ is therefore a commutative field with respect to the laws of composition (1) and (2). It is called *the field $\mathbb{C}$ of complex numbers.*

Since $(x_1, 0) + (x_2, 0) = (x_1 + x_2, 0)$ and $(x_1, 0)(x_2, 0) = (x_1 x_2, 0)$, the mapping $\mathbb{R} \to \mathbb{C}$, $x \mapsto (x, 0)$ is an *embedding of the field $\mathbb{R}$ into the field $\mathbb{C}$*. The real number $x$ is identified with the complex number $(x, 0)$. Thus $\mathbb{C}$ is a *field extension* of $\mathbb{R}$ with the unit element $e = (1, 0) = 1$. As $\mathbb{C}$ is a 2-dimensional real vector space, $\mathbb{C}$ is of degree 2 over $\mathbb{R}$, in the language of algebra.

The set $\mathbb{C} \setminus \{0\}$ of all non-zero complex numbers is denoted by $\mathbb{C}^{\times}$. $\mathbb{C}^{\times}$ *is an Abelian group with respect to multiplication in $\mathbb{C}$, whose neutral element is the unit element 1 (the multiplicative group of the field $\mathbb{C}$).*

One can motivate the particular definition of multiplication in (2) by the following considerations. In the $\mathbb{R}$-vector space $\mathbb{R}^2$ with the natural basis $(1, 0)$, $(0, 1)$ the first vector is to represent the unit element, and the second vector should have the property that its square is the negative of the unit element, in other words we require $(0, 1)^2 = -(1, 0)$. It then follows, if the ordinary laws are to hold, that

$$
\begin{aligned}
(x_1, y_1)(x_2, y_2) &= [x_1(1, 0) + y_1(0, 1)][x_2(1, 0) + y_2(0, 1)] \\
&= x_1 x_2 (1, 0) + (x_1 y_2 + y_1 x_2)(0, 1) + y_1 y_2 (0, 1)^2 \\
&= (x_1 x_2 - y_1 y_2)(1, 0) + (x_1 y_2 + y_1 x_2)(0, 1) \\
&= (x_1 x_2 - y_1 y_2, x_1 y_2 + y_1 x_2).
\end{aligned}
$$

*Note.* The motivation for (2) is rather different with HAMILTON: first he finds it suggestive to define products with real numbers by the rule $r(x_1, y_1) := (r x_1, r y_1)$ ($\mathbb{R}$ vector space structure). One then already has

$$
(x_1, y_1) = x_1 e + y_1 \varepsilon \quad \text{with} \quad e := (1, 0), \quad \varepsilon := (0, 1).
$$

Now if $e$ is to be the unit element and the distributive laws are to hold, then one must have

$$
(*) \qquad (x_1 e + y_1 \varepsilon)(x_2 e + y_2 \varepsilon) = x_1 x_2 e + (x_1 y_2 + y_1 x_2)\varepsilon + y_1 y_2 \varepsilon^2.
$$

The multiplication law is therefore determined as soon as $\varepsilon^2$, which must be of the form $pe + q\varepsilon$ is known. *There are however infinitely many ways of choosing $p$ and $q$ so that the resulting multiplication has an unique inverse.* (The reader may care to find examples.) HAMILTON therefore postulates (as he does later in the case of his quaternions, see, 6.E.2) the *product rule: the length of the product of two factors is equal to the product of the lengths of the factors, where the length $|z|$ of $z = (x, y)$ is defined as $+\sqrt{x^2 + y^2}$.* It is then only necessary to apply this product rule to

$$
\varepsilon^2 = pe + q\varepsilon \quad \text{and} \quad (e + \varepsilon)(e - \varepsilon) = e - \varepsilon^2 = (1 - p)e - q\varepsilon
$$

to deduce that $p = -1$, $q = 0$ (since $|\varepsilon^2| = |\varepsilon||\varepsilon| = 1$ and $|e + \varepsilon| = |e - \varepsilon| = \sqrt{2}$) so that $(*)$ becomes the same as equation (2).

On the product rule, see also 3.4.

**2. The Imaginary Unit $i$.** Traditionally one uses the notation which has been customary since the time of EULER and which became common practice through the influence of GAUSS

$$i := (0,1) \in \mathbb{C}.$$

This symbol is often called the *imaginary unit* of $\mathbb{C}$, and we have $i^2 = -1$. In the field $\mathbb{C}$ the real polynomial $X^2 + 1$ has the two zeros $i$ and $-i$. In the complex polynomial ring $X^2 + 1$ decomposes into linear factors.

For all $z = (x,y) \in \mathbb{C}$ the equation $(x,y) = (x,0) + (0,1)(y,0)$ holds and we therefore obtain the usual notation for complex numbers:

$$z = x + iy, \qquad x, y \in \mathbb{R}.$$

The *real and imaginary parts* of $z = x + iy$ are defined by $\operatorname{Re} z := x$, $\operatorname{Im} z := y$. Two complex numbers $z_1$, $z_2$ are equal if, and only if, they have equal real parts and equal imaginary parts:

$$z_1 = z_2 \Leftrightarrow \operatorname{Re} z_1 = \operatorname{Re} z_2 \quad \text{and} \quad \operatorname{Im} z_1 = \operatorname{Im} z_2.$$

A number $z \in \mathbb{C}$ is called *real* if $\operatorname{Im} z = 0$, and *purely imaginary* if $\operatorname{Re} z = 0$, so that in the latter case $z = iy$. The mappings $\operatorname{Re}: \mathbb{C} \to \mathbb{R}$, $\operatorname{Im}: \mathbb{C} \to \mathbb{R}$ are *linearly independent linear forms* of the $\mathbb{R}$-vector space $\mathbb{C}$.

**3. Geometric Representation.** Since the days of WESSEL, ARGAND and GAUSS (see 1.5 and 1.6) complex numbers have been visualized geometrically as points in the *plane* with a rectangular coordinate system (Fig. a). Addition of complex numbers is then represented by the familiar *vector addition*, in accordance with the parallelogram law illustrated in Fig. b.

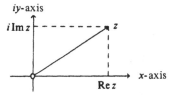

Fig. a

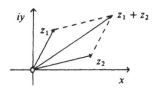

Fig. b

Multiplication of complex numbers is entirely governed by the *one* equation $i^2 = -1$. It follows automatically (see Para. 1) that

$$(x_1 + iy_1)(x_2 + iy_2) = (x_1 x_2 - y_1 y_2) + i(x_1 y_2 + y_1 x_2).$$

The geometrical interpretation of complex numbers in polar coordinates is no longer completely elementary and will be postponed until 6.2.          □

The unique representability of complex numbers in the form $x + iy$ together with the equation $i^2 = -1$, expressed in the language of algebra, says:

*The field $\mathbb{C}$ is a 2-dimensional (algebraic) extension of the field $\mathbb{R}$ and is isomorphic to the splitting field of the irreducible polynomial $X^2 + 1 \in \mathbb{R}[X]$.*

We are now already in a position to prove a first uniqueness theorem for $\mathbb{C}$.

**Theorem.** *Every 2-dimensional ring extension $K$ of $\mathbb{R}$ which has a unit and no divisors of zero is isomorphic to the field $\mathbb{C}$.*

**Proof.** Since $\dim_{\mathbb{R}} K = 2$ there exists a $u \in K \setminus \mathbb{R}$. Then $1 \in \mathbb{R} \subset K$ and $u$ together form a basis of the $\mathbb{R}$-vector space $K$. Consequently $u^2 = c + 2du$ with numbers $c, d \in \mathbb{R}$. For $v := u - d \notin \mathbb{R}$, it follows that $v^2 = r$ where $r := c + d^2 \in \mathbb{R}$. $r$ must be negative because otherwise $\sqrt{r}$ would belong to $\mathbb{R}$ and we should have $n = \pm\sqrt{r} \in \mathbb{R}$. Accordingly there exists an $s \in \mathbb{R}$ with $s^2 = -r^{-1}$. Hence for $w := sv \in K \setminus \mathbb{R}$, we have $w^2 = -1$. The mapping $\mathbb{C} \to K$, $x + iy \mapsto x + wy$ is now a field isomorphism.          □

The foregoing theorem will be significantly generalized in 4.3.5 using the fundamental theorem of algebra.

**4. Impossibility of Ordering the Field $\mathbb{C}$.** The field $\mathbb{R}$ of real numbers is an ordered field (see Chapter 2, §2). *The field of complex numbers,* on the other hand, *cannot be ordered,* that is to say it is impossible to define a relation "$> 0$", a relation of "being positive" in such a way that the following two rules are both satisfied:

1) *For every $z \in \mathbb{C}$, one and only one of the three relations $z > 0$, $z = 0$, $-z > 0$ is valid.*

2) *If $w > 0$ and $z > 0$ then $w + z > 0$ and $wz > 0$.*

**Proof.** Suppose there were such an ordering relation "$> 0$" in $\mathbb{C}$. Then, as in the real case, we should have $z^2 > 0$ for every non-zero $z$. In particular we should have $1^2 > 0$, $i^2 > 0$ and consequently $0 = i^2 + 1 > 0$, which is absurd.          □

The impossibility of ordering $\mathbb{C}$ is a further reason for the difficulties encountered in the 18th and 19th centuries with complex numbers. Eloquent evidence of this is afforded by the extracts from EULER's *Anleitung zur Algebra* quoted in 1.4.

**5. Representation by Means of $2 \times 2$ Real Matrices.** Instead of pairs of real numbers, real $2 \times 2$ matrices can be used for introducing complex numbers. With every complex number $c = a + ib$ we associate the $\mathbb{C}$-linear mapping

$$T_c : \mathbb{C} \to \mathbb{C}, \qquad z \mapsto cz = ax - by + i(bx + ay)$$

(the so-called left regular representation as defined in Algebra). This specifies more precisely, and generalizes, ARGAND's interpretation of complex numbers. Thus, for example, the linear transformation $z \to iz$ corresponding to $i$ is the counterclockwise rotation through one right angle, which sends 1 into $i$, $i$ into $-1$, and so on (see also 1.5). If one identifies $\mathbb{C}$ with $\mathbb{R}^2$ by $z = x + iy = \begin{pmatrix} x \\ y \end{pmatrix}$, then it follows that

$$T_c \begin{pmatrix} x \\ y \end{pmatrix} = \begin{pmatrix} ax - by \\ bx + ay \end{pmatrix} = \begin{pmatrix} a & -b \\ b & a \end{pmatrix} \begin{pmatrix} x \\ y \end{pmatrix}.$$

The linear transformation $T_c$ determined by $c = a + ib$ is thus described by the matrix $\begin{pmatrix} a & -b \\ b & a \end{pmatrix}$. One is thus led to consider the following mapping

$$F : \mathbb{C} \to \mathrm{Mat}(2, \mathbb{R}), \qquad c = a + ib \mapsto \begin{pmatrix} a & -b \\ b & a \end{pmatrix}$$

of the field $\mathbb{C}$ into the *non-commutative* ring $\mathrm{Mat}(2, \mathbb{R})$ of real $2 \times 2$ matrices (forgetting the motivation via $T_c$). This mapping is $\mathbb{R}$-*linear* and *multiplicative*, that is

$$F(rc + r'c') = rF(c) + r'F(c'), \quad F(cc') = F(c)F(c'), \quad r, r' \in \mathbb{R}; \ c, c' \in \mathbb{C}$$

where $F(c)F(c')$ is the *matrix product*. Clearly $F(1) = E := \begin{pmatrix} 1 & 0 \\ 0 & 1 \end{pmatrix}$, and it can be seen that:

*The set $C := \{ \begin{pmatrix} a & -b \\ b & a \end{pmatrix} : a, b \in \mathbb{R} \}$ is, with respect to the operation of matrix addition and matrix multiplication, a commutative field whose unit element is the unit matrix $E$. The $\mathbb{R}$-linear transformation*

$$F : \mathbb{C} \to C, \ a + bi \mapsto \begin{pmatrix} a & -b \\ b & a \end{pmatrix} \quad \text{with} \quad I := F(i) = \begin{pmatrix} 0 & -1 \\ 1 & 0 \end{pmatrix}, \ I^2 = -E,$$

*is a field isomorphism; the matrix $I$ is the "imaginary unit" in $C$.*

Introducing complex numbers through $2 \times 2$ matrices has the advantage over introducing them through ordered pairs of real numbers, that it is unnecessary to define an ad hoc multiplication. Current textbooks do not normally define complex numbers in terms of real $2 \times 2$ matrices; an exception is the book by COPSON, *An Introduction to the Theory of Functions of a Complex Variable* (Oxford: Clarendon Press, 1935).

There are *infinitely many other subfields, apart from C, isomorphic to $\mathbb{C}$ in* Mat$(2, \mathbb{R})$. The following theorem gives a complete picture of what they are.

**Theorem.** a) *For every invertible real $2 \times 2$ matrix $W$ the mapping*

$$g_W : \mathbb{C} \to \text{Mat}(2, \mathbb{R}), \qquad a + bi \mapsto W \begin{pmatrix} a & -b \\ b & a \end{pmatrix} W^{-1}$$

*is a monomorphism of real algebras (compare R.3).*

   b) *Every $\mathbb{R}$-linear homomorphism $g : \mathbb{C} \to \text{Mat}(2, \mathbb{R})$, $g \neq 0$, is of the form $g_W$.*

**Proof.** a) The case $W := E =$ the unit matrix was treated above. Since the mapping $T_W : \text{Mat}(2, \mathbb{R}) \to \text{Mat}(2, \mathbb{R})$, $A \to WAW^{-1}$ is an $\mathbb{R}$-algebra automorphism, a) follows from the fact that $g_W = T_W \circ g_E$.

   b) For $A := g(1)$, $B := g(i) \in \text{Mat}(2, \mathbb{R})$ we have $A^2 = A$, $BA = AB = B$, $B^2 = -A$. Since $\mathbb{C}$ is a field, $g$ is injective, and therefore $A \neq 0$. We choose a column vector $v \in \mathbb{R}^2$ such that $w := Av \neq 0$. Then

(∗)
$$Aw = A^2v = Av = w, \quad A(Bw) = BAw = Bw, \quad B^2w = -Aw = -w.$$

In view of the last equation, $w$, $Bw$ are linearly independent, because otherwise there would be an equation $Bw = \lambda w$, with $\lambda \in \mathbb{R}$, and this would lead to the contradiction $\lambda^2 = -1$. The matrix $W := (w, Bw) \in \text{Mat}(2, \mathbb{R})$ is thus invertible, and by (∗) it follows that $AW = W$, whence $A = E$, and furthermore $BW = (Bw, -w) = (w, Bw) \begin{pmatrix} 0 & -1 \\ 1 & 0 \end{pmatrix} = WI$. It has thus been shown that $g(1) = E = g_W(1)$, $g(i) = WIW^{-1} = g_W(i)$. From the $\mathbb{R}$-linearity of $g$ and $g_W$ it now follows that $g = g_W$. □

By way of example, for $W := \begin{pmatrix} 2 & 3 \\ 1 & 2 \end{pmatrix}$ we have

$$g_W(\mathbb{C}) = \left\{ \begin{pmatrix} a + 8b & -13b \\ 5b & a - 8b \end{pmatrix} : a, b \in \mathbb{R} \right\} \simeq \mathbb{C};$$

in this example $\begin{pmatrix} 8 & -13 \\ 5 & -8 \end{pmatrix}$ is the "imaginary unit."

## §3. ALGEBRAIC PROPERTIES OF THE FIELD $\mathbb{C}$

The field $\mathbb{C}$ possesses the *conjugation automorphism* $\mathbb{C} \to \mathbb{C}$, $z \mapsto \bar{z}$, which is fundamental in many contexts.

The *scalar product* $\langle w, z \rangle$ in $\mathbb{C}$, and the associated *absolute value function* $|z|$ can be introduced by

$$\langle w, z \rangle := \operatorname{Re}(w\bar{z}) = ux + vy, \qquad |z| := \sqrt{z\bar{z}} = \sqrt{x^2 + y^2},$$

where $w = u + iv$, $x = x + iy$. With the help of the function $|z|$ it will be shown in §5, by elementary arguments, that every quadratic equation $z^2 + az + b = 0$, $a, b \in \mathbb{C}$ is solvable in $\mathbb{C}$. This statement is a first indication that the field $\mathbb{C}$ is more "complete" than the field $\mathbb{R}$. The theorem on the solvability of all quadratic equations in $\mathbb{C}$ was already known long before EULER; it is a particular case of the famous and profound fundamental theorem of algebra which states that every non-constant polynomial with complex coefficients has at least one zero. This theorem will be discussed in Chapter 4.

**1. The Conjugation** $\mathbb{C} \to \mathbb{C}$, $z \mapsto \bar{z}$. As is well known, the field $\mathbb{R}$ has no automorphism apart from the identity (see Chapter 2, 5.3). In contrast with this the field $\mathbb{C}$ has an infinity of automorphisms. Among them is one which is distinguished from all others by the fact that it maps $\mathbb{R}$ onto itself, and sends $i$ into the second zero $-i$ (which, in principle, has precisely the same status) of the polynomial $X^2 + 1$.

For every complex number $z = x + iy$, $x, y \in \mathbb{R}$, the complex number

$$\bar{z} := x - iy = 2\operatorname{Re} z - z$$

is known as the *complex conjugate* of $z$.[10] In the Gaussian number plane $\bar{z}$ is represented by the reflection of $z$ in the real axis (see figure). We have

$$\operatorname{Re} z = \frac{1}{2}(z + \bar{z}), \quad \operatorname{Im} z = \frac{1}{2i}(z - \bar{z}), \quad z\bar{z} = x^2 + y^2 \in \mathbb{R}, \quad z\bar{z} > 0 \text{ for } z \neq 0.$$

In particular $z$ is real if and only if $z = \bar{z}$, and purely imaginary if and only if $z = -\bar{z}$.

Operations with complex conjugate numbers are governed by the following

---

[10] The term "conjugate" (conjugué) was introduced in 1821 by Cauchy in his *Cours d'analyse*.

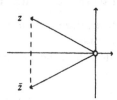

**Theorem.** *The conjugation mapping* $\mathbb{C} \to \mathbb{C}$, $z \mapsto \bar{z}$, *is an automorphism of the field* $\mathbb{C}$, *that is,* $\bar{1} = 1$ *and*

$$\overline{w + z} = \bar{w} + \bar{z}, \qquad \overline{wz} = \bar{w}\bar{z} \quad \text{for all} \quad w, z \in \mathbb{C}.$$

*The relation* $\bar{\bar{z}} = z$ *always holds. The fixed point set* $\{z \in \mathbb{C} : \bar{z} = z\}$ *is the field* $\mathbb{R}$.

All these statements follow without difficulty from the definition of $\bar{z}$; we shall content ourselves with verifying the multiplication rule. Let $w = u + iv$, $z = x + iy$. Then $wz = ux - vy + i(vx + uy)$ while

$$\overline{wz} = ux - vy - i(vx + uy) = (u - iv)(x - iy) = \bar{w}\bar{z}. \qquad \square$$

*Exercise.* Show that, for all $a, b, c, d \in \mathbb{C}$ with $a\bar{a} = b\bar{b} = c\bar{c}$, we have

$$(a - b)(c - d)(\bar{a} - \bar{d})(\bar{c} - \bar{b}) + i(c\bar{c} - d\bar{d})\text{Im}(c\bar{b} - c\bar{a} - a\bar{b}) \in \mathbb{R}. \qquad \square$$

The proof of the following criterion for linear independence is straightforward: *Two numbers* $w, z \in \mathbb{C}$ *are linearly dependent over* $\mathbb{R}$, *if and only if* $w\bar{z} \in \mathbb{R}$.

The conjugation transformation can be used advantageously to describe all $\mathbb{R}$-*linear transformations* $T : \mathbb{C} \to \mathbb{C}$. $\mathbb{R}$-linearity means that, for $z = x + iy$ we have $T(z) = xT(1) + yT(i)$. This immediately gives us:

*The following assertions about a transformation* $T : \mathbb{C} \to \mathbb{C}$ *are equivalent:*

i) $T$ *is* $\mathbb{R}$-*linear*.

ii) $T(z) = az + b\bar{z}$ *where* $a, b$ *are constants belonging to* $\mathbb{C}$.

*An* $\mathbb{R}$-*linear transformation* $T : \mathbb{C} \to \mathbb{C}$ *is* $\mathbb{C}$-*linear if and only if* $T(i) = iT(1)$; *this applies if and only if* $T(z) = az$.

The isomorphism $F : \mathbb{C} \to \mathcal{C}$ introduced in 2.5 has the property that

$$F(\bar{c}) = F(c)^t \quad \text{for} \quad c \in \mathbb{C}$$

where the transpose of a matrix $A$ is denoted by $A'$. Thus conjugation in $\mathbb{C}$ is nothing else but transposition in $\mathcal{C}$.

**2. The Field Automorphisms of $\mathbb{C}$.** The mapping $z \mapsto \bar{z}$ can be simply characterized.

**Theorem.** *The conjugation mapping is the only field automorphism of $\mathbb{C}$ which maps $\mathbb{R}$ into itself, and which is different from the identity mapping.*

**Proof.** If $f: \mathbb{C} \to \mathbb{C}$ is an automorphism with $f(\mathbb{R}) \subset \mathbb{R}$, then in the first place $f(x) = x$ for all $x \in \mathbb{R}$. It then follows that, for all $z = x + iy$, $x, y \in \mathbb{R}$

$$f(z) = f(x + iy) = f(x) + f(i)f(y) = x + f(i)y.$$

Since $i^2 = -1$, we have $f(i)^2 = f(i^2) = f(-1) = -1$, hence $f(i) = \pm i$. The case $f(i) = i$ gives $f = \mathrm{id}$, the case $f(i) = -i$ gives conjugation. □

At the beginning of this century (1901), no less famous an authority than DEDEKIND wrote: "die Zahlen des reellen Körpers scheinen mir durch die Stetigkeit so unlöslich miteinander verbunden zu sein, daß ich vermute, er könne außer der identischen gar keine andere Permutation [= Automorphismus] besitzen, und hieraus würde folgen, daß der Körper aller Zahlen [= Körper $\mathbb{C}$] nur die beiden genannten Permutationen besitzt. Nach einigen vergeblichen Versuchen, hierüber Gewißheit zu erlangen, habe ich diese Untersuchung aufgegeben; um so mehr würde es mich erfreuen, wenn ein anderer Mathematiker mir eine entscheidende Antwort auf diese Frage mitteilen wollte." (Math. Werke 2, S.277). [The numbers of the real field seem to me to be so inextricably connected to one another, that I would conjecture that this field has no automorphism other than the identity; and it would follow from this that the field of all numbers (the field $\mathbb{C}$) would possess only the two above-mentioned automorphisms. After a few unsuccessful attempts to establish this proposition on a rigorous basis, I have abandoned this investigation; I would therefore be all the more delighted if some other mathematician would let me have a decisive answer to this question.] It is now known that there are, in fact, infinitely many other automorphisms of $\mathbb{C}$ (which necessarily do *not* map $\mathbb{R}$ into itself). Such mappings are constructed by appealing to the axiom of choice. No one has yet actually seen such an automorphism. See *Grundwissen Mathematik, Vol. 2, Lineare Algebra und analytische Geometrie*, p. 44.

**3. The Natural Scalar Product $\mathrm{Re}(w\bar{z})$ and Euclidean Length $|z|$.** The *Euclidean* scalar product in the real vector space $\mathbb{C} = \mathbb{R}^2$ is given by

$$\langle w, z \rangle := \mathrm{Re}(w\bar{z}) = ux + vy, \quad \text{where} \quad w = u + iv, \quad z = x + iy.$$

As $z\bar{z} = x^2 + y^2$ is never negative, the nonnegative real square root

$$|z| := +\sqrt{\langle z, z \rangle} = +\sqrt{z\bar{z}} = \sqrt{x^2 + y^2}$$

always exists; it measures the *Euclidean distance* of the point $z$ from the origin in the Gaussian number plane, or in other words the length of the vector $z$. The number $|z|$ is known as the *absolute value*[11] of $z$. When $z$ is real, $|z|$ coincides with the absolute value, as defined in the usual way for real numbers. Clearly

$$|z| = |\bar{z}| \quad \text{for all} \quad z \in \mathbb{C}.$$

Since $z\bar{z} = |z|^2$ we have the following elegant representation of the inverse

$$z^{-1} = \frac{\bar{z}}{|z|^2} \quad \text{for all} \quad z \in \mathbb{C}^{\times}.$$

The mapping $\mathbb{C} \times \mathbb{C} \to \mathbb{R}$, $(w, z) \to \langle w, z \rangle$ is $\mathbb{R}$-*bilinear, symmetric, and positive definite*, that is, for all $w, w'z \in \mathbb{C}$ we have

$$\langle w + w', z \rangle = \langle w, z \rangle + \langle w', z \rangle; \quad \langle aw, z \rangle = a\langle w, z \rangle, \quad a \in \mathbb{R};$$

$$\langle w, z \rangle = \langle z, w \rangle; \quad \langle z, z \rangle > 0 \quad \text{whenever} \quad z \neq 0;$$

these rules follow immediately from the definition of $\langle \ , \ \rangle$.                              □

Two vectors $w, z$ are called *orthogonal* (*are perpendicular to one another*) when $\langle w, z \rangle = 0$. The vectors $iz$ and $z$ are always perpendicular to each other because $\text{Re}(iz\bar{z}) = |z|^2\text{Re}(i) = 0$. More generally, since $z\bar{z} \in \mathbb{R}$ we have the result:

*the vectors $z, cz \in \mathbb{C}^{\times}$ are orthogonal if and only if $c$ is purely imaginary.*

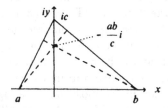

The reader may like to use this for a simple proof of the theorem that the altitudes of a triangle meet in a common point, the orthocenter (see the figure above where the orthocenter is $-\frac{ab}{c}i$).

It is amusing to interpret the scalar product $\text{Re}(w\bar{z})$ in the field $\mathcal{C}$ of real $2 \times 2$ matrices $\begin{pmatrix} a & -b \\ b & a \end{pmatrix}$. We set the following as an

*Exercise.* Show that

$$\langle A, B \rangle := \frac{1}{2} \text{trace}(A \cdot B^t), \qquad A, B \in \mathcal{C},$$

---

[11] Weierstrass used the term "absolute value" (absoluter Betrag) in his lectures: until then the usual expression had been "modulus."

is a *positive-definite, symmetric bilinear form.* Show that the isomorphism $F : \mathbb{C} \to C$ is length preserving, that is, $\langle F(w), F(z) \rangle = \langle w, z \rangle$. Show further that

$$\langle A, A \rangle = \det A.$$

**4. Product Rule and the "Two Squares" Theorem.** For calculating with absolute values we have the *product rule:*

$$|wz| = |w|\,|z| \quad \text{for all} \quad w, z \in \mathbb{C}.$$

To prove this we write $|wz|^2 = wz(\overline{wz}) = w\bar{w}z\bar{z} = |w|^2|z|^2$.      $\square$

The product rule contains a famous theorem, already known to DIO-PHANTUS OF ALEXANDRIA (Greek mathematician of the second half of the third century A.D.).

**"Two Squares" Theorem.** *For all $u, v, x, y \in \mathbb{R}$ we have*

$$(u^2 + v^2)(x^2 + y^2) = (ux - vy)^2 + (uy + vx)^2.$$

**Proof.** We apply the product rule to $w := u + iv$, $z := x + iy$.      $\square$

Here complex numbers serve only to discover the two-squares theorem. Once found it can easily be verified, by "multiplying out," that the identity is valid *for any commutative ring,* and in particular for the ring $\mathbb{Z}$ of integers. This fact is important in elementary number theory. Thus for example it shows that a natural number $n > 1$ is a sum of two squares of natural numbers if each of its prime factors has this property. It is shown in elementary number theory that the primes of the form $l^2 + m^2$, with $l, m \in \mathbb{N}$, are just the odd primes of the form $4k + 1$ and the prime 2.

Generalizations of the "two-squares" theorem will play an important role in the later chapters of this book (see, for example, 6.2.3, 8.2.4 and Chapter 9).      $\square$

The product rule implies the

*division rule*    $|w/z| = |w|/|z|$    *for all*   $w \in \mathbb{C}, \ z \in \mathbb{C}^{\times}.$

The product rule also implies, as an immediate corollary:

*The set $S^1 := \{z \in \mathbb{C} : |z| = 1\}$ of all complex numbers of unit length is a subgroup of $(\mathbb{C}^{\times}, \cdot)$ with respect to multiplication in $\mathbb{C}$.*

$S^1$ is represented in the Gaussian plane by the circumference of the unit circle centered on the origin. We shall call $S^1$ the *circle group,* and it will be used in 5.2 in defining the orthogonal group $O(\mathbb{C})$. It plays a decisive role in the introduction of polar coordinates in §6.

There is an important relationship between the three multiplicative groups $\mathbb{C}^\times$, $S^1$ and $\mathbb{R}_+^\times := \{x \in \mathbb{R}, x > 0\}$:

*The mapping $\mathbb{C}^\times \to \mathbb{R}_+^\times \times S^1$, $z \mapsto (|z|, z/|z|)$ is a (topological) isomorphism of the (topological) group $\mathbb{C}^\times$ onto the product of the (topological) groups $\mathbb{R}_+^\times$ and $S^1$.*

*Exercise.* Let $c \in S^1$. Show that there is a $w \in \{1, -1, i, -i\}$ such that $|c - w| < 1$. (See also 4.2.4 in this connection.)

**5. Quadratic Roots and Quadratic Equations.** To every real number $r \geq 0$, there is precisely one real number $s \geq 0$ such that $s^2 = r$; $s$ is called *the* nonnegative square root of $r$, and is written as $\sqrt{r}$ (this fact has already been used in the definition of $|z|$). It is not possible to extract a real square root from a negative real number. With complex numbers the situation is better.

**Existence Theorem.** *Let $c = a + ib$ where $a, b \in \mathbb{R}$, be any complex number. Let $\xi$ be defined by*

$$(1) \qquad \xi := \sqrt{\frac{1}{2}(|c| + a)} + i\eta\sqrt{\frac{1}{2}(|c| - a)}$$

*where $\eta := \pm 1$ with the sign chosen so that $b = \eta|b|$. Then $\xi^2 = c$.*

The proof is straightforward. We arrive at (1) automatically by starting from the equation $(x + iy)^2 = a + ib$ which is equivalent to the two equations $x^2 - y^2 = a$, $2xy = b$. It follows, since $x^2 + y^2 = |c|$, that $2x^2 = |c| + a$ and $2y^2 = |c| - a$, thus verifying (1). As in the real case, the number $\xi$ is called a *square root* of $c$ and is denoted by $\sqrt{c}$. Apart from $\xi$, the only other square root of $c$ is $-\xi$. The symbol $\sqrt{c}$ is therefore two-valued.

All quadratic equations, in standard form

$$z^2 + 2cz + d = 0, \qquad c, d \in \mathbb{C},$$

can now be solved immediately. Using the age-old trick of the Babylonians, the device of completing the square, the equation becomes

$$(z + c)^2 + d - c^2 = 0$$

whose two solutions $z_1, z_2$ can be read off at once:

$$z_1 := -c + \sqrt{c^2 - d}, \qquad z_2 := -c - \sqrt{c^2 - d},$$

where $\sqrt{c^2 - d}$ in both cases denote the *same* square root. One obtains the linear factorization

$$z^2 + 2cz + d = (z - z_1)(z - z_2)$$

and in particular the well-known rule taught at school.

*Vieta's Rule.*[12]   $z_1 + z_2 = -2c,$   $z_1 z_2 = d.$

In 6.3 we shall give the solution of quadratic equations in polar coordinates.

In 5.2 we shall use the proposition:
*To every number $c = a + ib \in S^1$ with $a \geq 0$ there exists a $\xi \in S^1$ such that*

$$(1') \qquad\qquad \xi^2 = c \quad and \quad |\operatorname{Im}\xi| \leq \frac{1}{\sqrt{2}}|b|.$$

**Proof.** Let $\xi$ be chosen to satisfy (1). Since $|\xi|^2 = |c| = 1$, $\xi \in S^1$. Since $1 = a^2 + b^2$ and $a \geq a^2$ as $0 \leq a \leq 1$, it follows by (1) that $2|\operatorname{Im}\xi|^2 = 1 - a \leq 1 - a^2 = b^2$, which is equivalent to (1'). $\qquad\square$

The existence theorem for square roots has some unsuspected consequences. We give a first sample in the next paragraph.

**6. Square Roots and $n$th Roots.** Let $n \geq 1$ be a natural number, and let $c \in \mathbb{C}$. Every complex number $\xi$ such that $\xi^n = c$ is called an *$n$th root of $c$*. The existence theorem 5 is so powerful that the existence of $n$th roots can be speedily deduced from it.

**Theorem.** *Every complex number $c$ has $n$th roots for $1 \leq n < \infty$.*

**Proof.** We use induction on $n$ and make use of the proposition

$(*)$ *Every real polynomial of odd degree has a real root, that is, it vanishes for some real value of the variable (by the intermediate value theorem) and in particular every number $r \in \mathbb{R}$ has a $(2m+1)$th root in $\mathbb{R}$, $m = 1, 2, \ldots$.*

By Theorem 5 the proposition is true for $n = 2$ (it is trivial for $n = 1$). Suppose $n > 2$. In the case $n = 2m$, there is in the first place an $\eta \in \mathbb{C}$

---

[12] François Vieta (1540–1603, Paris, Government official) introduced calculation with letters as symbols for numbers, using vowels for unknown and consonants for known quantities.

such that $\eta^2 = c$. Since $m < n$, there is then, by the inductive hypothesis, a $\xi \in \mathbb{C}$ such that $\xi^m = \eta$. It follows that $\xi^n = c$.

Now suppose $n$ to be odd. Because of $(*)$ we may assume that $c \notin \mathbb{R}$ and $|c| = 1$. We choose a $d \in \mathbb{C}$ such that $d^2 = c$. Then $\overline{d}d = 1$. Consider the polynomial

$$p(X) := i[\overline{d}(X + i)^n - d(X - i)^n] = i(\overline{d} - d)X^n + \text{lower order terms.}$$

Since $\overline{p(x)} = p(x)$ for all $x \in \mathbb{R}$, $p$ is a *real* polynomial. Since $d \notin \mathbb{R}$, $p$ has the odd degree $n$. By $(*)$ there is therefore a $\lambda \in \mathbb{R}$ such that $p(\lambda) = 0$. We conclude

$$\overline{d}(\lambda + i)^n = d(\lambda - i)^n, \quad \text{hence} \quad \left(\frac{\lambda + i}{\lambda - i}\right)^n = \frac{d}{\overline{d}} = d^2 = c. \qquad \square$$

The theorem can also be formulated as follows:

*Every polynomial in $\mathbb{C}(z)$ of the form $z^n - c$ of degee $n \geq 1$ has a complex zero.*

This is an important special case of the fundamental theorem of algebra.

*Historical Note.* The existence of $n$th roots is usually shown with the help of the complex exponential function, see 6.4, this method being particularly simple and economical. The fact that $n$th roots can be constructed in an elementary fashion without a knowledge of the exponential function had already been pointed out by DEDEKIND in a letter of 1878 to LIPSCHITZ (see LIPSCHITZ *Briefwechsel* ed. SCHARLAU, Vol. 2, Brunswick, Vieweg, 1986, p. 91). HURWITZ in 1911 beautifully demonstrated the power of the process of (iterated) square root extraction in his method of introducing the real logarithm function (see Über die Einführung der elementaren Funktionen in der algebraischen Analysis, *Math. Ann.* 70, 33–47; *Math. Werke* 1, 706–721).

We shall see in 4.2 that the existence of square roots in the final analysis leads to the fundamental theorem of algebra; we shall furthermore show in 7.4 that the famous GELFAND–MAZUR theorem in functional analysis is really based on nothing more than the *existence of square roots* and simple *topological properties of normed vector spaces*.

## §4. GEOMETRIC PROPERTIES OF THE FIELD $\mathbb{C}$

In this paragraph, the scalar product $\langle w, z \rangle$, the length function $|z|$, and the cross ratio of four points in $\mathbb{C}$, will constitute the focus of our attention.

We shall prove, among other things, PTOLEMY's famous theorem, now almost two thousand years old, on the diagonals of a cyclic quadrilateral, and the theorem on the WALLACE lines. We should like to make it clear that these particular geometric applications have been chosen on historical grounds. Many other equally striking and less well known applications could easily be found. We refer those interested to YAGLOM, *Complex Numbers in Geometry*, New York, Academic Press, 1968.

**1. The Identity** $\langle w, z \rangle^2 + \langle iw, z \rangle^2 = |w|^2 |z|^2$. Since we always have $\text{Re}(iz) = -\text{Im}\, z$, it follows that $\langle iw, z \rangle = -\text{Im}\, w\bar{z}$. We can therefore deduce, with the help of the product rule 3.4 the following useful identity

(1) $$\langle w, z \rangle^2 + \langle iw, z \rangle^2 = |w|^2 |z|^2, \qquad w, z \in \mathbb{C}.$$

**Proof.** $\langle w, z \rangle^2 + \langle iw, z \rangle^2 = (\text{Re}\, w\bar{z})^2 + (-\text{Im}\, w\bar{z})^2 = |w\bar{z}|^2 = |w|^2 |z|^2.$ □

As a corollary we obtain

**The CAUCHY–SCHWARZ Inequality.** $|\langle w, z \rangle| \le |w|\, |z|$ *for all* $w, z \in \mathbb{C}$ *with the equality sign applying if and only if* $w, z$ *are linearly dependent.*

**Proof.** The inequality is implicit in the identity (1), which also implies that there is equality when $\langle iw, z \rangle = -\text{Im}\, w\bar{z} = 0$, that is, when $w\bar{z} \in \mathbb{R}$. □

We give a second proof which uses the product rule and the inequalities $|\text{Re}\, z| \le |z|$, $|\text{Im}\, z| \le |z|$, $z \in \mathbb{C}$ which clearly follow from the respective definitions. We have $|\langle w, z \rangle| = |\text{Re}(w\bar{z})| \le |w\bar{z}| = |w|\, |\bar{z}| = |w|\, |z|$ from which we deduce that $|\text{Re}(w\bar{z})| = |w\bar{z}|$ if and only if $w\bar{z} \in \mathbb{R}$.

**2. Cosine Theorem and the Triangle Inequality.** Just as for every scalar product, we have

$$|w + z|^2 = |w|^2 + |z|^2 + 2\,\text{Re}\,(w\bar{z}) \qquad \text{(cosine)}$$

**Proof.** Thanks to the additivity and symmetry of $\langle w, z \rangle$ we have

$$|w + z|^2 = \langle w + z, w + z \rangle = \langle w, w \rangle + \langle w, z \rangle + \langle z, w \rangle + \langle z, z \rangle$$
$$= |w|^2 + 2\,\text{Re}(w\bar{z}) + |z|^2. \qquad \square$$

We shall return to the cosine theorem in 6.2, where the reason for the choice of name will be explained. With the help of the CAUCHY–SCHWARZ inequality one can prove the

**Triangle Inequality.** *For all* $w, z \in \mathbb{C}$, *we have* $|w + z| \le |w| + |z|$. *The equality sign applies if and only if* $w\bar{z} \ge 0$.

**Proof.** $|w + z|^2 = |w|^2 + 2\langle w, z \rangle + |z|^2 \le |w|^2 + 2|w|\, |z| + |z|^2 = (|w| + |z|)^2$. By the CAUCHY–SCHWARZ inequality $|\langle w, z \rangle| = |w|\, |z| \Leftrightarrow w\bar{z} \in \mathbb{R}$. Consequently the case $\langle w, z \rangle = |w|\, |z|$ applies if and only if $w\bar{z} \ge 0$.

A mapping $||: K \to \mathbb{R}$ of a (commutative) field $K$ into $\mathbb{R}$ is called a *valuation* of $K$, when, for all $w, z \in K$, the following relations hold:

1) $|z| \geq 0$, $\quad$ $|z| = 0 \Leftrightarrow z = 0$,

2) $|wz| = |w||z|$ $\quad$ (*Product rule*)

3) $|w + z| \leq |w| + |z|$ $\quad$ (*Triangle inequality*).

A field together with a valuation is called a *field with valuation*. The fields $\mathbb{Q}$ and $\mathbb{R}$ are fields with valuation. We have seen that $\mathbb{C}$ can be provided with a valuation, by means of the absolute value function $||: \mathbb{C} \to \mathbb{R}$, $z \to |z|$, and that this valuation is an extension of the valuation of $\mathbb{R}$ by means of the absolute value.

A subtle interplay between the absolute value function and the field operations is revealed in the following

**"Three-party" Theorem.** *Let $z_1$, $z_2$, $z_3$ be three distinct complex numbers such that $|z_1| = |z_2| = |z_3|$. Then the following statements are equivalent:*

i) $z_1$, $z_2$, $z_3$ *are the vertices of an equilateral triangle*

ii) $z_1 + z_2 + z_3 = 0$

iii) $z_1$, $z_2$, $z_3$ *are the roots of an equation $Z^3 = c$ where $c \in \mathbb{C}$.*

If one thinks of $z_1$, $z_2$, $z_3$ as political parties, interpreting equal in length as *equal in strength* then the implication (i) to (ii) provides the motivation for the name of the theorem.

The proof may be left to the reader. It can be reduced to the case $z_1 z_2 z_3 = 1$, and to prove ii) $\Rightarrow$ iii) one can consider the expression $z_1 z_2 z_3 (\bar{z}_1 + \bar{z}_2 + \bar{z}_3)$. $\qquad\qquad\qquad\qquad\qquad\qquad\qquad\qquad\qquad\qquad\qquad$ □

If one defines the *centroid* of a triangle with vertices $z_1$, $z_2$, $z_3$ as the point $\frac{1}{3}(z_1 + z_2 + z_3)$, the equivalence of i) and ii) asserts that the centroid of a triangle is at the center of its circumcircle if and only if the triangle is equilateral.

In analogy with the foregoing, if $z_1, z_2, z_3, z_4 \in \mathbb{C}$ and $|z_1| = \cdots = |z_4|$ the following three statements are equivalent:

i) $z_1, z_2, z_3, z_4$ *are the vertices of a rectangle.*

ii) $z_1 + z_2 + z_3 + z_4 = 0$.

iii) $z_1, \ldots, z_4$ *are the roots of an equation $(Z^2 - a^2)(Z^2 - b^2)$ with $|a| = |b| \neq 0$.*

**3. Numbers on Straight Lines and Circles. Cross-Ratio.** Two numbers $a, b \in \mathbb{C}$ lie on a straight line through 0, if and only if $a\bar{b} \in \mathbb{R}$ (see 3.1). More generally:

*Three numbers $a, b, c \in \mathbb{C}$, $a \neq b$, are collinear if and only if*

(1) $\qquad \dfrac{c-a}{b-a} \in \mathbb{R}, \qquad$ *that is, if and only if* $\quad c\bar{b} - c\bar{a} - a\bar{b} \in \mathbb{R}$.

The proof is trivial because the line through $a, b$ has the parametric representation $a + (b-a)s$, $s \in \mathbb{R}$. $\qquad\qquad\qquad\qquad\qquad\qquad\quad$ □

If $a, b, c, d \in \mathbb{R}$ with $a \neq d$, $b \neq c$, then the *cross-ratio* or *anharmonic ratio*, denoted by $CR(a, b, c, d)$ is defined by

(2) $\qquad CR(a, b, c, d) := \dfrac{a-b}{a-d} : \dfrac{c-b}{c-d} = \dfrac{(a-b)(c-d)}{(a-d)(c-b)}$

$$= \dfrac{(a-b)(c-d)(\bar{a}-\bar{d})(\bar{c}-\bar{b})}{|a-d|^2|c-b|^2} \in \mathbb{C}.$$

This number depends on the order of the four points, $a$, $b$, $c$, $d$. The reciprocal value is obtained when the points undergo a cyclic permutation:

$$CR(b, c, d, a) = CR(a, b, c, d)^{-1}.$$

We now prove:

**Theorem.** *Four numbers $a, b, c, d \in \mathbb{C}$, $a \neq d$, $b \neq c$, not all on the same straight line, lie on a circle if and only if their cross-ratio is real.*

**Proof.** Suppose say that $a, b, c$ are not collinear. Since this property and the cross-ratio are both *translation invariant*, we may assume that the center of the circumcircle of the triangle with vertices $a, b, c$ lies at the origin. Then $|a| = |b| = |c|$ and

$$(a-b)(c-d)(\bar{a}-\bar{d})(\bar{c}-\bar{b}) + i(|c|^2 - |d|^2)\text{Im}(c\bar{b} - c\bar{a} - a\bar{b}) \in \mathbb{R}$$

by exercise 3.1. Since $a, b, c$ are not collinear, $\text{Im}(c\bar{b} - c\bar{a} - a\bar{b}) \neq 0$ by (1). It follows therefore that

$$(a-b)(c-d)(\bar{a}-\bar{d})(\bar{c}-\bar{d}) \in \mathbb{R} \Leftrightarrow |c| = |d|$$

and by (2) this is what the theorem asserts. $\qquad\qquad\qquad\qquad\qquad\qquad$ □

In the theory of fractional linear transformations $z \mapsto \frac{az+b}{cz+d}$ the cross-ratio plays a central role. In this theory the argument $z$ is allowed to assume the value $\infty$. The cross-ratio is invariant under all fractional linear transformations, and this makes possible a new proof of the preceding theorem. See, for example CONWAY, *Functions of One Complex Variable*, Springer, 1978, p. 43.

**4. Cyclic Quadrilaterals and Cross-Ratio.** Any four distinct points $a, b, c, d \in \mathbb{C}$ define a quadrilateral $abcd$ in $\mathbb{C}$ with vertices $a$, $b$, $c$, $d$, whose sides are the line-segments joining $a$ to $b$, $b$ to $c$, $c$ to $d$ and $d$ to $a$. A quadrilateral is said to be *cyclic* when its vertices all lie on a circle and when two different sides intersect in a vertex, if they intersect at all. (The figure in the next paragraph illustrates a cyclic quadrilateral $abcd$; the quadrilateral $abcd$ which would be obtained by interchanging the vertices $b$ and $c$ would not be a cyclic quadrilateral.)

**Theorem.** *A quadrilateral $abcd$ is cyclic if and only if the cross-ratio $CR(a,b,c,d)$ is negative.*

**Proof** (using a continuity argument). Let $S^1$ be the given circle. The squares $Q$, $Q'$ whose vertices are respectively the points $1$, $i$, $-1$, $-i$ and $1$, $-i$, $-1$, $i$ are cyclic and the cross-ratio of their vertices is $-1$. It is "obvious" that a quadrilateral $V$ can be obtained from $Q$ or $Q'$ by a continuous displacement of the vertices along the circumference of $S^1$ in such a way that two vertices never coincide during the displacement.

Since the cross-ratio of four different points on $S^1$ is, by Theorem 3, a real number and since it is a continuous non-vanishing function of its arguments, it follows from the intermediate value theorem that a quadrilateral with vertices $a, b, c, d \in S^1$ is cyclic if and only if $CR(a, b, c, d) < 0$.

**5. PTOLEMY's Theorem.** The Egyptian mathematician Claudius Ptolemy (Alexandria, circa 150 A.D.) proved in his *Almagest*, Book 1, Chapter 10 the following theorem which is still occasionally discussed in school geometry:

*In any cyclic quadrilateral $abcd$ the sum of the products of the opposite sides is equal to the product of the diagonals*

$$|a - b| \cdot |c - d| + |a - d| \cdot |c - b| = |a - c| \cdot |b - d|.$$

Ptolemy made this theorem serve Astronomy and used it as a tool in the computation of his famous table of chords. If, in fact, one of the sides is a diameter, then it is an easy matter to derive the addition theorem

$$\sin(\alpha - \beta) = \sin \alpha \cos \beta - \cos \alpha \sin \beta.$$

Ptolemy proved his theorem by an elegant trick of elementary geometry. He constructs a point $e$ on the line $ac$ so that $\angle abe = \angle cbd$. The triangles $abe$ and $bcd$ are then similar, and a simple argument then leads to the desired conclusion.                                                                      □

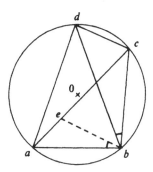

To prove PTOLEMY's theorem and more, with the aid of complex numbers, we assign to every quadrilateral $abcd$ in $\mathbb{C}$ the "PTOLEMY number"

$$P(abcd) := |(a - b)(c - d)| + |(a - d)(c - b)| - |(a - c)(b - d)|.$$

Since $(a-b)(c-d)-(a-d)(c-b) = (a-c)(b-d)$ holds for every commutative ring, and since $CR(a,b,c,d) = (a - b)(c - d)(a - d)^{-1}(c - b)^{-1}$, a direct verification shows that

$$P(abcd) = |(a - d)(b - c)|\,|CR(a,b,c,d)| + 1 - |CR(a,b,c,d) - 1|.$$

Since, by the triangle inequality $|w - 1| = |w| + 1$ if and only if $w$ is real and $\leq 0$, we have, thanks to Theorem 4, demonstrated

**Theorem.** *The following two statements about a quadrilateral $abcd$ in $\mathbb{C}$ are equivalent:*

i) *The assertion in* PTOLEMY*'s theorem holds for $abcd$: $P(abcd) = 0$.*

ii) *The quadrilateral $abcd$ is cyclic.*

The converse of PTOLEMY's theorem, that is the implication i) $\Rightarrow$ ii), was proposed in 1832 in CRELLE's *Journal*, Vol. 8, p. 320 as an exercise. Solutions are to be found in Volumes 10, p. 41; 11, 264–271 and 13, 233–236. CLAUSEN among others gave an elegant solution.

**6. WALLACE's Line.** Suppose $a, b, u \in \mathbb{C}$, $a \neq b$. The *foot* $v$ of the *perpendicular* from $u$ on to the line $L := \{z = a + s(b - a) : s \in \mathbb{R}\}$ through $a$ and $b$ is, since $i(b - a)$ is orthogonal to $(b - a)$, the point of intersection of $L$ with the line $L' := \{z = u + it(b - a)\}$, (see Fig. a). This gives for $s, t$ the condition $s - ti = (u - a)(b - a)^{-1}$, and thus $2s = (u - a)(b - a)^{-1} + (\bar{u} - \bar{a})(\bar{b} - \bar{a})^{-1}$ and therefore

$$v = \frac{1}{2}\left[a + u + (\bar{u} - \bar{a})\frac{b - a}{\bar{b} - \bar{a}}\right].$$

In the case $|a| = |b|$ we have $(b-a)(\bar{b}-\bar{a})^{-1} = -b(\bar{a})^{-1}$ and consequently

$$(*) \qquad\qquad v = \frac{1}{2}\left(a + b + u - \bar{u}\frac{ab}{|a|^2}\right), \quad \text{if} \quad |a| = |b|.$$

We make use of $(*)$ to prove a little-known statement about three "remarkable"[13] points of a triangle.

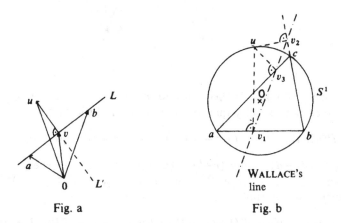

Fig. a                                    Fig. b

**Theorem.** *Let $a, b, c \in \mathbb{C}$ be the vertices of a triangle, and $v_1$, $v_2$, $v_3$ the feet of the perpendiculars from an arbitrary point $u \in \mathbb{C}$ onto the lines through the pair of points $a, b$; $b, c$; $c, a$ respectively. Then the following statements are equivalent (see Fig. b):*

i) *The points $v_1$, $v_2$, $v_3$ are collinear.*

ii) *The point $u$ lies on the circumcircle of the triangle whose vertices are $a, b, c$.*

**Proof.** We may assume that the circumcircle is $S^1$. We then have, by $(*)$, if we make the initial hypothesis that $v_2 \neq v_3$, $u \neq 0$.

$$\frac{v_1 - v_3}{v_2 - v_3} = \frac{b - c - \bar{u}ab + \bar{u}ac}{b - a - \bar{u}bc + \bar{u}ac} = \frac{(c-b)(\bar{u}a - 1)}{(a-b)(\bar{u}c - 1)} = \frac{c - b}{c - \bar{u}^{-1}} : \frac{a - b}{a - \bar{u}^{-1}}$$
$$= CR(c, b, a, \bar{u}^{-1}).$$

The equivalence i) $\Leftrightarrow$ ii) now follows from the results of Section 3, since $\bar{u}^{-1} \in S^1$ is equivalent to $u \in S^1$. The case $v_2 = v_3$ is, by virtue of $a \neq b$, possible only if $\bar{u}c = 1$, that is, if $u \in S^1$. In the case $u = 0$, we have

---

[13]The word "remarkable" is used in classical elementary geometry in the sense of "worthy of notice."

$(v_1 - v_3):(v_2 - v_3) = (c - b):(a - b)$, so that $v_1$, $v_2$, $v_3$ are not collinear because $a, b, c$ are not. □

In the case where $u$ lies on the circumcircle, the line through $v_1$, $v_2$, $v_3$ is called WALLACE's line,[14] after the self-taught Scottish mathematician William WALLACE (1768–1843) who, after having been a teacher in Perth, was Professor of Mathematics at Edinburgh University from 1819. This line is also sometimes (in fact more usually, if mistakenly) known as the SIMSON line, after the Scottish mathematician Robert SIMSON (1687–1768) who successfully sought to revive the study of ancient Greek geometry in England. However MACKAY showed, in two articles in the *Proceedings of Edinburgh Mathematical Society* 9, 1891, 83–91 and 23, 1905, 80–85, that no comparable result is to be found in the published works of SIMSON, whereas the implication ii) ⇒ i) appears, obviously for the first time, in an article by WALLACE in the *Mathematical Repository* 2, 1799–1800, p. 111.

## §5. THE GROUPS $O(\mathbb{C})$ AND $SO(2)$

In the following paragraphs we shall show, among other things, that the circle group $S^1$ is isomorphic to the orthogonal group $SO(2)$ of orthogonal $2 \times 2$ matrices with determinant 1, under the mapping $F: \mathbb{C} \to \mathcal{C}$, $a + bi \mapsto \begin{pmatrix} a & -b \\ b & a \end{pmatrix}$. We shall also obtain a classical parametric representation of the group $SO(2)$.

**1. Distance Preserving Mappings of $\mathbb{C}$.** A (not necessarily $\mathbb{R}$-linear) mapping $f: \mathbb{C} \to \mathbb{C}$ is called *distance preserving* (or isometric), if

$$|f(w) - f(z)| = |w - z| \quad \text{for} \quad w, z \in \mathbb{C}.$$

**Theorem.** *The following statements about $f: \mathbb{C} \to \mathbb{C}$ are equivalent:*

i) *$f$ satisfies $f(z) = f(0) + cz$ or $f(z) = f(0) + c\bar{z}$ with $c \in S^1$.*

ii) *$f$ is distance preserving.*

**Proof.** i) ⇒ ii). This is trivial since $f(w) - f(z) = c(w - z)$ or $= c(\bar{w} - \bar{z})$ respectively.

ii) ⇒ i). As $c := f(1) - f(0) \in S^1$, the mapping $g: \mathbb{C} \to \mathbb{C}$, $z \mapsto c^{-1}(f(z) - f(0))$ is certainly distance preserving. Since $g(0) = 0$ and $g(1) = 1$ we have $|g(z)|^2 = |z|^2$ and $|g(z) - 1|^2 = |z - 1|^2$. It follows from this that $\operatorname{Re} g(z) = \operatorname{Re} z$, and in particular that $g(i) = \pm i$. In the case where $g(i) = i$, then $\hat{g}(z) := -ig(iz)$ is distance preserving with $\hat{g}(0) = 0$, $\hat{g}(1) = 1$, and

---

[14] Not to be confused with the well-known Wallace line in geography and natural history.

therefore (from what has just been proved) $\mathrm{Re}(-ig(iz)) = \mathrm{Re}\,z$, that is, $\mathrm{Im}\,g(z) = \mathrm{Im}\,z$, whence $g(z) = z$ and $f(z) = f(0) + cz$. In the other case where $g(i) = -i$, it follows similarly with $\hat{g}(z) := ig(iz)$ that $\mathrm{Re}(ig(iz)) = \mathrm{Re}\,z$, that is $\mathrm{Im}\,g(z) = -\mathrm{Im}\,z$, and hence $f(z) = f(0) + c\bar{z}$.                     □

In particular every distance preserving mapping of $\mathbb{C}$ into itself which fixes the origin is $\mathbb{R}$-linear.

In linear algebra every distance preserving mapping of an Euclidean vector space $V$ into itself is called a motion (or displacement). The statement which we have just proved above is thus a special case of the general theorem that every (Euclidean) motion $f: V \to V$ has the form $x \to f(0) + h(x)$ where $h: V \to V$ is orthogonal.

**2. The Group $O(\mathbb{C})$.** An $\mathbb{R}$-*linear* mapping $f: \mathbb{C} \to \mathbb{C}$ is called *orthogonal* if $\langle f(w), f(z) \rangle = \langle w, z \rangle$ for all $w, z \in \mathbb{C}$. Every orthogonal mapping $f: \mathbb{C} \to \mathbb{C}$ is *length preserving*: $|f(z)| = |z|$, and therefore because of $\mathbb{R}$-linearity, also distance preserving.

**Theorem.** *A mapping* $f: \mathbb{C} \to \mathbb{C}$ *is orthogonal if and only if*

$$f(z) = cz \quad or \quad f(z) = c\bar{z} \quad with \quad c \in S^1.$$

**Proof.** The specified mappings are orthogonal. For example in the second case

$$\langle f(w), f(z) \rangle = \mathrm{Re}(c\bar{w}(\overline{c\bar{z}})) = |c|^2 \mathrm{Re}(\bar{w}z) = \langle w, z \rangle$$

since $c \in S^1$.

Conversely, if $f$ is orthogonal it is distance preserving and the statement follows from Theorem 1 because $f(0) = 0$.                     □

*Exercise.* Prove the theorem directly by using the characterization of $\mathbb{R}$-linear mappings in 3.1 and showing, by verification, that

$$|az + b\bar{z}| = |z| \text{ for all } z \in \mathbb{C} \Leftrightarrow a \in S^1 \text{ and } b = 0 \text{ or } a = 0 \text{ and } b \in S^1.$$

The orthogonal mappings of $\mathbb{C}$ form a non-Abelian *group*, under the operation of composition, the so-called *orthogonal group* $O(\mathbb{C})$. The orthogonal mappings of the form $T_c(z) = cz$, $c \in S^1$, are called *rotations*, and constitute a normal subgroup $SO(\mathbb{C})$ of $S(\mathbb{C})$. It follows from the foregoing considerations that:

*The mapping* $S^1 \to SO(\mathbb{C})$, $c \mapsto T_c$, *is a group isomorphism.*

In particular the group $SO(\mathbb{C})$ is *Abelian*. The mappings $f(z) = c\bar{z}$, $c \in S^1$ are called *reflections;* they constitute the only other coset in $O(\mathbb{C})$ relative to $SO(\mathbb{C})$.

**3. The Group $SO(2)$ and the Isomorphism $S^1 \to SO(2)$.** The set

$$(1) \qquad\qquad O(2) := \{A \in GL(2, \mathbb{R}): AA' = E\}$$

of all real *orthogonal* $2 \times 2$ matrices is an important subgroup of the group $GL(2, \mathbb{R})$ of all real *invertible* $2 \times 2$ matrices. Since $\det A = \det A^t$ we have $\det A = \pm 1$ for all $A \in O(2)$. The set

$$SO(2) := \{A \in O(2): \det A = 1\}$$

is a normal *subgroup of $O(2)$*, and is the group of all *proper orthogonal* $2 \times 2$ real matrices. Denoting by $\mathcal{C}$ the subfield of $\text{Mat}(2, \mathbb{R})$ which was introduced in 2.5, we then have the following:

**Theorem.** $SO(2) = \{A \in \mathcal{C}: \det A = 1\}$.

**Proof.** For $A = \begin{pmatrix} a & -b \\ b & a \end{pmatrix}$ we can verify immediately that $AA^t = (\det A)E$, from which it follows that $\{A \in \mathcal{C}: \det A = 1\} \subset SO(2)$.

For $A = \begin{pmatrix} a & b \\ c & d \end{pmatrix} \in SO(2)$ we have $A^{-1} = A^t = \begin{pmatrix} a & c \\ b & d \end{pmatrix}$ by (1). On the other hand since $A^{-1} = \begin{pmatrix} d & -b \\ -c & a \end{pmatrix}$ on account of $\det A = 1$, it follows that $d = a$, $c = -b$, or in other words $A \in \mathcal{C}$. $\qquad\square$

This immediately yields the:

**Isomorphism Theorem.** *The circle group $S^1$ is mapped isomorphically on to the group $SO(2)$ by the mapping $F: \mathbb{C} \to \mathcal{C}$, $a + bi \mapsto \begin{pmatrix} a & -b \\ b & a \end{pmatrix}$.*

**Proof.** The statement is clearly true since

$$F(S^1) = \left\{ A = \begin{pmatrix} a & -b \\ b & a \end{pmatrix} \in \mathcal{C}: \det A = a^2 + b^2 = 1 \right\}. \qquad\square$$

The orthogonal groups $SO(3)$ and $SO(4)$ will be described in Chapter 6, §3 with the help of quaternions.

**4. Rational Parametrization of Properly Orthogonal $2 \times 2$ Matrices.** The set $S^1 \setminus \{-1\}$ is mapped bijectively on to the imaginary axis, by mapping the point $\alpha + i\beta$ of $S^1$ onto $i\lambda$, the point of intersection between the line joining $-1$ to $\alpha + i\beta$ and the imaginary axis (see figure). A simple calculation (intercept theorem of THALES) gives:

$$(1) \qquad \alpha = \frac{1 - \lambda^2}{1 + \lambda^2}, \quad \beta = \frac{2\lambda}{1 + \lambda^2}, \quad \lambda = \frac{\beta}{1 + \alpha}.$$

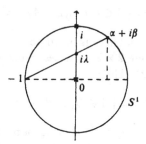

It follows that $\alpha + i\beta = \frac{1+i\lambda}{1-i\lambda}$, so that we have the *rational parametrization*

$$(2) \qquad S^1 \setminus \{-1\} = \left\{ \frac{1 + i\lambda}{1 - i\lambda} : \lambda \in \mathbb{R} \right\},$$

*where the real and imaginary parts of* $c := \frac{1+i\lambda}{1-i\lambda}$ *are rational, that is, belong to* $\mathbb{Q}$, *if and only if* $\lambda$ *is rational.*

In view of $F(S^1) = SO(2)$, this result can be expressed in the form

$$(3) \qquad SO(2) \setminus \{-E\} = \left\{ \frac{1}{1 + \lambda^2} \begin{pmatrix} 1 - \lambda^2 & -2\lambda \\ 2\lambda & 1 - \lambda^2 \end{pmatrix} : \lambda \in \mathbb{R} \right\},$$

*the matrix is rational if and only if* $\lambda$ *is rational.*

*Remark.* One can get rid of the exceptional role of $-1$ and $-E$ in the equations (2) and (3) if one replaces $\lambda$ by $\lambda/\kappa$ and simplifies. We then have, without any restriction

$$(2') \qquad S^1 = \left\{ \frac{\kappa + i\lambda}{\kappa - i\lambda} : (\kappa, \lambda) \in \mathbb{R}^2 \setminus \{0\} \right\}$$

$$= \left\{ \frac{1}{\kappa^2 + \lambda^2} [(\kappa^2 - \lambda^2) + 2\kappa\lambda i] : (\kappa, \lambda) \in \mathbb{R}^2 \setminus \{0\} \right\},$$

$$(3') \qquad SO(2) = \left\{ \frac{1}{\kappa^2 + \lambda^2} \begin{pmatrix} \kappa^2 - \lambda^2 & -2\kappa\lambda \\ 2\kappa\lambda & \kappa^2 - \lambda^2 \end{pmatrix} : (\kappa, \lambda) \in \mathbb{R}^2 \setminus \{0\} \right\}.$$

We shall make our acquaintance in 6.3.5 with EULER's famous rational parametric representation of the group $SO(3)$, which includes, as a special case, the representation $(3')$ of $SO(2)$.

The representation (3) for proper orthogonal $2 \times 2$ matrices is really nothing more than CAYLEY's representation

$(*)$   $A = (E - X)^{-1}(E + X)$, where $X \in \text{Mat}(2, \mathbb{R})$ is skew symmetric,

for *all* $2 \times 2$ skew symmetric matrices are of the form $\lambda \begin{pmatrix} 0 & -1 \\ 1 & 0 \end{pmatrix}$, $\lambda \in \mathbb{R}$, and since $X^2 = -\lambda^2 E$, the equation $(*)$ is the analogue of the equation $\alpha + i\beta = (1 - \lambda i)^{-1}(1 + \lambda i)$. Since $(E - X)^{-1} = (1 + \lambda^2)^{-1}(E + X)$ we have

$$A = (1 + \lambda^2)^{-1}(E + X)^2 = (1 + \lambda^2)^{-1}[(1 - \lambda^2)E + 2X]$$
$$= (1 + \lambda^2)^{-1} \begin{pmatrix} 1 - \lambda^2 & -2\lambda \\ 2\lambda & 1 - \lambda^2 \end{pmatrix}.$$

The equations (1) for the rational points on $S^1$ contain the so-called "Indian formulae" for Pythagorean triplets. A triplet of nonzero natural numbers $k, l, m$ is said to be *Pythagorean* if $k^2 + l^2 = m^2$. It is obvious that at least one of the numbers $k, l$ must be even. We shall show that:

*If $k, l, m$ is a Pythagorean triplet and $l$ is even, then there are nonzero natural numbers $r, s, t$ such that*

$$k = (r^2 - s^2)t, \quad l = 2rst, \quad m = (r^2 + s^2)t \quad \text{(the Indian formulae)}.$$

**Proof.** Corresponding to each $m^{-1}k + im^{-1}l \in S^1 \backslash \{-1\}$ there is a $\lambda = s/r$, with $r, s \in \mathbb{N} \backslash 0$ such that by (1)

$$k = (r^2 - s^2)\frac{m}{r^2 + s^2}, \quad l = 2rs\frac{m}{r^2 + s^2}.$$

If we now choose $r, s$ to be relatively prime, then $r^2 + s^2$, $rs$ also are relatively prime (the reader should prove this). As $\frac{1}{2}l = \frac{rsm}{r^2+s^2} \in \mathbb{N}$, it follows that $t := \frac{m}{r^2+s^2} \in \mathbb{N}$ which proves the proposition.   $\square$

## §6. POLAR COORDINATES AND $n$TH ROOTS

Polar coordinates are introduced in the complex number plane by writing every point $z \in \mathbb{C} = \mathbb{R}^2$ in the form $(r \cos \varphi, r \sin \varphi)$ as in the figure. Here $r := |z|$ is the distance of the point $z$ from the origin, and $\varphi$ is the angle in circular measure (radians) between the positive $x$-axis and the position vector of $z$. Every complex number $z \neq 0$ thus has the form

$$z = r(\cos \varphi + i \sin \varphi),$$

where the angle $\varphi$ is uniquely determined apart from an arbitrary integral multiple of $2\pi$.

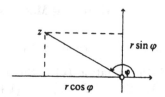

Although these things are clear enough intuitively, it is another matter to establish them precisely and a rigorous proof is *not trivial*. One needs properties of the sine and cosine function which despite being well known have proofs which lie rather deeper. In the treatment which follows we shall work mainly with the complex exponential function

$$\exp z = \sum_{0}^{\infty} \frac{z^{\nu}}{\nu!}$$

defined everywhere in $\mathbb{C}$.

We write $e^{i\varphi} := \exp(i\varphi)$ and appeal essentially to the

**Epimorphism Theorem.** *The mapping $p\colon \mathbb{R} \to S^1$, $\varphi \mapsto e^{i\varphi}$ is a group epimorphism of the (additive) group $\mathbb{R}$ onto the (multiplicative) circle group $S$. There is exactly one positive real number $\pi$ such that:*

a) *the group $2\pi\mathbb{Z}$ is the kernel $\{r \in \mathbb{R}\colon p(r) = 1\}$ of $p$; in particular:*

$$p(\varphi) = p(\psi) \Leftrightarrow \varphi - \psi \in 2\pi\mathbb{Z}; \qquad p([0, 2\pi)) = S^1.$$

b) $p(\pi/2) = i$.

It follows automatically from b) that $p(\pi) = -1$, $p(\frac{3}{2}\pi) = -i$. We call $p$ the *polar coordinate epimorphism*. The connection between $p$ and the trigonometrical functions

$$\cos z := \sum_{0}^{\infty} \frac{(-1)^{\nu}}{(2\nu)!} z^{2\nu}, \quad \sin z := \sum_{0}^{\infty} \frac{(-1)^{\nu}}{(2\nu + 1)!} z^{2\nu+1}, \quad z \in \mathbb{C},$$

is obtained by means of EULER's formula

$$\exp iz = \cos z + i \sin z$$

which obviously implies:

c) $p(\varphi) = e^{i\varphi} = \cos\varphi + i\sin\varphi$ *for all* $\varphi \in \mathbb{R}$.

EULER's formula and above all the epimorphism theorem are discussed at length in Chapter 5, see in particular 5.3.1 and 5.3.6.

**1. Polar Coordinates.** One of the consequences of the epimorphism theorem is the following:

**Theorem.** *Every complex number $z \in \mathbb{C}^{\times}$ can be written uniquely in the form*

$$(1) \qquad z = re^{i\varphi} = r(\cos\varphi + i\sin\varphi) \quad with \quad r := |z| \quad and \quad \varphi \in [0, 2\pi).$$

*For every other representation $z = \rho e^{i\psi} = \rho(\cos\psi + \sin\psi)$ with $\rho, \psi \in \mathbb{R}$, $\rho > 0$, the numbers $\rho, \varphi$ are given by $\rho = r$ and $\psi = \varphi + 2n\pi$ with $n \in \mathbb{Z}$.*

**Proof.** Since $r^{-1}z \in S^1$, there is a $\varphi \in [0, 2\pi)$ such that $p(\varphi) = r^{-1}z$. This means that $z = re^{i\varphi} = r(\cos\varphi + i\sin\varphi)$. If $z = \rho e^{i\psi}$ with $\rho > 0$, $\psi \in \mathbb{R}$, then $|z| = \rho$ since $e^{i\psi} \in S^1$. Hence $e^{i\varphi} = e^{i\psi}$, so that $\varphi - \psi \in 2\pi\mathbb{Z}$. $\qquad \square$

The equation (1) is called a *representation in polar coordinates*, the numbers $r$, $\varphi$, or more generally $r$, $\psi$, where $\psi = \varphi + 2n\pi$, are called *polar coordinates of $z$*. The number $\varphi \in [0, 2\pi]$ is known as the *argument* or *amplitude* of $z \in \mathbb{C}^{\times}$.

Polar coordinates were already used by NEWTON in 1671 in investigating plane spirals. The representation of complex numbers in polar coordinates first appears in EULER and D'ALEMBERT; the factor $\cos\varphi + i\sin\varphi$ is called by CAUCHY in 1821 (in his *Cours d'analyse*) an "expression réduite."

The numbers $1, i, -1, -i$ have the following polar coordinate representations

$$1 = 1 \cdot (\cos 0 + i\sin 0), \qquad i = 1 \cdot \left(\cos\frac{\pi}{2} + i\sin\frac{\pi}{2}\right),$$

$$-1 = 1 \cdot (\cos\pi + i\sin\pi), \qquad -i = 1 \cdot \left(\cos\frac{3\pi}{2} + i\sin\frac{3\pi}{2}\right);$$

so that we have the classical diagram illustrated below with the four values

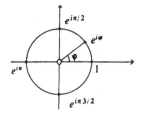

$$e^{i\pi/2} = i, \quad e^{i\pi} = -1, \quad e^{i3\pi/2} = -i, \quad e^{2\pi i} = 1;$$

these are particular cases of the identity

$$i^m = (e^{i\pi/2})^m = e^{im\pi/2}, \qquad m \in \mathbb{Z}.$$

The representation of conjugate complex numbers and of inverses is simple in polar coordinates. Since $\cos(-\varphi) = \cos\varphi$ and $\sin(-\varphi) = -\sin\varphi$, it follows that:

*If* $z = |z|e^{i\varphi} = |z|(\cos\varphi + i\sin\varphi)$, *then*

(2)

$$\bar{z} = |z|e^{-i\varphi} = |z|(\cos\varphi - i\sin\varphi), \quad z^{-1} = |z|^{-1}e^{-i\varphi} = |z|^{-1}(\cos\varphi - i\sin\varphi).$$

The second equation follows from the first since $z^{-1} = |z|^{-2}\bar{z}$.

The *real* polar coordinate mapping

$$\{r \in \mathbb{R}: r > 0\} \times \mathbb{R} \to \mathbb{C}^\times, \quad (r, \varphi) \mapsto (x, y) := (r\cos\varphi, r\sin\varphi)$$

is *differentiable arbitrarily often in the real domain.* We have

$$\det\begin{pmatrix} x_r & x_\varphi \\ y_r & y_\varphi \end{pmatrix} = \det\begin{pmatrix} \cos\varphi & -r\sin\varphi \\ \sin\varphi & r\cos\varphi \end{pmatrix} = r \neq 0,$$

and therefore there exists everywhere a real differentiable inverse mapping (which is given by

$$(x, y) \mapsto \left( \sqrt{x^2 + y^2}, \arccos \frac{x}{\sqrt{x^2 + y^2}} \right)$$

assuming the appropriate branch of the arccosine function is chosen).

**2. Multiplication of Complex Numbers in Polar Coordinates.** Since $e^{i\psi}e^{i\varphi} = e^{i(\psi+\varphi)}$ we have immediately for $w, z \in \mathbb{C}^\times$, the following.

**Theorem.** *If*

$$w = |w|e^{i\psi} = |w|(\cos\psi + i\sin\psi), \quad z = |z|e^{i\varphi} = |z|(\cos\varphi + i\sin\varphi),$$

*then*

(1)      $wz = |w|\,|z|e^{i(\psi+\varphi)} = |w|\,|z|(\cos(\psi + \varphi) + i\sin(\psi + \varphi)),$

*and hence also*

$$\frac{w}{z} = \frac{|w|}{|z|}e^{i(\psi-\varphi)} = \frac{|w|}{|z|}(\cos(\psi - \varphi) + i\sin(\psi - \varphi)).$$

The products and quotients of two complex numbers are therefore obtained by respectively multiplying and dividing their absolute values, and respectively adding and subtracting their amplitudes (see Fig. a). The equation (1) is fundamental and far more than simply a convenient calculating rule which makes the use of polar coordinates obviously advantageous in multiplying complex numbers. It is a profound and unexpected justification for the geometric interpretation of complex numbers in the plane. The mathematical power of this equation was already known to EULER.[15]

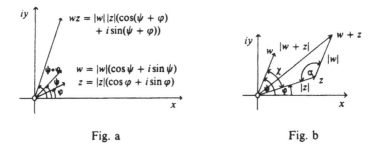

Fig. a                                    Fig. b

The scalar product $\langle w, z \rangle = \mathrm{Re}(w\bar{z})$ takes the well known form $\langle w, z \rangle = |w|\,|z|\cos\chi$, where $\chi := \psi - \varphi$ is the "angle between the vectors $w$ and $z$," as in seen by using the equation (1) in the form

$$w\bar{z} = |w|\,|z|(\cos(\psi - \varphi) + i\sin(\psi - \varphi))$$

(see Fig. b). It now becomes clear why the equation $|w + z|^2 = |w|^2 + |z|^2 + 2\,\mathrm{Re}(w\bar{z})$ was referred to as the *cosine theorem* in 4.2; since $\alpha + \chi = \pi$ (see Fig. b) we have $\cos\chi = -\cos\alpha$ and hence $|w + z|^2 = |w|^2 + |z|^2 - 2|w|\,|z|\cos\alpha$.

**3. de MOIVRE's Formula.** $(\cos\varphi + i\sin\varphi)^n = \cos n\varphi + i\sin n\varphi$ for $n \in \mathbb{Z}$. This is clear from $(e^{i\varphi})^n = e^{in\varphi}$; more generally, we have the following

**Theorem.** *For every complex number* $z = re^{i\varphi} = r(\cos\varphi + i\sin\varphi) \in \mathbb{C}^\times$ *the equation* $z^n = r^n e^{in\varphi} = r^n(\cos n\varphi + i\sin n\varphi)$ *holds for all* $n \in \mathbb{Z}$.

The French huguenot mathematician Abraham DE MOIVRE (1667–1754) emigrated to London after the revocation of the Edict of Nantes in 1685. He became a member of the Royal Society in 1697 and later of the Academies

---

[15] On page 154 of Cauchy's *Cours d'analyse* of 1821, we read however the sentence, so astounding to modern ears "L'équation $\cos(a + b) + \sqrt{-1}\sin(a + b) = (\cos a + \sqrt{-1}\sin a)(\cos b + \sqrt{-1}\sin b)$ elle-même, prise à la lettre, se trouve inexacte et n'a pas de sens."

in Paris and Berlin. His famous book on probability theory, the *Doctrine of chances* was published in 1718; he discovered the well known "Stirling's formula" $n! \approx \sqrt{2\pi n}(n/e)^n$ before Stirling; and in 1712 he was appointed by the Royal Society to adjudicate on the merits of the rival claims of NEWTON and LEIBNIZ in the discovery of the infinitesimal calculus. NEWTON in his old age, is said to have replied, when asked about anything mathematical "Go to Mr. DE MOIVRE; he knows these things better than I do." DE MOIVRE gave the first indication in 1707 of his "magic" formula by means of some numerical examples. By 1730 he seems to have been aware of the general formula

$$\cos \varphi = \frac{1}{2} \sqrt[n]{\cos n\varphi + i \sin n\varphi} + \frac{1}{2} \sqrt[n]{\cos n\varphi - i \sin n\varphi}, \quad n > 0.$$

In 1738 he describes (in a rather long-winded fashion) a procedure for finding roots of the form $\sqrt[n]{a + ib}$, which is equivalent as far as content goes, to the formula now known by his name. The formula in the form in which it is now usually expressed is first found in EULER in Chapter VIII of his *Introductio in analysin infinitorum* published in 1748. It was also EULER who, in 1749, gave the first valid proof of the formula for all $n \in \mathbb{Z}$ and who stripped DE MOIVRE's formula of all its mystery by the equation $(e^{i\varphi})^n = e^{in\varphi}$.

DE MOIVRE's formula provides a very simple method of expressing $\cos n\varphi$ and $\sin n\varphi$ as polynomials in $\cos \varphi$ and $\sin \varphi$, for all $n \geq 1$. Thus for example, we obtain for $n = 3$, by separating the real and imaginary parts:

$$\cos 3\varphi = \cos^3 \varphi - 3 \cos \varphi \sin^2 \varphi, \quad \sin 3\varphi = 3 \cos^2 \varphi \sin \varphi - \sin^3 \varphi.$$

The trigonometrical representation of the solutions of the quadratic equation $z^2 + az + b = 0$ foreshadowed in 3.5 arises in the following way: we write $\frac{1}{4}(a^2 - 4b) = r(\cos \varphi + i \sin \varphi)$ and the roots then take the form

$$z_1 = -\frac{1}{2}a + \sqrt{r}\left(\cos \frac{\varphi}{2} + i \sin \frac{\varphi}{2}\right), \quad z_2 = -\frac{1}{2}a - \sqrt{r}\left(\cos \frac{\varphi}{2} + i \sin \frac{\varphi}{2}\right).$$

**4. Roots of Unity.** As one of the most important applications of polar coordinates, we shall demonstrate the following.

**Lemma.** *Let $n \geq 1$ be a natural number. Then there are precisely $n$ different complex numbers $z$, such that $z^n = 1$, namely*

$$\zeta_\nu := \exp \frac{2\pi i}{n} \nu, \quad \nu = 0, 1, \ldots, n - 1.$$

*In particular $\zeta_\nu = \zeta^\nu$ where $\zeta := \zeta_1$.*

**Proof.** The equations $\zeta_\nu = \zeta^\nu$ and $\zeta_\nu^n = 1$ clearly hold (DE MOIVRE). Since

$$\zeta_\nu \zeta_\mu^{-1} = \exp \frac{2\pi i}{n}(\nu - \mu),$$

it follows that $\zeta_\nu = \zeta_\mu$ if and only if $\frac{1}{n}(\nu - \mu) \in \mathbb{Z}$ because the kernel of $p$ is $2\pi\mathbb{Z}$. Since $-n < \nu - \mu < n$ it follows that $\nu = \mu$, or in other words $\zeta_0, \zeta_1, \ldots, \zeta_{n-1}$ are all distinct from each other. For $z = |z|e^{i\varphi}$ we have $z^n = 1$ if and only if $|z| = 1$ and $e^{in\varphi} = 1$, that is, if $\varphi = \frac{2\pi k}{n}$ with $k \in \mathbb{Z}$. As $0 \le \varphi < 2\pi$, it follows that $k \in \{0, 1, \ldots, n-1\}$, that is $z = \zeta_k$. Accordingly there are no other complex numbers $z$, apart from $\zeta_0, \zeta_1, \ldots, \zeta_{n-1}$ satisfying the equation $z^n = 1$. □

The $n$ numbers $1, \zeta, \zeta^2, \ldots, \zeta^{n-1}$ are called *the $n$th roots of unity*. Geometrically, they represent the vertices of a regular $n$-sided polygon (the figure shows the fifth roots of unity). An $n$th root of unity is said to be *primitive* if all the other $n$th roots can be represented by one of its powers; the root $\zeta$ is always a primitive $n$th root, that is, for $n = 5$

$$\zeta = \frac{\sqrt{5} - 1}{4} + \frac{i}{4}\sqrt{2(5 + \sqrt{5})}.$$

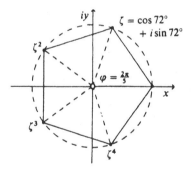

The lemma above can be immediately generalized. Writing

$$\xi := \sqrt[n]{|c|} \exp \frac{i\varphi}{n} \quad \text{for} \quad c = |c|e^{i\varphi} \in \mathbb{C}^\times,$$

where $\sqrt[n]{|c|}$ denotes the positive real $n$th root of $|c|$, we have the following:

**Existence and Uniqueness Theorem for $n$th Roots.** *Every complex number $c = |c|e^{i\varphi} \in \mathbb{C}^\times$ has precisely $n$ different complex $n$th roots, for every $n \in \mathbb{N}$, $n \ge 1$, namely the roots $\xi, \xi\zeta, \xi\zeta^2, \ldots, \xi\zeta^{n-1}$ where $\zeta := \exp \frac{2\pi i}{n}$.*

This provides a new proof of the theorem 3.6.

Realization of the many-valuedness of roots gradually developed during the 17th century. For example, the theorem that $n$th roots have $n$ distinct values was, by 1690, already very familiar to Michael ROLLE (1652–1719), a mathematician who worked in Paris and was a member of the Académie Française. Incidentally ROLLE found the well known theorem in the differential calculus which bears his name in the course of researches into the roots of polynomials, when he observed that between any two neighboring real roots of a real polynomial, there must always lie a root of the first derivative.

The British mathematician Roger COTES (1682–1716) who was a student and then Professor at Cambridge, and a friend of NEWTON, investigated in 1714 the factorization of the polynomials $Z^n - 1$ and $Z^{2n} + aZ^n + 1$ into real quadratic factors, in connection with his researches into the integration of rational functions by the method of decomposition into partial fractions. He was aware for example of the formula

$$Z^{2n} + 1 = \prod_{\nu=1}^{n} \left( Z^2 - 2Z \cos \frac{2\nu - 1}{2n} \pi + 1 \right).$$

COTES's results were first published posthumously in 1722 under the title *Harmonia mensurarum.* It was the desire to round off and improve upon these results which motivated DE MOIVRE among others.

# 4

# The Fundamental Theorem of Algebra

*R. Remmert*

Was beweisbar ist, soll in der Wissenschaft nicht ohne Beweis geglaubt werden (DEDEKIND 1887).

[In science, what is provable should never be believed without proof.]

We saw in 3.3.5 that every quadratic polynomial vanishes at two (possibly coincident) points in $\mathbb{C}$, the zeros of the polynomial, as they are often called. This statement is a special case of a far more general theorem, which GAUSS in 1849 (*Werke* 3, 73) called the *fundamental theorem* of the theory of algebraic equations, and which is now generally known in the literature as the so-called *fundamental theorem of algebra*.

*Every nonconstant complex polynomial has at least one zero in the field* $\mathbb{C}$.

In Algebra, a field is said to be *algebraically closed* if every polynomial $f \in K[X] \setminus K$ has a zero in $K$. The fundamental theorem can therefore also be stated in the form:

*The field* $\mathbb{C}$ *of complex numbers is algebraically closed.*

The designation of this statement as the fundamental theorem of algebra dates from a time when the word algebra was still understood as being broadly synonymous with the theory of polynomials with real or complex coefficients. This existence theorem, which is in fact nontrivial even for polynomials of the form $Z^n - a$ (see 3.3.6 and 3.6.4), will be discussed in some detail in this chapter, and proved in an "elementary" manner. It is equivalent to the theorem that every real polynomial can be expressed as a product of real linear and real quadratic factors.

The fundamental theorem of algebra is of outstanding significance in the history of the theory of complex numbers because it was the possibility of proving this theorem in the complex domain that, more than anything else, paved the way for a general recognition of complex numbers.

The genesis of the fundamental theorem will be fully explained in Section 1. In Section 2 we shall give what is possibly the simplest of all the proofs, one based on an old and beautiful idea used by ARGAND, which goes back to D'ALEMBERT. In Section 2 we shall give some first applications of the fundamental theorem, which will be called upon more and more in the later chapters on algebras. In particular we shall prove in 3.5 the theorem first published by HANKEL in 1867 on the uniqueness of the field $\mathbb{C}$.

In a supplementary paragraph we also discuss LAPLACE's elegant proof which is more "algebraic" than ARGAND's. The reader should consult the article by ZASSENHAUS, On the Fundamental Theorem of Algebra, *Amer. Math. Monthly*, **74**(1967), 485–497. A review of nearly a hundred classical proofs of the fundamental theorem was given in 1907 by NETTO and LE VAVASSEUR in their article "Les fonctions rationelles," *Enc. Sciences Math. Pures Appl.*, I, 2, 1–232, on pages 189–205.

## §1. ON THE HISTORY OF THE FUNDAMENTAL THEOREM

In this paragraph $f = a_0 + a_1X + \cdots a_nX^n \in \mathbb{R}[X]$ always denotes a *real polynomial of degree $n$* (and therefore $a_\nu \in \mathbb{R}$, $a_n \neq 0$). We consider only nonconstant polynomials, or in other words, we assume that $n \geq 1$. By a *zero* or *root* of $f$ we mean any element $c$ of any field $K$ which is an extension of $\mathbb{R}$, such that $f(c) = 0$. The element $c$ is also said to be a *solution of the polynomial equation* $f(x) = 0$. By *equation* we always mean a polynomial equation.

The most natural and straightforward way of showing that real equations always have complex solutions is to give an *explicit procedure for finding the roots which does not lead outside* $\mathbb{C}$. This happens with quadratic equations (see 3.3.5); it is what CARDAN succeeded in doing for cubic equations, and the same thing applies to biquadratic equations. We have formulae for the solutions which are "nested radical expressions" in which each radicand is a polynomial in the coefficients $a_0, \ldots, a_n$ and radical expressions of lower order. It can at once be verified without difficulty that the solutions constructed in this way are complex numbers (see VAN DER WAERDEN *Algebra* I, Berlin 1955, §59).

The situation is quite different with equations of the fifth and higher degrees. No method of solving such equations by radicals could be found;[1] until GAUSS all mathematicians *believed* in the existence of solutions in some sort of no-man's land (nowadays we would say in an unknown extension field of $\mathbb{C}$) and tried imaginatively to show that these solutions were

---

[1] N.H. Abel showed in 1826 in a paper published in the first volume of *Crelle's Journal* "Beweis der Unmöglichkeit, algebraische Gleichungen von höheren Graden, als den vierten, allgemein aufzulösen," 65–84 (see also *Oeuvres completes*, 1, 66–87) that it is fundamentally impossible to solve general equations of degree higher than the fourth by means of radicals.

in fact complex numbers.

We summarize below the main dates, starting from the first mystical appearance of the fundamental theorem to its present-day acceptance as a virtually self-evident truth. In addition to the references to the literature given in 3.1 we may also mention: *Abrégé d'histoire des mathématiques, I, sous la direction de Jean Dieudonné,* Paris, Hermann, 1978, especially Chapter IV.

**1. GIRARD (1595–1632) and DESCARTES (1596–1650).** Peter ROTH in 1608 stated that equations of the $n$th degree have at most $n$ solutions; VIETA (1540–1603), thanks to his theorem on the roots of equations, had been able to write down equations of the $n$th degree which actually have $n$ roots. It was the now forgotten Flemish mathematician Albert GIRARD who was the first to assert that there are always $n$ solutions. In his *L'invention en algèbre,* a work which appeared in 1629, he wrote "Toutes les équations d'algèbre reçoivent autant de solutions, que la dénomination de la plus haute quantité le démonstre ..." GIRARD gives no proof or any indication of one, but merely explains his proposition by some examples, including that of the equation $x^4 - 4x + 3 = 0$ whose solutions are 1, 1. $-1 + i\sqrt{2}$, $-1 - i\sqrt{2}$.

GIRARD does not assert that the solutions must always be of the form $a + b\sqrt{-1}$, $a, b \in \mathbb{R}$, apart from real solutions "(those that are $> 0$ and those that are $< 0$)" there are "autres enveloppées, comme celles qui ont des $\sqrt{-}$, comme $\sqrt{-3}$, ou autres nombres semblables." He thus leaves open the possibility of solutions which are not complex. In modern language he was putting forward the following proposition:

**GIRARD's Thesis.** *For every polynomial $f \in \mathbb{R}[X]$ of degree $n$ there exists a field $K$, an extension of $\mathbb{R}$, such that $f$ has exactly $n$ zeros (not necessarily distinct) in $K$. The field $K$ may perhaps be a proper overfield of $\mathbb{C}$.*

DESCARTES in 1637, in the third and last book of his *La géométrie,* gives a brief summary of what was then known about equations. He notes the important theorem[2] that a polynomial which vanishes at $c$ is always divisible by the factor $X - c$; he also described the so-called *"Descartes' rule of signs"* named after him. (See HAUPT, "Einführung in die Algebra," 2. Teil, Akad. Verl. Ges. Geest u. Portig 1954, S. 411.)

---

[2]This theorem was probably already known to Thomas Harriot (1560–1621) who in 1585 surveyed, on behalf of Sir Walter Raleigh, the colony of Virginia and was thus the first mathematician to live in North America.

DESCARTES takes a rather vague position on the thesis put forward by GIRARD (see 3.1.3).

**2. LEIBNIZ (1646–1716).** Through his efforts to integrate rational functions by decomposition into partial fractions, LEIBNIZ was led to consider the question of whether every real polynomial can be expressed as a product of factors of the first and second degrees. He put forward in 1702 in a work published in the *Acta Eruditorum* the view that this is not so, and supported this contention by pointing out that in the decomposition

$$X^4 + a^4 = (X^2 - a^2 i)(X^2 + a^2 i) = (X + a\sqrt{i})(X - a\sqrt{i})(X + a\sqrt{-i})(X - a\sqrt{-i})$$

the product of any two linear factors on the right is never a quadratic real polynomial. It does not seem to have occurred to LEIBNIZ that $\sqrt{i}$ could be of the form $a + bi$; because if he had seen that

$$\sqrt{i} = \frac{1}{2}\sqrt{2}(1 + i) \quad \text{and} \quad \sqrt{-i} = \frac{1}{2}\sqrt{2}(1 - i)$$

he would have noticed that the product of the first and third factors, and of the second and fourth factors are both real, and instead of his false assertion he would have obtained

$$X^4 + a^4 = (X^2 + a\sqrt{2}X + a^2)(X^2 - a\sqrt{2}X + a^2).$$

It is remarkable that he should not have been led to this factorization by the simple device of writing $X^4 + a^4 = (X^2 + a^2)^2 - 2a^2X^2$.

**3. EULER (1707–1783).** In a letter to Nikolaus BERNOULLI of the 1 November 1742 EULER enunciates the factorization theorem for real polynomials in precisely the form which LEIBNIZ had maintained was false. The presumed counter-example proposed by BERNOULLI, the polynomial $X^4 - 4X^3 + 2X^2 + 4X + 4$ with zeros

$$x_{1,2} = 1 \pm \sqrt{2 + i\sqrt{3}}, \quad x_{3,4} = 1 \pm \sqrt{2 - i\sqrt{3}}$$

was shown to be devoid of force, by proving that $(X - x_1)(X - x_3)$ and $(X - x_2)(X - x_4)$ are real polynomials, namely

$$X^2 - (2 + a)X + 1 + \sqrt{7} + a \quad \text{and} \quad X^2 - (2 - a)X + 1 + \sqrt{7} - a$$

with $a := \sqrt{4 + 2\sqrt{7}}$.

Soon afterwards, in a letter of the 15 February 1742 to his faithful correspondent GOLDBACH, EULER repeats his assertion but adds that he has not been able to prove it completely, but only "ungefähr, wie gewisse Fermatsche Sätze" [only roughly, as with certain theorems of Fermat]. In this letter he also mentions incidentally—something that seems perfectly clear to us nowadays and that has nothing to do with the problem of the existence of complex roots—that the imaginary roots of real polynomials can always be grouped together in pairs so as to produce real polynomials of the second degree after multiplication of the corresponding factors.[3] GOLDBACH remains sceptical even about this simple assertion and adduces as a counter-example the polynomial $Z^4 + 72Z^2 - 20$, which EULER immediately factorizes.

EULER's factorization theorem goes beyond GIRARD's thesis of which EULER must have been well aware. Since quadratic equations always have complex solutions, his statement is nothing else but the

**Fundamental Theorem of Algebra for Real Polynomials.** *Every polynomial of the nth degree $f \in \mathbb{R}[X]$ has precisely n zeros in the extension field $\mathbb{C}$.*

EULER was able to prove this theorem rigorously for all polynomials of degree $\leq 6$. In 1749 (Recherches sur les racines imaginaires des équations. *Histoire de l'Académie Royale des Sciences et Belles Lettres, Année MDCCXLIX*, Berlin 1751, 222–228, see also *Opera omnia* 6, 1 ser., 78–147) he attacked the general case. His idea was to decompose every monic polynomial $P$ of degree $2^n \geq 4$ into a product $P_1 P_2$ of two monic polynomials of degree $m := 2^{n-1}$. If this could be done then his theorem would be proved because an arbitrary polynomial $\neq 0$ can always be converted into such a polynomial by multiplication by $aX^d$ and iteration of the decomposition procedure finally yields a decomposition of $P$ into real quadratic polynomials.

EULER makes the initial assumption that $P$ is of the form

$$P(X) = X^{2m} + BX^{2m-2} + CX^{2m-3} + \cdots,$$

which is permissible since the coefficient $A$ of $X^{2m-1}$ can always be made to vanish by a translation $X \mapsto X - \frac{1}{2m}A$. This reduction had been known since the days of CARDANO (*Ars magna*, Chapter 17) if not earlier; VIETA had called the process "expurgatio." The polynomials $P_1$, $P_2$ now take the form

$$X^m + uX^{m-1} + \alpha X^{m-2} + \beta X^{m-3} + \cdots,$$

$$X^m - uX^{m-1} + \lambda X^{m-2} + \mu X^{m-3} + \cdots$$

because the coefficients of $X^{m-1}$ differ only in sign, in view of the vanishing of the coefficient of $X^{2m-1}$ in $P(X)$. By multiplying out and comparing coefficients, one obtains equations involving $B, C, \ldots$ and $u, \alpha, \beta, \ldots, \lambda, \mu, \ldots$. EULER asserts that $\alpha, \beta, \ldots, \lambda, \mu, \ldots$ are *rational functions* in $B, C, \ldots$ and $u$, and that by elimination of $\alpha, \beta, \ldots, \lambda, \mu, \ldots$ a monic real polynomial in

---

[3] This had already been remarked by Bombelli around 1560.

$u$ of degree $\binom{2m}{m}$ is obtained whose constant term is *negative*. Now this polynomial has a zero $u$, by the *intermediate value theorem* (BOLZANO–CAUCHY theorem) as EULER clearly saw. All this is carried out explicitly for $2m = 4$ (see *loc.cit.* pp. 93/94) but the proof in the general case is only sketchy (see pp. 105/106), and EULER passes over in silence many details (as GUASS was to criticize later—see Section 6).

EULER also stated his theorem in terms of complex numbers (*loc.cit.* p. 112):

*Si une équation algébrique, de degré qu'elle soit, a des racines imaginaires, chacune sera comprise dans cette formule générale $M + N\sqrt{-1}$, les lettres $M$ et $N$ marquant des quantités réelles.*

**4. d'ALEMBERT (1717–1783).** Three years before EULER, Jean le Rond D'ALEMBERT in 1746 made the *first* serious attempt to prove the factorization theorem (Recherches sur le calcul intégral, *Histoire de l'Academie Royale des Sciences et Belles Lettres, année MDCCXLVI*, Berlin 1748, 182–224). Accordingly this theorem has ever since been referred to in the French literature as D'ALEMBERT's theorem. The basic idea is simple, even if heavily concealed. *It is to try to minimize the absolute value of the polynomial $f$ by an appropriate choice of its argument.* D'ALEMBERT uses the following auxiliary proposition which he assumes without proof, and which was first correctly derived in 1851 by PUISEUX (on the implicit assumption of the Fundamental theorem!):

*To every pair $(b, c)$ of complex numbers with $f(b) = c$, there corresponds a natural number $q \geq 1$, and a series*

$$h(w) = b + \sum_{\nu=1}^{\infty} c_\nu (w - c)^{\nu/q},$$

*convergent in a neighborhood of $c$, such that for all numbers $w$ near $c$, $f(h(w)) = w$.*

D'ALEMBERT now starts from real numbers $b$, $c$ satisfying $f(b) = c$ (in fact he chooses $b$ so that the real function has a minimum at $b$) and then finds, if $c \neq 0$, with the help of his PUISEUX expansion, complex numbers $z_1$, $w_1$ with $|w_1| < c$, such that $f(z_1) = w_1$. Repetition of this process leads to smaller and smaller values for the absolute value of $f$, and by using a simple compactness argument (which D'ALEMBERT was unable to do), eventually to a zero of $f$.

The weaknesses in D'ALEMBERT's argument, which were inevitable in the prevailing circumstances, are subject to the criticisms which were rightly made by GAUSS (see paragraph 6). Nevertheless GAUSS also says, almost prophetically (*Werke* 3, p. 11): "Aus diesen Gründen vermag ich den d'Alembertschen Beweis nicht für ausreichend zu halten. Allein das verhindert mich nicht, daß mir der wahre Nerv des Beweises trotz aller Einwände unberührt zu sein scheint; ich glaube ..., daß man auf dieselben Grundlagen einen strengen Beweis unseres Satzes aufbauen kann." [For these reasons I am unable to regard the proof by d'Alembert as entirely satisfactory, but that does not prevent, in my opinion, the essential idea of the proof from being unaffected, despite all objections; I believe that ... a rigorous proof could be constructed on the same basis.]

This is precisely what ARGAND did in 1814 (see paragraph 8).

As a result of this work of D'ALEMBERT and EULER the view gradually came to prevail that it required only the existence of a single imaginary quantity $\sqrt{-1}$ in order to ensure that $n$ roots could be assigned to every algebraic equation of degree $n$ (GAUSS, *Werke* 10, 1, p. 404).

**5. LAGRANGE (1736–1813) and LAPLACE (1749–1827).** Already by 1772 Joseph Louis LAGRANGE in his memoir "Sur la forme des racines imaginaires des équations" (*Nouveaux mémoires de l'Académie Royale des Sciences et Belles Lettres, Année MDCCLXXVII*, Berlin 1774, 222–258 and *Oeuvres complètes*, 3, 477–516) had raised objections against EULER's proof. He remarked, among other things, that EULER's equation for $u$ could have undefined coefficients of the form $\frac{0}{0}$. LAGRANGE made a new attempt to demonstrate the existence of the factorization $P = P_1P_2$ sought by EULER. Thanks to his results on the permutation of roots of equations he succeeded to a large extent in closing the gaps in EULER's proof: but he also had to appeal to fictitious roots.

In the year 1795, Pierre Simon de LAPLACE[4] in his "Leçons de mathématiques données à l'Ecole Normale" (*Journal de l'Ecole Polytechnique, Septième et Huitieme cahier*, Tome II, 1–278, Paris, 1812, especially pp. 56–58; see also *Oeuvres complètes* 14, 10–111, especially 63–65) made an attempt to prove the Fundamental theorem, quite different from the EULER–LAGRANGE attempt. He uses ideas involving the discriminant of a polynomial. LAPLACE, like his predecessors, assumes that roots of polynomials "exist" in the platonic sense of the word. His extremely elegant proof has long been forgotten, and we reproduce it in modernized form as an appendix to this chapter.

---

[4]Laplace was appointed Minister of the Interior by Napoleon, who removed him from office after only six weeks because he brought the spirit of the infinitely small into the government [il portait enfin l'esprit des infiniment petits dans l'administration] (Napoleon I. *Mémoires pour servir à l'histoire de France, écrits à Sainte-Hélène, sous la dictée de l'empereur*, dicté au général Gourgaud, London 1823, Vol. 1, 111–112). After the restoration of the Bourbons, he was made a marquis and a peer of France.

**6. GAUSS's Critique.** In October 1797 GAUSS writes in his diary "Ae-
quationes habere radices imaginarias methodo genuina demonstratum" (see
*Math. Ann.* 57, p. 18, 1903). He published the above-mentioned proof of
the Fundamental theorem, which however by no means meets modern standards of rigor, in 1799 in his doctoral thesis "Demonstratio nova theorematis omnem functionem algebraicam rationalem integram unius variabilis
in factores reales primi vel secundi gradus resolvi posse" (*Werke* 3, 1.30)
which he submitted in absentia to PFAFF (1765–1825) at the University of
Helmstedt, and through which he obtained his doctorate. GAUSS begins his
dissertation by a detailed critical examination of all previous attempts to
prove the theorem known to him. This is not the place to discuss in detail
the objections raised by the twenty-two year old student against the proofs
of D'ALEMBERT, EULER, and LAGRANGE—and thus against the leading
mathematicians of the time—(the reader interested in this may refer, for
example, to TROPFKE, Vol. 1, 1980, 494–499). GAUSS's main objection was
that the existence of a point at which the polynomial takes the value zero is
always assumed and that this existence needs to be proved. Thus for example he reproaches EULER for using hypothetical roots (*Werke* 3, pp. 5, 14):[5]
"..., wenn man dann mit diesen unmöglichen Wurzeln so verfährt, als ob sie
etwas Wirkliches seien, und beispielsweise sagt, die Summe aller Wurzeln
der Gleichung $X^m + AX^{m-1} + \cdots = 0$ sei $= -A$, obschon unmögliche
unter ihnen sind (das heißt eigentlich: *wiewohl einige fehlen*), so kann ich
dies durchaus nicht billigen." [... if one carries out operations with these
impossible roots, as though they really existed, and says for example, the
sum of all the roots of the equation $x^m + AX^{m-1} + \cdots = 0$ is equal to $-A$,
even though some of them may be impossible (which really means: even if
some are *nonexistent and hence are missing*), then I can only say that I
thorougly disapprove of this type of argument.]

The improved proof by LAGRANGE is likewise disallowed. GAUSS writes
(*Werke* 3, p. 20):[6] "Dieser große Mathematiker bemühte sich vor Allem, die
Lücken in Eulers erstem Beweise auszufüllen, und wirklich hat er das, was
oben §8 den zweiten und den vierten Einwurf ausmacht, so tief durchforscht,
daß nichts Weiteres zu wünschen übrig bleibt. ... Den dritten Einwurf
dagegen berührt er überhaupt nicht; ja auch seine ganze Untersuchung
ist auf der Voraussetzung aufgebaut, jede Gleichung $m$-ten Grades habe
wirklich $m$ Wurzeln." [This great mathematician tried above all to fill in
the gaps in EULER's first proof, and indeed, as regards what constitutes
the second and fourth objections referred to in §8 above, he has pursued
his investigations so profoundly that nothing more remains to be desired.

---

[5] See next footnote.

[6] Citations based on the German translation in Ostwald's *Klassikern der Ex-
akten Wissenschaften*, No. 14. "Die vier Gaußschen Beweise für die Zerlegung
ganzer algebraischer Funktionen in reelle Faktoren ersten und zweiten Grades
(1799–1849)," made by Netto in 1899.

... On the other hand he has not touched at all the third objection; in fact his whole investigation is based on the assumption that every equation of the $m$th degree actually has $m$ roots.] And in 1815 (*Werke* 3, p. 105) he even talks in this connection of a "true petitio principii."

GAUSS in 1799 was not yet aware of LAPLACE's proof. However later on, even this attempt did not find favor in his eyes; he comments on it in 1815 in the *Göttingische gelehrten Anzeigen* (*Werke* 3, p. 105) writing "die scharf-sinnige Art, wie später LAPLACE diesen Gegenstand behandelt hat, [kann] gerade von dem Hauptvorwurfe, welcher alle jene versuchten Beweise trifft, nicht freigesprochen werden." [The ingenious way in which LAPLACE dealt with this matter cannot be absolved from the main objections affecting all these attempted proofs.]

We would now like to take another look at the situation from our modern point of view. In all the pre-Gaussian attempts, the question asked at the outset was not so much "do roots of an equation *exist*?" but rather "what *form* do they have?" and "are they of the form $a + b\sqrt{-1}$?" GIRARD's thesis is tacitly taken as an axiom, and no reasons of any kind are put forward in justification. It was even believed for a long time, that there existed a whole hierarchy of imaginary quantities–called by GAUSS in his dissertation (*Werke* 3, p. 14) "vera umbrae umbra" [veritable shadows of shadows]—of which the complex numbers $a + b\sqrt{-1}$, $a, b \in \mathbb{R}$ were the simplest. It was not until the 18th century when the idea had gained general acceptance that the solutions of polynomial equations were capable of being defined by "algebraic/analytical methods which never led outside the domain $\mathbb{C}$," that the following problem (which no longer seems so paradoxical knowing the background) began to be seriously considered:

*"Show that every imaginary quantity has the form $a + b\sqrt{-1}$."*

Interpreted with a little goodwill, the statement to be proved is nothing more than the assertion that the field $\mathbb{C}$ is *complete and not capable of any further algebraic extension.* In the work quoted in paragraph 3, the "Recherches sur les racines..." by EULER can be read (p. 147) the words: "Puisque donc toutes ces quantités imaginaires, qui sont formées par des opérations transcendantes, sont aussi comprises dans la forme générale $M + N\sqrt{-1}$, nous pourrons soutenir sans balancer, que généralement toutes les quantités imaginaires, quelques compliquées qu'elles puissent être, sont toujours réductibles à la forme $M + N\sqrt{-1}$." [Since all these imaginary quantities, produced by transcendental operations, are also comprized in the general form $M + N\sqrt{-1}$, we can maintain, without hesitation, that generally all imaginary quantities, no matter how complicated, are always reducible to the form $M + N\sqrt{-1}$.]

The Gaussian objection against the attempts of EULER–LAGRANGE and LAPLACE was invalidated as soon as Algebra was able to guarantee the existence of a splitting field for every polynomial. From that moment on, as Adolf KNESER already observed in 1888 (*Crelle's Journal* 102, p. 21), the attempted proofs became in effect fully valid. In 1907 FROBENIUS said (*Ges. Abhandl.* 3, p. 733) on the occasion of the official ceremony at Basle University to commemorate the bicentenary of Leonhard EULER's birth: "Für die Existenz der Wurzeln einer Gleichung führt er jenen am meisten algebraischen Beweis, der darauf fußt, daß jede reelle Gleichung unpaaren Grades eine reelle Wurzel besitzt. Ich halte es für unrecht, diesen Beweis ausschließlich GAUSS zuzuschreiben, der doch nur die letzte Feile daran gelegt hat." [He gave the most algebraic of the proofs of the existence of roots of an equation, the one which is based on the proposition that every real equation of odd degree has a real root. I regard it as unjust to ascribe this proof exclusively to GAUSS, who merely added the finishing touches.]

**7. GAUSS's Four Proofs.** The fundamentally new element in GAUSS's proof of 1799 is that he does not set out to calculate a root, but to *prove its existence*. To do this required, in the words of HANKEL (p. 97): "einen eminenten Aufwand von Schärfe des Gedankens und Productionskraft, wie beides in Gauß wunderbar vereinigt war." [a high degree of perspicacity of thought and fertility of invention which in GAUSS were wonderfully combined]. GAUSS in his doctoral dissertation does not however claim that he was the first to produce a correct proof of the Fundamental theorem, as is already made clear by the word "Nova" in the title, and as his remarks on D'ALEMBERT's attempted proof also bear witness (see paragraph 4), GAUSS gave, in all, *four* proofs of the Fundamental theorem of algebra, the fourth being published in 1849 in the year of the golden jubilee of his doctorate (see *Ostwald's classics* No. 14).

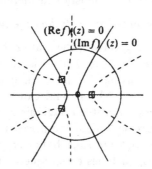

The first proof, of 1799, is topological, but has some significant gaps when judged in the light of present-day understanding. Let us take a closer look at the problem involved: the complex zeros of the real polynomial $f$ of degree $n$ are the *points of intersection* of the two *real algebraic curves* $(\operatorname{Re} f)(z) = 0$ and $(\operatorname{Im} f)(z) = 0$. If $R$ is sufficiently large, then exactly $2n$ points of each curve will lie on every circle $|z| = r$ for which $r > R$. Outside the circular disc $\{z \in \mathbb{C}: |z| \le R\}$ these points can each be associated with $2n$ *continuous* branches $A_\nu$ and $B_\nu$, $1 \le \nu \le 2n$, extending to infinity, and in fact these branches are so situated that between any two consecutive "branches" of the curve $(\mathrm{Re}\,f)(z) = 0$, there lies a branch of the curve $(\mathrm{Im}\,f)(z) = 0$ and vice versa. The figure illustrates the example $f(Z) := Z^3 + Z^2 - 2$ whose zeros are the points $1, -1 \pm i$. Gauss says (Art. 21): "Nun läßt sich aus der gegenseitigen Lage der in die Kreisscheibe eintretenden Zweige der Schluss, dass innerhalb des Kreises ein Schnitt eines Zweiges der ersten mit einem Zweige der zweiten Linie vorhanden sein müsse, auf so viele Arten ziehen, daß ich fast nicht weiss, welche Methode an erster Stelle vor den übrigen zu bevorzugen sei." [Now this alternation in the positions of the points of entry of the branches entering the disc allows us to draw the conclusion that a branch of the first curve must intersect with a branch of the second curve at some point in the interior of the circular disc. This conclusion can be drawn in so many different ways that I hardly know which method should be given pride of place.] In the subsequent geometrical argument on which he bases his proof, GAUSS uses results from higher geometry and in particular the theorem that "... if a (non-compact) branch of an algebraic curve enters a bounded space (here, a circular disc) it must necessarily emerge from this space." This theorem whose truth was taken for granted for over a hundred years, lies at the heart of the proof. Topologists have so far been able to prove it only by sophisticated arguments. GAUSS remarks in an explanatory footnote (*Werke* 3, p. 27, *Ostwald's classics* No. 14, p. 33): "Wie mir scheint, ist es wohl hinreichend sicher bewiesen, daß eine algebraische Curve weder plötzlich irgendwo abbricht, noch sich nach unendlich vielen Umläufen gewissermaßen in einem Punkt verlieren kann (wie die logarithmische Spiral)." [It seems to me that it can be taken as sufficiently securely established, that an algebraic curve can neither suddenly end abruptly anywhere, nor lose itself, so to speak, in a point after an infinity of circuits (as in the case of a logarithmic spiral).]

A careful and balanced criticism together with a completion of the first Gaussian proof was first given in 1920 by A. OSTROWSKI: ("Über den ersten und vierten Gaußschen Beweis des Fundamentalsatzes der Algebra," GAUSS Werke 10.2, Abh. 3). OSTROWSKI began with the words: "Während die im ersten Teil der Gaußschen Dissertation enthaltene Besprechung der früheren Beweisversuche des Fundamentalsatzes der Algebra sich durch ganz außerordentliche Sorgfalt auszeichnet, fällt daneben der im zweiten Teil entwickelte Beweis dieses Satzes etwas ab. Nicht etwa, weil dieser Beweis in geometrischer Einkleidung vorgetragen wird, sondern, weil bei ihm Eigenschaften der algebraischen Kurven verwendet werden, die weder in der *Dissertation* selbst, noch in der vorgaußschen Literatur bewiesen sind." [While the discussion, in the first part of GAUSS's dissertation, of the earlier attempts at proving the Fundamental theorem of algebra is distinguished by extraordinarily thorough and painstaking care, the proof of this theorem developed in the second half falls away somewhat from this high standard. Not so much because it is presented in a geometrical guise but rather because the proof makes use of geometrical properties of algebraic curves which are neither proved in the *dissertation* itself nor had been proved in the pre-Gaussian literature.]

In 1816 GAUSS gave a second proof of the Fundamental theorem which is almost completely algebraic. The only fact used taken from analysis is the theorem that *any real polynomial of odd degree always has a real zero.* GAUSS takes up the basic algebraic idea from EULER with a simplification proposed by DE FONCENEX in 1759, and uses the truly algebraic device of indeterminates, even though he does not have at his disposal the general concept of a field. He carries out mathematical operations which his predecessors had performed on *illegitimately* assumed roots, and which are perfectly valid in his case precisely because the operands are *legitimately* regarded as indeterminates. Such considerations still underlie the usual modern proof of the existence of a splitting field. GAUSS's second proof is, even by modern standards, absolutely correct.

GAUSS's third proof likewise dates from 1816; it is once more topological, but this time the idea is to count—by means of a double integral—the number of circuits which the image point $f(z)$ makes around the origin $0 \in \mathbb{C}$ when the point $z$ describes a closed curve around the origin $z = 0$. The basic idea of this proof is still to be found in the modern "function-theoretic" proofs based on evaluating the contour integral $(1/2\pi i) \int (f'(z)/f(z)) dz$ (ROUCHÉ's theorem).

Until 1849 all proofs, including those found in the intervening period by CAUCHY, ABEL, JACOBI and others, dealt with real polynomials only. It was only in his fourth proof, which is a variant of the first, that GAUSS in 1849, the time now being ripe for this step, allowed arbitrary complex polynomials. However this apparent generalization is not one of any real significance, because one can immediately switch from a *complex* polynomial $f \in \mathbb{C}[Z]$ to a *real* polynomial $g \in \mathbb{R}[Z]$, by means of $g(z) := \bar{f}(\bar{z})f(z)$. If $c$ is a zero of $g$, then $c$ or $\bar{c}$ is a zero of $f$. To modern eyes the proof for real polynomials is no simpler than for complex polynomials (and vice versa).

## 8. ARGAND (1768–1822) and CAUCHY (1798–1857).

What may well be the simplest of all the proofs of the Fundamental theorem of algebra was published in 1814 by R. ARGAND in his "Réflexions sur la nouvelle théorie d'analyse" *Annales de Mathématiques* 5, 197–209. ARGAND who had already sketched the essence of his proof in his essay on the representation of complex numbers, simplifies astonishingly the application of D'ALEMBERT's basic idea. He uses the general theorem on the existence of a minimum of a (continuous) function and so arrives at a completely new kind of proof. As ARGAND says nothing to justify the existence of the minimum, his elementary proof was not at first accepted. CAUCHY in 1820 gave what is essentially the same proof in his paper, "Sur les racines imaginaires des équations" (*Oeuvres* 1, 2, Ser., 258–263) but in a more accessible form, thereby contributing greatly to a wider dissemination of ARGAND's ideas.

Even with CAUCHY the proposition that $|f(z)|$ must somewhere attain its minimum is not properly established; it only became possible to do this after the general concept of the lower bound had been introduced. CAUCHY devotes a whole chapter (Chapitre X) of his *Cours d'analyse* to the fundamental theorem, but without mentioning ARGAND.

In the 19th century ARGAND's method of proof was adopted in various textbooks, e.g. in LIPSCHITZ's *Lehrbuch der Analysis*, Vol. 1 of 1877, and in the book published in 1886 by CHRYSTAL, *Algebra, An elementary textbook for higher classes of secondary schools and for colleges*. CHRYSTAL, whose textbook had an unusually great influence (see the discussion on CHRYSTAL's algebra by ABHYANKAR in *The mathematical intelligencer* 1, 1978, p. 37) called ARGAND's proof "both ingenious and profound" (p. 248).

ARGAND's proof has nowadays tended to fall into oblivion. Towards the end of the twenties, SCHREIER reproduced this proof in his Hamburg Lectures on *Analytical geometry and algebra;* it is given for example in the first volume of the *first* edition of the book by SCHREIER and SPERNER (Teubner Verlag, pp. 221 *et seq.*). LANDAU, in 1934, also presented a version of the ARGAND proof in his characteristic style (pp. 233 *et seq.*); the ARGAND proof is also to be found in the second volume of MANGOLDT and KNOPP (11th edn., Hirzel Verlag, Stuttgart 1958, pp. 546 *et seq.*). The ARGAND proof is reproduced in this chapter.

**9. The Fundamental Theorem of Algebra: Then and Now.** Nowadays one can only speculate about how mathematicians before the beginning of the nineteenth century had visualized the solutions of equations in their mind's eye. It is difficult for us to understand why, until the time of GAUSS, they had an unshakable belief in a kind of "extraterrestrial" existence of such solutions "somewhere or other," and then sought to show that these solutions were complex numbers. Still less can one conceive why it should be that, until far into the nineteenth century, algebra textbooks hardly ever troubled to enunciate this Fundamental theorem but juggled with it in a most amazing fashion (see HANKEL, 1867, p. 98). An honorable exception to this general attitude was the Göttingen mathematician and physicist Abraham Gotthelf KÄSTNER (1719–1800) who was GAUSS's predecessor at the Observatory (and who also wrote epigrams, satirical pieces, aphorisms and pointed comments on the latest literary novelties, and was friendly with GOTTSCHED). In 1767, KÄSTNER, in Article 210 of his *Anfangsgründe der endlichen Analysis* expressly *postulated* the Fundamental theorem as an *axiom*.

Nowadays the Fundamental theorem of algebra is one of the established propositions of algebra and of the theory of holomorphic functions respectively which students accept without protest. All proofs require, in the final analysis, the aid of *non-algebraic (analytic, transcendental)* methods and concepts. Either—like D'ALEMBERT, ARGAND and CAUCHY—one reduces successively the absolute value of the polynomial by a suitable choice of its argument, in which case one has to solve pure binomial equations and one needs to have available some theorem guaranteeing the existence of a minimum; or else—like EULER, LAGRANGE and LAPLACE—one splits off factors, and then the contribution needed from analysis can be kept more in the background. We require "only" the existence of square roots of complex numbers and the theorem that real polynomials of odd degree have a real zero.

Particularly favored are proofs which draw on results from CAUCHY's theory of functions: for instance the *maximum modulus principle* or the *open mapping theorem,* or LIOUVILLE's theorem to the effect that any function which is holomorphic and bounded throughout $\mathbb{C}$ is necessarily a constant. (See J. CONWAY, *Functions of One Complex Variable,* Springer-Verlag, 1978, p. 77.) Many mathematicians believe that there can be no purely algebraic proof, because the field $\mathbb{R}$, and consequently its extension field $\mathbb{C}$, is a construct belonging to analysis.

**10. Brief Biographical Notes on Carl Friedrich GAUSS.** He was born on the 30th April 1777 in Brunswick. He was a mathematician, astronomer, geodesist and physicist. In 1792 at the age of fifteen he had already conjectured the *Prime number theorem* (first proved a hundred years later) by counting from tables of primes and tables of logarithms which he had been given. He studied at Göttingen from 1795 to 1798 as holder of a special scholarship from the Duke of Brunswick. In 1796 he discovered thr ruler and compass construction of the regular 17-sided polygon. In 1799, he was awarded his doctorate *in absentia* by PFAFF at the University of Helmstedt which then belonged to the State of Brunswick. In 1801 he published the immortal, *Disquisitiones arithmeticae,* the "bible" of number theory. The same year he was appointed corresponding member of the St. Petersburg Academy. 1801 also saw his calculation of the orbit of Ceres by numerical analysis using only scanty observational data. In 1807 he was appointed Professor of Astronomy and Director of Göttingen Observatory and 1810 he refused the offer of a post in Berlin. In 1818 he began his work on the survey of the Kingdom of Hanover. In 1820 he invented the heliotrope—an instrument with a movable mirror for reflecting the sun's rays, used especially in geodesy. From 1821 to 1825 he directed survey work in the field. In 1828, he was guest of Alexander von HUMBOLDT in Berlin and made the acquaintance of Wilhelm WEBER. In 1841 he decided to learn Russian so as to be able to read the works of LOBACHEVSKY on non-Euclidean geometry which he had known about for a long time. In 1842 he was a founder member of the order "Pour le mérite" for the Arts and Sciences.[7] In 1845 he carried out long and wearisome calculations in

---

[7] Other founder members of the civilian division of the Pour le mérite order founded in 1842 by King William IV of Prussia were: J.I. Berzelius (chemist), F.W. Bessel (astronomer), J. Daguerre (painter and inventor of the daguer-

connection with the reorganization of the pension fund for the widows of Göttingen professors. He died in Göttingen the 23rd February 1855. Large parts of his mathematical knowledge were not made public until the papers which he left at his death were published; his motto was: *Pauca sed matura.* After his death medals were struck in the Kingdom of Hanover at the initiative of the King, on which he was described as "Princeps mathematicorum" a name by which he had already been called during his lifetime. By careful reading of foreign and other newspapers in a reading room in Göttingen and a systematic evaluation of the financial news, GAUSS managed to accumulate a considerable private fortune through stock exchange speculation. An obituary memoir *Gauss zum Gedächtnis* written by his friend Sartorius von WALTERSHAUSEN came out in 1856. A very stimulating book is the critical study by W.K. BÜHLER published in 1981 by Springer-Verlag, *GAUSS, a Bibliographical Study.*

## §2. PROOF OF THE FUNDAMENTAL THEOREM BASED ON ARGAND

ARGAND's proof makes use of three auxiliary propositions:

0) *Every complex polynomial is a continuous function in* $\mathbb{C}$.

1) *Every continuous function* $f: K \to \mathbb{R}$ *on a compactum $K$ in* $\mathbb{R}^2$ *assumes a minimum in $K$.*

2) *Every complex number has square roots.*

The first two statements belong to the foundations of analysis; statement 2) was proved in 3.3.5, and it was deduced therefrom in 3.3.6 (cf. also 3.6.4) that:

2') *Every complex number has kth roots,* $1 < k < \infty$.

We prove the theorem in three stages. First we show by a simple growth argument that the absolute value function (or modulus) $|f(x)|$ of any complex polynomial $f(z)$ in $\mathbb{C}$ always assumes a *minimum value;* this is the so-called Minimum theorem of CAUCHY. The D'ALEMBERT–GAUSS theorem now states that, for a nonconstant polynomial this minimum is always zero. The proof of this is given in three lines in 2.3 with the help of ARGAND's inequality, which provides a bound for the value of a complex

rotype), J.L. Gay-Lussac (chemist and physicist), J. Grimm (Germanist), F.H.A. v. Humboldt (Naturalist and geographer first chancellor of the order), C.G.J. Jacobi (mathematician), F. Liszt (musician), J.L.F. Mendelssohn- Bartholdy (composer), F. Rückert (poet and orientalist), A.W. v. Schlegel (poet), and L. Tieck (poet). Details taken from "Orden Pour Le Mérite für Wissenschaften Und Künste," Gebr. Mann Verlag, Berlin 1975.

polynomial. This inequality, which is the core of ARGAND's argument, will be derived in 3, and depends on a simple inequality for polynomials of the type $1 + bZ^k + Z^k g(Z)$ where $g(0) = 0$.

**1. CAUCHY's Minimum Theorem.** *For every polynomial $f(Z) = a_0 + a_1 Z + \cdots + a_n Z^n \in \mathbb{C}[Z]$ there is a $c \in \mathbb{C}$ such that $|f(c)| = \inf |f(\mathbb{C})|$.*

**Proof.** We can assume that $a_n \neq 0$ with $n \geq 1$. We need a *statement about growth:*

> (*) *there exists an $r \in \mathbb{R}$ such that $|f(z)| > |f(0)|$ for all $z \in \mathbb{C}$ with $|z| > r$.*

For $z \neq 0$, we have $|f(z)| = |z|^n |a_n + h(z^{-1})|$ with $h(W) := a_{n-1} W + \cdots + a_0 W^n \in \mathbb{C}[W]$. Since $h$ is continuous at 0, there is $\delta > 0$ such that $|h(w)| \leq \frac{1}{2}|a_n|$, whenever $|w| < \delta$. It follows that $|f(z)| \geq |z|^n(|a_n| - |h(z^{-1})|) \geq \frac{1}{2}|a_n| |z|^n$, when $|z| > \delta^{-1}$. It suffices therefore to choose $r > \delta^{-1}$ in order to ensure that $|a_n| r^n > 2|a_0|$.

After this preliminary work the proof of the minimum theorem can be swiftly concluded. Since $f(z)$ is continuous in $\mathbb{C}$, the same is true of $|f(z)|$ and therefore $|f(z)|$ assumes a minimum in the compact circle $K := \{z \in \mathbb{C} : |z| \leq r\}$ by reason of statement 1) of the introduction. There is therefore a $c \in K$ with $|f(c)| = \inf |f(K)|$. As $|f(c)| \leq |f(0)| \leq \inf |f(\mathbb{C} \setminus K)|$ by virtue of (*), it follows that $|f(c)| = \inf |f(\mathbb{C})|$.                             □

CAUCHY likewise drew upon the existence of the minimum in his *Cours d'analyse* of 1821 for a proof of the D'ALEMBERT–GAUSS theorem (Chapitre X). The existence of minima in compact sets, which we have taken without proof from real analysis, had of course not yet been proved in CAUCHY's time.

Some statement about the growth of polynomials, such as the one represented here by (*) is also needed in most of the function theoretical proofs.

**2. Proof of the Fundamental Theorem.** In addition to the minimum theorem we need:

ARGAND's Inequality: *Let $f(Z)$ be a nonconstant polynomial. Then for every point $c \in \mathbb{C}$ with $f(c) \neq 0$ there is another point $c' \in \mathbb{C}$ with*

$$|f(c')| < |f(c)|.$$

This inequality will be proved in the next paragraph, by the extraction of $k$th roots. It follows at once from the inequality that every nonconstant complex polynomial $f(Z) \in \mathbb{C}[Z]$ must have a zero $c$ in $\mathbb{C}$. For by the minimum theorem there exists a $c \in \mathbb{C}$ such that $|f(c)| \leq |f(z)|$ for all

$z \in \mathbb{C}$. If $f(c)$ were nonzero there would, by ARGAND's inequality, be a $c' \in \mathbb{C}$ with $|f(c')| < |f(c)|$, which would be absurd.

**3. Proof of ARGAND's Inequality.** The decisive role in the proof is played by the following.

**Lemma.** *Let $k$ be a natural number, not zero, and let*

$$h := 1 + bZ^k + Z^k g \quad with \quad b \in \mathbb{C}^\times, \ g \in \mathbb{C}[Z], \ g(0) = 0.$$

*Then there is a $u \in \mathbb{C}$ such that $|h(u)| < 1$.*

**Proof.** We choose a $k$th root $d \in \mathbb{C}$, of $-1/b$, so that $bd^k = -1$ (proposition $2'$ of the introduction). For all real $t$ with $0 < t \le 1$, we then have

$$|h(dt)| \le |1 - t^k| + |d^k t^k g(dt)| = 1 - t^k + t^k |d^k g(dt)|.$$

Since $g$, being a polynomial, is continuous at 0 (proposition 0 of the introduction), and since $g(0) = 0$, there exists a $\delta$, with $0 < \delta < 1$, such that $|d^k g(dt)| < \frac{1}{2}$ for all $t$ satisfying the inequality $0 < t < \delta$. For every such $t$, it then follows that $|h(dt)| \le 1 - t^k + \frac{1}{2}t^k < 1$.                    □

The reader will notice that, apart from $g(0) = 0$, the only property of $g: \mathbb{C} \to \mathbb{C}$ which has been used, is that of continuity at the origin. The lemma therefore holds for all such functions. The argument shows that $h$ assumes values less than 1 in an *arbitrarily small neighborhood of the origin.*

ARGAND's inequality now quickly follows: a nonconstant $f(Z)$ implies that $\hat{f}(Z) := f(c + Z)/f(c) \in \mathbb{C}[Z]$ is not constant. Now

$$\hat{f}(Z) = 1 + b_k Z^k + b_{k+1} Z^{k+1} + \cdots + b_n Z^n \quad with \quad b_k \ne 0, \ 1 \le k \le n.$$

Writing $g(Z) := b_{k+1} Z + \cdots + b_n Z^{n-k}$, we have $\hat{f} = 1 + b_k Z^k + Z^k g$ with $g(0) = 0$. By the Lemma there exists therefore an $u \in \mathbb{C}$, such that $|h(u)| < 1$. For $c' := c + u$, we then have

$$|f(c')| = |h(u)| \, |f(c)| < |f(c)|.$$                    □

In function theory ARGAND's inequality is a special case of the general "open mapping theorem" which asserts that nonconstant holomorphic functions always map open sets on to open sets. (See J. Conway, *Functions of One Complex Variable*, Springer-Verlag, 1978, p. 95.)

**4. Variant of the Proof.** We describe here a variant of the proof of the Fundamental theorem in which the existence of $k$th roots, with $k > 2$, is assumed only for positive real numbers, and their existence for arbitrary complex numbers is proved as a consequence. We use induction on the degree of the polynomial $f$, the initial step in the induction being clear. Since the polynomial $\hat{f}$ defined in the previous paragraph has the same degree as $f$ and since the truth of of Lemma 3 for all polynomials $h$ of degree $< n$ follows from the truth of the Fundamental theorem for all polynomials of degree $< n$ (via the ARGAND inequality) it suffices to show that:

*If the Fundamental theorem holds for all polynomials of degree $< n$, $n \geq 2$, then Lemma 3 holds for all polynomials $h$ of degree $n$.*

Let $h$ be any of the admissible polynomials of degree $n$ in Lemma 3. We distinguish three cases:

(1) $k < n$. Then by hypothesis the Fundamental theorem holds for all polynomials $z^k - a$, $a \in \mathbb{C}$; all $a \in \mathbb{C}$ therefore have $k$th roots and the lemma can be proved as in 3.

(2) $k = n$, with $n$ even. Then $h = 1 + bz^n$ with $b \neq 0$. Choose a square root $\eta$ of $-1/b$ and let $u$ be a $k/2$th root of $\eta$ (which is allowable since $k/2 < n$); it then follows that $h(u) = 0 < 1$.

(3) $k = n$, with $n$ odd. Again $h = 1 + bz^n$ with $b \neq 0$. One can then find a $u \in \mathbb{C}$ satisfying $|1 + bu^n| < 1$ in the following amusing way: for $c := -|b|/b \in S^1$ there is a $w \in \{1, -1, i, -i\}$ such that $|c - w| < 1$ (see Exercise 3.3.4). As $n$ is odd, the set $\{1, -1, i, -i\}$ is mapped onto itself by the transformation $x \mapsto x^n$, and there is therefore a $v \in \mathbb{C}$ such that $v^n = w$. For $u := v/\sqrt[n]{|b|} \in \mathbb{C}$ we have $|b| \cdot u^n = w$ and hence $bu^n = -w/c$. Since $|c| = 1$ it follows that

$$|1 + bu^n| = |1 - w/c| = |c - w| < 1. \qquad \square$$

The first inductive proof of this kind was given in 1941 by J.E. LITTLE-WOOD: "Mathematical notes (14): every polynomial has a root." *J. Lond. Math. Soc.*, 16, 95–98. An even simpler proof was given in 1956 by T. ES-TERMANN "On the fundamental theorem of algebra," *J. Lond. Math. Soc.*, 31, 238–240.

**5. Constructive Proofs of the Fundamental Theorem.** The ARGAND–CAUCHY proof is purely an existence proof and is *non-constructive*. As early as 1859 WEIERSTRASS in his note "Neuer Beweis des Fundamentalsatzes der Algebra" (*Math. Werke* 1, 247–256) had made the following start towards a constructive proof: given a polynomial $f(Z)$, a number $z_0 := c \in \mathbb{C}$ is chosen arbitrarily and the sequence $z_n := z_{n-1} - f(z_{n-1})$ defined recursively. WEIERSTRASS says (p. 247) "... it can be shown that when $n$ is increased indefinitely, $z_n$ under certain conditions, tends to a limit $z$ satisfying the equation $f(z) = 0$." More than 30 years later (1891, *Math. Werke* 3, 251–269) WEIERSTRASS once again discusses in detail the problem of a *constructive* proof by asking the following question:

"Is it possible for any given polynomial $f \in \mathbb{C}[Z]$, to produce a sequence $z_n$ of complex numbers by an effectively defined procedure, so that $|f(z_n)|$ is sufficiently small in relation to $|f(z_{n-1})|$ that it converges to a zero of $f$?" H. KNESER in 1940 in his paper entitled "Der Fundamentalsatz der Algebra und der Intuitionismus," *Math. Z.*, 46, 287–302, defined such a process which yields a constructive variant of the ARGAND–CAUCHY proof and which also satisfies the criticisms of the intuitionists. M. KNESER in 1981 further simplified his father's process in a paper entitled "Ergänzung zu einer Arbeit von Hellmuth KNESER über den Fundamentalsatz der Algebra," *Math. Z.*, 177, 285–287.

In 1979 HIRSCH and SMALE described a "sure fire algorithm" which produces, *for any* nonconstant polynomial $f(Z) \in \mathbb{C}[Z]$ and *any arbitrary* initial point $c \in \mathbb{C}$ a sequence $z_n$, with $z_0 = c$, which converges to a zero of $f$. More precisely it is shown that:

$$(*) \qquad\qquad |f(z_n)| \leq K^n |f(c)|, \qquad n = 0, 1, 2, \ldots$$

with a positive real constant $K < 1$, depending only on the degree of $f$, not on $f$ itself. For details, see the article "On algorithms for solving $f(x) = 0$" in *Comm. Pure Appl. Math.*, 32, 281–312 and in particular pp. 303 *et seq.* The inequality $(*)$, and with it a "sure fire algorithm" is already to be found in KNESER, *loc. cit.*, p. 292, formula (6), except that, to satisfy the demands of the intuitionists, $|f(c)|$ is replaced by $\text{Max}(1, |f(c)|)$.

## §3. APPLICATION OF THE FUNDAMENTAL THEOREM

The existence of *at least one* zero for every nonconstant complex polynomial already implies that complex polynomials *decompose* into linear and that real polynomials decompose into linear and quadratic factors. These consequences of the Fundamental theorem are completely elementary, and are a result of the simple fact that a polynomial with a zero at $c$ always has the factor $z - c$.

**1. Factorization Lemma.** *If $c \in \mathbb{C}$ is a zero of the polynomial $f \in \mathbb{C}[Z]$ of degree $n$, then there is just one polynomial $g \in \mathbb{C}[Z]$ of degree $n-1$, such that $f(Z) = (Z - c)g(Z)$.*

**Proof.** Let $f = a_0 + a_1 Z + \cdots + a_n Z^n$, $a_n \neq 0$. Since $Z^\nu - c^\nu = (Z - c)q_\nu(Z)$ with $q_\nu(Z) := Z^{\nu-1} + Z^{\nu-2}c + \cdots + c^{\nu-1}$ it follows that

$$f(Z) = f(Z) - f(c) = \sum_1^n a_\nu(Z^\nu - c^\nu) = (Z - c)g(Z),$$

where

$$g(Z) := \sum_1^n a_\nu q_\nu(Z).$$

It is clear that $f$ is of degree $n - 1$: since $g(z) = (z - c)^{-1}f(z)$, $z \neq c$, $g$ is uniquely determined by $f$ and $c$.                                     □

The factorization lemma holds for all commutative rings, provided that one gives up the uniqueness of $g$. By induction on $n$ we at once obtain the

**Corollary.** *A polynomial $f \in \mathbb{C}[Z]$ of degree $n$ has at most $n$ zeros.*

**2. Factorization of Complex Polynomials.** *Every complex polynomial $f \in \mathbb{C}[Z]$ of degree $n \geq 1$ is, disregarding the order of the factors, uniquely representable in the form*

$$(1) \qquad f(Z) = a(Z - c_1)^{n_1}(Z - c_2)^{n_2} \cdot \ldots \cdot (Z - c_r)^{n_r},$$

*where $a \in \mathbb{C}^\times$; $r \in \mathbb{N}$, $c_1, \ldots, c_r \in \mathbb{C}$ are distinct from one another, and $n_1, \ldots, n_r \in \mathbb{N} \setminus \{0\}$ with $n_1 + n_2 + \cdots + n_r = n$.*

**Proof.** We use induction on $n$, the case $n = 1$ being clearly true. Suppose $n > 1$. By the Fundamental theorem of algebra there exists a $c_1 \in \mathbb{C}$ for which $f$ vanishes. By lemma 1, $f(Z) = (Z - c_1)g(Z)$, where $g(Z) \in \mathbb{C}[Z]$ is of degree $n - 1$. By the inductive hypothesis there is a unique factorization

$$g(Z) = a(Z - c_1)^{n_1-1}(Z - c_2)^{n_2} \cdot \ldots \cdot (Z - c_r)^{n_r}$$

with $n_1 \geq 1, \ldots, n_r \geq 1$, $n_1 - 1 + n_2 + \cdots + n_r = n - 1$; $c_1, \ldots, c_r \in \mathbb{C}$ distinct from one another, and $a \in \mathbb{C}^\times$. Consequently (1) holds.           □

The theorem just proved is often stated in the form:

*Every complex polynomial of the nth degree has precisely n zeros* where each of the zeros $c_j$ is counted according to its multiplicity $n_j$.

**3. Factorization of Real Polynomials.** Every *real* polynomial $f = \Sigma a_\nu X^\nu$ is a complex polynomial satisfying the additional condition

$$\overline{f(z)} = f(\bar{z}) \quad \text{for all} \quad z \in \mathbb{C},$$

for since $\bar{a}_\nu = a_\nu$ it follows that $\overline{\sum a_\nu z^\nu} = \sum a_\nu \bar{z}^\nu$. In particular $\bar{c}$ is a zero of $f[X]$, whenever $c$ is. We easily deduce from this the

**Theorem.** *Every real polynomial $f \in \mathbb{R}[X]$ of degree $n \geq 1$ is (disregarding the order of the factors) uniquely representable in the form*

$$(1) \qquad f(X) = a(X - c_1)^{m_1} \cdot \ldots \cdot (X - c_s)^{m_s} q_1(X)^{n_1} \cdot \ldots \cdot q_t(X)^{n_t},$$

*where the following conditions hold:*

*(a) $a \in \mathbb{R}$, $a \neq 0$; $s, t \in \mathbb{N}$; $c_1, \ldots, c_s \in \mathbb{R}$ are distinct from one another; $m_1, \ldots, m_s, n_1, \ldots, n_t \in \mathbb{N} \setminus \{0\}$ with $m_1 + \cdots + m_s + 2n_1 + \cdots + 2n_t = n$.*

(b) $q_j(X) = X^2 - b_j X - a_j$ *with* $b_j^2 + 4a_j < 0$ *for* $j = 1, \ldots, t$; $q_1, \ldots, q_t$
*are distinct from one another.*

**Proof.** We regard $f$ as a complex polynomial and factorize it in accordance
with Theorem 2. We denote by $c_1, \ldots, c_s$ the *real* zeros. The other truly
complex zeros are taken in conjugate pairs to form real quadratic polyno-
mials $q(x) = (x - c)(x - \bar{c}) = x^2 - (c + \bar{c})x + c\bar{c} \in \mathbb{R}[x]$. Writing $b := c + \bar{c}$,
$a := -c\bar{c}$ we have $b^2 + 4a < 0$, for otherwise $q(x) = (x - \frac{1}{2}b)^2 - \frac{1}{4}(b^2 + 4a)$
would have a real zero. The assertion in the theorem now follows immedi-
ately.                                                                        □

Complex numbers no longer appear in the above enunciation of the preceding theorem. In the proof however they play an essential role as a *deus
ex machina*. GAUSS himself, incidentally, in his dissertation formulated the
fundamental theorem of algebra as a theorem on the factorization of real
polynomials, as its title already indicates (see 1.6). The latter form of the
theorem is used, among other places, in finding the indefinite integrals of
rational functions by partial fractions (see, for example, any standard Calculus text).

**4. Existence of Eigenvalues.** If $\varphi: E \to E$ is a $\mathbb{C}$-linear mapping of a
$\mathbb{C}$-vector space $E$ into itself, the $\lambda \in \mathbb{C}$ is called an *eigenvalue of* $\varphi$, if
there is a vector $v \neq 0$ in $E$ such that $\varphi(v) = \lambda v$. With the help of the
fundamental theorem of algebra we can prove the following:

**Theorem.** *If* $E \neq 0$ *is a finite dimensional* $\mathbb{C}$*-vector space, then every*
$\mathbb{C}$*-linear mapping* $\varphi: E \to E$ *has at least one eigenvalue.*

**Proof** (without using determinants). The set of all $\mathbb{C}$-linear mappings of
$E$ into itself is a *finite dimensional* $\mathbb{C}$-algebra, with respect to the composition of mappings (which is isomorphic to the algebra of all complex $n \times n$
matrices). The elements id, $\varphi, \varphi^2, \ldots, \varphi^k, \ldots$ are therefore linearly dependent, that is, there is a polynomial $f \in \mathbb{C}[Z]$, $f \neq 0$ such that $f(\varphi) = 0$. By
the factorization theorem 2 there exists an equation $f = a \prod_{\nu=1}^{r} (Z - c_\nu)^{n_\nu}$.
Consequently

$$(\varphi - c_1 \mathrm{id})^{n_1} \cdot (\varphi - c_2 \mathrm{id})^{n_2} \cdot \ldots \cdot (\varphi - c_r \mathrm{id})^{n_r} = 0.$$

Thus the mappings $\varphi - c_\nu \mathrm{id}: E \to E$ are not all invertible. Suppose that,
say, $\psi := \varphi - c_1 \mathrm{id}$ is not invertible, then since $E$ is finite dimensional, $\psi$
is not injective. There must therefore be a $v \neq 0$ in $E$ with $\psi(v) = 0$. It
follows that $\varphi(v) = c_1 v$, that is, $c_1$ is an eigenvalue of $\varphi$.                    □

A far-reaching generalization of this theorem will be found in 8.4.7.

**5. Prime Polynomials in $\mathbb{C}[Z]$ and $\mathbb{R}[X]$.** We reformulate the results of paragraphs 2 and 3 in a wider context. Let $K$ be any (commutative) field whatsoever. Then a polynomial $p \in K[X] \setminus K$ whose term of highest degree has the coefficient 1 is said to be a *monic prime polynomial*, if $p$ is not expressible as a product of two polynomials $g, h \in K[X] \setminus K$. All polynomials $X - c$, $c \in K$ are monic prime polynomials. We shall take from Algebra the following result:

*The polynomial ring $K[X]$ has unique factorization, that is to say, every polynomial $f \in K[X] \setminus \{0\}$ is (disregarding the order in which the factors are arranged) expressible uniquely in the form*

$$f = ap_1^{m_1} p_2^{m_2} \cdots\cdots p_r^{m_r} \quad with \quad r \in \mathbb{N};\ m_1, \ldots, m_r \in \mathbb{N} \setminus \{0\},$$

*where $a \in K \setminus \{0\}$ and $p_1, p_2, \ldots, p_r \in K[x]$ are monic prime polynomials distinct from one another.*

In the cases $K = \mathbb{C}$ and $K = \mathbb{R}$ the decomposition of polynomials into prime factors is described more precisely by theorems 2 and 3 respectively.

*In the polynomial ring $\mathbb{C}[Z]$ every monic prime polynomial $p$ is linear, that is, $p(Z) = Z - c$, $c \in \mathbb{C}$.*
*In the polynomial ring $\mathbb{R}[Z]$ every monic prime polynomial $p$ is either linear or quadratic: $p(X) = X - c$, $c \in \mathbb{R}$, or $p(X) = X^2 - bX - a$ with $b^2 + 4a < 0$.*

Each of the last two statements is equivalent to the Fundamental theorem of algebra. In arbitrary base fields $K$ there exist in general prime polynomials of arbitrarily high degree in $K[X]$. For example in $\mathbb{Q}[X]$ the polynomial $X^n - 2$ is a monic prime polynomial for every $n \geq 1$.

**6. Uniqueness of $\mathbb{C}$.** The choice of the field $\mathbb{C}$ of complex numbers is neither arbitrary nor haphazard. We have already become aware in 3.2.3 of one uniqueness result for $\mathbb{C}$. We shall now, with the help of the Fundamental theorem of algebra, establish a more general

**Uniqueness Theorem for $\mathbb{C}$.** *Let $K$ be a commutative extension ring of $\mathbb{R}$ without divisors of zero and with unit element 1 and such that every element of $K$ is algebraic over $\mathbb{R}$, that is, a zero of a real nonzero polynomial. Then $K$ is isomorphic to $\mathbb{R}$ or to $\mathbb{C}$.*

To prove this theorem we use the following simple lemma, based on the Fundamental theorem.

**Lemma.** *On the hypotheses of the Uniqueness theorem every element $v \in K \setminus \mathbb{R}$ satisfies an equation $v^2 = a + bv$ with $a, b \in \mathbb{R}$.*

**Proof.** By hypothesis there exists a nonzero polynomial $f$ such that $f(v) = 0$. As $K$ has no divisors of zero, it follows by Theorem 3 that there is also a polynomial $p$ of degree 1 or 2 which vanishes for the argument $v$. Since $p$ cannot be linear because $v \notin \mathbb{R}$, $p(X)$ must be of the form $p(X) = X^2 - bX - a$, that is, $v^2 = a + bv$.

We now come to the actual proof of the Uniqueness theorem. Suppose $K \neq \mathbb{R}$. We choose an element $v \in K \setminus \mathbb{R}$ and consider the 2-dimensional real vector space $V = \mathbb{R} + \mathbb{R}v$ generated by 1 and $v$. Since $v$, by the lemma, satisfies an equation $v^2 = a + bv$ with $a, b \in \mathbb{R}$, it follows that for any arbitrary elements $x_1 + y_1 v$, $x_2 + y_2 v \in V$:

$$(x_1 + y_1 v)(x_2 + y_2 v) = (x_1 x_2 + y_1 y_2 a) + (x_1 y_2 + y_1 x_2 + y_1 y_2 b)v \in V.$$

Thus $V$ is a commutative, 2-dimensional ring over $\mathbb{R}$ without zero divisors and with unit element, and is therefore, by theorem 3.2.3 isomorphic to $\mathbb{C}$.

It only remains to show that $K = V$. Let $u$ be any element of $K \setminus \mathbb{R}$. There is a real polynomial $f \neq 0$ with $f(u) = 0$. Over $\mathbb{C} \simeq V \subset K$, $f$ splits into linear factors $X - c$, $c \in V$. Since $K$ has no divisors of zero, one of these linear factors must vanish at $u$, that is, $u = c \in V$. We have therefore verified that $K = V \simeq \mathbb{C}$.                    □

The hypothesis in the Uniqueness theorem, that every element $w \in K$ is *algebraic* is always satisfied when $K$ is a finite dimensional vector space over $\mathbb{R}$: for the powers $1, w, w^2, \ldots, w^n, \ldots$ are then linearly dependent, that is, there is an equation $a_0 + a_1 w + \cdots + a_n w^n = 0$, in which the coefficients $a_v$ do not all vanish.

**7. The Prospects for "Hypercomplex Numbers."** The Uniqueness theorem asserts in particular:

*The field $\mathbb{C}$ is (up to isomorphism) the only proper commutative algebraic field over $\mathbb{R}$, and in particular there is no commutative algebraic extension field of $\mathbb{C}$ other than $\mathbb{C}$ itself.*

This theorem was presented by WEIERSTRASS in his Berlin lectures from 1863 onwards. It was published for the first time by HANKEL in his book *Theorie der complexen Zahlensysteme*. It is stated by HANKEL in the words (p. 107):

"*Ein höheres complexes Zahlensystem, dessen formale Rechenoperationen nach den Bedingungen des §28 bestimmt sind, und dessen Einheitsprodukte in's Besondere lineare Functionen der ursprünglichen Einheiten sind, und in welchem kein Product verschwinden kann, ohne dass einer seiner Factoren Null würde, enthält also in sich einen Widerspruch und kann nicht existieren.*" [*A higher complex number system, whose formal laws of* operation are determined by the conditions of §28[8] and whose products of units are in particular linear functions of the original units, and in which no product can vanish unless one of its factors is zero, is a contradiction of terms and *cannot exist.*]

HANKEL proudly declares (p. 107): "Damit ist die Frage beantwortet, deren Lösung 1831 GAUSS (Werke 2, S. 178) versprochen, aber nicht gegeben hat, "*warum die Relationen zwischen Dingen, die eine Mannigfaltigkeit von mehr als zwei Dimensionen darbieten, nicht noch andere in der allgemeinen Arithmetik zulässige Arten von Größen liefern können.*" [This answers a question whose solution GAUSS had promised in 1831 (*Werke*, 2, p. 178) but never gave: the question of *why relations between objects, which represent a manifold of more than two dimensions, cannot give rise to other permissible kinds of magnitudes in generalized arithmetic.*]

The hypothesis of commutativity is essential in the Uniqueness theorem. As is well known the *hypercomplex system of quaternions* described by HAMILTON in the year 1843 is a 4-*dimensional noncommutative* field extension of $\mathbb{R}$. Moreover there is also the 8-*dimensional hypercomplex system of octonions which is a further extension of* $\mathbb{R}$, *that is neither commutative nor associative, but yet has no divisors of zero.* We shall discuss these algebras in depth in Chapters 7 and 8 of this volume.

The hypothesis that the system must not contain divisors of zero is also an immediate condition for the validity of the Uniqueness theorem. For example the system $\mathbb{R} \times \mathbb{R}$ with a "ring-direct multiplication" defined by

$$(a, b)(c, d) := (ac, bd)$$

is a 2-dimensional commutative ring extension of $\mathbb{R}$ with unit element $e := (1, 1)$ which has, for example, $(1, 0)$ as a divisor of zero, and consequently is not isomorphic to $\mathbb{C}$. WEIERSTRASS (1884) and DEDEKIND (1885) showed that this example is significant and that *every finite dimensional, commutative ring extension of* $\mathbb{R}$ *with unit element but no nilpotent elements, is isomorphic to a ring direct sum of copies of* $\mathbb{R}$ *and* $\mathbb{C}$. (An element $x \neq 0$ is said to be nilpotent if there is an exponent $n \geq 2$ such that $x^n = 0$.)

### Appendix: Proof of the Fundamental Theorem, after LAPLACE

We shall discuss here the beautiful *algebraic* proof, which LAPLACE sketched in 1795 and which is somewhat different and perhaps simpler than the second proof that GAUSS gave in 1816. This proof is to be found in N. BOURBAKI's *Algèbre*, Chap. VI, 1952, pp. 40–41. In the *Note historique* BOURBAKI ascribes the proof to GAUSS (p. 150). Our source is an article

---

[8]The conditions of §28 state in effect that the system is a commutative ring with unit element, which is a finite dimensional vector space over $\mathbb{R}$.

by Hellmuth KNESER entitled "Laplace, Gauss und der Fundamentalsatz der Algebra" which was published in 1939 in *Deutsche Mathematik* 4, 318–322.

**1. Results Used.** We shall use the following well-known results.

1) *Every real polynomial of odd degree has at least one real zero (Corollary of the Intermediate value theorem).*

2) *Given any real polynomial $f$ which is not a constant, there exists an extension field $K$ of the field $\mathbb{R}$, such that $f$ splits in $K[X]$ into linear factors (existence of a splitting field).*

3) *Let $K$ be an extension field of $\mathbb{R}$, let $\zeta_1, \ldots, \zeta_n$ be elements of $K$, and let*

$$\eta_k := \sum_{1 \leq \nu_1 < \cdots < \nu_k \leq n} \zeta_{\nu_1} \cdot \ldots \cdot \zeta_{\nu_k}$$

*be the "elementary symmetric functions in $\zeta_1, \ldots, \zeta_n$ (so that $\eta_1 = \zeta_1 + \cdots + \zeta_n, \ldots, \eta_n = \zeta_1 \cdot \ldots \cdot \zeta_n$). Then (with $X$ as indeterminate)*

$$\prod_{\nu=1}^{n}(X - \zeta_\nu) = X^n - \eta_1 X^{n-1} + \eta_2 X^{n-2} - \cdots + (-1)^n \eta_n;$$

*and every polynomial symmetric[9] in $\zeta_1, \ldots, \zeta_n$ belonging to $\mathbb{R}[\zeta_1, \ldots, \zeta_n]$ is a real polynomial in $\eta_1, \ldots, \eta_n$ (Main theorem on symmetric functions).*

4) *Every quadratic complex polynomial splits into linear factors in $\mathbb{C}[Z]$.*

Of these four statements only the main theorem on symmetric functions, which was proved by NEWTON in 1673 would not necessarily be covered in a general mathematical education.

**2. Proof.** For ease in utilizing the statement (1.3) we shall write the coefficients of the given polynomial with alternating signs. The Fundamental theorem of algebra will have been proved as soon as it is shown that:

*Every polynomial $h = X^n - b_1 X^{n-1} + b_2 X^{n-2} - \cdots + (-1)^n b_n \in \mathbb{R}[X]$, $n \geq 1$, has a zero $c \in \mathbb{C}$.*

**Proof** (following LAPLACE). We write $n$ in the form $2^k q$, where $q \in \mathbb{N}$ is an odd number, and use induction on $k$. The start of the induction, $k = 0$, is clear, since the statement holds by virtue of 1). Suppose that

---

[9] A polynomial $p(\zeta_1, \ldots, \zeta_n)$ is said to be *symmetric*, if it is invariant under any permutation of the indices $1, \ldots, n$.

$k \geq 1$. By 2) there is a field $K \supset \mathbb{R}$ and elements $\zeta_1, \ldots, \zeta_n \in K$, such that $h = (X - \zeta_1)(X - \zeta_2) \cdot \ldots \cdot (X - \zeta_n) \in K[X]$. Using an artifice due to LAPLACE, we now form, for any real number $t$, the polynomial

$$L_t := \prod_{1 \leq \mu < \nu \leq n} (X - \zeta_\mu - \zeta_\nu - t\zeta_\mu\zeta_\nu) \in K[X].$$

When this polynomial is expanded in powers of $X$, all the coefficients are real *symmetric* polynomials in $\zeta_1, \ldots, \zeta_n$, because $L_t$, by its definition, remains invariant when the $\zeta_1, \ldots, \zeta_n$ are permuted in any way. By 3) these coefficients are real polynomials in the elementary symmetric functions of the $\zeta_1, \ldots, \zeta_n$, that is, in the real numbers $b_1, \ldots, b_n$. It follows that $L_t \in \mathbb{R}[X]$. Since $L_t$ is of degree $\frac{1}{2}n(n-1) = 2^{k-1}q(2^k q - 1)$ and as $q(2^k q - 1)$ is odd when $q$ is odd, because $k \geq 1$, it follows from the inductive hypothesis that $L_t$ has a zero in $\mathbb{C}$. The product form of $L_t$ now shows that for every $t \in \mathbb{R}$, there must be indices $\mu < \nu$, such that $\zeta_\mu + \zeta_\nu + t\zeta_\mu\zeta_\nu$ lies in $\mathbb{C}$. As there are only $\frac{1}{2}n(n-1)$ index pairs $(\mu, \nu)$ with $1 \leq \mu < \nu \leq n$ and infinitely many real numbers, it must always be possible to find $r, s \in \mathbb{R}$ with $r \neq s$ and $\kappa, \lambda$ with $1 \leq \kappa < \lambda \leq n$, such that

$$\zeta_\kappa + \zeta_\lambda + r\zeta_\kappa\zeta_\lambda \in \mathbb{C}, \qquad \zeta_\kappa + \zeta_\lambda + s\zeta_\kappa\zeta_\lambda \in \mathbb{C}.$$

Since $r \neq s$ it follows from this that

$$u := \zeta_\kappa\zeta_\lambda \in \mathbb{C}, \qquad v := \zeta_\kappa + \zeta_\lambda \in \mathbb{C}$$

and that $\zeta_\kappa, \zeta_\lambda$ are the roots of the polynomial

$$Z^2 - vZ + u \in \mathbb{C}[Z]$$

and that consequently, by 4), $\zeta_\kappa, \zeta_\lambda \in \mathbb{C}$.

## 3. Historical Note.

LAGRANGE said in 1797/98 about LAPLACE's proof that it "ne laisse rien à désirer comme simple démonstration" but held against it the fact that the calculations required would be virtually "impossible" to carry out in practice (*De la résolution des équations numériques de tous les degrés*, Paris, An VI, 1797/98, pp. 200–201). In the 2nd edition of this treatise by LAGRANGE, which appeared in 1808, no mention is made, incidentally, of GAUSS's first proof of 1799, and doubtless this was due to the limited circulation which the latter had enjoyed. H. KNESER commented in this connection "it is perhaps even more remarkable that in the third edition (which came out in 1828, after LAGRANGE's death, in a new version rearranged and edited by POINSOT) nothing had changed. Not only was POINSOT completely unaware of GAUSS's second and third proofs, which had appeared in 1816 in the *Göttinger Commentationes*, but he also expresses his complete satisfaction at what LAGRANGE and LAPLACE had achieved. Thus GAUSS's criticisms and ideas had not yet penetrated to Paris after nearly thirty years, twelve of which had been years of peace."

# 5

# What is $\pi$?

*R. Remmert*

> And he made a molten sea, ten cubits from the
> one brim to the other; it was round all about, and his
> height was five cubits, and a line of thirty cubits
> did compass it round about.
> (I KINGS, Chapter 7, verse 23).

There are many possible ways of introducing the number $\pi$, associated with the circle. We shall obtain $\pi$ from the *complex* exponential function

$$\exp z = 1 + \frac{z}{1!} + \frac{z^2}{2!} + \cdots .$$

*There is a (uniquely defined) real number $\pi > 0$, such that the numbers $2n\pi i$, $n \in \mathbb{Z}$, constitute the set of numbers mapped on to 1 by the exponential mapping $\exp z$; or, in other words, there is a unique number $\pi$ with the property that*

$$(1) \qquad \{w \in \mathbb{C} : \exp w = 1\} = 2\pi i \mathbb{Z}.$$

We shall take (1) as the definition of $\pi$, and deduce from it all its well-known properties. To go into more detail, we shall adopt the following procedure; after describing the history of the number $\pi$ in Section 1, we shall begin by developing the theory of the exponential function in the complex domain as far as is necessary for our purpose, and we shall assume that the reader has a certain familiarity with the basic ideas of real analysis. Absolutely convergent series are defined as in the real domain. The field $\mathbb{C}$ inherits the completeness of the field $\mathbb{R}$ so that CAUCHY's Multiplication theorem remains valid for absolutely convergent series of complex numbers. We shall use these elementary things without stopping to substantiate them afresh for the complex domain, and we shall also have nothing to say about the general limit concept for series of functions.[1] The central result of Section 2 is the Epimorphism Theorem 2.3, which describes the exponential function

---

[1] We justify this unsystematic procedure by appealing to a fundamental principle of applied didactics, which Schiller expressed in a letter to Goethe of the 5th February 1796 in the following words: "Wo es die Sache leidet, halte ich es immer für besser, nicht mit dem Anfang anzufangen, der immer das Schw-

as a *homomorphism* $\exp: \mathbb{C} \rightarrow \mathbb{C}^{\times}$, mapping the *additive* group $\mathbb{C}$ onto the *multiplicative* group $\mathbb{C}^{\times}$. This is quickly established once it is known that the image set $\exp(\mathbb{C})$ contains a neighborhood of the point 1. We give two proofs for this: a very short one based on differentiation, and a completely elementary one, which uses no differential calculus but merely the Intermediate value theorem for real continuous functions (see 2.3 and the appendix to §2).

Once the Epimorphism theorem is available, it is easy to verify the equation (1). After that the existence of the polar coordinate epimorphism, indispensable to the introduction of polar coordinates, can then be quickly established. This is the epimorphism $p : \mathbb{R} \rightarrow S^1, \varphi \mapsto e^{i\varphi}$ whose kernel is $2\pi\mathbb{Z}$. However the proof that $p(\pi/2) = i$ requires the use of the Intermediate Value Theorem (see 3.5 and 3.6).

"After ... exponentials ... the sine and cosine need to be considered, because they ... arise from exponential quantities as soon as these involve imaginary numbers." So wrote EULER in 1748 in §126 of his *Introductio in analysin infinitorum*. True to this sentiment we shall introduce in Section 3 the trigonometric functions by means of the exponential function. The famous EULER formulae

$$\cos z = \frac{1}{2}(e^{iz} + e^{-iz}), \qquad \sin z = \frac{1}{2i}(e^{iz} - e^{-iz})$$

are raised to the status of definitions. EULER's discovery of the relationship between the trigonometric functions and the exponential function completely recast the whole of analysis from its foundations. All the propositions of the elementary theory of the circular functions now follow almost by themselves and in particular the BALTZER–LANDAU characterization of $\pi$ (see 1.5 and 1.6). In Section 4 we discuss some classical formulae for $\pi$; we refer there also to the questions of irrationality and transcendence. The key to the solution of the problem of squaring of the circle lies in the fundamental relation $e^{2\pi i} = 1$.

## §1. ON THE HISTORY OF $\pi$

We summarize the important historical facts. Our sources are:

TROPFKE, J.: Geschichte der Elementar-Mathematik, 4, Ebene Geometrie, 3rd ed., pp. 260ff., De Gruyter, Berlin 1940

JUSCHKEWITSCH, A.P.: Geschichte der Mathematik im Mittelalter, Teubner-Verlag, Leipzig 1964

---

erste ist." [Whenever the subject allows, I always think it better not to begin at the beginning, which is always the most difficult.] This "theorem" to which mathematicians can hardly do justice in lectures and textbooks, was found, incidentally recorded in Riemann's posthumous papers right in the middle of some calculations.

RUDIO, F.: ARCHIMEDES, HUYGENS, LAMBERT, LEGENDRE. Vier Abhandlungen über die Kreismessung, Deutsch herausgegeben und mit
einer Übersicht über die Geschichte des Problems von der Quadratur
des Zirkels, von den ältesten Zeiten bis auf unsere Tage, Teubner Verlag,
Leipzig 1892. Reprint Dr. Martin Sändig OHG 1971
BECKMANN, P.: A history of $\pi$ (Pi), The Golem Press, Boulder, Colorado,
4th ed., 1977

**1. Definition by Measuring a Circle.** In any circle the ratio of the
circumference $C$ to the diameter, and the ratio of the area $A$ to the square
of the radius is constant. ARCHIMEDES (287–212 B.C.) recognized that in
each case the constant is the same. Since the time of EULER (1737) this
constant has been denoted by $\pi$, so that if we write $r$ for the radius, we
have:

$$C = 2\pi r, \qquad A = \pi r^2.$$

The letter $\pi$ appears for the first time in a book by the English mathematician W. OUGHTRED (1575–1660), who taught J. WALLIS, entitled
*Theorematum in libris Archimedis de sphaera et cylindro declaratio*, Oxo-
niae 1663. Whether EULER knew of OUGHTRED is difficult to determine,
but he may well have thought of this symbol as the initial letter of the
ordinary Greek word for circumference ($\pi\epsilon\rho\iota\phi\acute{\epsilon}\rho\epsilon\iota\alpha$). Until 1735 EULER
still wrote $p$ rather than $\pi$.

**2. Practical Approximations.** For the architects of the "molten sea"
in the courtyard of the temple of King Solomon, mentioned in the *Book
of Kings,* $\pi$ was 3. This value was also the one used in the main by the
Babylonians. A surprisingly good approximation is found in the Egyptian
arithmetic book of AHMES (*circa* 1900 B.C.) which gives the rule that
the area of a circle of diameter $d$ is $\left(d - \frac{d}{9}\right)^2$. This corresponds to an
approximation of $\pi$ by $\left(\frac{16}{9}\right)^2 \approx 3.16$. How this value was found is not
recorded.

In the Indian Śulbasūtras (literally "cord-rules," that is, rules for constructing altars of specified form by means of cords or ropes) are found two
rules:

1) to find a square equal in area to a given circle, deduct 2/15 from the
diameter, which leads to an approximate value for $\pi$ of $\left(\frac{26}{15}\right)^2 \approx 3.0044$;

2) to find a circle equal in area to a given square, take as radius the
line $MQ$ in the figure below, where $RQ = \frac{1}{3}RP$, which corresponds to
$\pi \approx 3.088$.

The Śalbasūtras were written down about 500 B.C. It is not known how long before then their content had been handed down by oral tradition.

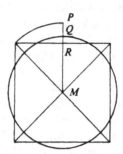

Albrecht DÜRER (1471–1528) of Nuremburg gives the following solution for the second problem:[2] divide the diagonal of the square into 10 parts and take 8 of them as the diameter of the circle. This amounts to saying that $1 \approx \left(\frac{2}{5}\sqrt{2}\right)^2 \pi$, that is, $\pi \approx 3\frac{1}{8}$. Thus DÜRER does not take the then generally accepted value of $3\frac{1}{7}$, which is probably to be explained by the fact that he liked to draw rather than calculate, and that there is no rational geometrical construction based on division which leads to $3\frac{1}{7}$. (The reader may care to try to prove this.)

According to K.R. POPPER (*The open society and its enemies, volume 1, the spell of Plato*, 5, revised ed., Routledge and Kegan Paul, London and Henley, 1966), PLATO (427–348/47) already knew a surprisingly good approximation for $\pi$; he is said to have given the estimate $\sqrt{2} + \sqrt{3} \approx 3.14626$ which has an error of less than 1.5 parts in a thousand.

**3. Systematic Approximation.** ARCHIMEDES was the first to give upper and lower bounds for $\pi$. He compared the circumference of the circle with the total length of the sides of the inscribed and circumscribed regular $n$-sided polygons and obtained for $n = 96$ the inequalities $3\frac{10}{71} < \pi < 3\frac{1}{7}$. The estimate $\pi > 3$ is trivial because an inscribed regular hexagon has a periphery of length $6r$. The value $\pi \approx 3\frac{1}{7} \approx 3.14$ is still used today as a sufficiently close approximation for many practical purposes.

With ARCHIMEDES's method it became possible to determine the value of $\pi$ more accurately. Already APOLLONIUS who was about 25 years younger than ARCHIMEDES calculated some better approximations. This is reported by EUTOCIUS in his commentary on ARCHIMEDES's *On the measurement of a circle* [ARCHIMEDES, *Opera*, Vol. 3, Leipzig, Teubner 1915, Reprint

---

[2]Underweysung der Messung mit dem Zirckel und Richtscheyt in Linien, Ebnen, und gantzen Corporen, Nuremberg 1525, [2]1528; end of the 2nd Book, figure 34 (facsimile edition published by A. Jaeggli und Chr. Papesch, Zürich 1966).

1972, pp. 258–9]; unfortunately he gives no numbers. PTOLEMY (around 150 A.D.) chose a mean between the two values of ARCHIMEDES, namely $\pi \approx 3\frac{17}{120} \approx 3.14166\ldots$ (Handbuch der Astronomie, Deutsch von K. MANITIUS, 2nd ed., Leipzig 1963, pp. 384–5).

Since then astronomers in all nations strove to find improved values for $\pi$. The Chinese knew of some already in the first century A.D. Thus the astronomer and philosopher ZHANG HENG (78–139) worked with the value $\sqrt{10} \approx 3.162$; while the scholar and warlord WANG FAN (died 267) was aware of the better fractional approximation $\frac{142}{45} \approx 3.155$. LIU HUI calculated (circa 263) from a regular polygon of 192 sides the bounds $3.14\frac{64}{625} < \pi < 3.14\frac{169}{625}$ and later from one of 3,072 sides an approximate value corresponding to the decimal fraction 3.14159. Finally, from ZU CHONG-ZHI (430–501) came the approximation $\pi \approx \frac{355}{113}$, which is accurate to the first six decimal places. This approximation, as is well known, is one of the convergents in the expansion of $\pi$ as a regular continued fraction (see 5.6). This fraction was rediscovered by the Dutchman Valentin OTHO towards the end of the 16th century. Whether the Chinese had learnt anything from the discoveries of ARCHIMEDES or PTOLEMY is not known, but anyhow there were already cultural contacts at the time because Chinese silk was being sold in Rome.

In the Indian astronomical work, the Sūryasiddhānta (circa 400 A.D.) $\sqrt{10}$ is used, ĀRAYBHATA gives $\frac{62832}{20000}$ in 498 A.D. This value also appears in the works of al-HwĀRIZMĪ (Baghdad, beginning of the 9th century A.D.). The height of achievement of the Islamic astronomers in such calculations was reached, though much later, by al-KĀŠĪ, who was an astronomer at the observatory in Samarkand founded by ULUG BEG. He calculated the circumference of a circle of unit radius by means of a regular polygon of $3.2^{28}$ sides and thus found $2\pi$ in the form of a sexagesimal fraction 6; 16, 59, 28, 1, 34, 51, 46, 14, 50 with an error of less than a quarter-unit in the last place. He then converted this to the decimal fraction 6.283 185 307 179 586 5 (one of the earliest appearances of decimal fractions).

Rules for the mensuration of circles, equivalent to taking a value of $3\frac{1}{7}$ for $\pi$, seem to have spread through the western world through the activities of Roman surveyors and the writings of BOETHIUS (circa 480–524 A.D.). LEONARDO of PISA (circa 1170–1240?) who made himself master of the mathematical knowledge of the time in the course of his travels in the Orient, calculated $\pi$ from a 96-sided polygon to obtain $\pi \approx \frac{864}{275} \approx 3.141818\ldots$ (La pratica di geometria. In Scritti di Leonardo Pisano, B. Boncompagni, ed., Vol. 2, Rome 1862, pp. 90 et seq.); LUDOLPH VAN CEULEN (1540–1610, Leyden) gave the value correct to 35 decimal places, and so $\pi$ is often called after him, LUDOLPH's number. The first twenty correct decimals are as follows:

$$\pi = 3.14159\,26535\,89793\,23846\ldots.$$

The House of Representatives of the State of Indiana in the U.S.A. unanimously passed in 1897 an "Act introducing a new mathematical truth," which proposed two values for $\pi$, namely 4 and 3.2. The Senate of Indiana postponed "indefinitely" the adoption of this measure. Fortunately for the people of Indiana, the "indefinitely" still continues (see D. SINGMASTER, The legal values of pi, *Math. Intelligencer*, 7(2), 1985, 69–72).

**4. Analytical Formulae.** The first analytical representation of $\pi$ was found by VIETA in 1579 in the form of the infinite product

$$\frac{2}{\pi} = \sqrt{\frac{1}{2}} \cdot \sqrt{\frac{1}{2} + \frac{1}{2}\sqrt{\frac{1}{2}}} \cdot \sqrt{\frac{1}{2} + \frac{1}{2}\sqrt{\frac{1}{2} + \frac{1}{2}\sqrt{\frac{1}{2}}}} \cdots$$

This is probably the very first infinite product in the history of mathematics. WALLIS in 1655 discovered, in the course of investigations to do with integration, his famous product

$$\frac{\pi}{2} = \frac{2 \cdot 2}{1 \cdot 3} \cdot \frac{4 \cdot 4}{3 \cdot 5} \cdot \frac{6 \cdot 6}{5 \cdot 7} \cdots \frac{2n \cdot 2n}{(2n-1) \cdot (2n+1)} \cdots$$

It is remarkable that these first formulae for $\pi$ are not infinite series.

The next great advances towards an understanding of the number $\pi$ had to await the development of the infinitesimal calculus and the theory of infinite series. In 1671 James GREGORY gave the classical series representation

$$\frac{\pi}{4} = 1 - \frac{1}{3} + \frac{1}{5} - \frac{1}{7} + \cdots,$$

which was rediscovered in 1674 by LEIBNIZ, but which, like WALLIS's product, is unsuitable for numerical calculations, because of the slowness of its convergence. NEWTON by putting $z := \frac{1}{2}$ in the arc sin series

$$\arcsin z = z + \frac{1}{2}\frac{z^3}{3} + \frac{1}{2} \cdot \frac{3}{4}\frac{z^5}{5} + \cdots + \frac{1 \cdot 3 \cdot \ldots \cdot (2n-1)}{2 \cdot 4 \cdot \ldots \cdot 2n}\frac{z^{2n+1}}{2n+1} + \cdots$$

obtained, around 1665, the representation

$$\frac{\pi}{6} = \frac{1}{2} + \frac{1}{2} \cdot \frac{1}{3} \cdot \frac{1}{8} + \frac{1}{2} \cdot \frac{3}{4} \cdot \frac{1}{5} \cdot \frac{1}{32} + \frac{1}{2} \cdot \frac{3}{4} \cdot \frac{5}{6} \cdot \frac{1}{7} \cdot \frac{1}{128} + \cdots,$$

which enabled him to calculate with great ease the first 14 decimal places of $\pi$.

**5. BALTZER's Definition.** If one wishes to express the geometric definitions of $\pi$ given in paragraph 1 in an analytical form, one has to use integrals. The unit circle may be described by $x^2 + y^2 = 1$; the arc length of its upper half and its total area are given by

$$\int_{-1}^{1} y\, dx = \int_{-1}^{1} \sqrt{1 - x^2}\, dx = \frac{\pi}{2}$$

and
$$\int_{-1}^{1} \sqrt{1 + (y')^2}\, dx = \int_{-1}^{1} \frac{1}{\sqrt{1 - x^2}}\, dx = \pi$$
respectively.

These equations can be elevated to the status of definitions of $\pi$. It is worth pointing out here that WEIERSTRASS as early as 1841 in his function theoretic proof of the Expansion theorem now usually known as the LAURENT series theorem, had already introduced the idea of defining $\pi$ by the improper integral
$$\pi := \int_{-\infty}^{\infty} \frac{dx}{1 + x^2}$$
(*Math. Werke* 1, p. 53).

In lectures and books on the infinitesimal calculus, integrals are not normally used to define $\pi$, because as a general rule the integral calculus is not treated until after the differential calculus, while $\pi$ and $\frac{1}{2}\pi$ need to be introduced at an early stage as zeros of the sine and cosine functions respectively. It is more usual therefore to define $\frac{1}{2}\pi$ as the *smallest positive* zero of the cosine function defined by its power series; the existence of such a zero being proved with the help of the Intermediate value theorem. This method of introducing the number $\pi$ was already used by Richard BALTZER (1818–1887), who was a professor at Giessen from 1869 onwards, and a friend of KRONECKER. In the first volume of his *Elemente der Mathematik* one reads (see, for example, 5th ed., 1875, p. 195) "Während $x$ den realen Weg von 1 bis 2 zurücklegt, geht cos $x$ ohne Unterbrechung der Continuität aus dem Positiven ins Negative:
$$\cos 1 = 1 - \frac{1}{2} + \frac{1}{4!}\left(1 - \frac{1}{5 \cdot 6}\right) + \cdots > 0,$$
$$\cos 2 = -\frac{1}{3} - \frac{2^6}{6!}\left(1 - \frac{2^2}{7 \cdot 8}\right) - \cdots < 0$$
also giebt es zwischen 1 und 2 einen realen Werth $x$, bei welchem cos $x$ null ist. Dieser Werth ... wird durch $\frac{1}{2}\pi$ bezeichnet." [While $x$ travels along the real path from 1 to 2, cos $x$ goes without any break in continuity from a positive value to a negative value:
$$\cos 1 = 1 - \frac{1}{2} + \frac{1}{4!}\left(1 - \frac{1}{5 \cdot 6}\right) + \cdots > 0$$
$$\cos 2 = 1 - \frac{1}{3} - \frac{2^6}{6!}\left(1 - \frac{2^2}{7 \cdot 8}\right) - \cdots < 0$$
so that there is a real value of $x$ between 1 and 2 for which cos $x$ has the value zero. This value ... is denoted by $\frac{1}{2}\pi$.]

**6. LANDAU and His Contemporary Critics.** BALTZER's method of introducing $\pi$ is not geometrical, but it is probably the most convenient way of arriving rapidly at $\pi$ in the real domain. Edumund LANDAU (1877–1938) advocated and publicized this approach in his Göttingen lectures and his *Einführung in die Differentialrechnung und Integralrechnung* (Verlag Noordoff, Groningen) published in 1934, and written in his characteristic "telegraphic" style. On page 193 of this book can be read *"Die Weltkonstante aus* Satz 262 *werde dauernd mit* $\pi$ *bezeichnet."* [The universal constant in Theorem 262 will always be denoted by $\pi$.] LANDAU, who was a pupil of FROBENIUS, was appointed in 1909 Professor of Mathematics in Göttingen as successor to MINKOWSKI. In 1933 he was dismissed on racial grounds. There is an obituary notice by K. KNOPP in *Jahresber. DMV*, 54, 1951, 55–62.

The definition of $\frac{1}{2}\pi$ as the smallest positive zero of $\cos x$ is now commonplace. It is therefore all the more incomprehensible to us nowadays that this particular *method* of defining $\pi$ should have unleashed in 1934 an academic dispute for which the epithet "disgraceful" would be far too mild a description. A highly distinguished colleague in Berlin attacked LANDAU savagely. It will be enough to quote two of his sentences: "Uns Deutsche läßt eine solche Rumpftheorie unbefriedigt" (Sonderausg. Sitz. Ber. Preuss. Akad. Wiss., Phys.-Math. Kl. XX, p. 6); und weitaus deutlicher: "So ist ... die mannhafte Ablehnung, die ein großer Mathematiker, Edmund LANDAU, bei der Göttinger Studentenschaft gefunden hat, letzten Endes darin begründet, daß der undeutsche Stil dieses Mannes in Forschung und Lehre deutschem Empfinden unerträglich ist. Ein Volk, das eingesehen hat, ... wie Volksfremde daran arbeiten, ihm fremde Art aufzuzwingen, muß Lehrer von einem ihm fremden Typus ablehnen." (Persönlichkeitsstruktur und mathematisches Schaffen, Forsch. u. Fortschr., 10. Jahrg. Nr. 18, 1934, p. 236.) [Such a tail-end of a theory leaves us Germans quite unsatisfied] and more specifically: [Thus ... the valiant rejection by the Göttingen student body which a great mathematician, Edmund LANDAU, has experienced is due in the final analysis to the fact that the un-German style of this man in his research and teaching is unbearable to German feelings. A people who have perceived, ... how members of another race are working to impose ideas foreign to its own must refuse teachers of an alien culture.]

Such abstruse, outrageous, and monstrous opinions were immediately and sharply rejected by the British mathematician G.H. HARDY in August 1934 in his note "The $J$-type and the $S$-type among the mathematicians" (*Collected Papers*, 7, 1979, 610–611) he wrote: "There are many of us, many Englishmen and many Germans, who said things during the War which we scarcely meant and are sorry to remember now. Anxiety for one's own position, dread of falling behind the rising torrent of folly, determination at all costs not to be outdone, may be natural if not particularly heroic excuses. Prof. Bieberbach's reputation excludes such explanations of his utterances; and I find myself driven to the more uncharitable conclusion that he really believes them true."

## §2. THE EXPONENTIAL HOMOMORPHISM $exp: \mathbb{C} \to \mathbb{C}^\times$

The *exponential series*, first written down for real arguments by NEWTON in a letter to LEIBNIZ of the 24th October 1676 (see *Math. Schriften*, ed. GERHARDT, vol. 1, p. 138)

$$\exp z = 1 + z + \frac{z^2}{2!} + \frac{z^3}{3!} + \cdots + \frac{z^n}{n!} + \cdots = \sum_0^\infty \frac{z^\nu}{\nu!}$$

is *absolutely convergent* for all $z \in \mathbb{C}$. This can be proved in exactly the same way as in the real case. We have thus defined in the whole of $\mathbb{C}$ a complex function exp: $\mathbb{C} \to \mathbb{C}$, which is called the *(complex) exponential function*, and which is the natural extension of the real exponential function into the complex domain. This function plays, ever since the days of EULER, a dominant role among the so-called "elementary transcendental functions." We shall derive the Addition theorem which is of fundamental importance for the theory of the function $\exp z$, from CAUCHY's theorem on the product of two series:

Let $\sum_0^\infty a_\mu$, $\sum_0^\infty b_\nu$ be absolutely convergent series. Then their "Cauchy product" $\sum_0^\infty p_\lambda$, where $p_\lambda := \sum_{\mu+\nu=\lambda} a_\mu b_\nu$, is absolutely convergent, and

$$\left( \sum_0^\infty a_\mu \right) \left( \sum_0^\infty b_\nu \right) = \sum_0^\infty p_\lambda.$$

The reader will find a proof of this theorem in the famous *Cours d'analyse* of CAUCHY, which appeared in Paris in 1821 (see, for example, *Oeuvres*, 3, Ser. 2, p. 237) and also in any modern Advanced Calculus text. The Addition theorem states that the mapping

$$\exp: \mathbb{C} \to \mathbb{C}^\times, \qquad z \mapsto \exp z$$

is a *homomorphism* of the *additive group* $\mathbb{C}$ into the multiplicative group $\mathbb{C}^\times$. Whenever a mathematician sees a group homomorphism $\sigma: G \to H$, he immediately looks for the image group $\sigma(G)$, and the kernel

$$\operatorname{Ker} \sigma := \{ g \in G : \sigma(g) = \text{neutral element of } H \}.$$

We shall show that, for the exponential homomorphism,

$$\exp(\mathbb{C}) = \mathbb{C}^\times, \qquad \operatorname{Ker}(\exp) = 2\pi i \mathbb{Z},$$

where $\pi$ is a positive real number. To prove that $\exp(\mathbb{C}) = \mathbb{C}^\times$ we use a simple

**Convergence Lemma.** *Corresponding to any $w \in \mathbb{C}^\times$ there is a sequence in $\mathbb{C}^\times$ with $w_n^{2^n} = w$ and $\lim w_n = 1$.*

We shall prove this straightaway. We write $w = |w|c$ with $c \in S^1$. There is a $c_1 = a_1 + ib_1 \in S^1$, with $a_1 \geq 0$ and $c_1^2 = c$. In view of the concluding remark in 3.3.5 we can now find a succession of numbers $c_n = a_n + ib_n \in S^1$, such that $c_n^2 = c_{n-1}$, $a_n \geq 0$, $|b_n| \leq \frac{1}{\sqrt{2}}|b_{n-1}|$. We see that $c_n^{2^n} = c$ and

$$|b_n| \leq \left(\frac{1}{\sqrt{2}}\right)^{n-1} |b_1| \leq \left(\frac{1}{\sqrt{2}}\right)^{n-1}$$

Since $\sqrt{2} > 1$ it follows that $\lim b_n = 0$, and therefore $\lim a_n^2 = \lim(1 - b_n^2) = 1$, so that $\lim a_n = 1$ since $a_n \geq 0$. It is thus clear that $\lim c_n = \lim(a_n + ib_n) = 1$. For the sequence defined by $w_n := \sqrt[2^n]{|w|}c_n$ we now have $w_n^{2^n} = w$ and $\lim w_n = 1$, because as is well known $\lim \sqrt[n]{r} = 1$ for any $r > 0$.                                                                    □

Apart from the convergence lemma, we shall also make use of two elementary facts taken from the theory of functions (see J. Conway, *Functions of One Complex Variable*, Springer-Verlag, 1978, p. 37).

1) *Any power series* $f(z) = \sum_0^\infty a_\nu z^\nu$ *is holomorphic inside its circle of convergence, and within this circle* $f'(z) = \sum_1^\infty \nu a_\nu z^{\nu-1}$.

2) *If* $f$ *is holomorphic and* $f'$ *vanishes at every point inside some circle, then* $f$ *is constant.*

## 1. The Addition Theorem. $(\exp w)(\exp z) = \exp(w + z)$. To prove this we write

$$(\exp w)(\exp z) = \sum_{\lambda=0}^\infty p_\lambda \quad \text{with} \quad p_\lambda := \sum_{\mu+\nu=\lambda} \frac{w^\mu z^\nu}{\mu! \, \nu!} = \sum_{\nu=0}^\lambda \frac{1}{(\lambda-\nu)!\nu!} w^{\lambda-\nu} z^\nu$$

by CAUCHY's theorem on the multiplication of series. Now

$$\frac{1}{(\lambda-\nu)!}\frac{1}{\nu!} = \frac{1}{\lambda!}\binom{\lambda}{\nu},$$

and consequently, by the Binomial theorem,

$$p_\lambda = \frac{1}{\lambda!}\sum_{\nu=0}^\lambda \binom{\lambda}{\nu} w^{\lambda-\nu} z^\nu = \frac{1}{\lambda!}(w+z)^\lambda,$$

so that

$$(\exp w)(\exp z) = \sum_{\lambda=0}^\infty \frac{(w+z)^\lambda}{\lambda!} = \exp(w+z).$$                □

The Addition theorem asserts that the exponential function obeys the "rule for powers." To bring out this point more clearly one often prefers to write

$$e^z := \exp z, \quad \text{where} \quad e := 1 + \frac{1}{1!} + \frac{1}{2!} + \cdots \quad \text{(EULER 1728)}.$$

If one uses this notation, which is not without its dangers, then the Addition theorem takes the suggestive form of the

**Power Rule.** $e^w e^z = e^{w+z}$ *for all* $w, z \in \mathbb{C}$.

If one puts $w := -z$ in the Addition theorem, it follows from $\exp 0 = 1$, that

$$(\exp z)^{-1} = \exp(-z) \quad \text{for all} \quad z \in \mathbb{C};$$

and in particular that the function $\exp(z)$ has no zeros, so that it maps $\mathbb{C}$ into $\mathbb{C}^\times$. The Addition theorem now states that

*The mapping* $\exp: \mathbb{C} \to \mathbb{C}^\times$ *is a homomorphism of the additive group* $\mathbb{C}$ *into the multiplicative group* $\mathbb{C}^\times$.

**2. Elementary Consequences.** The conjugation of convergent sequences is compatible with the formation of the limit. Consequently $\overline{\exp z} = \exp \bar{z}$ from which it follows that

(1) $$|\exp z| = \exp(\operatorname{Re} z) \quad \text{for all} \quad z \in \mathbb{C}.$$

**Proof.** Since $z + \bar{z} = 2 \operatorname{Re} z$, we have, by the addition theorem

$$|\exp z| = \left|\exp \frac{1}{2}z\right|^2 = \left(\exp \frac{1}{2}z\right)\overline{\left(\exp \frac{1}{2}z\right)} = \left(\exp \frac{1}{2}z\right)\left(\exp \frac{1}{2}\bar{z}\right)$$

$$= \exp\left[\frac{1}{2}(z + \bar{z})\right] = \exp(\operatorname{Re} z). \qquad \square$$

Since it is clear, from the form of the exponential series, that $\exp x > 1$ for $x > 0$, it follows that $\exp x = (\exp(-x))^{-1} < 1$ for $x < 0$. The statement (1) therefore implies

(2) $$|\exp z| = 1 \Leftrightarrow z \in \mathbb{R}i;$$

and in particular $y \mapsto \exp(iy)$ is a mapping of $\mathbb{R}$ into the unit circle $S^1$. As regards the behavior of its functional values we shall show that

(3) $$\operatorname{Im}(\exp(iy)) > 0 \quad \text{for} \quad 0 < y < \sqrt{6}.$$

**Proof.** Since $\exp(iy) = \sum_0^\infty \frac{1}{\nu!}(iy)^\nu$ and since $(iy)^{2n} \in \mathbb{R}$, we have

$$\operatorname{Im}(\exp(iy)) = y - \frac{1}{3!}y^3 + \cdots + \frac{(-1)^n}{(2n+1)!}y^{2n+1} - \cdots$$

(the sine series; cf. 3.1(2)). By writing this in the form

$$y\left(1 - \frac{1}{6}y^2\right) + \frac{1}{5!}y^5\left(1 - \frac{1}{6 \cdot 7}y^2\right) + \cdots$$

we deduce at once that $\mathrm{Im}(\exp(iy)) > 0$ for $0 < y < \sqrt{6}$.                □

Since $\exp(-iy) = (\exp(iy))^{-1}$, we deduce directly from (3) the following.

**Lemma.** *The only point in the open interval* $(-1, +1)$ *at which the function* $\mathbb{R} \to S^1$, $y \mapsto \exp(iy)$ *has a real value is the point* $y = 0$.

The continuity of $\exp z$ is easily deduced from the Addition theorem. Writing $q := 1 + 1/2! + 1/3! + \cdots$ we have $|\exp w - 1| \leq |w|\,|1 + w/2! + w^2/3! + \cdots| \leq |q|\,|w|$ for all $w \in \mathbb{C}$ with $|w| \leq 1$. It follows therefore for any $c \in \mathbb{C}$ and all $z \in \mathbb{C}$ with $|z - c| \leq 1$, that:

$$|\exp z - \exp c| = |\exp c|\,|\exp(z - c) - 1| \leq q|\exp c|\,|z - c|$$

and hence $|\exp z - \exp c| \leq \varepsilon$, provided that $|z - c| \leq \min(1, |q \exp c|^{-1}\varepsilon)$.
□

It follows from the continuity of $\exp z$ with the help of the Intermediate value theorem, that, since $\exp s > 1 + s$ for $s > 0$ and $\exp(-x) = 1/\exp x$, we have

(4)                                       $\exp \mathbb{R} = \{r \in \mathbb{R} : r > 0\}$.

**3. Epimorphism Theorem.** *The exponential homomorphism* $\exp \mathbb{C} \to \mathbb{C}^\times$ *is an epimorphism, that is, it is surjective.*

Our proof is based on the following.

**Lemma.** *There is a neighborhood* $U$ *of the point* $1 \in \mathbb{C}$ *such that* $U \subset \exp(\mathbb{C})$.

**Proof.** The logarithmic series $\lambda(z) := z - \dfrac{z^2}{2} + \dfrac{z^3}{3} - + \cdots$ converges for $|z| < 1$ and is therefore holomorphic with $\lambda'(z) = 1 - z + z^2 - z^3 + - \cdots = (1 + z)^{-1}$ for $|z| < 1$, by the first statement in the Introduction. Similarly $\exp z$ is holomorphic in $\mathbb{C}$, and $(\exp z)' = \exp z$ in $\mathbb{C}$. Consequently $f(z) := (1 + z)\exp(-\lambda(z))$ also is holomorphic inside the unit circle, and it follows by the chain rule that $f'(z)$ is identically zero for all $|z| < 1$. Consequently, by statement 2 of the Introduction, $f(z)$ is a constant, and since $f(0) = 1$, we must have $\exp \lambda(z) = 1 + z$ for all $|z| < 1$. It now directly follows that the disc $U := \{z \in \mathbb{C} : |z - 1| < 1\}$ lies in $\exp(\mathbb{C})$, because for any $a \in \mathbb{C}$ satisfying $|a - 1| < 1$ the number $b := \lambda(a - 1)$ is well defined, and $\exp b = a$.                □

The proof of the Epimorphism theorem can now be quickly completed. By the convergence lemma in the introduction there exists, corresponding to every $w \in \mathbb{C}^\times$, a sequence $w_n \in \mathbb{C}$ with $w_n^{2^n} = w$ and $\lim w_n = 1$. By the lemma in the preceding paragraph there exists also an index $m \geq 1$ and a $\hat{z} \in \mathbb{C}$ such that $w_m = \exp \hat{z}$. For $z := 2^m \hat{z}$ we then have, by virtue of the Addition theorem, $\exp z = (\exp \hat{z})^{2^m} = w$, and we see that $\exp(\mathbb{C}) = \mathbb{C}$. $\square$

We sketch a second proof of the Epimorphism theorem which works without the sequence $w_n$. For any $w \in \mathbb{C}^\times$ the set of points $W := \{wz : z \in U\}$ is a neighborhood of $w$ in $\mathbb{C}^\times$. If $w \in \exp(\mathbb{C})$ then $W \subset \exp(\mathbb{C})$ by the lemma above and the group property of $\exp(\mathbb{C})$. Thus $\exp(\mathbb{C})$ is an open subgroup of the *connected* group $\mathbb{C}^\times$. However, by an elementary theorem of the general theory of topological groups a connected group $G$ has no open subgroups other than $G$.

In the proof of the above lemma the identity $\exp \lambda(z) = 1 + z$ plays the decisive role. It is possible to prove this identity in an *elementary* way without use of the differential calculus. I am indebted to M. KNESER for the following argument. The proof is based on using the formulae:

$$(1) \qquad \lambda(z) = \lim_{n \to \infty} n[(1+z)^{1/n} - 1], \quad z \in \mathbb{E},$$

$$(2) \qquad \exp w = \lim_{n \to \infty} \left(1 + \frac{w_n}{n}\right)^n, \quad \text{if} \ \lim w_n = w,$$

where

$$(1+z)^a := \sum_0^\infty \binom{a}{n} z^n, \qquad z \in \mathbb{E}.$$

From (1) and (2) may be deduced immediately, by taking $w_n := n[(1+z)^{1/n} - 1]$

$$\exp \lambda(z) = \lim_{n \to \infty} [(1+z)^{1/n}]^n.$$

The statement asserted by the theorem now follows, if one also remembers that

$$(3) \qquad (1+z)^a (1+z)^b = (1+z)^{a+b}, \qquad z \in \mathbb{E}.$$

The statements (1)–(3) can be proved by elementary arguments, thus (3) is equivalent to identities involving binomial coefficients $\binom{a}{k}$, $\binom{b}{l}$ and $\binom{a+b}{m}$ which hold for all natural numbers $a$, $b$ and hence generally, by the binomial theorem.

In an appendix to this section we shall give another elementary proof of the lemma.

## 4. The Kernel of the Exponential Homomorphism. Definition of $\pi$.

With the help of the equation $\exp(\mathbb{C}) = \mathbb{C}^\times$ we can determine without trouble the kernel, $\mathrm{Ker}(\exp) = \{w \in \mathbb{C} : \exp w = 1\}$.

**Theorem.** *There is a uniquely determined positive real number $\pi$ with the property that*

$$\mathrm{Ker}(\exp) = 2\pi i \mathbb{Z}.$$

**Proof.** We put $K := \mathrm{Ker}(\exp)$. Now $K \neq 0$ because $\exp \colon \mathbb{C} \to \mathbb{C}^{\times}$ *is not an isomorphism as* $\mathbb{C}^{\times}$ contains the element $-1$ of order $2$ whereas the *additive* group $\mathbb{C}$ has no elements of finite order. For any $c \in K$, we have $1 = |\exp c|$, so that it follows by 2(2) that $K \subset \mathbb{R}i$. There are therefore numbers $s > 0$ such that $si \in K$ (note that $-c$ is in $K$ whenever $c$ is). Since, by Lemma 2 there is no nonzero number $iy$, with $y \in (-1,1)$, that belongs to $K$, there must, by reason of the *continuity* of $\exp z$, be a *least positive real* number $\pi$ such that $2\pi i \in K$. Then $2\pi i \mathbb{Z} \subset K$ is trivial. Conversely if $r \in \mathbb{R}$ with $ri \in K$, there must, since $\pi > 0$, be an $n \in \mathbb{Z}$ such that $2n\pi \leq r < 2(n+1)\pi$. As $K$ is an additive subgroup of $\mathbb{C}$, it follows that $i(r - 2n\pi) \in K$. Since $0 \leq r - 2n\pi < 2\pi$, it follows that $r = 2n\pi$ because of the choice of $\pi$ as the least number with the specified property. We have thus shown that $K = 2\pi i \mathbb{Z}$, and the uniqueness of $\pi$ is clear. $\qquad\square$

*In what follows we shall use the statement asserted by this theorem as the definition of $\pi$.*

Since $e^{2\pi i} = 1$ and $\pi i \notin \mathrm{Ker}(\exp)$ we must have $e^{i\pi} = -1$. In the one equation

$$0 = 1 + e^{i\pi}$$

the five fundamental numbers $0, 1, i, e, \pi$ are "interwoven in a truly wonderful manner" which has given rise on occasions to metaphysical speculation.

**Appendix. Elementary Proof of Lemma 3.** We use Lemma 2 and the following proposition, which is a consequence of the Intermediate value theorem of the infinitesimal calculus.

1) *If $I$ is a compact interval in $\mathbb{R}$, and if $f \colon I \to \mathbb{R}$ is continuous, then the image set $f(I)$ is a compact interval.*

To prove the lemma we now define $u(y) := \mathrm{Re}\,\exp(iy)$ and $v(y) := \mathrm{Im}\,\exp(iy)$. Since

$$(*) \qquad \exp z = e^x u(y) + ie^x v(y) \quad \text{for} \quad z = x + iy,$$

and since continuous complex functions necessarily have continuous real and imaginary parts and absolute value functions, it follows from the continuity of $\exp z$ that $u(y) = (\mathrm{Re}\,\exp z)/|\exp z|$ is continuous throughout $\mathbb{R}$ and that the function $h(y) := (\mathrm{Im}\,\exp z)/(\mathrm{Re}\,\exp z) = u(y)/v(y)$ is continuous wherever $v(y)$ does not vanish. (Of course it is known that $u(y) = \cos y$, $v(y) = \sin y$ and $h(y) = \tan y$, but this is irrelevant to our argument.) Since $u(0) = 1$ and $u$ is continuous, there exists an $\varepsilon > 0$ such that $u$ is positive in the closed interval $I := [-\varepsilon, \varepsilon]$. Thus $h$ is well defined and continuous in $I$ and we assert:

*The image $h(I)$ is an interval $[c, d]$ with $c < d$; the image $\exp(\mathbb{R} \times I)$ is the "sector" $S := \{s(1 + it) \colon s > 0,\, t \in h(I)\}$ (see figure).*

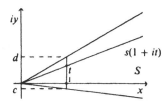

**Proof.** By 1), $h(I)$ is a compact interval $[c, d]$ in $\mathbb{R}$. Since $v(0) = 0$ we have $0 = h(0) \in I$. If $c$ were equal to $d$, $h$ and thus $v$ would vanish identically in $I$. In view of $(*)$ this would mean that $e^{iy} \in \mathbb{R}$ for all $y \in I$, which is impossible by Lemma 2. Consequently $c < d$.

For $z = (x, y) \in \mathbb{R} \times I$ we have $\exp z = e^x u(y)[1 + ih(y)]$. Since $u(y) > 0$ for any $y \in I$, and since $e^x$ runs through all positive real numbers, it follows that $\exp(\mathbb{R} \times I) = S$. Since $1 \in \mathbb{C}$ is the image of $(0, 0) \in \mathbb{R} \times I$, $S \subset \exp(\mathbb{C})$ is a neighborhood of the point 1 with the required property.

## §3. CLASSICAL CHARACTERIZATIONS OF $\pi$

In this section we show that the number $\pi$ defined in 2.4 has all the properties one normally learns in real analysis. The characterization of $\pi$ and $\frac{1}{2}\pi$ as the least positive zero of the sine and cosine function respectively is a simple matter with the help of the results of the preceding sections, if one makes use of the relationship between the exponential function and the trigonometrical functions

$$e^{iz} = \cos z + i \sin z$$

which was discovered by EULER, and which remains invisible as long as one is confined to the real domain. To determine the circumference and area of a circle in terms of $\pi$, we take over the basic definitions from analysis.

**1. Definitions of cos $z$ and sin $z$.** We define the *complex cosine* and *sine functions* throughout $\mathbb{C}$ by:

$$(1) \qquad \cos z := \frac{e^{iz} + e^{-iz}}{2}, \quad \sin z := \frac{e^{iz} - e^{-iz}}{2i}, \quad z \in \mathbb{C},$$

and note immediately that these are the well-known trigonometrical functions, usually defined directly through their power series

$$(2) \qquad \cos z = \sum_0^\infty \frac{(-1)^\nu}{(2\nu)!} z^{2\nu}, \quad \sin z = \sum_0^\infty \frac{(-1)^\nu}{(2\nu+1)!} z^{2\nu+1}, \quad z \in \mathbb{C},$$

For, writing $s_n(z) := \sum_0^n z^\nu / \nu!$, we have for all $m \in \mathbb{N}$

$$s_{2m+1}(\pm iz) = \sum_{\mu=0}^m (-1)^\mu \frac{z^{2\mu}}{(2\mu)!} \pm i \sum_{\mu=0}^m (-1)^\mu \frac{z^{2\mu+1}}{(2\mu+1)!}, \quad z \in \mathbb{C}$$

and the equations (2) follow immediately by addition and subtraction respectively, since $e^z = \lim_{n\to\infty} s_n(z)$. $\qquad\qquad\qquad\qquad\qquad\square$

From (1), the classical Eulerian formula

$$\exp(iz) = \cos z + i \sin z, \qquad z \in \mathbb{C}$$

is obtained by addition.

For real arguments $z = x$, we have $\cos x$, $\sin x \in \mathbb{R}$; consequently

$$\exp(ix) = \cos x + i \sin x, \qquad x \in \mathbb{R},$$

is the *decomposition of* $\exp(ix)$ *into its real and imaginary parts*. This representation was repeatedly used in 3, §6. It follows now, for example, since $e^{2\pi i} = 1$ and $e^{i\pi} = -1$ that

$$\cos 2\pi = 1, \quad \sin 2\pi = 0; \quad \cos \pi = -1, \quad \sin \pi = 0.$$

It also follows at once from (1), that the cosine function is *even* and the sine function *odd*, that is, that

$$\cos(-z) = \cos z, \quad \sin(-z) = -\sin z.$$

**2. Addition Theorem.** *For all* $w, z \in \mathbb{C}$

$$\cos(w+z) = \cos w \cos z - \sin w \sin z,$$
$$(1)$$
$$\sin(w+z) = \sin w \cos z + \cos w \sin z.$$

**Proof.** We start from the identity

$$e^{i(w+z)} = e^{iw} \cdot e^{iz} = (\cos w + i \sin w)(\cos z + i \sin z)$$
$$= \cos w \cos z - \sin w \sin z + i(\sin w \cos z + \cos w \sin z).$$

Writing $-w$ and $-z$ in place of $w$ and $z$, we obtain:

$$e^{-i(w+z)} = \cos w \cos z - \sin w \sin z - i(\sin w \cos z + \cos w \sin z).$$

The equations (1) are then obtained by addition and subtraction respectively. □

Innumerable other formulae, for example, $\cos^2 z + \sin^2 z = 1$ and the "halving formulae"

$$\sin z = 2 \sin \frac{1}{2} z \cos \frac{1}{2} z, \quad \cos^2 \frac{1}{2} z = \frac{1}{2}(1 + \cos z)$$

can be derived from the Addition theorem.

In Section 4 we shall make essential use of the relations:

$$\cos w - \cos z = -2 \sin \tfrac{1}{2}(w + z) \sin \tfrac{1}{2}(w - z),$$

(2)

$$\sin w - \sin z = 2 \cos \tfrac{1}{2}(w + z) \sin \tfrac{1}{2}(w - z).$$

**Proof.** It follows from the equations (1), by subtraction, that

$$\cos(w + z) - \cos(w - z) = -2 \sin w \sin z,$$
$$\sin(w + z) - \sin(w - z) = 2 \cos w \sin z.$$

(2) then follows by writing $\frac{1}{2}(w + z)$, $\frac{1}{2}(w - z)$ instead of $w$, $z$ respectively.

**3. The Number $\pi$ and the zeros of cos $z$ and sin $z$.** In contrast to $\exp z$, the function $\cos z$ and $\sin z$ have zeros.

**Theorem on Zeros.** *The only (real or complex) zeros of $\sin z$ are the real numbers $n\pi, n \in \mathbb{Z}$, and the only (real or complex) zeros of $\cos z$ are the real numbers $\frac{1}{2}\pi + n\pi, n \in \mathbb{Z}$.*

**Proof.** In view of $e^{i\pi} = -1$ we have

$$2i \sin z = e^{-iz}(e^{2iz} - 1), \quad 2 \cos z = e^{i(\pi - z)}(e^{2i(z - \frac{1}{2}\pi)} - 1).$$

Theorem 2.4 now gives:

$$\sin w = 0 \Leftrightarrow 2iw \in \mathrm{Ker}(\exp) = 2\pi i \mathbb{Z} \Leftrightarrow w = n\pi, \quad n \in \mathbb{Z},$$
$$\cos w = 0 \Leftrightarrow 2i\left(w - \tfrac{1}{2}\pi\right) \in 2\pi i \mathbb{Z} \quad \Leftrightarrow w = \tfrac{1}{2}\pi + n\pi, \quad n \in \mathbb{Z}. \quad □$$

We see that $\pi$ and $\frac{1}{2}\pi$ are in fact the least positive zeros of $\sin z$ and $\cos z$ respectively. Even though all the real zeros of cos and sin are known from the real theory, it still had to be shown that there are no properly complex zeros in the extended domain of the argument.

**4. The Number $\pi$ and the Periods of exp $z$, cos $z$ and sin $z$.** A function $f: \mathbb{C} \to \mathbb{C}$ is said to be *periodic*, when there is a complex number

$\omega \neq 0$, such that $f(z+\omega) = f(z)$ for all $z \in \mathbb{C}$. The number $\omega$ is then called a *period of f*. If $f$ is periodic, then the set

$$\mathrm{Per}(f) := \{\omega \in \mathbb{C} : \omega \text{ is a period of } f\} \cup \{0\}$$

of all periods of $f$, including 0, is *an additive subgroup of* $\mathbb{C}$.

**Periodicity Theorem.** *The functions* exp, cos *and* sin *are periodic and*

$$\mathrm{Per}(\exp) = \mathrm{Ker}(\exp) = 2\pi i \mathbb{Z}, \qquad \mathrm{Per}(\cos) = \mathrm{Per}(\sin) = 2\pi \mathbb{Z}.$$

**Proof.** For a number $\omega \in \mathbb{C}$, the function $\exp(z+\omega) = \exp z \exp \omega$ coincides with the function $\exp z$, if and only if $\exp \omega = 1$. This proves, by Theorem 2.4 that $\mathrm{Per}(\exp) = \mathrm{Ker}(\exp) = 2\pi i \mathbb{Z}$.

Since $\cos(z + \omega) - \cos z = -2 \sin\left(z + \frac{\omega}{2}\right) \sin \frac{\omega}{2}$ we have $\omega \in \mathrm{Per}(\cos)$ if and only if $\sin \frac{\omega}{2} = 0$, that is, whenever $\omega \in 2\pi \mathbb{Z}$. The statement for the sine function follows similarly from $\sin(z + \omega) - \sin z = 2 \cos\left(z + \frac{\omega}{2}\right) \sin \frac{\omega}{2}$. $\square$

There is an essential difference between the behavior of the exponential function in the real and complex domain. In the real domain it assumes, since $\mathrm{Ker}(\exp) \cap \mathbb{R} = \{0\}$, *every positive real value once and once only*, whereas in the complex domain it possesses the purely imaginary "minimal period" $2\pi i$ (not seen in the real case) and assumes *every* value $c \neq 0$— including negative real values—countably often.

We see also that the number $2\pi$ can be characterized as the smallest positive number that is a period of both the functions cos and sin. Even though one knows that cos and sin as real functions have $2\pi$ as least common period, one still needs to show that $2\pi$ remains a period in the complex domain and that there are no new additional properly complex periods.

**5. The Inequality sin $y > 0$ for $0 < y < \pi$ and the Equation $e^{i\frac{\pi}{2}} = i$.** The equation $e^{i\pi} = -1$ naturally raises the question of the values of $\xi := e^{i\frac{\pi}{2}}$ and $\eta := e^{i\frac{\pi}{4}}$ and so on. Since $\xi^2 = e^{i\pi} = -1$, $\eta^2 = \xi$ there are respectively two and four possible values to be considered, namely $\xi = \pm i$ and $\eta = \pm\frac{1}{2}\sqrt{2}(1 \pm i)$. To determine the sign we note that

(1) $$\sin y > 0 \quad \text{for} \quad 0 < y < \pi.$$

**Proof.** The sine function is continuous in $\mathbb{R}$ and by 2.2(3) it is positive in the interval $(0, \sqrt{6})$. If $\sin y$ were negative anywhere in $(0, \pi)$ it would, by the *Intermediate value theorem* vanish for some $r$, with $0 < r < \pi$, and this would contradict the theorem on zeros in Section 3. $\square$

It follows from (1) that only the plus sign in the expression for $\xi$ and $\eta$ can be valid, that is, that

(2) $$e^{i\frac{\pi}{2}} = i, \qquad e^{i\frac{\pi}{4}} = \frac{1}{2}\sqrt{2}(1 + i).$$

The first equation here was already known to BERNOULLI in 1702; in the form

$$\ln i = i\frac{\pi}{2} \quad \text{or} \quad \pi = 2\frac{\ln i}{i}$$

it played an important part in the controversy between LEIBNIZ and BERNOULLI over the true values of the natural logarithms of $-1$ and $i$. □

It must clearly be said that without appealing to the Intermediate value theorem, the minus sign cannot be excluded in the formula $e^{i\frac{\pi}{2}} = \pm i$. All the conclusions drawn so far apply equally well to the function $\exp(-z)$ which assumes the value $-i$ at $i\frac{\pi}{2}$, as the reader may care to check for himself. To arrive at the important equation $e^{i\frac{\pi}{2}} = i$ recourse to the Intermediate value theorem is therefore again essential.

*The equations (2) assert, when written in real form, that*

$$\cos\frac{\pi}{2} = 0, \quad \sin\frac{\pi}{2} = 1, \quad \cos\frac{\pi}{4} = \sin\frac{\pi}{4} = \frac{1}{2}\sqrt{2}.$$

With the help of the Intermediate value theorem and $\cos 0 = 1$, it can be shown, as above, that

$$(1') \qquad\qquad \cos y > 0 \quad \text{for} \quad -\frac{1}{2}\pi < y < \frac{1}{2}\pi.$$

**6. The Polar Coordinate Epimorphism $p: \mathbb{R} \to S^1$.** In Chapter 3, §6 polar coordinates were introduced. The statements made at the time without proof are now clear. From the Epimorphism theorem 2.3 and the fact that $|\exp z| = 1$ holds if and only if $z \in \mathbb{R}i$ (see 2.2(2)), it can be deduced that $\exp(i\mathbb{R}) = S^1$, where $S^1$ again denotes the (multiplicative) circle group (see 3.3.4), and from this the theorem which was decisively used in 3.6.1 namely the

**Epimorphism Theorem.** *The mapping $p: \mathbb{R} \to S^1$, $\varphi \mapsto e^{i\varphi}$ is a group epimorphism whose kernel is the group $2\pi\mathbb{Z}$, and we have: $p\left(\frac{\pi}{2}\right) = i$.*

**Proof.** For $\varphi, \psi \in \mathbb{R}$ we have $p(\varphi+\psi) = \exp(i\varphi+i\psi) = (\exp i\varphi)(\exp i\psi) = p(\varphi)p(\psi)$; consequently $p$ is an epimorphism, since $p(\mathbb{R}) = \exp(i\mathbb{R}) = S^1$. As $\operatorname{Ker}(\exp) = 2\pi i\mathbb{Z}$ it also follows that $\operatorname{Ker} p = \{t \in \mathbb{R}: it \in \operatorname{Ker}(\exp)\} = \{t \in \mathbb{R}: t \in 2\pi\mathbb{Z}\}$. This last statement was proved in the previous section. □

**7. The Number $\pi$ and the Circumference and Area of a Circle.** A mapping $\gamma: I \to \mathbb{C}$, $t \mapsto z(t) = x(t) + iy(t)$ of a closed interval $[a,b]$ of $\mathbb{R}$

into $\mathbb{C}$ is called a *continuously differentiable path* in $\mathbb{C}$, if the functions $x(t)$ and $y(t)$ are continuously differentiable in $I$. For such paths $\gamma$ the integral

$$L(\gamma) := \int_a^b |z'(t)|dt \text{ exists where } z'(t) := x'(t) + iy'(t).$$

Since $|z'(t)| = \sqrt{x'(t)^2 + y'(t)^2}$ this expression—as is shown in analysis—represents the (Euclidean) length $L(\gamma)$ of the path $\gamma$.

If $c \in \mathbb{C}$ is a point and $r > 0$, the continuously differentiable path $\gamma_\psi : [0, \psi] \to \mathbb{C}$, $\varphi \mapsto z(\varphi) := c + re^{i\varphi}$, where $0 < \psi \le 2\pi$, is a *circular arc* of center $c$ and radius $r$, which runs from $c + r$ to $c + re^{i\psi}$ (see figure).

As $z'(\varphi) = ire^{i\varphi}$, we have $|z'(\varphi)| = r$ and consequently $L(\gamma_\psi) = \int_0^\psi |z'(\varphi)|d\varphi = \psi r$. The length of the circular arc is thus $\psi r$. Since $\gamma_{2\pi}$ is the full circumference, it follows in particular that:

*The circumference of a circle of radius $r$ is $2\pi r$.*      □

If $f : [a, b] \to \mathbb{R}$ is continuous, the integral $\int_a^b f(x)dx$ exists; it measures the *area* under the graph of $f$. As the semicircle about the origin of radius $r > 0$ is represented by the function $\sqrt{r^2 - x^2}$, $x \in [-r, r]$, the area $I$ of the whole circle is given by

$$I := 2 \int_{-r}^r \sqrt{r^2 - x^2}\, dx.$$

Substituting $r \cos \varphi$ for $x$, and using the identity $\sin^2 \varphi = \frac{1}{2}(1 - \cos 2\varphi)$ we obtain

$$I = 2r \int_{-\pi}^0 \sin \varphi(-r \sin \varphi)d\varphi = 2r^2 \int_0^\pi \sin^2 \varphi\, d\varphi = r^2 \left(\varphi - \frac{1}{2}\sin 2\varphi\right)\Big|_0^\pi = \pi r^2.$$

*The area of a circle of radius $r$ is $r^2\pi$.*

## §4. Classical Formulae for $\pi$

From the countless formulae for $\pi$ (whose sheer number makes it impossible to review them all) we shall select those of LEIBNIZ, VIETA, WALLIS and EULER which stand out because of their special historical significance.

In Section 5 we give a representation of $\pi$ by an integral, which WEIER-STRASS in 1841 in a work written in his youth entitled "Darstellung einer analytischen Function einer complexen Veränderlichen, deren absoluter Betrag zwischen zwei gegebenen Grenzen liegt" (*Math. Werke* 1, 51–66) used precisely for the purpose of defining $\pi$. Finally we discuss some expressions for $\pi$ in the form of continued fractions, and the transcendence problem for $\pi$.

**1. LEIBNIZ's Series for $\pi$.** The tangent function $\tan x$ is a monotonic strictly increasing function in the interval $(-\frac{1}{2}\pi, \frac{1}{2}\pi)$, since its derivative $1/\cos^2 x$ is strictly positive; moreover, $\tan x$ assumes all real values. There exists therefore an inverse function $\arctan: \mathbb{R} \to (-\frac{1}{2}\pi, \frac{1}{2}\pi)$ whose derivative is given by

$$\arctan'(x) = \frac{1}{\tan'(\arctan x)} = \cos^2(\arctan x) = \frac{1}{1+x^2};$$

the last equation being obtained by putting $y := \arctan x$ and noting that $x^2 = \tan^2 y = (1/\cos^2 y) - 1$. The geometric series $(1 + t^2)^{-1} = \sum_0^\infty (-1)^\nu t^{2\nu}$ which is uniformly convergent for $|t| < 1$, yields, on term by term integration, after interchanging the order of integration and summation, which is valid in these conditions, the *arctangent series*

(1)
$$\arctan x = \int_0^x \frac{dt}{1+t^2} = \sum_0^\infty (-1)^\nu \int_0^x t^{2\nu} dt$$

$$= x - \frac{x^3}{3} + \frac{x^5}{5} - \frac{x^7}{7} + \cdots, \quad |x| < 1.$$

By ABEL's theorem on the limit of a power series (see L. AHLFORS, *Complex Analysis*, 2nd ed., McGraw-Hill, 1966, p. 42) and K. KNOPP, *Theorie und Anwendung der unendlichen Reihen*, Springer-Verlag, 4, Aufl. 1947, p. 179) the expansion is also valid for $x := 1$, its value there being $\arctan 1 = \frac{\pi}{4}$. In this way we obtain the "only just" convergent series

$$\frac{\pi}{4} = 1 - \frac{1}{3} + \frac{1}{5} - \frac{1}{7} + - \cdots = \sum_{\nu=0}^\infty \frac{(-1)^\nu}{2\nu+1}.$$

This is LEIBNIZ's series for $\pi$, which he discovered by geometrical considerations. It is a formula which so to speak "yields the number $\pi$ by purely arithmetic operations. It is as though, by this representation, the veil had been lifted from this strange number" as KNOPP pointed out (*loc cit.* p. 220).

LEIBNIZ's series, which interestingly enough, was already known in India around 1500, is entirely unsuitable for practical calculation. To calculate

$\pi$ to an accuracy of $10^{-k}$, about $\frac{1}{4}10^k$ terms would be needed; in fact for $n \geq 1$:

$$\frac{1}{4}\frac{1}{2n+1} < \frac{\pi}{4} - \sum_{\nu=0}^{2n-1} \frac{(-1)^\nu}{2\nu+1} < \frac{1}{4}\frac{1}{2n-1}.$$

## 2. VIETA's Product Formula for $\pi$.

The halving-formula $\sin z = 2\sin\frac{z}{2}\cos\frac{z}{2}$ for the sine function gives, by induction,

$$\sin z = 2^n \sin \frac{z}{2^n} \prod_{\nu=1}^{n} \cos \frac{z}{2^\nu}, \quad z \in \mathbb{C}, \quad n = 1, 2, \ldots .$$

Since $\lim_{n\to\infty} 2^n \sin \frac{z}{2^n} = z$, we obtain the infinite products

$$\sin z = z \prod_{\nu=1}^{\infty} \cos \frac{z}{2^\nu}, \quad z \in \mathbb{C},$$

$$(*) \qquad \frac{2}{\pi} = \cos \frac{\pi}{4} \cdot \cos \frac{\pi}{8} \cdot \cos \frac{\pi}{16} \cdot \ldots \cos \frac{\pi}{2^{\nu+1}} \cdot \ldots \qquad \left(\text{for } z := \frac{1}{2}\pi\right).$$

This, as it stands, is almost VIETA's formula. Since $\cos^2 \frac{z}{2} = \frac{1}{2}(1 + \cos z)$ by 3.2 and since $\cos x \geq 0$ for $x \in [0, \frac{1}{2}\pi]$ by 3.5(1'), we have $\cos \frac{z}{2} = \sqrt{\frac{1}{2} + \frac{1}{2}\cos x}$ for such $x$; and in particular therefore:

$$\cos \frac{\pi}{4} = \sqrt{\frac{1}{2}}, \quad \cos \frac{\pi}{8} = \sqrt{\frac{1}{2} + \frac{1}{2} \cdot \sqrt{\frac{1}{2}}}, \quad \cos \frac{\pi}{16} = \sqrt{\frac{1}{2} + \frac{1}{2} \cdot \sqrt{\frac{1}{2} + \frac{1}{2} \cdot \sqrt{\frac{1}{2}}}}, \ldots,$$

so that $(*)$ becomes VIETA's formula:

$$\frac{2}{\pi} = \sqrt{\frac{1}{2}} \cdot \sqrt{\frac{1}{2} + \frac{1}{2} \cdot \sqrt{\frac{1}{2}}} \cdot \sqrt{\frac{1}{2} + \frac{1}{2} \cdot \sqrt{\frac{1}{2} + \frac{1}{2} \cdot \sqrt{\frac{1}{2}}}} \cdots .$$

The "VIETA's sequence" $v_n := \prod_{\nu=1}^{n} \cos \frac{\pi}{2^{\nu+1}} = (2^n \sin \frac{\pi}{2^{n+1}})^{-1}$ converges rapidly:

$$(1) \qquad 0 < v_n - \frac{2}{\pi} < \frac{1}{48}\sqrt{2}\pi^2 \frac{1}{4^n} < \frac{3}{10}\frac{1}{4^n}.$$

**Proof.** Since $v_n$ decreases monotonically $v_n > 2\pi^{-1}$. As $\sin x > x - \frac{x^3}{3!}$ for $0 < x < \sqrt{42}$ (by estimation of the real Taylor series) we have $2\pi^{-1}v_n^{-1} > 1 - \frac{\pi^2}{24}\frac{1}{4^n}$ and the inequalities (1) then follow on multiplying by $v_n$, since $v_n \leq v_1 = \frac{1}{2}\sqrt{2}$, $\pi^2 < 10$, $\sqrt{2} < 1,44$. $\qquad \square$

The following numerical examples show the good convergence:

| $n$ | $v_n$ | $2v_n^{-1}$ |
|---|---|---|
| 5 | 0.6368755077217... | 3.140331156954... |
| 15 | 0.6366197726114... | 3.141592652386... |
| 21 | 0.6366197723676... | 3.141592653589... |

The last value is already correct to the 12th decimal.

The estimation (1) can, by the way, easily be improved to the equation

$$\lim 4^n \left( v_n - \frac{2}{\pi} \right) = \frac{\pi}{12}.$$

### 3. EULER's Product for the Sine and WALLIS's Product for $\pi$.

The French mathematician J. HADAMARD (1865–1963) is supposed to have said "*Le plus court chemin entre deux énoncés réels passe par le complexe.*" As an example of this principle that the "shortest way is via the complex," we shall deduce EULER's product formula for the sine function, and with it WALLIS's formula for $\pi$.

From de MOIVRE's formula $(\cos t + i \sin t)^k = \sin kt + i \sin kt$, $t \in \mathbb{R}$, we obtain, on separating real and imaginary parts:

$$\sin kt = \sin t \left[ k \cos^{k-1} t - \binom{k}{3} \cos^{k-3} t \sin^2 t + \cdots \right], \qquad k \in \mathbb{N},$$

and hence, since $\cos^{2k} t = (1 - \sin^2 t)^k$:

*the function $\sin kt$ is, for odd $k = 2n + 1$, a rational polynomial $p(\sin t)$ in $\sin t$ of degree $k$.*

From now on everything stays in $\mathbb{R}$: as $p(\sin t) = \sin kt$ has the $k$ distinct real zeros $\sin \frac{\nu \pi}{k}$, $\nu = 0, \pm 1, \ldots, \pm n$ it follows that:

$$\sin kt = C \prod_{\nu=-n}^{n} \left( \sin t - \sin \frac{\nu \pi}{k} \right),$$

where the constant $C$ is determined by dividing through by $t$ and taking the limit as $t$ tends to zero, so that we have

$$k = C \prod_{\nu=-n}^{n} {}' \left( - \sin \frac{\nu \pi}{k} \right)$$

(where the dash attached to the product sign indicates the omission of the term corresponding to $\nu = 0$). If we now write $x$ instead of $kt$, we obtain

$$\sin x = k \sin x/k \prod_{\nu=-n}^{n} {}' \left( 1 - \frac{\sin x/k}{\sin \nu \pi/k} \right) = k \sin x/k \prod_{\nu=1}^{n} \left( 1 - \frac{\sin^2 x/k}{\sin^2 \nu \pi/k} \right),$$

where $n = \frac{1}{2}(k-1)$. Since

$$\lim_{k \to \infty} k \sin x/k = x, \qquad \lim_{k \to \infty} \frac{\sin x/k}{\sin \nu\pi/k} = \frac{x}{\nu\pi},$$

this yields after a "naive" passage to the limit,

**EULER's Product Formula:** $\sin x = x \prod_{1}^{\infty} \left(1 - \frac{x^2}{\nu^2\pi^2}\right)$; of course, this last step involving the proof of the convergence of the infinite product, needs to be, and can easily be, justified. $\qquad\qquad\qquad\qquad\qquad\qquad$ $\square$

If we now put $x := \frac{\pi}{2}$, we obtain after slight rearrangement

**WALLIS's Formula.**

$$\frac{\pi}{2} = \lim_{n \to \infty} \frac{2}{1} \cdot \frac{2}{3} \cdot \frac{4}{3} \cdot \frac{4}{5} \cdot \frac{6}{5} \cdot \frac{6}{7} \cdot \ldots \cdot \frac{2n}{2n-1} \cdot \frac{2n}{2n+1}.$$

It follows from this, for example, that

$$\sqrt{\pi} = \lim_{n \to \infty} \frac{2}{1} \cdot \frac{4}{3} \cdot \ldots \cdot \frac{2n}{2n-1} \cdot \frac{1}{\sqrt{n}};$$

which can also be expressed as an *asymptotic* equation for the binomial coefficient $\binom{2n}{n}$,

$$\binom{2n}{n} \simeq \frac{2^{2n}}{\sqrt{n\pi}}.$$

The monotonically increasing "WALLIS's sequence"

$$w_n := \frac{2^2 4^2 \cdot \ldots \cdot (2n)^2}{3^2 5^2 \cdot \ldots \cdot (2n-1)^2} \cdot \frac{1}{2n+1}$$

converges very poorly; an elementary computation shows that

$$\frac{1}{3}\frac{1}{n+1} < \frac{\pi}{2} - w_n < \frac{1}{2}\frac{1}{n}, \qquad \lim_{n \to \infty} n\left(\frac{\pi}{2} - w_n\right) = \frac{\pi}{8}.$$

With slightly more trouble it can be shown that

$$\frac{\pi}{2} - w_n = \frac{1}{4}w_n\left(\frac{1}{n} - \frac{3}{4}\frac{1}{n^2} + \frac{M_n}{n^3}\right),$$

where $M_n$ is a *bounded* sequence. For the modified WALLIS's sequence $\hat{w}_n := w_n(1 + \frac{1}{4n})$ we therefore have

$$\lim_{n \to \infty} n^2\left(\hat{w}_n - \frac{\pi}{2}\right) = \frac{3}{32}\pi.$$

The convergence of $w_n$ and $\hat{w}_n$ to $\pi/2$ thus involve error terms of the order of $1/n$ and $1/n^2$ respectively. The following table illustrates the slowness of the convergence:

| $n$ | $2w_n$ | $2\hat{w}_n$ |
|---|---|---|
| 10 | 3.067703807... | 3.144396403... |
| $10^2$ | 3.133787491... | 3.141621960... |
| $10^3$ | 3.140807746... | 3.141592948... |
| $10^4$ | 3.141514119... | 3.141592658... |
| $10^5$ | 3.141584800... | 3.141592655... |

the values in the last row being correct only to 4 and 8 decimal places respectively.

**4. EULER's Series for $\pi^2$, $\pi^4$,....** EULER in his *De summis serierum reciprocarum* (*Opera omnia*, Ser. 1, XIV, 73–86) succeeded in 1734 in deriving, from his product formula for the sine, the famous series

$$(*) \qquad \frac{\pi^2}{6} = \sum_1^\infty \frac{1}{\nu^2}, \quad \frac{\pi^4}{90} = \sum_1^\infty \frac{1}{\nu^4}, \quad \frac{\pi^6}{945} = \sum_1^\infty \frac{1}{\nu^6}, \quad \frac{\pi^8}{9450} = \sum_1^\infty \frac{1}{\nu^8}, \ldots ;$$

Jacob and John BERNOULLI had long sought in vain to find the sum of the series $1 + \frac{1}{4} + \frac{1}{9} + \frac{1}{16} + \cdots$. EULER obtained the formulae $(*)$ from the identity

$$(1) \qquad 1 - \frac{1}{3!}\pi^2 x^2 + \frac{1}{5!}\pi^4 x^4 - \cdots = \frac{\sin \pi x}{\pi x} = \prod_1^\infty \left(1 - \frac{x^2}{\nu^2}\right),$$

which holds, by virtue of 3.1(2) and his product formula, for all $x \in \mathbb{R}$, $x \neq 0$. His method was to compare coefficients of the powers of $x$ after expanding the product on the right. In his *Introductio in analysin infinitorum* he describes the process as follows (Chapter 10, §165):

"If $1 + Az + Bz^2 + Cz^3 + Dz^4 + \cdots = (1+\alpha z)(1+\beta z)(1+\gamma z)(1+\delta z)\ldots$ then *these factors*, whether *finite or infinite* in number, must *reproduce the expression* $1 + Az + Bz^2 + Cz^3 + Dz^4 + \cdots$ when actually multiplied out by one another." This gives him

$$A = \alpha + \beta + \gamma + \delta + \cdots, \quad B = \alpha\beta + \alpha\gamma + \alpha\delta + \beta\gamma + \beta\delta + \gamma\delta + \cdots,$$

etc. Application to (1) gives immediately $\frac{1}{3!}\pi^2 = \sum_1 \frac{1}{\nu^2}$ and then with this, similarly

$$\frac{1}{5!}\pi^4 = \frac{1}{2} \sum_{\mu \neq \nu}^\infty \frac{1}{\mu^2} \cdot \frac{1}{\nu^2} = \frac{1}{2} \sum_{\nu=1}^\infty \frac{1}{\nu^2} \left(\sum_{\mu=1}^\infty \frac{1}{\mu^2} - \frac{1}{\nu^2}\right)$$

$$= \frac{1}{2} \sum_{\nu=1}^\infty \frac{1}{\nu^2} \left(\frac{\pi^2}{6} - \frac{1}{\nu^2}\right),$$

which yields the second formula in (*). EULER shows by his method that every sum $\sum_1^\infty \frac{1}{\nu^{2k}}$ is a rational multiple of $\pi^{2k}$; more precisely

$$\sum_1^\infty \frac{1}{\nu^{2k}} = (-1)^{k-1} \frac{(2\pi)^{2k}}{2(2k)!} B_{2k}, \qquad k = 1, 2, \ldots ,$$

where $B_2, B_4, B_6, \ldots$ are the BERNOULLI numbers. Proofs of this general formula will be found in (Sansone and Gerretsen, *Lectures on the Theory of Functions of a Complex Variable*, P. Noordhoff–Groningen, 1960, p. 143).

The convergence of the series $\sum_1 \frac{1}{\nu^k}$ is very poor: about 100 million terms of EULER's first series are needed to give $\pi^2/6$ correctly to the first seven places of decimals.

## 5. The WEIERSTRASS Definition of $\pi$. The integral formula

$$(1) \qquad\qquad \int_{S^1} \frac{dz}{z} = 2\pi i$$

is fundamental for the theory of functions. It can be obtained immediately, if $S^1$ is described by $z(\varphi) := e^{i\varphi}$, $0 \le \varphi \le 2\pi$, and one notes that $z'(\varphi) = iz(\varphi)$ so that:

$$\int_{S^1} \frac{dz}{z} = \int_0^{2\pi} \frac{z'(\varphi)}{z(\varphi)} d\varphi = \int_0^{2\pi} i\, d\varphi = 2\pi i.$$

WEIERSTRASS, in 1841 in his proof of the theorem on the expansion of a function in a LAURENT series, calculated the integral (1) as follows (*Math. Werke* 1, pp. 52–53). He defines $S^1$ by

$$z(\lambda) := \frac{1 + i\lambda}{1 - i\lambda}, \qquad -\infty < \lambda < \infty;$$

this is the rational parametrization of $S^1$ discussed in 3.5.4. Since $z'(\lambda) = \frac{2i}{(1-i\lambda)^2}$, it follows that $\frac{z'(\lambda)}{z(\lambda)} = \frac{2i}{1+\lambda^2}$ and consequently, if one also notes that

$$\int_1^\infty \frac{d\lambda}{1 + \lambda^2} = \int_0^1 \frac{d\tau}{1 + \tau^2}$$

(using the substitution $\lambda := \tau^{-1}$), we have:

$$\int_{S^1} \frac{dz}{z} = \int_{-\infty}^\infty \frac{z'(\lambda)}{z(\lambda)} d\lambda = 2i \int_{-\infty}^\infty \frac{d\lambda}{1 + \lambda^2} = 4i \int_0^\infty \frac{d\lambda}{1 + \lambda^2} = 8i \int_0^1 \frac{d\lambda}{1 + \lambda^2}.$$

If the formula (1) is available, then we also know that:

$$(2) \qquad\qquad \pi = \int_{-\infty}^\infty \frac{d\lambda}{1 + \lambda^2} = 4 \int_0^1 \frac{d\lambda}{1 + \lambda^2}.$$

This identity is pointed out by WEIERSTRASS as a possible definition for $\pi$.

More generally $\int_0^x \frac{d\lambda}{1+\lambda^2} = \text{arc tan } x$, so that (2) is nothing more than the statement that arc tan $1 = \pi/4$, or in other words, $\tan \pi/4 = 1$, which is clear if one knows that $\sin \frac{\pi}{4} = \cos \frac{\pi}{4} \; (= \frac{1}{2}\sqrt{2})$.

**6. The Irrationality of $\pi$ and Its Continued Fraction Expansion.**
The statement that the circumference and diameter of a circle are incommensurable had already been asserted by ARISTOTLE, but the first proof of the irrationality of $\pi$ was given in 1766 by Johann Heinrich LAMBERT (1728–1777) in his *Vorläufige Kenntnisse für die, so die Quadratur und Rectification des Circuls suchen* (*Werke* 1, 194–212), a sort of manual for would-be circle squarers, written in highly original language. The proof was based on the theory of continued fractions. He found the infinite continued fraction

$$\tan z = \cfrac{z}{1 - \cfrac{z^2}{3 - \cfrac{z^2}{5 - \cfrac{z^2}{7 - \cdots}}}}$$

and deduced from it the irrationality of $\tan(z)$ for all real rational arguments $z \neq 0$, and in particular the result that $\pi \notin \mathbb{Q}$ since $\tan \frac{1}{4}\pi = 1$. However LAMBERT's proof is not completely rigorous because it lacks a lemma on the irrationality of certain infinite continued fractions (having particularly good convergence). This lemma was proved in 1806 by Adrien-Marie LEGENDRE (1752–1833) in the 6th edition of his *Elements de géometric, Note IV*. LEGENDRE also shows there that

$\pi^2$ *is irrational.*

LAMBERT's continued fraction for $\sqrt{q} \tan \sqrt{q}$ is, in fact, irrational, by virtue of LEGENDRE's lemma, for all $q \in \mathbb{Q}$, $q \neq 0$, and therefore $\pi = \sqrt{q}$; $q \in \mathbb{Q}$ is impossible, because $\tan \pi = 0$. $\qquad\square$

LAMBERT's and LEGENDRE's work on the subject is readily accessible in the article by RUDIO mentioned in the introduction to this chapter. Perhaps the simplest modern proof of the irrationality of $\pi^2$ runs as follows. We introduce the polynomial $p_n(x) := \frac{1}{n!}x^n(1-x)^n$, $n \geq 1$, and begin by noting that

1) $0 < p_n(x) < \frac{1}{n!}$ for $0 < x < 1$; $p_n^{(\nu)}(0)$, $p_n^{(\nu)}(1) \in \mathbb{Z}$ for all $\nu$.

2) For $P_n(x) := b^n\{\pi^{2n}p_n(x) - \pi^{2n-2}p_n''(x) + \pi^{2n-4}p_n^{(4)}(x) - \cdots + (-1)^n p_n^{(2n)}(x)\}$ we have

$$\frac{d}{dx}(P_n'(x)\sin \pi x - \pi P_n(x)\cos \pi x) = b^n \pi^{2n+2} p_n(x)\sin \pi x, \quad b \in \mathbb{R}.$$

The inequalities in 1) are trivial; the statement that $p_n^{(\nu)}(1)$ is an integer follows, by induction on $n$, from the equation $p_n'(x) = (1 - 2x)p_{n-1}(x)$; the statement 2) is easily proved by first noting that the derivative on the left is simply $(P_n''(x) + \pi^2 P_n(x))\sin \pi x$.

If now $\pi^2$ were rational, say $\pi^2 = a/b$ with $a, b \geq 1$ natural numbers, then the values of $P_n(0)$ and $P_n(1)$ formed with this $b$ would, by 1), be rational integers. Consequently we should have, by 2), since $b^n\pi^{2n+2} = a^n\pi^2$

$$\pi a^n \int_0^1 p_n(x)\sin \pi x \, dx = [\pi^{-1}P_n'(x)\sin \pi x - P_n(x)\cos \pi x]_0^1$$

$$= P_n(0) + P_n(1) \in \mathbb{Z}.$$

On the other hand, since $0 < \sin \pi x \leq 1$ for $0 < x < 1$, we deduce from 1) that:

$$0 < \pi a^n \int_0^1 p_n(x)\sin \pi x \, dx < \pi \frac{a^n}{n!} < 1 \quad \text{for large } n,$$

since $\lim_{n \to \infty} \frac{a^n}{n!} = 0$ for every $a \in \mathbb{R}$, in view of the convergence of the exponential series.

Thus, for all large enough $n$, we should have

$$0 < P_n(0) + P_n(1) < 1 \quad \text{in contradiction to} \quad P_n(0) + P_n(1) \in \mathbb{Z}. \qquad \square$$

This proof is based on an idea of I. NIVEN: *A simple proof that $\pi$ is irrational*, in *Bull. Amer. Math. Soc.*, 53, 1947, 509. The extension to $\pi^2$ is due to Y. IWAMOTO: *A proof that $\pi^2$ is irrational*, in *J. Osaka Inst. Sci. Tech.*, 1, 1949, 147–148. The reader should also compare the proof given in the book by G.H. HARDY and E.M. WRIGHT *An introduction to the theory of numbers*, 3rd edn., Oxford, Clarendon Press, 1954, especially p. 47. Also compare a paper by J. Hančl, "A simple proof of the irrationality of $\pi^4$," *Amer. Math. Monthly* 93 (1986), 374–375. $\qquad \square$

The following two continued fractions among others exist for the number $\pi$

$$\frac{\pi}{4} = \cfrac{1}{1 + \cfrac{1^2}{2 + \cfrac{3^2}{2 + \cfrac{5^2}{2 + 7^2}}}}, \qquad \pi = 3 + \cfrac{1}{7 + \cfrac{1}{15 + \cfrac{1}{1 + \cfrac{1}{292 + 1}}}}$$

The expansion on the left was found in 1656 by Lord BROUNCKER (1620–1684), the first President of the Royal Society, by transforming WALLIS's product. EULER, in §369 of his *Introductio,* used the LEIBNIZ series instead. The expansion on the right is the so-called *regular* continued fraction for $\pi$. Every positive real number has a unique representation in the form of a regular continued fraction, in which only positive integers appear, and in which all the "numerators in the denominators" are 1. There is no known law governing the successive terms in the regular continued fraction for $\pi$. The successive integers $3, 7, 15, 1, 292, \ldots$ shown are simply calculated from the decimal representation of $\pi$ by using the continued fraction algorithm.

BROUNCKER's continued fraction has a very poor convergence. Regular continued fractions on the other hand have excellent convergence. The first few convergents to $\pi$, for example, give the approximations $3$, $\frac{22}{7}$, $\frac{333}{106}$, $\frac{355}{113}$, $\frac{103993}{33102}$; the approximation in 1.3 given by ZU CHONG-ZHI is thus the fourth convergent. For further details on the relation between $\pi$ and continued fractions we refer the reader to the two volume work by O. PERRON: *Die Lehre von den Kettenbrüchen,* Stuttgart, Teubner Verlag, 3rd edn., 1954-1957. The approximation of $\pi$ and $\pi^2$ by rational numbers $p/q$ has some fundamental limitations. For example, a result of M. MIGNOTTE, *Approximations rationelles de $\pi$ et quelques autres nombres,* Bull. Soc. Math. France, Mem. 37, 121–132 shows that:

$$\left| \pi - \frac{p}{q} \right| > \frac{1}{q^{20.6}} \quad \text{for} \quad q > 1, \qquad \left| \pi^2 - \frac{p}{q} \right| > \frac{1}{q^{18}} \quad \text{for} \quad q \gg 0.$$

**7. Transcendence of $\pi$.** The problem of constructing a square equal in area to a given circle by means of a ruler and compass construction had already engaged the attention of the ancient Greeks. This is the problem usually referred to as "squaring the circle." It is shown in Algebra that a real number can be constructed by these means if and only if it lies in a finite extension of the field $\mathbb{Q}$ formed by successive adjunction of square roots. In particular therefore the numbers constructible by ruler and compass are at most those which are *algebraic* (over $\mathbb{Q}$), that is to say which annihilate a polynomial $p \in \mathbb{Q}[X] \setminus \{0\}$.

The problem of squaring the circle is equivalent to the question of whether $\pi$ is constructible by ruler and compass. In view of the foregoing remarks, $\pi$ would then have to be an algebraic number. EULER, LAMBERT and LEGENDRE were already of the opinion that this is not so. Thus LEGENDRE at the end of his paper on the irrationality of $\pi^2$ says quite clearly (see RUDIO, p. 59) "It is probable that $\pi$ is not even contained among the algebraic irrationals, in other words it cannot be the root of an algebraic equation with a finite number of terms, and rational coefficients. However it seems difficult to prove this theorem rigorously."

Numbers which are not algebraic are called "transcendental" (omnem rationem transcendunt). Thus LEGENDRE in 1806 conjectured that $\pi$ is transcendental. This was an extraordinarily bold conjecture, because at that time, no one even knew that there were such things as transcendental numbers (in contrast to irrational numbers, such as, for example, $\sqrt{2}$, whose existence had been known to the Greeks). It was not until 1844 that Joseph LIOUVILLE (1809–1882) first proved that all (irrational) numbers having "very good" rational approximations, such as for example the number

$$10^{-1!} + 10^{-2!} + 10^{-3!} + \cdots = 0.11000\,10000 \cdots ,$$

are transcendental. In 1874 Georg CANTOR (1845–1918) gave his sensational proof, using an enumerative argument, that there are only countably many algebraic numbers, but uncountably many transcendental numbers (see on this, for example, O. PERRON: *Irrationalzahlen*, Berlin, de Gruyter, 1960, 174–181).

The great breakthrough in the theory of transcendental numbers came in 1873 when the French mathematician Charles HERMITE (1822–1901) developed methods by which he was able to prove that *the number $e$ is transcendental*. By a refinement of HERMITE's argument the German mathematician Carl Louis Ferdinand von LINDEMANN (1852–1939) who had taught HILBERT and HURWITZ in Königsberg, and subsequently went to Munich in 1893, proved in 1882 in a short paper "Über die Zahl $\pi$," published in *Math. Ann.*, 20, 213–225, his famous theorem that:

$\pi$  *is transcendental.*

In this way the thousand-year-old question about the quadrature of the circle was finally answered in the negative. Oblivious to this fact, amateur mathematicians still try to tackle this problem as they did before; they often find good approximation processes, and in most cases it is difficult to convince them that their "solution" does not contradict the transcendentality of $\pi$.                                                                        □

LINDEMANN himself seems to have been quite surprised at having been able to solve a thousand-year-old problem. Thus we read in the introduction to his paper (p. 213): "Man wird sonach die Unmöglichkeit der Quadratur des Kreises darthun, wenn man nachweist, dass *die Zahl $\pi$ überhaupt nicht Wurzel einer algebraischen Gleichung irgend welchen Grades mit rationalen Coefficienten sein kann.* Den dafür nöthigen Beweis zu erbringen, ist im Folgenden versucht worden." [The impossibility of the quadrature of the circle will thus have been established when one has proved that *the number $\pi$ can never be the root of any algebraic equation of any degree with rational coefficients.* We seek to prove this in the following pages.] The propositions of HERMITE and LINDEMANN are included in the following general theorem.

**The LINDEMANN–WEIERSTRASS Theorem** (see WEIERSTRASS: Zu Lindemann's Abhandlung: "Über die Ludolph'sche Zahl," in *Math. Werke*, 2, 341–462, particularly 360–361). *Let $c_1, \ldots, c_n \in \mathbb{C}$ be pairwise distinct algebraic numbers belonging to $\mathbb{C}$. Then there exists no equation*

$a_1 e^{c_1} + \cdots + a_n e^{c_n} = 0$ *in which* $a_1, \ldots, a_n$ *are algebraic numbers and are not all equal to zero.*

If, in this theorem, we put $n := 2$, $c_1 := c$, $c_2 := 0$, we obtain the result: *for every algebraic number* $c \in \mathbb{C}^\times$ *the number* $a := e^c$ *is transcendental.*

The case $c := 1$ proves the transcendence of $e$, and since $1 = e^{2\pi i}$ the transcendence of $\pi$ follows as well.                    □

Meanwhile it is also known that $e^\pi = i^{-2i}$ is transcendental (GELFOND, 1929). As for the number $\pi^e$ nothing is known for certain, and on the whole our knowledge about transcendental numbers is still extremely limited. As $e$ is transcendental, the numbers $e\pi$ and $e + \pi$ cannot both be algebraic; but it is still not known whether $e\pi$ or $e + \pi$ is rational.

## FURTHER READING

[1] BECKMANN, PETER, A History of Pi. Boulder, Col.: Golem, 1970

[2] BORWEIN, J. and PETER BORWEIN, Pi and the AGM, New York: John Wiley, 1987

[3] DAVIS, P.J., The Long, Long Trail of Pi. In The Lore of Large Numbers, Chapter 17. New York: Random House, 1961

[4] DIEUDONNÉ, J., Abrégé d'histoire des mathématiques I, Paris: Hermann, 1978 (especially pp. 283 ff)

[5] SCHNEIDER, Th., Einführung in die transzendenten Zahlen, Grundl. Math. Wiss., Springer-Verlag, 1957

[6] SIEGEL, C.L., Transzendente Zahlen, BI Hochschultaschenbuch 137, Mannheim, 1967

# 6

# The $p$-Adic Numbers

*J. Neukirch*

## §1. NUMBERS AS FUNCTIONS

The $p$-adic numbers were invented at the beginning of the twentieth century
by the German mathematician Kurt HENSEL (1861–1941). The aim was to
make the methods of power series expansions, which play such a dominant
role in the theory of functions, available to the theory of numbers as well.
The idea sprang from the observation that numbers behave in many ways
just like functions, and in a certain sense numbers may also be regarded as
functions on a topological space.

To explain this, we begin by considering polynomials

$$f(z) = a_0 + a_1 z + \cdots + a_n z^n$$

with complex coefficients $a_i \in \mathbb{C}$, which we can regard as functions on the
complex plane. This characteristic property can be formulated in purely
algebraic terms in the following way. Let $a \in \mathbb{C}$ be a point of the complex
plane. The totality of all functions belonging to the polynomial ring $\mathbb{C}[z]$
that vanish at the point $a$ forms a maximal ideal of $\mathbb{C}[z]$, namely, the prime
ideal

$$\mathfrak{p} = (z - a) = \{(z - a)g(z) \mid g(z) \in \mathbb{C}[z]\}.$$

Thus, there is a one-to-one correspondence between points of the complex
plane and maximal ideals $\mathfrak{p}$ of $\mathbb{C}[z]$.

We denote the set of all these ideals by

$$X = \mathrm{Max}(\mathbb{C}[z]).$$

If we regard $X$ as a new space, we can interpret the elements $f = f(z)$ of
the ring $\mathbb{C}[z]$ as functions on $X$, by defining the *value* of $f$ at the point
$\mathfrak{p} = (z - a) \in X$, as the residue class

$$f(\mathfrak{p}) := f \bmod \mathfrak{p}$$

in the residue class field $\kappa(\mathfrak{p}) = \mathbb{C}[z]/\mathfrak{p}$. (This definition is justified because
of the canonical isomorphism

$$\mathbb{C}[z]/\mathfrak{p} \cong \mathbb{C},$$

by which the residue class $f \bmod \mathfrak{p}$ is mapped onto $f(a)$.) The topology on $\mathbb{C}$ cannot be carried algebraically onto $X$; all that can be defined by algebraic means are the point sets defined by the equations

$$f(z) = 0.$$

These finite sets are defined to be closed.

The promised interpretation of numbers as functions is now based on an analogy between the ring $\mathbb{Z}$ and the ring $\mathbb{C}[z]$, in which the prime numbers $p \in \mathbb{Z}$ are the analogues of the prime elements $z - a \in \mathbb{C}[z]$ and the elements $f \in \mathbb{Z}$, the analogues of the elements $f(z) \in \mathbb{C}[z]$. Accordingly we form the set

$$X = \operatorname{Max}(\mathbb{Z})$$

of all maximal ideals $(p) = p\mathbb{Z}$ of $\mathbb{Z}$, that is, the set of all prime numbers $p$. We regard $X$ as a topological space, by defining the closed sets in $X$ to be its finite subsets. For the elements $f \in \mathbb{Z}$, which are now to take over the role of functions on $X$, we define their "value" at the point $p \in X$ to be

$$f(p) := f \bmod p.$$

$f(p)$ is an element of the residue class field $\kappa(p) = \mathbb{Z}/p\mathbb{Z} = \mathbb{F}_p$, and thus the values of $f$ do not all lie in one and the same field.

This way of looking at things at once raises the further question of whether, in addition to the "value" of the number at $p$ one could not also define the higher derivatives of $f$ in some meaningful way. In the case of the polynomials $f(x) \in \mathbb{C}[z]$ the higher derivatives at the point $z = a$ are given (almost) by the coefficients of the expansion

$$f(z) = a_0 + a_1(z - a) + \cdots + a_n(z - a)^n$$

and more generally, in the case of rational functions $f(z) = \frac{g(z)}{h(z)} \in \mathbb{C}(z)$ by the Laurent series expansion

$$f(z) = \sum_{\nu = -m}^{\infty} a_\nu (z - a)^\nu.$$

We are now led to the concept of the $p$-adic number by observing that every rational number $f \in \mathbb{Q}$ can be given an analogous expansion with respect to every prime element $p$ of $\mathbb{Z}$. First of all, every natural number $f \in \mathbb{N}$ possesses a $p$-*adic expansion*

$$f = a_0 + a_1 p + \cdots + a_n p^n,$$

in which the coefficients $a_i$ lie in $\{0, 1, \ldots, p - 1\}$, that is, in a fixed representative system of the "field of values." This representation is clearly unique. It is found by repeatedly dividing by $p$, using the algorithm:

$$
\begin{aligned}
f &= a_0 &&+ pf_1 \\
f_1 &= a_1 &&+ pf_2 \quad \text{etc.} \\
f_{n-1} &= a_{n-1} + pf_n \\
f_n &= a_n.
\end{aligned}
$$

In these equations $a_i \in \{0, 1, \ldots, p-1\}$ denotes the representative of $f_i \bmod p \in \mathbb{Z}/p\mathbb{Z}$. In concrete cases the number $f$ is often simply denoted by the $a_0, a_1 \ldots a_n$, thus for example

$$
\begin{aligned}
216 &= 0.0011011 &&\text{(2-adic)} \\
216 &= 0.0022 &&\text{(3-adic)} \\
216 &= 1.331 &&\text{(5-adic)}
\end{aligned}
$$

If we now wish to find a $p$-adic expansion for negative and even for fractional numbers, then we are forced to consider infinite series of the form

$$
\sum_{\nu=-m}^{\infty} a_\nu p^\nu .
$$

This is initially meant in a purely formal sense, that is, $\sum_{\nu=-m}^{\infty} a_\nu p^\nu$ simply denotes the sequence of the partial sums

$$
s_n = \sum_{\nu=-m}^{n-1} a_\nu p^\nu , \qquad n = 1, 2, \ldots .
$$

**Definition.** *Let $p$ be a fixed prime. A $p$-adic number is a formal infinite series*

$$
a_{-m}p^{-m} + \cdots + a_{-1}p^{-1} + a_0 + a_1 p + a_2 p^2 + \cdots ,
$$

*in which $a_i \in \{0, 1, \ldots, p-1\}$.*
   *The $p$-adic integers are the series*

$$
a_0 + a_1 p + a_2 p^2 + \cdots .
$$

*The complete set of all $p$-adic numbers is denoted by $\mathbb{Q}_p$, and that of all $p$-adic integers by $\mathbb{Z}_p$.*

*The $p$-adic expansion of an arbitrary rational number $f$ results from the following theorem on residue classes in $\mathbb{Z}/p^n\mathbb{Z}$.*

**Theorem 1.** *The residue classes $a \bmod p^n \in \mathbb{Z}/p^n\mathbb{Z}$ are expressible uniquely in the form*

$$
a \equiv a_0 + a_1 p + a_2 p^2 + \cdots + a_{n-1}p^{n-1} \bmod p^n
$$

*where $0 \le a_i < p$ for $i = 0, \ldots, n-1$.*

**Proof** (by induction). The complete theorem is obviously true for $n = 1$. If we assume the statement to be true for $n - 1$, then we have an unique representation

$$a = a_0 + a_1 p + a_2 p^2 + \cdots + a_{n-2} p^{n-2} + g p^{n-1}$$

with some integer $g$. If $g = a_{n-1} \bmod p$ with $0 \le a_{n-1} < p$, then $a_{n-1}$ is uniquely defined, and the asserted congruence therefore holds.     □

Every integer $f$, and more generally every rational number $f = g/h$ whose denominator $h$ is not divisible by $p$, now defines a sequence of residue classes

$$\bar{s}_n = f \bmod p^n \in \mathbb{Z}/p^n\mathbb{Z}, \qquad n = 1, 2, \ldots$$

and by the theorem above, we have

$$\begin{aligned}
\bar{s}_0 &= a_0 \bmod p \\
\bar{s}_1 &= a_0 + a_1 p \bmod p^2 \\
\bar{s}_2 &= a_0 + a_1 p + a_2 \bmod p^3 \\
&\ \vdots \quad \vdots
\end{aligned}$$

with uniquely defined and unchanging coefficients

$$a_0, a_1, a_2, \ldots \in \{0, 1, \ldots, p-1\}.$$

The number sequence

$$s_n = a_0 + a_1 p + a_2 p^2 + \cdots + a_{n-1} p^{n-1}, \qquad n = 1, 2, \ldots$$

defines a $p$-adic integer

$$\sum_{\nu=0}^{\infty} a_\nu p^\nu \in \mathbb{Z}_p.$$

We call this the *$p$-adic expansion* (or $p$-adic representation) of $f$. If, more generally, $f \in \mathbb{Q}$ is an arbitrary rational number, we write

$$f = \frac{g}{h} p^{-m} \quad \text{with} \quad (gh, p) = 1,$$

and if

$$a_0 + a_1 p + a_2 p^2 + \cdots \in \mathbb{Z}_p$$

is the $p$-adic expansion of $g/h$, then we assign to $f$ the $p$-adic number

$$a_0 p^{-m} + a_1 p^{-m+1} + \cdots + a_m + a_{m+1} p + \cdots \in \mathbb{Q}_p$$

as its $p$-adic expansion.

In this way we obtain a canonical mapping

$$\mathbb{Q} \to \mathbb{Q}_p$$

which maps $\mathbb{Z}$ into $\mathbb{Z}_p$, and which by virtue of the uniqueness statement in Theorem 1, is injective. We now identify $\mathbb{Q}$ with its image in $\mathbb{Q}_p$, so that $\mathbb{Q} \subseteq \mathbb{Q}_p$ and $\mathbb{Z} \subseteq \mathbb{Z}_p$, thus obtaining for every rational number $f \in \mathbb{Q}$ an equation

$$f = \sum_{\nu=-m}^{\infty} a_\nu p^\nu$$

and thereby establishing the analogue which we sought of the power series expansion in the theory of functions.

*Examples.* a) $-1 = (p-1) + (p-1)p + (p-1)p^2 + \cdots$.
   We have

$$-1 = (p-1) + (p-1)p + \cdots + (p-1)p^{n-1} - p^n,$$

also

$$-1 \equiv (p-1) + (p-1)p + \cdots + (p-1)p^{n-1} \bmod p^n.$$

b) $\dfrac{1}{p-1} = 1 + p + p^2 + \cdots$.
   We have

$$1 = (1 + p + \cdots + p^{n-1})(1-p) + p^n,$$

also

$$\frac{1}{1-p} \equiv 1 + p + \cdots + p^{n-1} \bmod p^n.$$

Addition and multiplication can be defined for $p$-adic numbers, whereby $\mathbb{Z}_p$ becomes a ring whose quotient field is $\mathbb{Q}_p$. A straightforward attempt to define the sum and product by adopting the usual "carry" rules to which we are accustomed in ordinary decimal operations leads, however, to some significant complications. These disappear if we make use of a slightly different representation of the $p$-adic numbers $f = \sum_{\nu=0}^{\infty} a_\nu p^\nu$ in which we regard them, not as a sequence of the integer partial sums

$$s_n = \sum_{\nu=0}^{n-1} a_\nu p^\nu \in \mathbb{Z},$$

but as a sequence of the residue classes

$$\bar{s}_n = s_n \bmod p^n \in \mathbb{Z}/p^n\mathbb{Z}.$$

The terms of this sequence lie in different rings $\mathbb{Z}/p^n\mathbb{Z}$, but they are all related to one another through the canonical projections

$$\mathbb{Z}/p\mathbb{Z} \xleftarrow{\lambda_1} \mathbb{Z}/p^2\mathbb{Z} \xleftarrow{\lambda_2} \mathbb{Z}/p^3\mathbb{Z} \xleftarrow{\lambda_3} \cdots$$

and the relation $\lambda_n(\bar{s}_{n+1}) = \bar{s}_n$ holds.

We now consider, in the direct product

$$\prod_{n=1}^{\infty} \mathbb{Z}/p^n\mathbb{Z} = \{(x_n)_{n\in\mathbb{N}} \mid x_n \in \mathbb{Z}/p^n\mathbb{Z}\}$$

all those elements $(x_n)_{n\in\mathbb{N}}$ having the property

$$\lambda_n(x_{n+1}) = x_n \quad \text{for all} \quad n = 1, 2, \dots.$$

This set is called the *inverse limit* of the ring $\mathbb{Z}/p^n\mathbb{Z}$ and is denoted by

$$\varprojlim_n \mathbb{Z}/p^n\mathbb{Z} = \left\{(x_n) \in \prod_{n=1}^{\infty} \mathbb{Z}/p^n\mathbb{Z} \mid \lambda_n(x_{n+1}) = x_n, \ n = 1, 2, \dots\right\}.$$

The modified representation of the $p$-adic numbers to which we referred earlier is now obtained through the following.

**Theorem 2.** *If we associate with each p-adic integer*

$$f = (s_n)_{n\in\mathbb{N}}, \qquad s_n = \sum_{\nu=0}^{n-1} a_\nu p^\nu,$$

*the sequence* $(\bar{s}_n)_{n\in\mathbb{N}}$ *of residue classes* $\bar{s}_n = s_n \bmod p^n \in \mathbb{Z}/p^n\mathbb{Z}$, *we obtain a bijection*

$$\mathbb{Z}_p \xrightarrow{\sim} \varprojlim_n \mathbb{Z}/p^n\mathbb{Z}. \qquad\qquad \square$$

The proof is an immediate consequence of Theorem 1. The projective limit $\varprojlim_n \mathbb{Z}/p^n\mathbb{Z}$ now has the advantage of being a ring, in a direct fashion, namely a subring of the direct product $\prod_{n=1}^{\infty} \mathbb{Z}/p^n\mathbb{Z}$, in which addition and multiplication are defined componentwise. If we identify $\mathbb{Z}_p$ with $\varprojlim_n \mathbb{Z}/p^n\mathbb{Z}$, then $\mathbb{Z}_p$ also becomes a ring, *the ring of p-adic integers*.

As every element $f \in \mathbb{Q}_p$ has a representation

$$f = p^{-m}g$$

with $g \in \mathbb{Z}_p$, addition and multiplication in $\mathbb{Z}_p$ can be extended to $\mathbb{Q}_p$, and $\mathbb{Q}_p$ becomes the quotient field of $\mathbb{Z}_p$.

In $\mathbb{Z}_p$ we were able to rediscover the rational integers $a \in \mathbb{Z}$ in the guise of those $p$-adic numbers whose expansions $a_0 + a_1 p + a_2 p^2 + \cdots$ were derived from the congruences

$$a \equiv a_0 + a_1 p + \cdots + a_{n-1}p^{n-1} \bmod p^n,$$

$0 \le a_i < p$. Through the identification

$$\mathbb{Z}_p = \varprojlim_n \mathbb{Z}/p^n\mathbb{Z}$$

$\mathbb{Z}$ therefore goes over into the set of tuples

$$(a \bmod p, a \bmod p^2, a \bmod p^3, \ldots)$$

and thus becomes a subring of $\mathbb{Z}_p$. Similarly $\mathbb{Q}$ becomes a subfield of $\mathbb{Q}_p$, the field of $p$-adic numbers.

In §3 we shall give a new definition of the $p$-adic numbers closely imitating that of the real numbers, which will bring out in an entirely straightforward way the ring and field structure of $\mathbb{Z}_p$ and $\mathbb{Q}_p$.

Corresponding to the familiar results on the decimal representation of rational numbers, we have, for $p$-adic numbers the following expansion theorem.

**Theorem 3.** *A p-adic number* $a = \sum_{\nu=-m}^{\infty} a_\nu p^\nu \in \mathbb{Q}_p$ *is rational if and only if the sequence of digits* $(a_\nu)$ *is periodic from some point onwards (that is, a finite number of digits before the beginning of the first period is allowed).*

**Proof.** We may obviously assume that $m = 0$ and $a_0 \neq 0$. Let the sequence of digits $(a_\nu)$ be periodic, that is to say, of the form

$$(a_0, a_1, \ldots) = (b_0, b_1, \ldots, b_{h-1}, \overline{c_0, c_1, \ldots, c_{n-1}}),$$

where the line above the letters $c$ indicates the principal period. We write

$$b = b_0 + b_1 p + \cdots + b_{h-1} p^{h-1}$$
$$c = c_0 + c_1 p + \cdots + c_{n-1} p^{n-1},$$

so that

$$a = b + cp^h (1 + p^n + p^{2n} + \cdots) = b + c \frac{p^h}{1 - p^n} \in \mathbb{Q}.$$

Conversely let $a$ be rational. To prove the periodicity of the $p$-adic representation of $a$, it suffices to bring $a$ into the above form, namely

$$a = \frac{b(p^n - 1) - cp^h}{p^n - 1}$$

with $b, c$ being integers such that

$$0 \leq b < p^h, \qquad 0 \leq c < p^n.$$

For we then have

$$b = b_0 + b_1 p + \cdots + b_{h-1} p^{h-1}, \qquad 0 \leq b_i < p,$$
$$c = c_0 + c_1 p + \cdots + c_{n-1} p^{n-1}, \qquad 0 \leq c_i < p,$$

and the substitution of these $p$-adic representations gives us, by the argument above, the (non-periodic) pre-period $b_0, b_1, \ldots, b_{h-1}$ and the principal period $c_0, c_1, \ldots, c_{n-1}$.

Since $m = 0$, the denominator $f$ of $a$ is prime to $p$ and thus $p^n \equiv 1 \bmod f$ for a suitable $n$. We can therefore write

$$a = \frac{g}{p^n - 1}$$

with $g \in \mathbb{Z}$. We choose a power $p^h$ such that

$$0 \le g < p^h \quad \text{or} \quad -p^h \le g < 0$$

depending on whether $a \ge 0$ or $a < 0$. Since $(p^n - 1, p^h) = 1$, we can put

$$g = b(p^n - 1) - cp^h$$

with $b, c \in \mathbb{Z}$, and at the same time prescribe that $c$ shall belong to any arbitrarily specified system of representatives $\bmod(p^n - 1)$. We stipulate

$$0 \le c \le p^n - 2 \quad \text{or} \quad 1 \le c \le p^n - 1,$$

depending on whether $a \ge 0$ or $a < 0$. In both cases $0 \le c < p^n$, as required, and it follows from

$$b(p^n - 1) = g + cp^h$$

in both cases that $0 \le b < p^h$ as required.                                            □

## §2. The Arithmetic Significance of the $p$-Adic Numbers

Despite being colored by their function theoretic origin, the $p$-adic numbers fulfill their true destiny in the realms of arithmetic, and indeed in one of its classical heartlands, the theory of Diophantine equations. A *Diophantine problem* is one in which we are given an equation

$$F(x_1, \ldots, x_n) = 0$$

where $F$ is a given polynomial in one or more variables, $x_1, \ldots, x_n$ and are asked for its solutions in integers. This difficult problem can be attacked by weakening the question and considering instead the set of congruences for all $m$:

$$F(x_1, \ldots, x_n) = 0 \bmod m$$

or, what amounts to the same thing because of the Chinese Remainder Theorem, the set of congruences

$$F(x_1, \ldots, x_n) \equiv 0 \bmod p^\nu$$

for all prime powers. We might hope that from the existence or nonexistence of solutions to the congruences, we might be able to draw corresponding conclusions about the original equation. For a fixed prime $p$, the infinite set of these congruences can now, with the help of the $p$-adic numbers, again be expressed as a single equation. This comes from the following.

**Theorem 4.** *Let $F(x_1, \ldots, x_n)$ be a polynomial whose coefficients are rational integers and $p$ a fixed prime. The congruence*

$$F(x_1, \ldots, x_n) \equiv 0 \bmod p^\nu$$

*is solvable for arbitrary $\nu \geq 1$ if and only if the equation*

$$F(x_1, \ldots, x_n) = 0$$

*is solvable in $p$-adic integers.*

**Proof.** We interpret the ring $\mathbb{Z}_p$, as in §1, as the inverse limit

$$\mathbb{Z}_p = \varprojlim_\nu \mathbb{Z}/p^\nu \mathbb{Z} \subseteq \prod_{\nu=1}^\infty \mathbb{Z}/p^\nu \mathbb{Z}.$$

The equation $F = 0$ factorizes, in the ring on the right, into components over the individual rings $\mathbb{Z}/p^\nu \mathbb{Z}$, and thus into the congruences

$$F(x_1, \ldots, x_n) \equiv 0 \bmod p^\nu.$$

If now

$$(x_1, \ldots, x_n) = (x_1^{(\nu)}, \ldots, x_n^{(\nu)})_{\nu \in \mathbb{N}} \in \mathbb{Z}_p^n,$$

$(x_i^{(\nu)})_{\nu \in \mathbb{N}} \in \mathbb{Z}_p = \varprojlim_\nu \mathbb{Z}/p^\nu \mathbb{Z}$, is a $p$-adic solution of $F(x_1, \ldots, x_n) = 0$, then the congruences are solved by

$$F(x_1^{(\nu)}, \ldots, x_n^{(\nu)}) \equiv 0 \bmod p^\nu, \qquad \nu = 1, 2, \ldots.$$

Conversely, let us suppose that for every $\nu \geq 1$ we are given a solution $(x_1^{(\nu)}, \ldots, x_n^{(\nu)})$ of the congruence

$$F(x_1, \ldots, x_n) \equiv 0 \bmod p^\nu.$$

If the elements $(x_i^{(\nu)})_{\nu \in \mathbb{N}} \in \prod_{\nu=1}^\infty \mathbb{Z}/p^\nu \mathbb{Z}$ aleady lay in $\varprojlim_\nu \mathbb{Z}/p^\nu \mathbb{Z}$ for all $i = 1, \ldots, n$, then we would have a $p$-adic solution of the equation $F = 0$. Since this is not automatically the case, we shall form the sequence $(x_1^{(\nu)}, \ldots, x_n^{(\nu)})$ a subsequence that meets our wishes. To keep the notation simple we shall deal only with the case of one variable ($n = 1$) and write $x_\nu$ for $x_1^{(\nu)}$. The general case can be proved in exactly the same way. As $\mathbb{Z}/p\mathbb{Z}$

is finite, there are infinitely many terms of $x_\nu$ which are congruent modulo $p$ to a fixed element $y_1 \in \mathbb{Z}/p\mathbb{Z}$. We can therefore choose a subsequence $\{x_\nu^{(1)}\}$ of $\{x_\nu\}$ with

$$x_\nu^{(1)} \equiv y_1 \bmod p \quad \text{and} \quad F(x_\nu^{(1)}) \equiv 0 \bmod p.$$

In the same way we can select from $\{x_\nu^{(1)}\}$ a subsequence $\{x_\nu^{(2)}\}$ with

$$x_\nu^{(2)} \equiv y_2 \bmod p^2 \quad \text{and} \quad F(x_\nu^{(2)}) \equiv 0 \bmod p^2,$$

where $y_2 \in \mathbb{Z}/p^2\mathbb{Z}$ since obviously $y_2 = y_1 \bmod p$. If we continue in this way we obtain, for every $k \geq 1$, a subsequence $\{x_\nu^{(k)}\}$ of $\{x_\nu^{(k-1)}\}$, whose terms satisfy the congruences

$$x_\nu^{(k)} \equiv y_k \bmod p^k \quad \text{and} \quad F(x_\nu^{(k)}) \equiv 0 \bmod p^k$$

with certain $y_k \in \mathbb{Z}/p^k\mathbb{Z}$, for which

$$y_k \equiv y_{k-1} \bmod p^{k-1}.$$

The $y_k$ thus define a $p$-adic number $y = (y_k)_{k \in \mathbb{N}} \in \varprojlim_k \mathbb{Z}/p^k\mathbb{Z} = \mathbb{Z}_p$ such that

$$F(y_k) \equiv 0 \bmod p^k$$

for all $k \geq 1$, that is, $F(y) = 0$.                                                    $\square$

*Example.* Consider for $\nu \geq 1$ the congruences

$$x^2 \equiv 2 \bmod 7^\nu.$$

For $\nu = 1$ the congruence has the solutions

(1)                                    $x_0 \equiv \pm 3 \bmod 7.$

Now let $\nu = 2$. Clearly

(2)                                    $x^2 \equiv 2 \bmod 7^2$

implies $x^2 = \bmod y$, and thus a solution of (2) must be of the form $\pm 3 + 7t$, so that a solution of (2) must be of the form $+3 + 7t$. If we substitute $x_1 = 3 + 7t_1$ in (2), we obtain

$$(3 + 7t_1)^2 \equiv 2 \bmod 7$$
$$9 + 6 \cdot 7t_1 + 7^2 t_1^2 \equiv 2 \bmod 7$$
$$1 + 6t_1 \equiv 0 \bmod 7$$
$$t_1 \equiv 1 \bmod 7$$

and we thus get as a solution to $x^2 \equiv 2 \bmod 7^2$,

$$x_1 \equiv 3 + 1 \cdot 7 \bmod 7^2.$$

For $\nu = 3$, we find $x_2 = x_1 + 7^2 t_2$, the value $t_2 \equiv 2 \bmod 7^3$, and thus, for the congruence

$$x^3 \equiv 2 \bmod 7^3$$

the solution

$$x_2 \equiv 3 + 1 \cdot 7 + 2 \cdot 7^2 \bmod 7^3.$$

It is easily seen that this process can be continued indefinitely, so that one obtains a 7-adic solution

$$x = 3 + 1 \cdot 7 + 2 \cdot 7^2 + \cdots \in \mathbb{Z}_7$$

of the equation $x^2 = 2$. This is denoted by $\sqrt{2}$ but is nevertheless to be strictly distinguished from the square root of 2 lying in the field $\mathbb{R}$.

If the polynomial $F(x_1, \ldots, x_n)$ is homogeneous and of degree $d \geq 1$, then the equation $F = 0$, obviously always has the trivial solution $(0, \ldots, 0)$, and the question of interest is whether it has any nontrivial solutions, and if so what they are. The proof of Theorem 4 can now be modified slightly to show that the congruences

$$F(x_1, \ldots, x_n) \equiv 0 \bmod p^\nu$$

have a nontrivial solution for all $\nu \geq 1$ if and only if the equation

$$F(x_1, \ldots, x_n) = 0$$

has a nontrivial $p$-adic solution.

At the beginning of this section we mentioned the question of whether from the solvability of an equation $F = 0$ in $\mathbb{Z}_p$ for all primes $p$ (that is, the existence of a common solution to all the congruences $F \equiv 0 \bmod m$) one can deduce the solvability of $F = 0$ in rational integers. This deduction can very seldom be made (that is, the condition mentioned, though obviously necessary is rarely sufficient). However, in the case of quadratic forms we have the following so-called "local-global principle" of Minkowski–Hasse, which we state here without giving any proof (see [1], §7).

**Theorem 5.** *Let $F(x_1, \ldots, x_n)$ be a quadratic form with rational coefficients. The equation*

$$F(x_1, \ldots, x_n) = 0$$

*has a nontrivial solution in $\mathbb{Q}$, if and only if it has a nontrivial solution in $\mathbb{R}$ and in $\mathbb{Q}_p$ for all primes $p$.*

## §3. The Analytical Nature of p-Adic Numbers

The series representation

$$(1) \qquad a_0 + a_1 p + a_2 p^2 + \cdots, \qquad 0 \le a_i < p$$

of a p-adic integer bears a close similarity to the representation of a real number between 0 and 10 as a decimal, that is, as the sum of a series of decimal fractions

$$a_0 + a_1 \left(\frac{1}{10}\right) + a_2 \left(\frac{1}{10}\right)^2 + \cdots, \qquad 0 \le a_i < 10.$$

However, unlike the latter, the p-adic series does not converge. Despite this nonconvergence however, the field $\mathbb{Q}_p$ of the p-adic numbers can be constructed from the field $\mathbb{Q}$ in virtually the same way as the field $\mathbb{R}$ is constructed from $\mathbb{Q}$. This is done by replacing the usual absolute value $|\ |$ in $\mathbb{Q}$ by a new "p-adic" absolute value $|\ |_p$ which has the effect of making the series (1) converge, and which enables the p-adic numbers to be regarded, in the usual way, as limits of Cauchy sequences of rational numbers.

The p-adic absolute value $|\ |_p$ is defined as follows. Let $a = \frac{b}{c} \in \mathbb{Q}$ be a nonzero rational number with $b, c \in \mathbb{Z}$ and $\frac{b}{c} \in \mathbb{Q}$. We divide $b$ and $c$ by the prime $p$ as many times as is possible, so that

$$(2) \qquad a = p^m \frac{b'}{c'}, \qquad (b'c', p) = 1,$$

and define

$$|a|_p = \frac{1}{p^m}.$$

The p-adic value is thus no longer a measure of the absolute magnitude of a number $a \in \mathbb{N}$, but rather has the property of being small when $a$ is divisible by a high power of $p$. In particular the partial sums associated with a p-adic series $a_0 + a_1 p + a_2 p^2 + \cdots$ form a convergent sequence with respect to the valuation $|\ |_p$.

The exponent $m$ in the representation (2) of the number $a$ is denoted by $v_p(a)$, and one writes formally $v_p(0) = \infty$. We have thus obtained a function

$$v_p : \mathbb{Q} \to \mathbb{Z} \cup \{\infty\}$$

with the following three properties, which are easily verified:

1) $v_p(a) = \infty \Leftrightarrow a = 0$,

2) $v_p(ab) = v_p(a) + v_p(b)$,

3) $v_p(a + b) \ge \min\{v_p(a), v_p(b)\}$,

where $\infty$ is a symbol satisfying the relations $x + \infty = \infty$ and $\infty > x$ for all $x \in \mathbb{Z}$. The function $v_p$ is called the *p-adic exponential valuation of* $\mathbb{Q}$.

The $p$-adic absolute value is given by

$$| \ |_p : \mathbb{Q} \to \mathbb{R}, \qquad a \mapsto |a|_p = p^{-v_p(a)},$$

and, in view of the relations 1), 2), 3) it satisfies the conditions for a *norm* on $\mathbb{Q}$, namely:

1) $|a|_p = 0 \Leftrightarrow a = 0$

2) $|ab|_p = |a|_p |b|_p$

3) $|a + b|_p \leq \max\{|a|_p, |b|_p\} \leq |a|_p + |b|_p$.

It can be shown that with $| \ |_p$ and $| \ |$ we have essentially exhausted the norms which can exist on $\mathbb{Q}$, in the sense that any other norm is a power $| \ |_p^s$ or $| \ |^s$ of one of these, where $s$ is a positive real number. The ordinary absolute value $| \ |$ is, for good reasons which we shall not go into here, denoted by $| \ |_\infty$. Along with the absolute values $| \ |_p$ it satisfies the following important *closure relation*.

**Theorem 6.** *For every nonzero rational integer* $a$

$$\prod_p |a|_p = 1$$

*where* $p$ *runs through all the primes, including the so-called infinite prime.*

**Proof.** In the canonical factorization of $a$

$$a = \pm \prod_{p \neq \infty} p^{v_p}$$

the exponent $v_p$ of $p$ is simply the exponential valuation $v_p(a)$ and the sign is equal to $\frac{a}{|a|_\infty}$. The equation can therefore be written in the form

$$a = \frac{a}{|a|_\infty} \prod_{p \neq \infty} \frac{1}{|a|_p},$$

so that in fact $\prod_p |a|_p = 1$.                                            □

We shall now redefine the field $\mathbb{Q}_p$ of $p$-adic numbers, following the same procedure as in the construction of the field of real numbers. We shall then go on to show that this new analytical definition is completely equivalent to the Hensel definition which was motivated by ideas from the theory of functions.

By a *Cauchy sequence*, with respect to $|\ |_p$, we mean a sequence $\{x_n\}$ of rational numbers such that, to every $\varepsilon > 0$ there corresponds a natural number $n_0$ for which

$$|x_n - x_m|_p < \varepsilon \quad \text{for all} \quad n, m > n_0.$$

*Example.* Any formal series

$$\sum_{\nu=0}^{\infty} a_\nu p^\nu, \qquad 0 \le a_\nu < p$$

provides, through its partial sums

$$x_n = \sum_{\nu=0}^{n-1} a_\nu p^\nu$$

an example of a Cauchy sequence, since for all $n > m$,

$$|x_n - x_m|_p = \left| \sum_{\nu=m}^{n-1} a_\nu p^\nu \right|_p \le \max_{m \le \nu < n} \{|a_\nu p^\nu|_p\} \le \frac{1}{p^m}.$$

A sequence $\{x_n\}$ in $\mathbb{Q}$ is called a *null sequence* w.r.t. $|\ |_p$, if $|x_n|_p$ is a sequence converging to zero in the usual sense.

*Example.* $1, p, p^2, p^3, \ldots$.

The Cauchy sequences form a ring $R$; the null sequences a maximal ideal $\mathfrak{m}$ in $R$. We define the field of $p$-adic numbers as the residue class field

$$\mathbb{Q}_p := R/\mathfrak{m}.$$

We can embed $\mathbb{Q}$ in $\mathbb{Q}_p$, by assigning to every $a \in \mathbb{Q}$ the residue class represented by the constant sequence $(a, a, a, \ldots)$. The $p$-adic absolute value $|\ |_p$ can be extended from $\mathbb{Q}$ to $\mathbb{Q}_p$ by defining, for any element $x = \{x_n\} \bmod \mathfrak{m} \in R/\mathfrak{m}$ the value

$$|x|_p := \lim_{n \to \infty} |x_n|_p \in \mathbb{R}.$$

The limit exists, because $|x_n|_p$ is a Cauchy sequence in $R$, and it is independent of the choice of the sequence $\{x_n\}$ in its residue class mod $\mathfrak{m}$, because for any $p$-adic null sequence $\{y_n\} \in \mathfrak{m}$ the relation $\lim_{n \to \infty} |y_n|_p = 0$ certainly holds.

The exponential valuation $v_p$ of $\mathbb{Q}$ can also be extended to the exponential valuation

$$v_p \colon \mathbb{Q}_p \to \mathbb{R} \cup \{\infty\}$$

by defining $v_p(x) = -\log_p |x|_p$, or what amounts to the same thing, by defining $v_p(x) = \lim_{n \to \infty} v_p(x_n)$, where $x$ is the class of rational Cauchy sequences $\{x_n\}$. We again have

$$|x|_p = p^{-v_p(x)}.$$

Since the image of $\mathbb{Q}^*$ under the mapping $v_p$ is the discrete set $\mathbb{Z}$, the same is true of the image of $\mathbb{Q}_p^*$, that is, $v_p$ is a surjective homomorphism

$$v_p : \mathbb{Q}_p^* \to \mathbb{Z}.$$

As with the real numbers, it can be proved that the field $\mathbb{Q}_p$ is *complete* with respect to the $p$-adic absolute value $|\ |_p$, that is, every Cauchy sequence in $\mathbb{Q}_p$ is convergent with respect to $|\ |_p$. Accordingly, for each prime number $p$, we can associate, alongside the field $\mathbb{R}$ of real numbers, a new complete field $\mathbb{Q}_p$. Out of the field $\mathbb{Q}$ has arisen an infinite family of fields

$$\mathbb{Q}_2, \ \mathbb{Q}_3, \ \mathbb{Q}_5, \ \mathbb{Q}_7, \ \mathbb{Q}_{11}, \ldots, \mathbb{Q}_\infty = \mathbb{R}.$$

An important peculiarity of the $p$-adic valuation $|\ |_p$ is that it not only satisfies the usual triangle inequality, but also the stronger inequality

$$|x + y|_p \leq \max\{|x|_p, |y|_p\}.$$

From this can be deduced a remarkable result.

**Theorem 7.** *The set*

$$\mathbb{Z}_p := \{x \in \mathbb{Q}_p \mid |x|_p \leq 1\}$$

*is a subring of $\mathbb{Q}_p$ whose units form the group*

$$\mathbb{Z}_p^* = \{x \in \mathbb{Q}_p \mid |x|_p = 1\}. \qquad \qquad \square$$

The elements of $\mathbb{Z}_p$ are called $p$-adic *integers*. The connection with the Hensel definition given in §1 is made clear by the following:

**Theorem 8.** (i) *Every p-adic number $x \in \mathbb{Q}_p^*$ has a unique representation*

$$x = p^m u \quad \text{with} \quad m \in \mathbb{Z}, \ u \in \mathbb{Z}_p^*$$

(ii) *The ideal $p\mathbb{Z}_p$ is a maximal ideal with residue class*

$$\mathbb{Z}_p / p\mathbb{Z}_p \cong \mathbb{Z}/p\mathbb{Z}$$

(iii) *The complete set of ideals of $\mathbb{Z}_p$ is given by $p^n \mathbb{Z}_p$, $n \geq 1$, and by*

$$\mathbb{Z}_p / p^n \mathbb{Z}_p \cong \mathbb{Z}/p^n \mathbb{Z}.$$

**Proof.** If $x \in \mathbb{Q}_p^*$ and $v_p(x) = m \in \mathbb{Z}$, then $v_p(xp^{-m}) = 0$, and hence $|xp^{-m}|_p = 1$, so that $u = xp^{-m} \in \mathbb{Z}_p^*$. This proves (i), while (ii) is a special case of (iii). Suppose $\hbar \neq 0$, $\mathbb{Z}_p$ to be an ideal of $\mathbb{Z}_p$. Let $x = p^m u$, $u \in \mathbb{Z}_p^*$, be an element of $\hbar$ with the smallest $m$ (since $|x|_p < 1$ $m$ must be greater than 0). Then $\hbar = p^m \mathbb{Z}_p$, because if $y = p^n u' \in \hbar$, $u' \in \mathbb{Z}_p^*$, then $n \geq m$ and thus $y = (p^{n-m} u') p^m \in p^m \mathbb{Z}_p$.                              $\square$

We now consider the homomorphism

$$\mathbb{Z} \to \mathbb{Z}_p / p^n \mathbb{Z}_p, \qquad a \mapsto a \bmod p^n \mathbb{Z}_p,$$

whose kernel is $p^n \mathbb{Z}$. This homomorphism is surjective. To prove this one can easily see that the numbers $x \in \mathbb{Z}_p$ are already limits of rational integers, and if $a \in \mathbb{Z}$ with

$$|x - a|_p \leq \frac{1}{p^n}$$

then $v_p(x - a) = m \geq n$, that is, $x - a = p^m u \in p^n \mathbb{Z}_p$, and so $a = x \bmod p^n \mathbb{Z}_p$. The homomorphism is therefore in fact an isomorphism

$$\mathbb{Z}/p^n\mathbb{Z} \overset{\sim}{\to} \mathbb{Z}_p/p^n\mathbb{Z}_p.$$

In §1, we defined $p$-adic integers as formal series

$$\sum_{\nu=0}^{\infty} a_\nu p^\nu, \qquad 0 \leq a_\nu < p$$

and identified them with the sequences

$$\bar{s}_n = s_n \bmod p^n, \qquad n = 1, 2, \ldots$$

where $s_n$ runs through the partial sums defined by

$$s_n = \sum_{\nu=0}^{n-1} a_\nu p^\nu.$$

These sequences define the inverse limit

$$\varprojlim_n \mathbb{Z}/p^n\mathbb{Z} = \left\{ (x_n)_{n \in \mathbb{N}} \in \prod_{n=1}^{\infty} \mathbb{Z}/p^n\mathbb{Z} \mid x_{n+1} \mapsto x_n \right\}$$

and we looked upon the $p$-adic integers as the elements of this ring. Since

$$\mathbb{Z}_p/p^n\mathbb{Z}_p \cong \mathbb{Z}/p^n\mathbb{Z}$$

we obtain, for each $n \geq 1$, a surjective homomorphism

$$\mathbb{Z}_p \to \mathbb{Z}/p^n\mathbb{Z},$$

and it is clear that the family of these homomorphisms gives us a homomorphism

$$\mathbb{Z}_p \longrightarrow \varprojlim_n \mathbb{Z}/p^n\mathbb{Z}.$$

The identification of the new analytical definition of $\mathbb{Z}_p$ (and thus of $\mathbb{Q}_p$) with the older Hensel definition can now be made.

**Theorem 9.** *The homomorphism*

$$\mathbb{Z}_p \longrightarrow \varprojlim_n \mathbb{Z}/p^n\mathbb{Z}$$

*is an isomorphism.*

**Proof.** If $x \in \mathbb{Z}_p$ is mapped onto zero, this means that $x \in p^n\mathbb{Z}_p$ for all $n$, that is, $|x|_p \leq \frac{1}{p^n}$ for all $n$, and hence $|x|_p = 0$ so that $x = 0$. This proves injectivity.

An element $\varprojlim_n \mathbb{Z}/p^n\mathbb{Z}$ is given by a sequence of partial sums

$$s_n = \sum_{\nu=0}^{n-1} a_\nu p^\nu, \qquad 0 \leq a_\nu < p^\nu.$$

We saw earlier that this sequence is a Cauchy sequence in $\mathbb{Z}_p$ and thus converges to an element

$$x = \sum_{\nu=0}^{\infty} a_\nu p^\nu \in \mathbb{Z}_p.$$

Since

$$x - s_n = \sum_{\nu=n}^{\infty} a_\nu p^\nu \in p^n\mathbb{Z}_p$$

it follows that $x \equiv s_n \bmod p^n$ for all $n$, that is, that $x$ is mapped onto the element of $\varprojlim_n \mathbb{Z}/p^n\mathbb{Z}$ corresponding to the sequence $(s_n)_{n\in\mathbb{N}}$ defined above. This proves surjectivity. $\qquad\square$

We emphasize that the elements of the right-hand side of

$$\mathbb{Z}_p \longrightarrow \varprojlim_n \mathbb{Z}/p^n\mathbb{Z}$$

are given formally by the sequence of partial sums

$$s_n = \sum_{\nu=0}^{n-1} a_\nu p^\nu, \qquad n = 1, 2, \ldots .$$

On the left-hand side, however, these sequences considered with respect to their absolute values converge and represent in the familiar fashion the elements of $\mathbb{Z}_p$ as convergent infinite series

$$x = \sum_{\nu=0}^{\infty} a_\nu p^\nu.$$

The isomorphism $\mathbb{Z}_p \cong \varprojlim \mathbb{Z}/p^n\mathbb{Z}$ gives us additional information about the topology on $\mathbb{Q}_p$, defined by the absolute value $|\ |_p$. The direct product

$$\prod_{\nu=1}^{\infty} \mathbb{Z}/p^\nu\mathbb{Z}$$

in fact, has the product topology, in which the individual factors are regarded as topological spaces endowed with the discrete topology. Since these factors are compact, the product is compact as well (by Tychonoff's theorem).

It can now easily be shown that the inverse limit $\varprojlim_{\nu} \mathbb{Z}/p^\nu\mathbb{Z}$ is a closed subset of this product, and is likewise a compact space. It is also not difficult to verify that the ring isomorphism

$$\mathbb{Z}_p \to \varprojlim_{\nu} \mathbb{Z}/p^\nu\mathbb{Z}$$

is also a homeomorphism between topological spaces. Consequently, $\mathbb{Z}_p$ is a compactum, and since

$$\mathbb{Z}_p = \{x \in \mathbb{Q}_p \mid |x|_p < 2\}$$

also an open subset of $\mathbb{Q}_p$. Every element $a \in \mathbb{Q}_p$ therefore possesses, in $a + \mathbb{Z}_p$, an open compact neighborhood. We have therefore proved the following.

**Theorem 10.** *The field $\mathbb{Q}_p$ is locally compact.*                                    □

The considerations in this section appear to release the $p$-adic numbers from their original role, modeled on that of the analytic functions, and to bring them into a closer analogy with the complex numbers themselves. It is particularly remarkable that in recent time a $p$-adic theory of analytic functions has been developed, in which $p$-adic numbers have replaced complex numbers both as arguments of the functions and as functional values. This theory was initiated by the American mathematician J. Tate, and has been widely developed by the two German mathematicians H. Grauert and R. Remmert.

## §4. The p-Adic Numbers

A far-reaching theory can be built up on the basis provided by the $p$-adic numbers, namely, the theory of algebraic extensions of the field $\mathbb{Q}_p$, or, to express it in another way, the theory of algebraic equations

$$f(x) = a_n x^n + a_{n-1} x^{n-1} + \cdots + a_0 = 0$$

in one variable. We saw, in Section 2, that the solvability of such an equation in the ring $\mathbb{Z}_p$ is equivalent to the solvability of the congruences $f(x) \equiv 0 \bmod p^{\nu}$ for all $\nu$. Of fundamental importance here is the fact that a sufficient condition for this is that the congruence

$$f(x) \equiv 0 \bmod p$$

should be solvable, as long as one restricts oneself to simple zeros. More generally, we have the important result.

**Hensel's Lemma.** *If a polynomial $f(x) \in \mathbb{Z}_p[x]$ has the decomposition modulo $p$*

$$f(x) \equiv g_0(x) h_0(x) \bmod p$$

*where the polynomials $g_0, h_0 \in \mathbb{Z}_p[x]$ are coprime modulo $p$, and if $g_0$ is monic, then there exists a decomposition over $\mathbb{Z}_p$*

$$f(x) = g(x) h(x)$$

*with polynomials $g, h \in \mathbb{Z}_p[x]$, such that $g(x)$ is monic and*

$$g(x) \equiv g_0(x) \bmod p, \quad h(x) \equiv h_0(x) \bmod p.$$

**Proof.** Let $d = \deg(f)$, $m = \deg(g_0)$ and without loss of generality let us suppose that $\deg(h_0) \leq d - m$. We then put the polynomials $g$ and $h$, which have to be determined, into the form

$$g = g_0 + y_1 p + y_2 p^2 + \cdots$$
$$h = h_0 + z_1 p + z_2 p^2 + \cdots$$

with polynomials $y_i, z_i \in \mathbb{Z}_p[x]$ of degrees $< m$ and $\leq d - m$ respectively. We now determine the polynomials

$$g_n = g_0 + y_1 p + \cdots + y_{n-1} p^{n-1}$$
$$h_n = h_0 + z_1 p + \cdots + z_{n-1} p^{n-1}$$

successively, in such a way that

(*)                                  $f \equiv g_n h_n \bmod p^n$

holds for each $n$ in turn. The equation $f = gh$ will then hold by a passage to the limit. For $n = 1$, the congruence $(*)$ is the hypothesis stated in the lemma, and we assume that its truth has been established for $n$. The requirement for $g_{n+1}, h_{n+1}$, in view of

$$g_{n+1} = g_n + y_n p^n, \quad h_{n+1} = h_n + z_n p^n$$

then becomes

$$f - g_n h_n \equiv (g_n z_n + h_n y_n) p^n \bmod p^{n+1}$$

or, after division by $p^n$

$$g_n z_n + h_n y_n \equiv g_0 z_n + h_0 y_n \equiv f_n \bmod p,$$

where $f_n = \frac{1}{p^n}(f - g_n h_n) \in \mathbb{Z}_p[x]$. Since $g_0$ and $h_0$ are coprime in $\mathbb{F}_p[x]$, there must be polynomials $z_n, y_n \in \mathbb{Z}_p[x]$ of the required kind, and $y_n$ can be chosen to be reduced to its minimum residue modulo $g_0$, so that $\deg(y_n) < m$. Since $\deg(h_0) \leq d - m$ and $\deg(f_n) \leq d$, it follows that $\deg(g_0 z_n) < d$ and hence $\deg(z_n) \leq d - m$ as required. $\qquad \square$

*Example.* The polynomial $x^{p-1} - 1$ splits into separate linear factors in the residue class field $\mathbb{Z}_p/p\mathbb{Z}_p = \mathbb{F}_p$. By (repeated) application of Hensel's lemma, therefore, it also splits into linear factors in $\mathbb{Q}_p$ (that is, linear factors whose coefficients belong to $\mathbb{Q}_p$) and we obtain the surprising result that $\mathbb{Q}_p$ contains the $(p-1)$th roots of unity.

We now consider the finite algebraic extensions of $\mathbb{Q}_p$. In contrast to the field $R$, the field $\mathbb{Q}_p$ possesses many such extensions. However, just as in the case $\mathbb{C}/\mathbb{R}$ the topological structure of the ground field $\mathbb{Q}_p$ is extended on each extension field. More precisely, we have the following.

**Theorem 11.** *Let $K/\mathbb{Q}_p$ be a finite extension of degree $n$. Then the absolute value $|\ |_p$ of $\mathbb{Q}_p$ can be extended to an absolute value $|\ |_{\mathfrak{p}}$ on $K$, namely, by defining*

$$|\alpha|_{\mathfrak{p}} = \sqrt[n]{|N(\alpha)|_p}$$

*where $N$ denotes the norm of $K/\mathbb{Q}_p$. The field $K$ is likewise complete with respect to $|\ |_{\mathfrak{p}}$.*

**Proof.** The properties $|\alpha|_{\mathfrak{p}} = 0 \Leftrightarrow \alpha = 0$ and

$$|\alpha\beta|_{\mathfrak{p}} = |\alpha|_{\mathfrak{p}} |\beta|_{\mathfrak{p}}$$

clearly hold, the latter because of the multiplicativity of the norm. We shall prove the stronger version of the triangle inequality

$$|\alpha + \beta|_{\mathfrak{p}} \leq \max\{|\alpha|_{\mathfrak{p}}, |\beta|_{\mathfrak{p}}\}$$

with the help of Hensel's lemma. After dividing by $\alpha$ or $\beta$, this reduces to checking that

$$|\alpha|_{\mathfrak{p}} \le 1 \Rightarrow |\alpha - 1|_{\mathfrak{p}} \le 1.$$

In view of the transitivity of the norm we may assume for this purpose that $K = \mathbb{Q}_p(\alpha)$. If therefore

$$f(x) = x^n + a_1 x^{n-1} + \cdots + a_n,$$

is the minimal polynomial of $\alpha$, then $N(\alpha) = \pm a_n$ and

$$f(x + 1) = x^n + \cdots + (1 + a_1 + \cdots + a_n)$$

is the minimal polynomial of $\alpha - 1$, that is, $N(\alpha - 1) = \pm(1 + a_1 + \cdots + a_n)$. We have to show therefore that $|1 + a_1 + \cdots + a_n|_{\mathfrak{p}} \le 1$ if $|a_n|_{\mathfrak{p}} \le 1$, or in other words

$$a_n \in \mathbb{Z}_p \Rightarrow 1 + a_1 + \cdots + a_n \in \mathbb{Z}_p.$$

We shall in fact show that $a_n \in \mathbb{Z}_p$ implies that $a_1, \ldots, a_n \in \mathbb{Z}_p$. Assume for the sake of argument that $a_1, \ldots, a_n$ were not all in $\mathbb{Z}_p$.

We then multiply $f(x)$ by $p^m$, the smallest positive power of $p$ required to ensure that all the coefficients $b_i$ of

$$f_0(x) = p^m f(x) = b_0 x^n + b_1 x^{n-1} + \cdots + b_{n-1} x + b_n$$

lie in $\mathbb{Z}_p$. These coefficients have the property that

$$b_0, b_n \equiv 0 \bmod p, \quad \text{while} \quad b_1, \ldots, b_{n-1}$$

are not all congruent to zero mod $p$. Among those coefficients is therefore a last nonzero coefficient $b_r$ satisfying $b_r \not\equiv 0 \bmod p$. This means that there is now a factorization

$$f_0(x) \equiv (b_0 x^r + b_1 x^{r-1} + \cdots + b_r) x^{n-r} \bmod p$$

into factors which are relatively prime modulo $p$. By Hensel's lemma it follows that $f(x)$ is reducible, which contradicts the definition of $f(x)$ thus disproving the assumption.

The completeness of $K$ is established by the familiar arguments, just as with $\mathbb{R}$-vector spaces, by choosing a basis $\omega_1, \ldots, \omega_n$ of $K/\mathbb{Q}_p$ and showing that a sequence

$$\alpha_i = a_{1i}\omega_1 + \cdots + a_{ni}\omega_n, \qquad a_{ji} \in \mathbb{Q}_p,$$

is a Cauchy sequence in $K$ if and only if the coefficient sequences $\{a_{1i}\}, \ldots, \{a_{ni}\}$ are Cauchy sequences in $\mathbb{Q}_p$.

To prove the uniqueness of the extension let $| \; |$ be any arbitrary extension of $| \; |_{\mathfrak{p}}$ on $K$. Then, for

$$|\alpha| < 1 \Leftrightarrow | \; | - \lim_{\nu \to \infty} \alpha^\nu = 0.$$

If we write, in terms of the basis $\omega_1, \ldots, \omega_n$ of $K$,

$$\alpha^\nu = a_{1\nu}\omega_n + \cdots + a_{n\nu}\omega_n,$$

then this is equivalent to the statement that $\lim_{\nu \to \infty} a_{i\nu} = 0$ in $\mathbb{Q}_p$ for all $i = 1, \ldots, n$. The inequality $|\alpha| < 1$ thus does not depend on the choice of the extension. In other words if $|\ |_1$ and $|\ |_2$ are any two extensions

$$(*) \qquad\qquad\qquad |\alpha|_1 < 1 \Leftrightarrow |\alpha|_2 < 1.$$

Suppose now that $\alpha_0 \in K$ is a fixed element with $0 < |\alpha_0|_1 < 1$. For an arbitrary $\alpha \neq 0$ we now consider all $k, l \in \mathbb{Z}$, $l \neq 0$ such that $|\alpha_0^k|_1 < |\alpha^l|_1$, or, in other words, such that

$$\frac{k}{l} > \frac{\log|\alpha|_1}{\log|\alpha_0|_1}.$$

By virtue of $(*)$ the fractions $\frac{k}{l}$ are at the same time all rational numbers satisfying

$$\frac{k}{l} > \frac{\log|\alpha|_2}{\log|\alpha_0|_2}.$$

It follows from this that

$$\frac{\log|\alpha|_1}{\log|\alpha_0|_1} = \frac{\log|\alpha|_2}{\log|\alpha_0|_2}, \quad \text{that is,} \quad \frac{\log|\alpha|_1}{\log|\alpha|_2} = \frac{\log|\alpha_0|_1}{\log|\alpha_0|_2} = s,$$

so that $|\alpha|_1 = |\alpha|_2^s$. As $|\ |_1$ and $|\ |_2$ coincide on $\mathbb{Q}_p$, $s$ must be equal to 1, and hence $|\ |_1 = |\ |_2$. This completes the proof of Theorem 11. $\qquad\square$

The field $\mathbb{Q}_p$ of the $p$-adic numbers passes on many of its properties to its finite extension $K$. The subset

$$\mathcal{O} = \{\alpha \in K \mid |\alpha|_{\mathfrak{p}} \leq 1\}$$

again forms, just like $\mathbb{Z}_p$, a ring with the group of units

$$\mathcal{O}^* = \{\alpha \in K \mid |\alpha|_{\mathfrak{p}} = 1\}$$

and the single maximal ideal

$$\mathfrak{p} = \{\alpha \in K \mid |\alpha|_{\mathfrak{p}} < 1\}.$$

The residue class field $\kappa(\mathfrak{p}) = \mathcal{O}/\mathfrak{p}$ is a finite extension of the residue class field $\kappa(p) = \mathbb{Z}_p/p\mathbb{Z}_p = \mathbb{F}_p$ and consequently, a finite field $\mathbb{F}_q$. For these reasons $K$ is known as a $\mathfrak{p}$-adic number field and its elements are known as $\mathfrak{p}$-adic numbers.

As $\mathbb{Q}_p$ is locally compact, it is clear from the basis representation $K = \mathbb{Q}_p\omega_1 + \cdots + \mathbb{Q}_p\omega_n$ that every finite extension $K$ of $\mathbb{Q}_p$ is likewise locally compact. Conversely, it can be shown that the finite extensions of the fields $\mathbb{Q}_2$, $\mathbb{Q}_3$, $\mathbb{Q}_5$, ..., $\mathbb{Q}_\infty = \mathbb{R}$ constitute precisely the totality of all the nondiscrete locally compact topological fields of characteristic zero (see [8], Ch. I §3, Th. 9).

An important objective of the theory of numbers is that of obtaining an overall view of the finite extensions of the field $\mathbb{Q}_p$. One of the most beautiful and profound theorems gives a complete answer to this question, as long as we confine ourselves to *Abelian* extensions, that is, to finite Galois extensions whose Galois group is commutative. In these circumstances we can take as ground field an arbitrary p-adic number field $K$ instead of $\mathbb{Q}$. If $L|K$ is a finite extension, then we may take $N_{L|K}(L^*) \subseteq K^*$ to be its norm group.

**Theorem.** *Let $K$ be a p-adic number field. The mapping*

$$L|K \mapsto N_{L|K}(L^*)$$

*is a one-to-one correspondence between the Abelian extension $L$ of $K$ and the subgroups $I$ of $K^*$ of finite index. With this relationship we even have, for Galois group $G(L|K)$, a canonical isomorphism*

$$G(L|K) \cong K^*/N_{L|K}(L^*).$$

This theorem, which reflects the structure of the Abelian extensions $L/K$ in the structure of the multiplicative group $K^*$ of the ground field $K$, is known as the fundamental theorem of local class field theory (see [5]). In a certain sense the classification of all finite extensions $L|K$ has recently been achieved. These extensions correspond in fact under Galois theory in one-to-one fashion to the open subgroups of the Galois group $G_K = G(\bar{K}/K)$ of the algebraic closure $\bar{K}$ of $K$, and this group $G_K$ was explicitly defined in terms of generators and relations in 1982, by the two German mathematicians Uwe JANNSEN and Kay WINGBERG. Another classification of the extensions of $K$ is being attempted with the help of "LANGLAND's conjecture," which seeks to put them into a close relationship with the representations of the groups $GL_n(K)$.

## REFERENCES

[1] BOREVIČ, Z.I., ŠAFAREVIČ, I.R.: Number Theory. Translated by N. Greenleaf, New York: Academic Press, 1967

[2] HASSE, H.: Number Theory. Translated by H.G. Zimmer. New York: Springer-Verlag, 1980

[3] HENSEL, K.: Theorie der algebraischen Zahlen. Teubner Verlag Leipzig und Berlin, 1908

[4] JANNSEN, U., WINGBERG, K.: Die Struktur der absoluten Galois-gruppe p-adischer Zahlkörper. Invent. Math. 70, S. 71–98 (1982)

[5] NEUKIRCH, J.: Class Field Theory. Springer-Verlag, Berlin, Heidelberg, New York, Tokyo 1986

[6] KOCH, H., PIEPER, H.: Zahlentheorie. VEB Deutscher Verlag der Wissenschaften, Berlin 1976

[7] SERRE, J.-P.: Corps locaux. Hermann, Paris 1962 [English translation. Local fields, in: GTM, 67, Springer, 1979]

[8] WEIL, A.: Basic Number Theory. Springer-Verlag, New York, 1967

# Part B

# Real Division Algebras

# Introduction

*M. Koecher, R. Remmert*

> Erst durch die Behandlung der gewöhnlichen
> imaginären Zahlen ... in Gemeinschaft mit
> den höheren complexen Zahlen kann ihre
> wahre Bedeutung in das volle Licht gesetzt werden
> (HANKEL 1867).
>
> [It is not until the ordinary imaginary numbers are
> treated ... in common with the higher complex num-
> bers that their true meaning can be brought
> into full daylight.]

**1.** GAUSS in 1831 was convinced that, outside the system of complex numbers, there were no "hypercomplex" number systems in which the basic
properties of complex numbers persist; however, he expressed himself in
thoroughly sibylline utterances (see 4.3.6). The Uniqueness theorem for the
field $\mathbb{C}$ appears to be a convincing pointer in support of GAUSS's thesis. In
the 1880's, a friendly dispute arose between WEIERSTRASS and DEDEKIND
about the proper interpretation of GAUSS's words. Described in modern
language, the controversy revolved around the question of characterizing
all finite-dimensional, commutative and associative $\mathbb{R}$ algebras with unit
element, *divisors of zero* being allowed.

In the year 1843 HAMILTON discovered his quaternions, and shortly afterwards GRAVES and CAYLEY constructed their octaves. These new hypercomplex systems are no longer fields—in the case of quaternions the
commutative law of multiplication no longer holds, while in the case of
octaves even the associative law of multiplication is abandoned—but every
non-zero element still has an inverse. *Division can be performed and remains unambiguous;* this property of the ordinary (rational) numbers was
regarded as indispensable by the founding fathers of the theory. Divisors
of zero, or even nilpotent elements, which are nowadays encountered by
first-year students learning about matrices, were not allowed.[1] Indeed the

---

[1] Weierstrass was the first to introduce, in 1883, in his "Zur Theorie der aus
$n$ Haupteinheiten gebildeten komplexen Größen" (*Math. Werke* 2, 311–339) the
concept of a "divisor of zero" (p. 314); he also struggles with nilpotent elements
(p. 319). The significance of the property of "absence of divisors of zero" had
already been clearly perceived by Hankel in 1867 (see 4.3.6).

idea that it was worthwhile concerning one's self with hypercomplex numbers was by no means undisputed. As late as 1890 E. STUDY, in his article "Über Systeme complexer Zahlen und ihre Anwendungen in der Theorie der Transformationsgruppen" (*Monatsh. Math. u. Phys.* 1, 283–355, specially pp. 341/42) wrote: "In weiten Kreisen, namentlich in Deutschland, ist die Ansicht verbreitet, dass die Systeme von complexen Zahlen oder ähnliche Algorithmen überhaupt gar keinen Nutzen hätten, ausgenommen allein die gewöhnlichen complexen Zahlen; und man begründet dies damit, dass durch sie nichts geleistet werden könnte, was nicht 'ebenso gut' auch ohne sie zu leisten wäre." [In a number of circles, particularly in Germany, there is a widely held view that systems of complex numbers or similar algorithms have actually been of hardly any real use, with the single exception of the ordinary complex numbers. The reason given as justification for this attitude has been that no results could ever be provided by these systems that could not equally well have been provided without their help.]

**2.** Since the beginning of the 20th century hypercomplex systems of numbers have been (loosely but more succinctly) called *real algebras*. If division can be performed unambiguously, we speak of a *division algebra*. We shall adopt an historical approach focusing our attention on division algebras. The classical division algebra is the four-dimensional quaternion algebra. We shall deal with quaternions in detail in Chapter 7. We prove the famous theorem of FROBENIUS on the uniqueness of quaternions in Chapter 8, and we shall also establish there the beautiful theorem of HOPF that every finite-dimensional commutative division algebra with unit element, other than $\mathbb{R}$, must be isomorphic to $\mathbb{C}$. In that same Chapter 8 we shall also give an "elementary" proof of the celebrated GELFAND–MAZUR theorem, which states that every normed, commutative, associative, real, division algebra $\neq 0$ is isomorphic to $\mathbb{R}$ or $\mathbb{C}$.

The eight-dimensional division algebra of CAYLEY numbers will be studied in Chapter 9, and ZORN's theorem on the uniqueness of the CAYLEY numbers will be proved. We deal with composition algebras in Chapter 10 and we shall discuss their characterization by HURWITZ; by way of application, we shall determine the class of all real vector product algebras. In Chapter 11, written by F. HIRZEBRUCH, we shall use topological methods to obtain a deep result due to KERVAIRE and MILNOR which asserts that division algebras are possible only in 1, 2, 4, and 8 dimensions.

To enable us to formulate our results precisely, we begin with a preliminary section in which we summarize the basic concepts and facts from the general theory of algebras. This is largely in the form of a repertory, and subsequent references to this chapter will be indicated by the letter $R$.

# Repertory. Basic Concepts from the Theory of Algebras

*M. Koecher, R. Remmert*

Die größten und fruchtbarsten Fortschritte in der
Mathematik sind vorzugsweise durch die Schöpfung neuer
Begriffe gemacht, nachdem die häufige Wiederkehr
zusammengesetzter Erscheinungen dazu gedrängt hat
(R. DEDEKIND, Was sind und was sollen
die Zahlen? 1888).

[The greatest and most fruitful advances in Mathematics are chiefly made by the creation of new ideas
and concepts, after the frequent reoccurrence of com-
posite (or complex) phenomena has driven us to this.
(R. DEDEKIND: What are numbers, and what
are they good for?]

We take $\mathbb{R}$ as the basic field, though in place of $\mathbb{R}$ one could equally well have chosen any commutative field $K$. Real numbers will always be denoted in Chapters 7 to 11 by small Greek letters. Every $n$-dimensional $\mathbb{R}$-vector space is isomorphic to the number space $\mathbb{R}^n$ of $n$-tuples $x = (\xi_1, \ldots, \xi_n)$.

**1. Real Algebras.** A vector space $V$ over $\mathbb{R}$ with a *"product mapping"* (or *multiplication*) $V \times V \to V$, $(x, y) \mapsto xy$ is said to be an *algebra over* $\mathbb{R}$, or an $\mathbb{R}$-*algebra* or (*real*) *algebra*, if the two *distributive laws*

$$(\alpha x + \beta y)z = \alpha(xz) + \beta(yz), \qquad x(\alpha y + \beta z) = \alpha(xy) + \beta(xz)$$

hold for all $\alpha, \beta \in \mathbb{R}$ and all $x, y, z \in V$ (*bilinearity of the product*). In particular, the relations $\alpha(xy) = (\alpha x)y = x(\alpha y)$ are always valid. If the *associative law* $x(yz) = (xy)z$ holds for all $x, y, z \in V$, then the algebra is said to be *associative;* if the commutative law $xy = yx$ holds for all $x, y \in V$, then we speak of a *commutative* algebra. Under these definitions an $\mathbb{R}$-algebra is, in general, *neither* associative *nor* commutative.

An element $e \in V$ is called an *identity element* (or unit element) of the algebra, if $ex = xe = x$ for all $x \in V$, and it can be seen at once that every algebra has at most one identity element.

To distinguish between different algebras defined on $V$, the multiplication symbol is often indicated explicitly as part of the notation, so that one writes $\mathcal{A} := (V, \cdot)$. The dimension of the $\mathbb{R}$-vector space $V$ is called the *dimension of the algebra;* $\dim \mathcal{A} := \dim V$.

In every algebra, powers are defined inductively by $x^m := x \cdot x^{m-1}$. Great care is needed in calculating with powers; thus, for example, in general $x \cdot x^2 \neq x^2 \cdot x$, and even in the commutative case it is not possible to show that $x^4 = (x^2)^2$ necessarily holds. An algebra $\mathcal{A}$ is said to be *power-associative* if the

**exponential rule** $x^m x^n = x^{m+n}$ *for all* $x \in \mathcal{A}$ *and all* $m \geq 1$, $n \geq 1$ *always holds.*

Every associative algebra is power associative.

An element $x$ of an algebra $\mathcal{A}$ is said to be a *divisor of zero in* $\mathcal{A}$ if there is an element $y \neq 0$ in $\mathcal{A}$ such that $xy = 0$ or $yx = 0$. An algebra is said to have no zero divisors, if it contains no divisors of zero. In this case the equation $xy = 0$ holds if and only if $x = 0$ or $y = 0$.

**2. Examples of Real Algebras.** We give seven instructive examples.

0) The fields $\mathbb{R}$ and $\mathbb{C}$ are associative and commutative $\mathbb{R}$-algebras of dimensions 1 and 2 respectively, each with identity element and without zero divisors.

1) The $\mathbb{R}$-vector space $\mathrm{Mat}(n, \mathbb{R})$ of all real $n \times n$ matrices is an $n^2$-dimensional, associative $\mathbb{R}$-algebra with identity element (the unit matrix) with respect to matrix multiplication.

2) The $\mathbb{R}$-vector space $\mathrm{Mat}(n, \mathbb{C})$ of all complex $n \times n$ matrices is a $2n^2$-dimensional, associative $\mathbb{R}$-algebra with identity element, with respect to matrix multiplication. The algebras $\mathrm{Mat}(n, \mathbb{R})$ and $\mathrm{Mat}(n, \mathbb{C})$ are *noncommutative* when $n > 1$.

3) For any two vectors $a = (\alpha_1, \alpha_2, \alpha_3)$, $b = (\beta_1, \beta_2, \beta_3) \in \mathbb{R}^3$ we may define the *vector product* by

$$a \times b := (\alpha_2 \beta_3 - \alpha_3 \beta_2, \alpha_3 \beta_1 - \alpha_1 \beta_3, \alpha_1 \beta_2 - \alpha_2 \beta_1) \in \mathbb{R}^3.$$

The vector space $\mathbb{R}^3$ thus becomes a three-dimensional $\mathbb{R}$-algebra which is *non-associative* and *anti-commutative*. This algebra is the simplest non-trivial example of a LIE-*algebra*. Such algebras play an important role in many parts of modern mathematics (see also 6.1.4 and 9.3).

4) The $\mathbb{R}$-vector space $\mathrm{Sym}(n, \mathbb{R})$ of all real symmetric $n \times n$ matrices is, with respect to the symmetrical matrix product $(A, B) \mapsto \frac{1}{2}(AB + BA)$, a commutative algebra which is not associative when $n > 1$.

5) Any $\mathbb{R}$-vector space $V \neq 0$ can be made into an *associative* and *commutative* $\mathbb{R}$-algebra with *identity element*. We fix a nonzero element $e \in V$,

choose any supplementary space $U$ to the line $\mathbb{R}e \subset V$ and define for arbitrary vectors $x = \alpha e + u$, $x' = \alpha' e + u' \in \mathbb{R}e \oplus U$ a multiplication by $xx' := (\alpha\alpha')e + \alpha u' + \alpha' u$. Then $e$ is an identity element, and $uu' = 0$ for all $u, u' \in U$. In this algebra *every* element belonging to $U$ is a divisor of zero.

6) If $\mathcal{A}_1 = (V_1, \cdot), \ldots, \mathcal{A}_s = (V_s, \cdot)$ are real algebras, then we can define in the vector space $V := V_1 \oplus \cdots \oplus V_s$ the direct sum of these vector spaces, a product by the rule

$$xy := x_1 y_1 + \cdots + x_s y_s \quad \textit{(component-wise multiplication)}$$

where $x = x_1 + \cdots x_s$, $y = y_1 + \cdots + y_s \in V$, and $x_i, y_i \in V_i$. The algebra $\mathcal{A} := (V, \cdot)$ obtained in this way is called the *direct sum* of the algebras $\mathcal{A}_1, \ldots, \mathcal{A}_s$, and we write $\mathcal{A} = \mathcal{A}_1 \oplus \cdots \oplus \mathcal{A}_s$. If $\mathcal{A}_1, \ldots, \mathcal{A}_s$ are all commutative or all associative, then so is $\mathcal{A}$. In the case where $s > 1$, $\mathcal{A}$ always has divisors of zero. If $e_i$ is an identity element of $\mathcal{A}_i$, $1 \leq i \leq s$, then $e := e_1 + \cdots + e_s$ is the identity element of $\mathcal{A}$.

All the algebras in 0)–5) are power associative.

## 3. Subalgebras and Algebra Homomorphisms.
A real subspace $U$ of an $\mathbb{R}$-algebra $\mathcal{A} = (V, \cdot)$ is said to be an $\mathbb{R}$-subalgebra of $\mathcal{A}$, if $xy \in U$ for all $x, y \in U$.

*Examples.* 1) The set $\left\{ \begin{pmatrix} \alpha & -\beta \\ \beta & \alpha \end{pmatrix} : \alpha, \beta \in \mathbb{R} \right\}$ is an $\mathbb{R}$-subalgebra of $\mathrm{Mat}(2, \mathbb{R})$ (see Chapter 3.2.5).

2) The sets of upper triangular matrices form in each case $\mathbb{R}$-subalgebras of $\mathrm{Mat}(n, \mathbb{R})$ and $\mathrm{Mat}(n, \mathbb{C})$ of dimension $\frac{1}{2}n(n+1)$ and $n(n+1)$ respectively.

If $\mathcal{A} = (V, \cdot)$ and $\mathcal{B} = (W, \cdot)$ are any two algebras, an $\mathbb{R}$-*linear mapping* $f: V \to W$ is said to be an $\mathbb{R}$-algebra homomorphism, if

$$f(xy) = f(x)f(y) \quad \text{for all} \quad x, y \in V.$$

One speaks of a *mono-, epi-, iso-, endo-* or *auto-morphism* when the $\mathbb{R}$-linear mapping $f: V \to W$ is a morphism of the corresponding type.

*Example.* The mapping $f: \mathbb{C} \to \mathrm{Mat}(2, \mathbb{R})$, $\alpha + \beta i \mapsto \begin{pmatrix} \alpha & -\beta \\ \beta & \alpha \end{pmatrix}$ is an algebra monomorphism.

*Remark.* If $\mathcal{A}$ is an algebra with identity element $e$, then $f: \mathbb{R} \to \mathcal{A}$, $\alpha \mapsto \alpha e$, is an algebra monomorphism. In particular every one-dimensional real algebra with identity element is isomorphic to $\mathbb{R}$.

## 4. Determination of All One-Dimensional Algebras.
Every real vector space $V$ trivially becomes an algebra, if one chooses as multiplication

$V \times V \to V$, the zero mapping $(x, y) \mapsto 0$. We shall show that, in the one-dimensional case, this pathological behavior is the only exception.

**Theorem.** *Any one-dimensional algebra, whose multiplication is not the zero mapping, is isomorphic to the algebra* $\mathbb{R}$.

**Proof.** In view of the remark in 3 above, it suffices to show that $\mathcal{A}$ has an identity element. Clearly $\mathcal{A} = \mathbb{R}a$ with $a \in \mathcal{A} \setminus \{0\}$. Since $xy = 0$ does not always hold, it follows that $a^2 \neq 0$, and hence also $\mathcal{A} = \mathbb{R}a^2$. Consequently there is an equation $a = \varepsilon a^2$ with $\varepsilon \in \mathbb{R}$, and therefore $e := \varepsilon a$ is an identity element of $\mathcal{A}$.                                                      $\square$

**5. Division Algebras.** Since the time of HAMILTON (finite-dimensional) division algebras have played a central role. An algebra $\mathcal{A} \neq 0$ is said to be a *division algebra*, if for all $a, b \in V$, $a \neq 0$, the two equations $ax = b$ and $ya = b$ *have unique solutions* in $\mathcal{A}$.

The fields $\mathbb{R}$ and $\mathbb{C}$ are associative and commutative division algebras of dimensions 1 and 2, respectively. The matrix algebras $\mathrm{Mat}(n, \mathbb{R})$ and $\mathrm{Mat}(n, \mathbb{C})$ are not division algebras when $n > 1$. The $\mathbb{R}$-vector space $\mathbb{C}$ is a 2-dimensional *commutative, non-associative* division algebra *without identity element* with respect to the multiplication $w \circ z := \overline{wz}$.

**Lemma.** *If $\mathcal{A}$ is an associative division algebra, then $G := \mathcal{A} \setminus \{0\}$ is a group with respect to the multiplication in $\mathcal{A}$. The neutral element of $G$ is the identity element of $\mathcal{A}$.*

**Proof.** Since within $G$ every equation $ax = b$ and $ya = b$ has a unique solution, $G$ must be a group.                                                     $\square$

*Every division algebra is without zero divisors.* As regards a converse, we have merely the following:

**Criterion.** *The following statements about a finite-dimensional algebra $\mathcal{A}$ are equivalent:*

i) *$\mathcal{A}$ is a division algebra,*

ii) *$\mathcal{A}$ is without zero divisors.*

**Proof.** We have only to show that ii) $\Rightarrow$ i). Let $a \in \mathcal{A} \setminus \{0\}$. The mapping $\mathcal{A} \to \mathcal{A}$, $x \mapsto ax$ is injective by hypothesis, and in fact, since $\dim \mathcal{A} < \infty$, is actually bijective. Thus every equation $ax = b$ has an unique solution. Similarly, by considering the mapping $\mathcal{A} \to \mathcal{A}$, $y \mapsto ya$ we see that the equation $ya = b$ has an unique solution.                                        $\square$

It is not a trivial matter to give an example of a real division algebra other than $\mathbb{R}$ and $\mathbb{C}$. The simplest such algebra is the Hamiltonian algebra of quaternions, described in the next chapter.

**6. Construction of Algebras by Means of Bases.** There is a simple process whereby a real $n$-dimensional vector space $V$ is made into an algebra $A = (V, \cdot)$. We first take a base $e_1, e_2, \ldots, e_n$ in $V$. If $x = \sum_1^n \alpha_\mu e_\mu$, $y = \sum_1^n \beta_\nu e_\nu \in V$ are arbitrary vectors of $V$, then, for any product $(x, y) \mapsto xy$ we have, by virtue of the distributive law,

$$xy = \sum_{\mu,\nu=1}^{n} (\alpha_\mu \beta_\nu) e_\mu e_\nu.$$

A multiplication in $V$ is therefore already completely defined once the $n^2$ individual products $e_\mu e_\nu$ have been assigned. Their values can be arbitrarily chosen in $V$, and in this way every possible $\mathbb{R}$-algebra on $V$ can be obtained. Most of these algebras are of no interest at all. If one wishes to construct algebras with an identity element then one may conveniently postulate that $e_1$ should be this element. It will then follow that $e_1 e_\nu = e_\nu e_1 = e_\nu$ for all $\nu = 1, \ldots, n$; but the remaining $(n-1)^2$ products $e_\mu e_\nu$, $2 \le \mu, \nu \le n$, can be assigned freely.

If in addition to $A = (V, \cdot)$ another $\mathbb{R}$-algebra $B = (W, \cdot)$ is given, then the $\mathbb{R}$-linear mapping $f: V \to W$ is an algebra homomorphism, if and only if $f(e_\mu e_\nu) = f(e_\mu) f(e_\nu)$ for all $\mu, \nu = 1, 2, \ldots, n$.

There is an obvious criterion for associativity and commutativity:

*The algebra $A = (V, \cdot)$ is associative if and only if $(e_\lambda e_\mu) e_\nu = e_\lambda (e_\mu e_\nu)$ for all $\lambda, \mu, \nu = 1, 2, \ldots, n$; it is commutative if and only if $e_\mu e_\nu = e_\nu e_\mu$ for all $\mu, \nu = 1, 2, \ldots, n$.*

It is extremely tedious to verify by practical calculation that the $n^3$ associativity conditions are satisfied. Even in the case where $e_1$ is an identity element, there are still $(n-1)^3$ equations to test. For this reason algebras are hardly ever defined today by specifying the products $e_\mu e_\nu$ of the base vectors. Nevertheless, it was this classical procedure that HAMILTON used to define his quaternions. DEDEKIND and WEIERSTRASS also used bases (the so-called principal units) in the commutative and associative case.

ADDITIONAL READING (for Chapters 7–11):

[1] I.L. KANTOR and A.S. SOLODOVNIKOV, Hypercomplex Numbers: An Elementary Introduction to Algebras, Springer-Verlag (1989)

[2] K.H. PARSHALL: In pursuit of the finite division algebra theorem and beyond: Joseph H.M. Wedderburn, Leonard E. Dickson and Oswald Veblen, *Arch. Internat. Hist. Sci.* 33, no. 111 (1983), 274–299.

# 7

# Hamilton's Quaternions

*M. Koecher, R. Remmert*

> Love of fame moves and cheers great mathe-
> maticians (W.R. HAMILTON).

## INTRODUCTION

**1.** Sir William Rowan Hamilton was born in Dublin in 1805, and at the age
of five was already reading Latin, Greek and Hebrew. He entered Trinity
College Dublin in 1823, and while still an undergraduate was, in 1827, appointed Andrewes Professor of Astronomy at that university, and Director
of the Dunsink Observatory with the title "Royal Astronomer of Ireland."
In that same year he began to develop geometric optics on extremal principles and in 1834/35 extended these ideas to dynamics, with the introduction
of the principle of least action, the Hamiltonian function, and his canonical
equations of motion. He was knighted in 1835 and was President of the
Royal Irish Academy from 1837 to 1845. His great discovery of quaternions
was made in 1843. He died in 1865 at Dunsink.

One of HAMILTON's earlier achievements in 1835 had been to legitimize
the traditional use of complex numbers in mathematics. He showed that
calculating with complex numbers $x + iy$ was logically equivalent to performing operations on ordered pairs $(x, y)$ of real numbers in accordance
with certain postulated rules (see 3.1.8). This was the origin of his interest
in the question of whether the geometrical interpretation of addition, and
more particularly of multiplication of complex numbers in the plane $\mathbb{R}^2$,
might not somehow—through the creation of *hypercomplex* numbers—have
an analogue in the three dimensional space $\mathbb{R}^3$ of our visual intuition.

HAMILTON had been hoping for many years to find a satisfactory form
of multiplication for real number triples with the right properties. Shortly
before his death in 1865 he wrote to his son (*Math. Papers* **3**, p. XV):
"Every morning, on my coming down to breakfast, you used to ask me:
'Well, Papa, can you multiply triplets?' Whereto I was always obliged to
reply, with a sad shake of the head: 'No, I can only add and subtract them'."

It is easy enough to see nowadays that there can be no $\mathbb{R}$-linear multiplication of all real number triples $(\alpha, \beta, \gamma)$ in $\mathbb{R}^3$ which simply extends
the multiplication in $\mathbb{C} = \mathbb{R}^2 \subset \mathbb{R}^3$ of the pairs $(\alpha, \beta)$. For if $e := (1, 0, 0)$,
$i := (0, 1, 0)$, $j := (0, 0, 1)$ be the canonical base of $\mathbb{R}^3$, then $ij$ would have

to be of the form $\rho e + \sigma i + \tau j$. It would then follow, if one assumes $i^2 = -e$ and $i(ij) = ii(j) = -j$, that

$$-j = \rho i - \sigma e + \tau ij = \rho i - \sigma e + \tau(\rho e + \sigma i + \tau j) = (\tau\rho - \sigma)e + (\tau\sigma + \rho)i + \tau^2 j,$$

and thus (since $e, i, j$ are linearly independent) that $\tau^2 = -1$, which would imply $\tau \notin \mathbb{R}$.[1]

**2.** HAMILTON's efforts are at first unsuccessful: He is looking for a multiplication with triplets in which, as with number pairs, the usual rules would still apply (in other words he assumes a principle of permanence). He begins by trying

$$\alpha + \beta i + \gamma j \quad \text{with} \quad i^2 = j^2 = -1,$$

(in which the existence of a neutral element is already implied) and considers the simplest case

$$(*) \qquad (\alpha + \beta i + \gamma j)^2 = \alpha^2 - \beta^2 - \gamma^2 + 2i\alpha\beta + 2j\alpha\gamma + 2ij\beta\gamma,$$

where the expression on the right is calculated in the ordinary way using the commutative laws.

The "touchstone" which he uses to test the value of the product of two vectors is, as in the case of $\mathbb{C}$ (where we have the modulus law) the principle that the length of the "product" of two vectors should be equal to the product of their individual lengths; the length of $\alpha + \beta i + \gamma j$ being its "Euclidean" length $\sqrt{\alpha^2 + \beta^2 + \gamma^2}$. The sum of the squares of the coefficients of 1, $i$ and $j$ on the right hand side of $(*)$ yields

$$(\alpha^2 - \beta^2 - \gamma^2)^2 + (2\alpha\beta)^2 + (2\alpha\gamma)^2 = (\alpha^2 + \beta^2 + \gamma^2)^2;$$

and thus HAMILTON has established the fact that the product rule will certainly hold provided $ij$ is made equal to zero. But he does not like this. And then he notices that the term on the right of $(*)$ should really be $ij + ji$ rather than $2ij$. This has to vanish so that $ji = -ij$; and so he is led to sacrifice the commutative law. One can see all this very clearly from a letter which HAMILTON wrote to John GRAVES on the 17th October 1843 (*Math. Papers* **3**, 106–110): "Behold me therefore tempted for a moment to

---

[1]In fact, one can prove the better

**Theorem.** *Every real division algebra $\mathcal{A}$ of odd dimension with unit element $e$ is isomorphic to $\mathbb{R}$, and therefore has dimension 1.*

**Proof.** Let $a \in \mathcal{A}$. The "left-multiplication" $L_a: \mathcal{A} \to \mathcal{A}$, $x \mapsto ax$ is a vector space endomorphism. Since dim $\mathcal{A}$ is odd, $L_a$ has a real eigenvalue (by the Bolzano–Cauchy intermediate value theorem). If $v \neq 0$ is an associated eigenvector, then $av = \lambda v$, that is $(a - \lambda e)v = 0$. Since $\mathcal{A}$ is a division algebra, it follows that $a = \lambda e$, or in other words $a \in \mathbb{R}e$, from which we see that $\mathcal{A} = \mathbb{R}e$.

fancy that $ij = 0$. But this seemed odd and uncomfortable, and I perceived that the same suppression of the term which was *de trop* might be attained by assuming what seemed to me less harsh, namely that $ji = -ij$. I made therefore $ij = k$, $ji = -k$, reserving to myself to inquire whether $k$ was 0 or not."

And now HAMILTON hit upon the ingenious idea that gave a new and decisive direction to the whole problem: he "jumped with $k$ into a fourth dimension." In other words, he took $k$ to be linearly independent of 1, $i$ and $j$. In his letter to GRAVES he wrote (*loc. cit.*) "and there dawned on me the notion that we must admit, in some sense, a *fourth dimension* of space for the purpose of calculating with triplets."

HAMILTON now carefully investigates what $k^2$ should be. If one were to use the associative law it would be immediately apparent that

$$k^2 = (ij)(ij) = i(ji)j = -i(ij)j = -i^2j^2 = -1;$$

but he does not use this argument, because he is not sure whether his multiplication is associative (his notes on this point are to be found in *Math. Papers* **3**, 103–105).

Later on he brings out clearly the validity of the associative law; thus he writes (*Math. Papers* **3**, p. 114): "... the commutative character is lost .... However it will be found that another important property of the old multiplication is preserved, or extended to the new, namely, that which may be called the *associative* character of the operation ...." This could well be the first introduction of the word "associative" in Mathematics.

**3.** The breakthrough came to HAMILTON on the 16th October 1843 on his way to a meeting of the Royal Irish Academy; during that meeting he announced his discovery of quaternions. He devoted the remainder of his life exclusively to their further exploration. He himself described in 1858 the moment of discovery in the following words (*North British Rev.* **14**, 1858): "...Tomorrow will be the fifteenth birthday of the Quaternions. They started into life, or light, full grown, on the 16th of October, 1843, as I was walking with Lady Hamilton to Dublin, and came up to Brougham Bridge. That is to say, I then and there felt the galvanic circuit of thought closed, and the sparks which fell from it were the fundamental equations between $i, j, k$ *exactly such* as I have used them ever since. I pulled out, on the spot, a pocketbook, which still exists, and made an entry, on which, *at the very moment*, I felt that it might be worth my while to expend the labour of at least ten (or it might be fifteen) years to come. But then it is fair to say that this was because I felt a *problem* to have been at that moment *solved*, an intellectual *want relieved*, which had *haunted* me for at least *fifteen years* before..." And in the letter, which we have already mentioned, to his son, he says, referring to that memorable October day: "Nor could I resist the impulse—unphilosophical as it may have been—to

cut with a knife on a stone of Brougham Bridge the fundamental formula
with the symbols $i, j, k$:

$$i^2 = j^2 = k^2 = ijk = -1."$$

With great delight HAMILTON verifies the validity of the product rule for
his quaternion multiplication, and writes (*Math. Papers* **3**, p. 108): "But I
considered it essential to try whether [my] equations were consistent with
the law of moduli,..., without which consistence being verified, I should
have regarded the whole speculation as a failure."

Neither HAMILTON nor anyone else at the time was aware that EULER
had already been in possession of the characteristic laws applying to quaternions, as early as 1748. In a letter to GOLDBACH on the 4th of May he gives
the product rule in the form of the "four squares theorem" (see 2.3 on this
point). GAUSS also knew about the rules for calculating with quaternions;
he wrote in 1819 a short note (not published at the time) on "Mutations
of space," in which the quaternion formulae appear (*Werke* **8**, 357–362).

**4.** HAMILTON regarded the creation of his quaternions as being on a par
with the creation of the infinitesimal calculus. He acknowledged no contemporary mathematicians other than GAUSS and GRASSMANN as having
played any part. F. ENGEL, on page 208 of his very readable account of
GRASSMANN's life, wrote: "Graßmann teilt sich mit Gauß in die Ehre, daß
Hamilton ihm zutraut, er könne die Quaternionen gefunden haben, und
sich immer von Neuem freut, daß es allem Anscheine nach doch nicht der
Fall ist" ("Graßmanns Leben," Teubner Verlag, Leipzig 1911). [GRASS-
MANN shares with GAUSS the honor, accorded to him by HAMILTON, that
he (GRASSMANN) could have discovered quaternions, and that he (HAMIL-
TON) is always delighted with the news that to all appearances this is not
the case.]

HAMILTON believed that his quaternions would play a key role in physics.
With missionary zeal he strove to get them accepted by the mathematical world. Thus in Dublin, quaternions became an official examination
subject; a "cosmic" significance was attributed to them. Felix KLEIN in
his well-known *Vorlesungen über die Entwicklung der Mathematik im 19.
Jahrhundert* (Vol. 1, p. 184) [Lectures on the development of mathematics
in the 19th century] gave a very harsh judgement when he wrote: "Hamilton selbst gestaltete sie [= Quaternionen] für sich zu einer Art orthodoxer
Lehre des mathematischen Credo, in die er alle seine geometrischen und
sonstigen Interessen hineinzwang, je mehr sich gegen Ende seines Lebens
sein Geist vereinseitigte und ...." [Hamilton himself regarded quaternions
as a kind of orthodox doctrine of the mathematical Credo, into which all
his geometrical and other interests were forced, and this tendency became

more pronounced as towards the end of his life his mind set and he became obsessed with a single idea...]

**5.** In Ireland and England, HAMILTON became the figurehead of a school of "quaternionists" who "outdid their master in intolerance and rigidity." At the center stood a mystic formalism treated with due reverence by the initiated. One dreamt of a quaternionistic theory of functions and expected to gain new and profound insights into the whole realm of mathematics. To promote these utopian aims there was even founded, in 1895, an "International Association for promoting the study of quaternions," at Yale University in New Haven, Connecticut. Even now there are still faint echos from the great days of the quaternionists in Ireland. Thus Eamon de VALERA, the President of Ireland from 1959 to 1973, during his period of office, would occasionally attend a mathematical colloquium in Dublin, whenever the announcement of the discourse contained the word "quaternions."

The history of algebra has shown that the significance of quaternions was vastly overestimated in the last century. Nowadays it has become clear that the quaternion algebra is only a particular algebra of complex 2 × 2 matrices (see §1). It was not the discovery of quaternions which was the great achievement, but rather the recognition which came about as a result of that discovery, of the great freedom which one has available, to construct hypercomplex systems. Lord KELVIN (1824–1907) the famous Scottish physicist and writer on thermodynamics, commented caustically: "Quaternions came from Hamilton after his really good work had been done; and though beautifully ingenious, have been an unmixed evil to those who have touched them in any way."

In contrast to this opinion is a well-known saying by Thomas HILL (who was a student of B. PEIRCE, the President of Harvard in 1862): "In the great mathematical birth of 1843, the Quaternions of HAMILTON, there is as much real promise of benefit to mankind as in any event of Victoria's reign."

We refer readers who would like further historical details to:

CROWE, M.J.: *A History of Vector Analysis,* University of Notre Dame Press, Notre Dame, London 1967

ROTHE, H.: Die Hamiltonschen Quaternionen und ihre Verallgemeinerungen, *Encykl. Math. Wiss.* III, 1.2, 1300–1423, Teubner Verlag, Leipzig 1914–1931.

VAN DER WAERDEN, B.L.: Hamiltons Entdeckung der Quaternionen, Veröffentlichungen der Joachim Jungius Gesellschaft der Wissenschaften, Vandenhoeck u. Ruprecht, Göttingen 1973, 14 Seiten

## §1. THE QUATERNION ALGEBRA $\mathbb{H}$

We introduce quaternions in §1.1, following HAMILTON's example, by means of the multiplication table for the natural basis. In §1.2 quaternions are represented as special complex $2 \times 2$ matrices. A subalgebra $\mathcal{H}$ of $\mathrm{Mat}(2, \mathbb{C})$ and a natural isomorphism $F: \mathbb{H} \to \mathcal{H}$ of the quaternion algebra $\mathbb{H}$ onto $\mathcal{H}$ is constructed, which was already known to CAYLEY in 1858. With this isomorphism it becomes obvious among other things that $\mathbb{H}$ is an associative division algebra over $\mathbb{R}$. HAMILTON had to find a direct verification of the associativity of $\mathbb{H}$, because in the year of the discovery 1843, matrices were as yet unknown. It was not until 1858, that CAYLEY introduced matrices and the matrix calculus in his "A memoir on the theory of matrices" (*Math. Papers* **2**, 475–496), which includes the quaternion calculus as a special case. The algebra $\mathcal{H}$ and the isomorphism $F$ can be usefully applied throughout the whole of this chapter.

In paragraphs §1.3 to §1.7 the basic algebraic properties of the quaternions will be discussed.

**1. The Algebra $\mathbb{H}$ of the Quaternions.** In the four-dimensional $\mathbb{R}$-vector space $\mathbb{R}^4$ of ordered real number quadruples, we choose the standard basis

$$e_1 := (1,0,0,0), \quad e_2 := (0,1,0,0), \quad e_3 := (0,0,1,0), \quad e_4 := (0,0,0,1).$$

We now introduce the so-called *Hamiltonian multiplication*. Let $e_1$, be the unit element; then the nine products $e_\mu e_\nu$, $2 \leq \mu, \nu \leq 4$, still have to be specified, and we define them by the following relations

$$\left. \begin{array}{lll} e_2 e_2 := -e_1, & e_2 e_3 := e_4, & e_2 e_4 := -e_3 \\ e_3 e_2 := -e_4, & e_3 e_3 := -e_1, & e_3 e_4 := e_2 \\ e_4 e_2 := e_3, & e_4 e_3 := -e_2, & e_4 e_4 := -e_1 \end{array} \right\} \quad \text{(HAMILTON relations)}$$

This is often set out in the form of a multiplication table

|       | $e_2$   | $e_3$   | $e_4$   |
|-------|---------|---------|---------|
| $e_2$ | $-e_1$  | $e_4$   | $-e_3$  |
| $e_3$ | $-e_4$  | $-e_1$  | $e_2$   |
| $e_4$ | $e_3$   | $-e_2$  | $-e_1$  |

The four-dimensional real $\mathbb{R}$-algebra constructed in this way is called the *quaternion algebra* and denoted by $\mathbb{H}$. The elements of $\mathbb{H}$ were given the name of *quaternions* by HAMILTON.[2] Since $e_2 e_3 \neq e_3 e_2$ it is clear that *the quaternion algebra $\mathbb{H}$ is not commutative.*

---

[2] The word means any group of four persons or things, and was used, for example, in the New Testament to describe the four groups of four soldiers used by King Herod to guard Peter. (*Acts of the apostles*, 12, 4): "... he put him in prison, and delivered him to four quaternions of soldiers to keep him" (see Temple *100 years of mathematics*, London, Duckworth, 1981, p. 46).

The validity of the 27 equations $(e_\lambda e_\mu)e_\nu = e_\lambda(e_\mu e_\nu)$, $2 \le \lambda, \mu, \nu \le 4$, can be checked directly from the multiplication table, thus verifying that the quaternion algebra is associative. We refrain from doing this because associativity and more will emerge in the next paragraph in a more elegant way. Traditionally $e_1, e_2, e_3, e_4$ are denoted by $e, i, j, k$ respectively so that

$$i^2 = j^2 = k^2 = ijk = -e, \qquad ij = -ji = k.$$

The other products are derived from these by cyclic interchange of $i, j, k$. Using the distributive law we thus obtain the

**Product formula**:

$$(\alpha e + \beta i + \gamma j + \delta k)(\alpha' e + \beta' i + \gamma' j + \delta' k)$$
$$= (\alpha\alpha' - \beta\beta' - \gamma\gamma' - \delta\delta')e + (\alpha\beta' + \beta\alpha' + \gamma\delta' - \delta\gamma')i$$
$$+(\alpha\gamma' - \beta\delta' + \gamma\alpha' + \delta\beta')j + (\alpha\delta' + \beta\gamma' - \gamma\beta' + \delta\alpha')k.$$

The classical method of writing quaternions with the symbols $i, j, k$ has certain hidden dangers, for example, if we try to deal with quaternions with complex instead of real numbers as coefficients.

$\mathbb{R}e$ is an $\mathbb{R}$-subalgebra of $\mathbb{H}$. In contrast to our practice with $\mathbb{C}$, we do not however identify $\mathbb{R}e$ with $\mathbb{R}$, and therefore we consistently write $e$ and not $1$ for the unit element of $\mathbb{H}$.

**2. The Matrix Algebra $\mathcal{H}$ and the Isomorphism $F: \mathbb{H} \to \mathcal{H}$.** The set $\mathcal{C}$ of all real $2 \times 2$ matrices $\begin{pmatrix} \alpha & -\beta \\ \beta & \alpha \end{pmatrix}$, $\alpha, \beta \in \mathbb{R}$, is an $\mathbb{R}$-subalgebra of $\mathrm{Mat}(2, \mathbb{R})$, and the mapping $\alpha + \beta i \mapsto \begin{pmatrix} \alpha & -\beta \\ \beta & \alpha \end{pmatrix}$ is an $\mathbb{R}$-algebra isomorphism $\mathbb{C} \to \mathcal{C}$ (see 3.2.5). In analogy with this, we have the following.

**Theorem.** *The set* $\mathcal{H} := \left\{ \begin{pmatrix} w & -z \\ \bar{z} & \bar{w} \end{pmatrix} : w, z \in \mathbb{C} \right\}$ *is an $\mathbb{R}$-subalgebra of* $\mathrm{Mat}(2, \mathbb{C})$, *with unit element* $E := \begin{pmatrix} 1 & 0 \\ 0 & 1 \end{pmatrix}$. *Every matrix* $A = \begin{pmatrix} w & -z \\ \bar{z} & \bar{w} \end{pmatrix} \in \mathcal{H}$ *satisfies, over $\mathbb{R}$, the quadratic equation*

(1)   $A^2 - (\mathrm{trace}\, A)A + (\det A)E = 0$
      *where* $\mathrm{trace}\, A = 2\,\mathrm{Re}\, w$, $\det A = |w|^2 + |z|^2$

$\mathcal{H}$ *is a 4-dimensional, associative division algebra.*

**Proof.** 1) It is easily verified, by direct calculation, that $\mathcal{H}$ is a four-dimensional $\mathbb{R}$-vector subspace of $\mathrm{Mat}(2, \mathbb{C})$ which is closed under matrix multiplication. The matrix equation $A^2 - (\mathrm{trace}\, A)A + (\det A)E = 0$ can be checked in the same way.

2) The algebra $\mathcal{H}$ is associative because $\mathrm{Mat}(2,\mathbb{C})$ is. To see that $\mathcal{H}$ is a division algebra we need to use the criterion R.5. Accordingly suppose $A, B \in \mathcal{H}$ and $AB = 0$. It then follows that $\det A \cdot \det B = 0$, and hence $\det A = 0$ or $\det B = 0$. As $\det \begin{pmatrix} w & -z \\ \bar{z} & \bar{w} \end{pmatrix} = |w|^2 + |z|^2$ vanishes only for $w = z = 0$, the required statement follows.    $\square$

The equation (1) is the statement of the so-called CAYLEY theorem (or of the CAYLEY–HAMILTON theorem for the special case of $2 \times 2$ matrices) (See S. Lang, *An Introduction to Linear Algebra*, 2nd ed., Springer-Verlag.)

**Lemma.** *The mapping*

$$F: \mathbb{H} \to \mathcal{H}, \qquad (\alpha, \beta, \gamma, \delta) \mapsto \begin{pmatrix} \alpha + \beta i & -\gamma - \delta i \\ \gamma - \delta i & \alpha - \beta i \end{pmatrix},$$

*is an $\mathbb{R}$-algebra isomorphism, and*

$$F(e_1) = E, \qquad\qquad F(e_2) = \begin{pmatrix} i & 0 \\ 0 & -i \end{pmatrix} =: I,$$

$$F(e_3) = \begin{pmatrix} 0 & -1 \\ +1 & 0 \end{pmatrix} =: J, \quad F(e_4) = \begin{pmatrix} 0 & -i \\ -i & 0 \end{pmatrix} =: K.$$

**Proof.** The mapping $F$ is obviously $\mathbb{R}$-linear and bijective. It remains to be shown (see R.6) that $F(e_\mu)F(e_\nu) = F(e_\mu e_\nu)$ for $\mu, \nu = 1, 2, 3, 4$. This however is clear, because the matrices $E, I, J, K$ are the images under $F$ of $e_1, e_2, e_3, e_4$ and satisfy the same laws of multiplication as $e_1, e_2, e_3, e_4$. The relations $I^2 = J^2 = -E$, $IJ = -JI = K$ are easily checked and the remaining relations can be derived from the associative law, for example, $K^2 = (IJ)(-JI) = -IJ^2I = I^2 = -E$.    $\square$

**Corollary.** *The Hamiltonian algebra $\mathbb{H}$ is an associative division algebra.*

By Lemma R.5, $\mathcal{H} \setminus \{0\}$ is a group with respect to multiplication. One can immediately verify that:

*The set $\{E, -E, I, -I, J, -J, K, -K\}$ is a noncommutative subgroup of $\mathcal{H} \setminus \{0\}$, each of whose elements other than $\pm E$ is of order 4.*

This group, and any group isomorphic to it, is known, in the literature, as the *(finite) quaternion group*.

The representation of quaternions by complex $2 \times 2$ matrices which we have used here was already familiar to CAYLEY in 1858. In his famous "Memoir on the theory of matrices" (*Math. Papers* **2**, p. 491) he writes: "It may be noticed in passing, that if $L, M$ are skew convertible matrices of the order 2, and if these matrices are also such that $L^2 = -1$, $M^2 = -1$, then putting $N = LM = -ML$, we obtain

$$L^2 = -1, \qquad M^2 = -1, \qquad N^2 = -1,$$

$$L = MN = -NM, \quad M = NL = -NL[\text{sic}], \quad N = LM = -ML,$$

which is a system of relations precisely similar to that in the theory of quaternions." CAYLEY does not however give explicit examples for $L, M$.

As calculations with complex matrices can be performed more elegantly than with quaternions, it is often preferable—as above—to prove theorems about $\mathbb{H}$, by first proving them for the algebra $\mathcal{H}$, and then using the isomorphism $F : \mathbb{H} \to \mathcal{H}$ to "lift" them to $\mathbb{H}$. We shall use this principle again later. □

As with complex numbers earlier, there are many possible ways of representing the quaternion algebra $\mathbb{H}$ as an $\mathbb{R}$-subalgebra of $\mathrm{Mat}(2, \mathbb{C})$. One can choose three matrices $I_2, I_3, I_4 \in \mathrm{Mat}(2, \mathbb{C})$, in any way one likes as long as the nine Hamiltonian conditions are satisfied. The mapping

$$\mathbb{H} \to \mathrm{Mat}(2, \mathbb{C}), \qquad (\alpha, \beta, \gamma, \delta) \mapsto \alpha E + \beta I_2 + \gamma I_3 + \delta I_4$$

is then an $\mathbb{R}$-*algebra monomorphism*. It can be shown, as a generalization of the theorem in 3.2.5, that:

*If $g : \mathbb{H} \to \mathrm{Mat}(2, \mathbb{C})$ is an $\mathbb{R}$-algebra monomorphism, then there is an invertible matrix $W \in \mathrm{Mat}(2, \mathbb{C})$, such that the associated "inner automorphism" $\iota_W : \mathrm{Mat}(2, \mathbb{C}) \to \mathrm{Mat}(2, \mathbb{C})$, $A \mapsto W^{-1}AW$ has the property $g = \iota_W \circ F$.*

**3. The Imaginary Space of $\mathbb{H}$.** We use the standard basis $e, i, j, k$. The three-dimensional vector subspace

$$(1) \qquad\qquad \mathrm{Im}\,\mathbb{H} := \mathbb{R}i + \mathbb{R}j + \mathbb{R}k$$

of $\mathbb{H}$ is called—in analogy to the complex numbers—the imaginary space of $\mathbb{H}$. Its elements are called "purely imaginary." $\mathbb{H}$ is a direct sum of the vector spaces $\mathbb{R}e$ and $\mathrm{Im}\,\mathbb{H}$

$$(2) \qquad\qquad \mathbb{H} = \mathbb{R}e \oplus \mathrm{Im}\,\mathbb{H}.$$

The line $\mathbb{R}e$ is defined invariantly by the unit element $e$. The definition of $\mathrm{Im}\,\mathbb{H}$ is initially dependent on the basis. In order to characterize $\mathrm{Im}\,\mathbb{H}$ invariantly, we note that the quaternion $x = \alpha e + \beta i + \gamma j + \delta k$ satisfies, by Theorem 2, the quadratic equation

$$(3) \qquad\qquad x^2 = 2\alpha x - (\alpha^2 + \beta^2 + \gamma^2 + \delta^2)e.$$

As $x \in \text{Im}\,\mathbb{H}$ if and only if $\alpha = 0$, we obtain the basis-free representation

(4)                 $\text{Im}\,\mathbb{H} = \{x \in \mathbb{H}: x^2 \in \mathbb{R}e \text{ and } x \notin \mathbb{R}e \setminus \{0\}\}.$

$\text{Im}\,\mathbb{H}$ is *not* an $\mathbb{R}$-subalgebra of $\mathbb{H}$. We note that:

*For purely imaginary quaternions $u, v$, the following relations hold: $u^2 = -we$ with $\omega \geq 0$ and $uv + vu \in \mathbb{R}e$.*

**Proof.** If $u = \beta i + \gamma j + \delta k$ then $u^2 = -(\beta^2 + \gamma^2 + \delta^2)e$ with $\beta^2 + \gamma^2 + \delta^2 \geq 0$. Since $u$, $v$, $u + v$ all belong to $\text{Im}\,\mathbb{H}$, it follows that $uv + vu = (u+v)^2 - u^2 - v^2 \in \mathbb{R}e$.                           $\square$

In particular for every $u \in \text{Im}\,\mathbb{H}$, $u \neq 0$, a scalar $\rho$ (namely $\rho := \sqrt{\omega^{-1}}$)) can be found such that $(\rho u)^2 = -e$ (normalization).

The imaginary space $\text{Im}\,\mathbb{H}$ plays a dominant role in the theory of quaternions. Its elements are also called *vectorial* (or pure) quaternions. The expression "vector" first appears in HAMILTON's writings in 1845, (*Q. Jl. Math.* 1, p. 56). In the long drawn-out war of resistance against the vector calculus, Lord KELVIN was even in 1896 expressing the opinion that: "Vector is a useless survival, or offshoot from quaternions, and has never been of the slightest use to any creature."

By (2) every quaternion $x$ can be expressed uniquely in the form

(5)              $x = \alpha e + u$   with   $\alpha \in \mathbb{R}$   and   $u \in \text{Im}\,\mathbb{H}.$

In this expression $\alpha e$ is sometimes called the *scalar part* (or *real part*) and $u$ the *vector(ial) part* (or *imaginary part*) of $x$.

Every plane in $\mathbb{H}$, containing the straight line $\mathbb{R}e$, is a subalgebra of $\mathbb{H}$, isomorphic to $\mathbb{C}$. It is however fundamentally impossible to make $\mathbb{H}$ "somehow or other" into a $\mathbb{C}$-algebra.[3]

**4. Quaternion Product, Vector Product and Scalar Product.** For vectorial quaternions $u = \beta i + \gamma j + \delta k$, $v = \rho i + \sigma j + \tau k$ we have

(1)   $uv = -(\beta\rho + \gamma\sigma + \delta\tau)e + (\gamma\tau - \delta\sigma)i + (\delta\rho - \beta\tau)j + (\beta\sigma - \gamma\rho)k.$

Here the "scalar part" is, apart from sign, the canonical Euclidean *scalar product* $\langle u, v \rangle$ of the vectors $u = (\beta, \gamma, \delta)$, $v = (\rho, \sigma, \tau) \in \mathbb{R}^3$; the "vectorial

---

[3]In fact we can improve on this with the following.

**Theorem.** *Every finite dimensional complex division algebra with unit element is isomorphic to $\mathbb{C}$.*

This is proved in the same way as the theorem in the footnote on page 190: the left-multiplication $L_a$ now has a complex eigenvalue $\lambda$ (Fundamental theorem of algebra).

part" of $uv$ is the *vector product* (*cross product*) of these two vectors. We thus obtain the aesthetically pleasing formula

$$(2) \qquad uv = -\langle u, v \rangle e + u \times v, \qquad u, v, u \times v \in \operatorname{Im} \mathbb{H},$$

The mapping $(u, v) \mapsto u \times v$ is by definition bilinear and anticommutative:

$$(3) \qquad u \times v = -v \times u, \qquad u, v \in \operatorname{Im} \mathbb{H}.$$

From (3) we get immediately using (2)

$$(4) \quad u \times v = \frac{1}{2}(uv - vu), \qquad \langle u, v \rangle e = -\frac{1}{2}(uv + vu) \quad \text{for all } u, v \in \operatorname{Im} \mathbb{H}.$$

The vector product is not associative. We note that

$$(5) \qquad u \times (v \times w) = \frac{1}{2}(uvw - vwu), \qquad u, v, w \in \operatorname{Im} \mathbb{H}.$$

**Proof.** Since $uvw = -\langle v, w \rangle u + u(v \times w)$ and $vwu = -\langle v, w \rangle u + (v \times w)u$ by (2), the identity (5) follows from $u(v \times w) - (v \times w)u = 2u \times (v \times w)$. $\square$

*Exercise.* Show that every quaternion $a \in \mathbb{H}$ can be represented (in infinitely many different ways) as the product $a = bc$ of two purely imaginary quaternions $b, c$.

As a substitute to some extent for the associative law we have the

**GRASSMANN Identity:** $u \times (v \times w) = \langle u, w \rangle v - \langle u, v \rangle w.$

**Proof.** This follows from (5) with the help of (3), since

$$uvw - vwu = (uv + vu)w - v(uw + wu) = -2\langle u, v \rangle w + 2\langle u, w \rangle v. \qquad \square$$

If we introduce $u, v, w$ cyclically in (5) or in the GRASSMAN identity, and add, we obtain the

**JACOBI Identity:** $u \times (v \times w) + v \times (w \times u) + w \times (u \times v) = 0.$

This identity and (4) assert that the span $\mathbb{R}^3 \cong \operatorname{Im} \mathbb{H}$, with the vector product, is a LIE algebra (see R.2.3).

It follows directly from (1), or from (2) with the product rule 2.2(4), that:

$$(6) \qquad \langle u, v \rangle^2 + |u \times v|^2 = |u|^2 |v|^2.$$

This is a strengthened version of the CAUCHY–SCHWARZ inequality. If we write $\langle u, v \rangle = |u| \, |v| \cos \varphi$ with $\varphi \in [0, \pi]$, we obtain

$$|u \times v| = |u| \, |v| \sin \varphi.$$

Thus $|u \times v|$ is the area of the parallelogram spanned by the vectors $u, v$. The equation (6) plays a central role in the theory of vector product algebras (see 10.3.1).

The triple (scalar) product of three vectors $u, v, w \in \operatorname{Im} \mathbb{H}$ is the real number $\langle u \times v, w \rangle$. Since $\mathbb{R}e$ and $\operatorname{Im} \mathbb{H}$ are orthogonal, it follows from (2) that

$$\langle u \times v, w \rangle = \langle uv, w \rangle, \qquad u, v, w \in \operatorname{Im} \mathbb{H}.$$

We can immediately deduce from (1) that $\langle u \times v, u \rangle = 0$. After replacing $u$ by $u + v$, and taking account of (4), we get the

**Interchange Rule:** $\langle u \times v, w \rangle = \langle u, v \times w \rangle$, $u, v, w \in \operatorname{Im} \mathbb{H}$.

It is at once clear from this that the mapping $\langle u, v, w \rangle \mapsto \langle u \times v, w \rangle$ is a determinant function of the vector space $\operatorname{Im} \mathbb{H}$.

*Exercise.* Show that, for $2 \times 2$ matrices $A, B, C$

$$\begin{aligned}
[A, [B, C]] = {} & \{2\sigma(AB) - \sigma(A)\sigma(B)\}C - \{2\sigma(AC) - \sigma(A)\sigma(C)\}B \\
& + \{\sigma(B)\sigma(AC) - \sigma(C)\sigma(AB)\}E,
\end{aligned}$$

where $[A, B] := AB - BA$, $\sigma(A) :=$ trace $A$, $E :=$ unit matrix. Show also that, in the case where $A, B, C \in F(\operatorname{Im} \mathbb{H})$, this is the GRASSMANN identity.

*Historical Remarks.* Vector multiplication was discovered by H. GRASS-MANN in 1844 (one year after HAMILTON's discovery of quaternions) as a special case of a much more general so-called "exterior product." The algebra of vectors in $\mathbb{R}^3$ first became popular however in the eighties of the last century through the works of the American physicist and mathematician Josiah Willard GIBBS (1839–1903) who was a professor at Yale University. GIBBS maintained amongst other arguments—what seems to us nowadays almost self-evident—that the scalar product $\langle u, v \rangle$ and the vector product $u \times v$ have their own independent meaning and that the quaternion product $uv$ in which these two products are combined with one another has no essential significance in many problems. GIBBS was an opponent of the quaternionists, and it was because of this controversy that a colleague of GIBBS founded in 1895 at Yale the association for the worldwide promotion of quaternions, mentioned earlier in the introduction.

**5. Noncommutativity of $\mathbb{H}$. The Center.** The fact that $\mathbb{H}$ is not commutative leads to many unusual consequences. Thus polynomials can have more zeros than is indicated by their degree. For example, the quadratic polynomial $X^2 + e$ has, as zeros, *all* purely imaginary quaternions $u = \beta i + \gamma j + \delta k$, whose "length" $\beta^2 + \gamma^2 + \delta^2$ equals one. These quaternions represent the surface of the unit sphere in the three-dimensional space $\mathbb{R}^3$ or the real number triples $(\beta, \gamma, \delta)$.[4]

Another statement we can make is that:

*There are cubic polynomials over $\mathbb{H}$, for example, $X^2iXi + iX^2iX - iXiX^2 - XiX^2i$, which assume the value zero for all quaternions.*

As every quaternion satisfies an equation $X^2 = \alpha X + \beta e$ the truth of this statement can be proved by substituting $\alpha X + \beta e$ for $X^2$ in the above polynomial. □

Since $\mathbb{H}$ is not commutative, the naive definition of determinants fails. For example neither

$$\det \begin{pmatrix} a & b \\ c & d \end{pmatrix} := ad - bc \qquad \text{nor} \qquad \det \begin{pmatrix} a & b \\ c & d \end{pmatrix} := ad - cb$$

would be a suitable definition. In the first case, we would have

$$\det \begin{pmatrix} i & j \\ i & j \end{pmatrix} = ij - ji = 2k \neq 0,$$

and in the second case we would have

$$\det \begin{pmatrix} i & i \\ j & j \end{pmatrix} = ij - ji \neq 0,$$

so that neither determinant would vanish even though the first has equal columns and the second equal rows.

To measure the departure from commutativity of an algebra $\mathcal{A}$, we consider its *center*

$$Z(\mathcal{A}) := \{z \in \mathcal{A} : zx = xz \quad \text{for all} \quad x \in \mathcal{A}\}.$$

If $\mathcal{A}$ is associative, $Z(\mathcal{A})$ is a subalgebra of $\mathcal{A}$, and $Z(\mathcal{A}) = \mathcal{A}$ if and only if $\mathcal{A}$ is commutative. For algebras with a unit element $e$, $\mathbb{R}e \in Z(\mathcal{A})$. The extreme case $Z(\mathcal{A}) = \mathbb{R}e$ can occur.

*For the algebra $\mathbb{H}$ we have $Z(\mathbb{H}) = \mathbb{R}e = \{x \in \mathbb{H} : xu = ux \text{ for all } u \in \operatorname{Im}\mathbb{H}\}$.*

This is included in the following statement:

*For all $u \in \mathbb{H} \setminus \mathbb{R}e$, $\{x \in \mathbb{H} : xu = ux\} = \mathbb{R}e + \mathbb{R}u$.*

**Proof.** Since $\{x \in \mathbb{H} : xu = ux\} = \{x \in \mathbb{H} : xv = vx\}$ for $v := u - (\mathbb{R}e\, u)e$ one can assume that $u \in \operatorname{Im}\mathbb{H}$, $u \neq 0$. One can even assume $u^2 = -e$ and

---

[4]The reason for this phenomenon is that polynomials over $\mathbb{H}$ no longer factor in the usual way. For example $(X - x)(X - y) = X^2 - xX - Xy + xy$, and the linear terms cannot be combined into $-(x + y)X$.

$x^2 = -e$ (we pass from $x$ to $x - \alpha e$ and normalize). It then follows that $(x - u)(x + u) = x^2 - ux + xu - u^2 = 0$, so that $x = \pm u$. $\quad\square$

The noncommutativity of $\mathbb{H}$ is also the reason why $\mathbb{H}$ has many $\mathbb{R}$-algebra automorphisms: every $a \in \mathbb{H}$, $a \neq 0$ induces a so-called *inner automorphism* $h_a : \mathbb{H} \to \mathbb{H}$, $x \mapsto axa^{-1}$. Since $Z(\mathcal{A}) = \mathbb{R}e$, we have $h_a = h_b$, if and only if $b^{-1}a \in \mathbb{R}e$. We shall show in 3.2 that the $\mathbb{R}$-algebra $\mathbb{H}$ has no other automorphisms.

*Exercise.* Show that, for any two elements $a, b \in \mathbb{H}$, the following statements are equivalent:

i) $ab = ba$.

ii) $e, a, b$ are linearly independent.

iii) there is a subalgebra of $\mathbb{H}$, isomorphic to $\mathbb{C}$, and containing $a$ and $b$.

## 6. The Endomorphisms of the $\mathbb{R}$-Vector Space $\mathbb{H}$.

For any two quaternions $a, b$ the mapping $\mathbb{H} \to \mathbb{H}$, $x \mapsto axb$ is an $\mathbb{R}$-linear mapping of $\mathbb{H}$ into itself (*an endomorphism*). We denote by $\operatorname{End}\mathbb{H}$ the $\mathbb{R}$-vector space of all endomorphisms of $\mathbb{H}$.

**Theorem.** *If $a_1, \ldots, a_4$ is a basis of $\mathbb{H}$, the mapping $\mathbb{H}^4 \to \operatorname{End}\mathbb{H}$,*

$$(b_1, b_2, b_3, b_4) \mapsto f \in \operatorname{End}\mathbb{H} \quad \text{with} \quad f(x) := \sum_1^4 a_\nu x b_\nu$$

*is $\mathbb{R}$-linear and bijective.*

**Proof.** The $\mathbb{R}$-linearity is obvious. Since $\dim\mathbb{H}^4 = \dim(\operatorname{End}\mathbb{H}) = 16$, as $\dim\mathbb{H} = 4$, it only remains to prove the injectivity of the mapping in question. This is the case $n = 4$ of the following auxiliary proposition:

*Let $n = 1, 2, 3$ or $4$ and suppose $\sum_1^n a_\nu x b_\nu = 0$ for all $x \in \mathbb{H}$ then $b_1 = \cdots = b_n = 0$.*

We argue by induction, the case $n = 1$ being clear. Suppose $n > 1$, then if $b_1$ were not zero, we should have

$$(*) \qquad a_1 x + \sum_2^n a_\nu x q_\nu = 0 \quad \text{with} \quad q_\nu := b_\nu b_1^{-1}.$$

If we now multiply this equation on the right by $y$, and subtract from it the equation obtained by replacing $x$ by $xy$ (in the original equation) we obtain

$$\sum_2^n a_\nu x (q_\nu y - y q_\nu) = 0 \quad \text{and hence} \quad q_\nu y = y q_\nu \quad \text{for all} \quad y \in \mathbb{H}$$

by the inductive hypothesis. Since $Z(\mathbb{H}) = \mathbb{R}e$ it follows that $q_\nu = \alpha_\nu e$, $\alpha_\nu \in \mathbb{R}$. We now deduce from $(*)$ that

$$\left(a_1 + \sum_2^n \alpha_\nu a_\nu\right) x = 0, \quad \text{that is} \quad a_1 + \sum_2^n \alpha_\nu a_\nu = 0,$$

or in other words $a_1, \ldots, a_4$ would be linearly dependent. It follows that $b_1 = 0$, and similarly $b_2 = b_3 = b_4 = 0$.      $\square$

*Example.* The conjugation $x \mapsto \bar{x}$ (see §2.1) belongs to End $\mathbb{H}$, and with respect to the basis $1$, $i$, $j$, $k$, we have:

$$\bar{x} = -\frac{1}{2}(x + ixi + jxj + kxk).$$

The theorem proved here is to be found in HAMILTON's work *Elements of quaternions*, which was published by his son in 1866. The analogue of this theorem does not hold for the field $\mathbb{C}$, where (see 3.3.1) the $\mathbb{R}$-linear mappings $\mathbb{C} \to \mathbb{C}$ have the form $z \mapsto az + b\bar{z}$. The fact that one can work without conjugate quaternions is really tied in with the fact that $\mathbb{H}$ has a one-dimensional center, whereas this is not true of $\mathbb{C}$.

**7. Quaternion Multiplication and Vector Analysis.** HAMILTON applied quaternion multiplication to derive important formulae in vector analysis in an elegant fashion. He introduced the "Nabla" operator

$$\nabla := \frac{\partial}{\partial x}i + \frac{\partial}{\partial y}j + \frac{\partial}{\partial z}k$$

(he chose the word nabla because of the similarity of the shape of the symbol to that of an Hebrew musical instrument of that name). The application of $\nabla$ to a differentiable function $f(x, y, z)$ of three real variables, gives the gradient of $f$

$$\nabla f := \frac{\partial f}{\partial x}i + \frac{\partial f}{\partial y}j + \frac{\partial f}{\partial z}k = \text{grad } f.$$

Application of $\nabla$ to a "differentiable quaternion field" $F(x, y, z) = u(x, y, z)i + v(x, y, z)j + w(x, y, z)k$ gives, when formally expanded:

$$\nabla F = -\left(\frac{\partial u}{\partial x} + \frac{\partial v}{\partial y} + \frac{\partial w}{\partial z}\right)e + \left(\frac{\partial w}{\partial y} - \frac{\partial v}{\partial z}\right)i$$
$$+ \left(\frac{\partial u}{\partial z} - \frac{\partial w}{\partial x}\right)j + \left(\frac{\partial v}{\partial x} - \frac{\partial u}{\partial y}\right)k.$$

The real part, up to sign, is the *divergence* of $F$, and the imaginary part is the *curl F*, of the vector field $F$:

$$\nabla F = -\text{div } F + \text{curl } F.$$

Applying the operator $\nabla$ twice to a function $f$ leads to the well-known LAPLACE *operator* of potential theory, more precisely:

$$\nabla^2 f = -\Delta f = -\left(\frac{\partial^2 f}{\partial x^2} + \frac{\partial^2 f}{\partial y^2} + \frac{\partial^2 f}{\partial z^2}\right).$$

All this works amazingly well. Felix KLEIN, in the first volume of his *Vorlesungen über die Entwicklung der Mathematik im 19. Jahrhundert* (p. 188) writes: "Die Leichtigkeit und Eleganz ist in der Tat überraschend, und es läß sich wohl von hier aus die alles andere ablehnende Begeisterung der Quaternionisten für ihr System begreifen, die bald über vernünftige Grenzen hinauswuchs, in einer weder der Mathematik als Ganzem noch der Quaternionentheorie selbst förderlichen Weise." [The ease and elegance is indeed astonishing, and may well account for the enthusiasm of the quaternionists for their system and their rejection of all others; an enthusiasm which soon outgrew all reasonable bounds and advanced neither the theory of quaternions itself nor mathematics as a whole.]

**8. The Fundamental Theorem of Algebra for Quaternions.** It is easily seen that

*Every polynomial $X^n - a$, $a \in \mathbb{H}$, of degree $n > 0$ has zeros in every plane in $\mathbb{H}$ containing 0, e, and a.*

**Proof.** Every such plane $E$ is a subalgebra of $\mathbb{H}$, isomorphic to $\mathbb{C}$, and so $X^n - a$, by the fundamental theorem of algebra for complex numbers, always has zeros in $E$.                                                           □

The number of zeros of $X^n - a$ can be infinite, for example, for $X^2 + e$. The reader may care to show that:

*If $\operatorname{Im} a \neq 0$, then $X^n - a$ has exactly $n$ zeros in $\mathbb{H}$.*                    □

The exponential series

$$\exp x := \sum_0^\infty \frac{x^n}{n!}, \qquad x \in \mathbb{H}$$

converges *absolutely*, and uniformly on compact subsets (that is, w.r.t. *the norm*) of $\mathbb{H}$. By multiplication of series, or by reduction to the complex case (see the proof above and Exercise 7.1.5), one obtains the

**Addition Theorem.** $(\exp x)(\exp y) = \exp(x + y)$, *if* $xy = yx$.

Quaternions have a "representation in polar coordinates":

$$a = |a|(\exp u) \quad \text{with} \quad u \in \operatorname{Im}\mathbb{H}.$$

To see this, we again consider a plane in $\mathbb{H}$, containing $0$, $e$, and $a$, and transfer to it what we know about $\mathbb{C}$. With this representation, the roots of $X^n - a$ can be given an explicit solution:

*if $a = |a|(\exp u)$, then $b := \sqrt[n]{|a|}(\exp \frac{1}{n} u)$ is a zero of $X^n - a$.* □

It is not immediately apparent that quadratic polynomials $X^2 + aX + b$, $a, b \in \mathbb{H}$, always have zeros in $\mathbb{H}$ (the reduction to pure polynomials by "completing the square" only works if $ab = ba$). Nevertheless as we shall see a fundamental theorem of algebra does in fact hold for $\mathbb{H}$ as well. We first define, inductively, the concept of a monomial (over $\mathbb{H}$). Any constant $a \neq 0$ is a monomial of degree 0. The "indeterminate" $X$ is a monomial of degree 1. If $m_1$ and $m_2$ are monomials of degree $k_1$ and $k_2$ respectively, their product $m_1 m_2$ is a monomial of degree $k_1 + k_2$. The general monomial of the $n$th degree accordingly has the form

$$a_1 X^{n_1} a_2 X^{n_2} a_3 \cdots a_r X^{n_r} a_{r+1}, a_\rho \in \mathbb{H} \setminus \{0\}, \quad n = n_1 + n_2 + \cdots + n_r.$$

Any finite sum of monomials is said to be a polynomial (over $\mathbb{H}$). Every polynomial $p$ defines a continuous mapping $p: \mathbb{H} \to \mathbb{H}$, $x \mapsto p(x)$.

**Fundamental Theorem of Algebra for Quaternions.** *Let $p$ be a polynomial over $\mathbb{H}$ of degree $n > 0$ of the form $m + q$ where $m$ is a monomial of degree $n$ and $q$ a polynomial of degree $< n$. Then the mapping $p: \mathbb{H} \to \mathbb{H}$ is surjective, and in particular $p$ has zeros in $\mathbb{H}$.*

*Remark.* The hypothesis that only *one* monomial of highest degree is present in $p$, is essential to the validity of this theorem. Thus, for example, the linear polynomial $iX - Xi + 1$ has no zero in $\mathbb{H}$ (because for any $a \in \mathbb{H}$, the polynomial $aX - Xa$ assumes values only in $\operatorname{Im} \mathbb{H}$, since for all $a = \alpha e + u$, $x = \xi e + v$, $u, v \in \operatorname{Im} \mathbb{H}$, we have $ax - xa = 2u \times v$). □

The usual proofs of the fundamental theorem for $\mathbb{C}$ do not carry over to $\mathbb{H}$. With the help of the more powerful methods of topology a proof can be given as follows. Since in $p$, *the monomial of the $n$th degree dominates the remaining terms* for large $x \in \mathbb{H}$, we have the *growth equation:* $\lim_{x \to \infty} |p(x)| = \infty$. Thus $p$ can be extended to a *continuous* mapping $\hat{p}: S^4 \to S^4$ of the four-dimensional sphere into itself with $\hat{p}(\infty) := \infty$. ($S^4$ is regarded as the compactification of $\mathbb{H} \cong \mathbb{R}^4$ by the addition of a point $\infty$). It can now be shown that the mapping $\hat{p}$ has degree $n$ (in the sense of topology). Since $n \neq 0$, it follows from a general theorem of topology that $\hat{p}$ is surjective, and this means that $p(\mathbb{H}) = (\mathbb{H})$.

*Historical Note.* The fundamental theorem was proved in 1944 by EILEN-BERG and NIVEN, using the notion of the degree of a mapping, in the paper: The "fundamental theorem of algebra for quaternions," in *Bull. AMS* **50**, 246–248, after NIVEN had already resolved the special case in which all terms in $p$ have the form $aX^k$ and the indeterminate $X$ commutes with $a$ (see I. NIVEN: Equations in quaternions, in *Am. Math. Monthly* **48**, 654–661). The topological proof is also given in the textbook of S. EILENBERG and N. STEENROD: *Foundations of algebraic topology*, Princeton University Press, 1952, 306–311. It is rather surprising that, until the year 1941, the subject of the fundamental theorem for quaternions was never treated in the literature.

## §2. THE ALGEBRA $\mathbb{H}$ AS A EUCLIDEAN VECTOR SPACE

If $V$ is a real vector space, then a *bilinear form* $V \times V \to \mathbb{R}$, $(x,y) \mapsto \langle x,y \rangle$, is said to be a *scalar product*, if it is *symmetric* and *positive definite*, that is

$$\langle x,y \rangle = \langle y,x \rangle \quad \text{and} \quad \langle x,x \rangle > 0 \quad \text{for} \quad x \neq 0.$$

$V$ together with a scalar product is called a *Euclidean* vector space. The number $|x| := +\sqrt{\langle x,x \rangle} \geq 0$ is called the (*Euclidean*) *length*, or the *norm*, of the vector $x \in V$. Two vectors $x, y \in V$ are said to be *orthogonal* (or to be *perpendicular to each other*), if $\langle x,y \rangle = 0$.

The object of this section is to introduce a scalar product in the quaternion algebra $\mathbb{H}$, which fits in well with the multiplication in $\mathbb{H}$. In $\mathbb{C} = \mathbb{R}^2$, the scalar product $\langle w,z \rangle = \operatorname{Re}(w\bar{z})$ is an optimal choice which is compatible with multiplication in $\mathbb{C}$, as the product rule $|wz| = |w|\,|z|$ shows (see 3.3.4). We shall see that an analogous situation applies to $\mathbb{H} = \mathbb{R}^4$, if one defines, for any two quaternions $x = \alpha e + \beta i + \gamma j + \delta k$, $x' = \alpha' e + \beta' i + \gamma' j + \delta' k \in \mathbb{H}$, the *canonical scalar product*

$$(1) \qquad\qquad \langle x,x' \rangle := \alpha\alpha' + \beta\beta' + \gamma\gamma' + \delta\delta' \in \mathbb{R}.$$

Then it is clear that $e, i, j, k$ constitute an *orthonormal basis* of $\mathbb{H}$. By (1) the length $|x|$ of $x$ is given by

$$(2) \qquad\qquad |x|^2 := \langle x,x \rangle = \alpha^2 + \beta^2 + \gamma^2 + \delta^2.$$

**1. Conjugation and the Linear Form $\mathbb{R}e$.** By 1.3(5) every quaternion $x$ has the basis-independent representation $x = \alpha e + u$, $u \in \operatorname{Im}\mathbb{H}$. We shall discuss the $\mathbb{R}$-*linear conjugation* (mapping) defined (by analogy with conjugation in $\mathbb{C}$) by

$$(1) \qquad\qquad \mathbb{H} \to \mathbb{H}, \qquad x \mapsto \bar{x} := \alpha e - u.$$

We then have

$$(2) \qquad\qquad \bar{\bar{x}} = x, \qquad \operatorname{Im}\mathbb{H} = \{x \in \mathbb{H} : \bar{x} = -x\},$$

and the *fixed point set* is the *straight line* $\mathbb{R}e$. It is also clear that

$$(3) \qquad\qquad |\bar{x}| = |x|, \qquad x \in \mathbb{H} \qquad (\text{preservation of length}).$$

We shall continually make use of the multiplication rule

(4) $$\overline{xy} = \bar{y}\bar{x};$$

which follows, for example, from the product formula 1.1, though one need verify it only for the basis quaternions $e, i, j, k$, because its general validity is then a consequence of the fact that the mapping $(x, y) \mapsto \overline{xy} - \bar{y}\bar{x}$ is bilinear. In view of the identities (2) and (4) the mapping $x \mapsto \bar{x}$ is called an *involution* of the quaternion algebra $\mathbb{H}$.

We also simulate the *real part mapping* in $\mathbb{C}$, and introduce the $\mathbb{R}$-*linear form*

(5)     $\mathrm{Re}: \mathbb{H} \to \mathbb{R}, \quad x \mapsto \mathrm{Re}(x) := \alpha, \quad \text{where} \quad x = \alpha e + u, \quad u \in \mathrm{Im}\,\mathbb{H}.$

Clearly Re is characterized by the properties

$$\mathrm{Re}(e) = 1 \quad \text{and} \quad \text{kernel Re} = \mathrm{Im}\,\mathbb{H}.$$

It is also clear from the definition that (analogously to 3.3.1)

(6) $$x + \bar{x} = 2\,\mathrm{Re}(x)e \quad \text{and} \quad \mathrm{Re}(\bar{x}) = \mathrm{Re}(x).$$

The important quadratic equation (3) in 1.3 can now be written as

(7) $$x^2 = 2\,\mathrm{Re}(x)x - |x|^2 e.$$

Since $(x, y) \mapsto \mathrm{Re}(xy) - \mathrm{Re}(yx)$ is bilinear,

(8) $$\mathrm{Re}(xy) = \mathrm{Re}(yx),$$

holds generally, because it obviously holds for $e, i, j, k$. Incidentally it may be mentioned that $\mathrm{Re}(xy)$ is the bilinear form of the *Lorentz metric* in $\mathbb{R}^4$, because (by the product formula 1.1)

$$\mathrm{Re}(xy) = \alpha\alpha' - \beta\beta' - \gamma\gamma' - \delta\delta'$$

for

$$x = \alpha e + \beta i + \gamma j + \delta k, \qquad x' = \alpha' e + \beta' i + \gamma' j + \delta' k \in \mathbb{H}.$$

*Remark.* The proofs of the rules (4) and (8) become more readily understandable if one makes use of the algebra isomorphism introduced in 1.2, namely

$$F: \mathbb{H} \to \mathcal{H}, \qquad x = (\alpha, \beta, \gamma, \delta) \mapsto F(x) = \begin{pmatrix} w & -z \\ \bar{z} & \bar{w} \end{pmatrix},$$

$$w := \alpha + i\beta, \qquad z := \gamma + i\delta \in \mathbb{C},$$

and works in the matrix algebra $\mathcal{H}$. Thus, writing $A^t$ for the transpose of a matrix $A$, so that

$$F(\bar{x}) = \overline{F(x)^t}, \qquad \mathrm{Re}\, x = \frac{1}{2}\mathrm{trace}\, F(x),$$

the familiar rules of the matrix calculus give us

$$F(\overline{xy}) = \overline{F(xy)^t} = (\overline{F(x)F(y)})^t = (\overline{F(x)}\,\overline{F(y)})^t = \overline{F(y)^t}\,\overline{F(x)^t}$$
$$= F(\bar{y})F(\bar{x}) = F(\bar{y}\bar{x}),$$

and therefore $\overline{xy} = \bar{y}\bar{x}$ since $F$ is injective. Since $\mathrm{Re}(xy) = \frac{1}{2}\mathrm{trace}(F(x)F(y))$ (8) follows immediately from the commutativity of the trace: $\mathrm{trace}(AB) = \mathrm{trace}(BA)$.

**2. Properties of the Scalar Product.** In the introduction, the scalar product $\langle x, x' \rangle$ was defined by (1) in terms of the basis $e, i, j, k$ of $\mathbb{H}$. It is easy to describe it in terms independent of the basis by means of conjugation. We first verify

(1)    $x\bar{x} = \bar{x}x = \langle x, x \rangle e,$  *and in particular* $x^{-1} = |x|^{-2}\bar{x}$ *for* $x \neq 0$.

Writing $x + y$ in place of $x$, we have

(2)    $$\langle x, y \rangle e = \frac{1}{2}(x\bar{y} + y\bar{x})$$

in view of the bilinearity of $\langle x, y \rangle$. We deduce at once from (2) the

**Orthogonality Criterion:** $\langle x, y \rangle = 0 \Leftrightarrow x\bar{y} = -y\bar{x} \Leftrightarrow x\bar{y} \in \mathrm{Im}\,\mathbb{H}.$

The scalar product in $\mathbb{C}$ is given by $\mathrm{Re}(w\bar{z})$. The same formula applies for $\mathbb{H}$:

(3)    $\langle x, y \rangle = \mathrm{Re}(x\bar{y}) = \mathrm{Re}(\bar{x}y),$  *in particular* $\langle x, e \rangle = \mathrm{Re}(x).$

If one wishes to avoid the straightforward but tedious deduction of (3) from the product formula, one can argue as follows. Since $x\bar{y} + \overline{xy} = 2\,\mathrm{Re}(x\bar{y})e$ by 1(6), and since $\overline{xy} = y\bar{x}$ by 1(4), the relation (3) follows from (2). A second proof of (3) is contained in the remark that the mapping $\mathbb{H} \times \mathbb{H} \to \mathbb{R}$, $(x, y) \mapsto \mathrm{Re}(x\bar{y})$, is easily seen to be bilinear, symmetric and positive definite and that $e, i, j, k$ form an orthonormal basis.                                $\square$

The fundamental property is

(4)    $$|xy| = |x|\,|y| \qquad (\textit{product rule}).$$

**Proof.** Using (1), 1(4), the associative law, and then (1) a second time, we have

$$|xy|^2 e = \langle xy, xy \rangle e = (\overline{xy})(xy) = \bar{y}(\bar{x}x)y = \langle x, x \rangle \bar{y}y$$
$$= \langle x, x \rangle \langle y, y \rangle e = |x|^2 |y|^2 e.$$                                $\square$

Finally we prove yet another formula, which will prove useful in 3.2, and which expresses in a surprising way a triple product of the form $yxy$ as a linear combination of $y$ and $\bar{x}$:

(5)     $yxy = 2\langle \bar{x}, y\rangle y - \langle y, y\rangle \bar{x},$     $x, y \in \mathbb{H}$     (*triple product identity*).

**Proof.** The identity (2) is equivalent to $2\langle \bar{x}, y\rangle e = \bar{x}\bar{y} + yx$. Right-multiplication by $y$ now gives the required result, since $\bar{y}y = \langle yy\rangle e$.     □

*Remark.* If we consider, in the algebra $\mathcal{H}$, the mapping

$$\mathcal{H} \times \mathcal{H} \to \mathbb{R}, \qquad (A, B) \mapsto \langle A, B\rangle := \frac{1}{2}\mathrm{trace}(A\bar{B}^t),$$

it is clear, from the remark in §2.1, that the algebra isomorphism $F \colon \mathbb{H} \to \mathcal{H}$ has the property $\langle F(x), F(y)\rangle = \mathrm{Re}(x\bar{y})$. Formula (3) says therefore that $\langle F(x), F(y)\rangle = \langle x, y\rangle$. This means that $\langle A, B\rangle$ is a scalar product in $\mathcal{H}$ (which could of course be verified directly) and that $F \colon \mathbb{H} \to \mathcal{H}$ is an orthogonal mapping (see 3.1 for this concept). Since (4) $\mathrm{trace}(\bar{A}A^t) = 2 \det A$, it follows that $\det F(x) = |x|^2$, so that the product rule (4) translates into the product rule for determinants.

**3. The "Four Squares Theorem."** In 3.3.4 we deduced the "two-squares" theorem from the product rule for $\mathbb{C}$. In the same way we deduce, from the product rule for $\mathbb{H}$, the famous

**Four Squares Theorem.** *For all* $\alpha, \beta, \gamma, \delta, \alpha', \beta', \gamma', \delta' \in \mathbb{R}$ *we have:*

$$(\alpha^2 + \beta^2 + \gamma^2 + \delta^2)(\alpha'^2 + \beta'^2 + \gamma'^2 + \delta'^2)$$
$$= (\alpha\alpha' - \beta\beta' - \gamma\gamma' - \delta\delta')^2 + (\alpha\beta' + \beta\alpha' + \gamma\delta' - \delta\gamma')^2$$
$$+ (\alpha\gamma' + \gamma\alpha' + \delta\beta' - \beta\delta')^2 + (\alpha\delta' + \delta\alpha' + \beta\gamma' - \gamma\beta')^2.$$

**Proof.** The identity follows from the product rule 2(4) and the product formula in 1.1.     □

The "four squares theorem" was discovered by Euler in 1748 (letter to Goldbach of the 4th May; see "Correspondance entre Leonhard Euler et Chr. Goldbach 1729–1763," in *Correspondance mathématique et physique de quelques célèbres géomètres du XVIIIèm siècle*, ed. P.-H. Fuss, St. Petersbourg 1843, vol. 1, p. 452). Euler was trying to prove the theorem, which had already been stated by Fermat in 1659, that every natural number is the sum of four squares of natural numbers; by means of his identity he was able to reduce this theorem to the corresponding assertion for primes. The first complete proof of the theorem stated by Fermat was given in 1770 by Lagrange (further information on this will be found in the book by W. Scharlau and H. Opolka: *From Fermat to Minkowski*, Springer-Verlag, 1985).

GAUSS remarked (in an unpublished manuscript found after his death, *Werke* 3, 383–4) that, if complex numbers are used, the "four squares theorem" is contained in the identity

$$(|u|^2 + |v|^2)(|w|^2 + |z|^2) = |uw + vz|^2 + |u\bar{z} - v\bar{w}|^2, \qquad u, v, w, z \in \mathbb{C},$$

which is nothing else than the theorem on the product of determinants applied to the matrices $\begin{pmatrix} u & v \\ -\bar{v} & \bar{u} \end{pmatrix}$ and $\begin{pmatrix} w & -\bar{z} \\ z & \bar{w} \end{pmatrix}$ in $\mathcal{H}$.

HAMILTON, as we have already pointed out in the introduction to this chapter, elevated the "four squares theorem" into a "touchstone" to test the value of his quaternions. Once the four squares formula has been found, it is obvious (as in the case of two squares) that it must be true in *any commutative* ring.

## 4. Preservation of Length, and of the Conjugacy Relation Under Automorphisms.

The excellent interplay within $\mathbb{H}$ between the operations of conjugation, multiplication, and the formation of the scalar product is again underlined by the following.

**Theorem.** *Every $\mathbb{R}$-algebra automorphism $h: \mathbb{H} \to \mathbb{H}$ has the following two properties:*

$$(1) \qquad\qquad h(\bar{x}) = \overline{h(x)}, \qquad |h(x)| = |x|, \qquad x \in \mathbb{H},$$

*which assert that the mapping $h$ preserves conjugacy and length respectively.*

**Proof.** Since $h(e) = e$ and $\mathrm{Im}\,\mathbb{H} = \{x \in \mathbb{H} : x^2 = -\omega e$ with $\omega \geq 0\}$ it follows that $h(\mathrm{Im}\,\mathbb{H}) \subset \mathrm{Im}\,\mathbb{H}$. Hence, for $x = \alpha e + u \in \mathbb{H}$, $\alpha \in \mathbb{R}$, $u \in \mathrm{Im}\,\mathbb{H}$, it also follows that $h(x) = \alpha e + h(u)$ with $h(u) \in \mathrm{Im}\,\mathbb{H}$. This implies $\overline{h(x)} = \alpha e - h(u) = h(\alpha e - u) = h(\bar{x})$. Moreover $|h(x)|^2 e = h(x)\overline{h(x)} = h(x\bar{x}) = |x|^2 e$, that is $|h(x)| = |x|$.  $\square$

In the theorem we have just proved, the bijectivity of $h$ is used nowhere. In fact the statement holds good for all $\mathbb{R}$-algebra endomorphisms $h \neq 0$ of $\mathbb{H}$, because we always have kernel $h = 0$, as $\mathbb{H}$ is a division algebra, and $h(e) = e$. The above theorem was used in 3.2 to prove that all automorphisms of $\mathbb{H}$ are of the form $x \mapsto axa^{-1}$, $a \neq 0$. For the $\mathbb{R}$-algebra $\mathbb{C}$ the corresponding statement is trivial because $\mathbb{C}$ has only two $\mathbb{R}$-automorphisms, namely, the identity mapping and the conjugation mapping (see 3.3.2).

## 5. The Group $S^3$ of Quaternions of Length 1.

As with complex numbers (see 3.3.4) the product rule gives us immediately

*The set* $S^3 := \{x \in \mathbb{H} \colon |x| = 1\}$ *of all quaternions of length 1 constitutes a group with respect to multiplication in* $\mathbb{H}$*, which is a subgroup of the multiplicative group* $\mathbb{H}^\times := (\mathbb{H} \setminus \{0\}, \cdot)$.

As $e, i, j, k \in S^3$ it is clear that the group $S^3$ is *not abelian*. In $\mathbb{R}^4 \simeq \mathbb{H}$, the set $S^3$ is the "surface of the unit (hyper)sphere" whose center is at the origin. $S^3$ is *compact*, it is also called the *three-dimensional sphere;* topologically $S^3$ can be obtained from $\mathbb{R}^3$, the familiar space of our physical intuition, by compactification through the addition of a point at infinity. The group $S^3$ will play a central role in the next section when we come to study the orthogonal mappings of $\mathbb{H}$ and of $\mathrm{Im}\,\mathbb{H}$.

The group $S^3$ is its own commutator subgroup. In particular, for every $x \in S^3$ there are elements $u, v \in S^3 \cap \mathrm{Im}\,\mathbb{H}$ with $x = uvu^{-1}v^{-1}$.

**Proof.** For any such $x$ there is a $y \in S^3$ with $y^2 = x$. From Exercise 7.1.4 there are elements $u, v \in \mathrm{Im}\,\mathbb{H}$ with $y = uv$. We may assume that $u, v \in S^3$. Then $u^{-1} = -u$ and $v^{-1} = -v$. Therefore $x = (uv)^2 = uvu^{-1}v^{-1}$. $\quad\square$

In the space $\mathbb{R}^{n+1}$ of $(n+1)$-tuples $x = (\xi_0, \ldots, \xi_n)$, $y = (\eta_0, \ldots, \eta_n)$ with the scalar product $\langle x, y \rangle = \sum_0^n \xi_\nu \eta_\nu$ we define the "$n$-dimensional sphere," by $S^n := \{x \in \mathbb{R}^{n+1} \colon |x| = 1\}$. A nontrivial theorem states that $S^1$ and $S^3$ are the only spheres with a "continuous" group structure.

The following relationship exists between the multiplicative groups $\mathbb{H}^\times$, $S^3$ and $\mathbb{R}_+^\times := \{x \in \mathbb{R}, x > 0\}$:

*The mapping* $\mathbb{H}^\times \to \mathbb{R}_+^\times \times S^3$, $x \mapsto (|x|, x/|x|)$ *is a (topological) isomorphism of the (topological) group* $\mathbb{H}^\times$ *onto the product of the (topological) groups* $\mathbb{R}_+^\times$ *and* $S^3$.

For every quaternion $x \neq 0$, $x\bar{x}^{-1} \in S^3$. One can verify directly that:

*The mapping* $h \colon \mathbb{H} \setminus \{0\} \to S^3$, $x \mapsto x\bar{x}^{-1} = |x|^{-1}x^2$, *is surjective:*

(1) $\quad h\left(e + \dfrac{b}{1+\alpha}\right) = \alpha + b$, *if* $\alpha e + b \in S^3 \setminus \{-e\}$, $\quad \alpha \in \mathbb{R}$, $b \in \mathrm{Im}\,\mathbb{H}$;

$$h(i) = -e.\qquad\qquad\square$$

If one puts $x = \kappa e + b$, $b \in \mathrm{Im}\,\mathbb{H}$, then $b^2 = -|b|^2 e$ and hence

(2) $$h(x) = \frac{\kappa^2 - |b|^2}{\kappa^2 + |b|^2}e + \frac{2\kappa}{\kappa^2 + |b|^2}b;$$

we have thus obtained the following *parametric representation* for the group $S^3$

(3) $\quad S^3 = \left\{ \dfrac{1}{\kappa^2 + |b|^2}[(\kappa^2 - |b|^2)e + 2\kappa b] \colon (\kappa, b) \in (\mathbb{R} \times \mathrm{Im}\,\mathbb{H}) \setminus \{0\} \right\}$

as a generalization of the parametric representation 3.5.4(2') of the circle group $S^1$. The equations (1) and (2) also yield the result (whose proof is left as an exercise):

*Every "rational" quaternion* $\alpha e + \beta_1 i + \beta_2 j + \beta_3 k \in S^3 \setminus \{e\}$, $\alpha, \beta_\nu \in \mathbb{Q}$, *has the form*

$$(4) \qquad \alpha = \frac{1-q^2}{1+q^2}, \quad \beta_\nu = \frac{2q_\nu}{1+q^2} \quad \text{with} \quad q_\nu := \frac{\beta_\nu}{1+\alpha} \in \mathbb{Q}, \quad 1 \le \nu \le 3,$$

$$q^2 := q_1^2 + q_2^2 + q_3^2 \in \mathbb{Q}.$$

This representation can be utilized (by analogy with 3.5.4) to parametrize *Pythagorean quintuplets,* that is to say 5-tuples $(k, l, m, n, p)$ of nonzero natural numbers satisfying the equation $k^2 + l^2 + m^2 + n^2 = p^2$. The reader interested in this may care to work through the simple calculations.

**6. The Special Unitary Group $SU(2)$ and the Isomorphism $S^3 \to SU(2)$.** The set

$$(1) \qquad\qquad U(2) := \{U \in GL(2, \mathbb{C}) : U\bar{U}^t = E\}$$

of all *unitary* $2 \times 2$ matrices is an important subgroup of the group $GL(2, \mathbb{C})$ of all complex, *nonsingular* $2 \times 2$ matrices. Since $\det \bar{A}^t = \overline{\det A}$ we have $|\det U| = 1$ for all $U \in U(2)$. The *special unitary group* $SU(2)$ is the *normal subgroup* of the group $U(2)$ defined by

$$SU(2) := \{U \in U(2) : \det U = 1\}.$$

In terms of the subalgebra $\mathcal{H}$ of $\text{Mat}(2, \mathbb{C})$ defined as in 1.2, we now have the

**Theorem.** $SU(2) = \{A \in \mathcal{H} : \det A = 1\}$, *and in particular* $SU(2) \subset \mathcal{H}$.

**Proof.** The equation $A\bar{A}^t = (\det A) \cdot E$ can be immediately verified for

$$A = \begin{pmatrix} w & -z \\ \bar{z} & \bar{w} \end{pmatrix} \in \mathcal{H}$$

and from this follows the inclusion relation $\{A \in \mathcal{H} : \det A = 1\} \subset SU(2)$.

For $U = \begin{pmatrix} a & b \\ c & d \end{pmatrix} \in SU(2)$ we have $U^{-1} = \bar{U}^t = \begin{pmatrix} \bar{a} & \bar{c} \\ \bar{b} & \bar{d} \end{pmatrix}$ by (1).

Since however $U^{-1} = \begin{pmatrix} d & -b \\ -c & a \end{pmatrix}$, it follows that $d = \bar{a}$, $c = -\bar{b}$, that is $U \in \mathcal{H}$. $\qquad\qquad\square$

This immediately yields the

**Isomorphism Theorem.** *The algebra isomorphism* $F : \mathbb{H} \to \mathcal{H}$ *maps the group* $S^3$ *of all quaternions of length* 1 *isomorphically onto the special unitary group* $SU(2)$.

**Proof.** Since $|x|^2 = \det F(x)$ we have $F(S^3) = \{A \in \mathcal{H} : \det A = 1\} = SU(2)$. $\qquad\qquad\square$

An "Eulerian parametrization" of the group $SU(2)$ can now be obtained from 5(3)

$$SU(2) =$$
$$\left\{ \frac{1}{\kappa^2 + \lambda^2 + \mu^2 + \nu^2} \begin{pmatrix} \kappa^2 - \lambda^2 - \mu^2 - \nu^2 + 2\kappa\lambda i & 2\kappa\mu + 2\kappa\nu i \\ -2\kappa\mu + 2\kappa\nu i & \kappa^2 - \lambda^2 - \mu^2 - \nu^2 - 2\kappa\lambda i \end{pmatrix} \right\},$$

where $\kappa, \lambda, \mu, \nu$ run through all real quadruples $\neq 0$.

The reader should compare the results discussed in this section with those considered in 5.3 and 5.4 of Chapter 3.

## §3. THE ORTHOGONAL GROUPS $O(3)$, $O(4)$ AND QUATERNIONS

HAMILTON tried for many years to find an algebraic structure in the space of our physical world, with whose help the Euclidean geometry of the $\mathbb{R}^3$ would be more easily understood. We have seen that the structure of a division algebra cannot be realized until we have embedded the $\mathbb{R}^3$ in an $\mathbb{R}^4$, and that there are interesting connections between quaternion multiplication and the natural scalar product in $\mathbb{R}^4$. It now turns out that with the "purely imaginary quaternions" one can also give a very elegant interpretation of rotations in $\mathbb{R}^3$ in terms of quaternion multiplication.

Already in 1844, that is within a year after the discovery of quaternions, HAMILTON and CAYLEY were aware that *every properly orthogonal mapping* of $\mathbb{R}^3$ has the form

$$\operatorname{Im}\mathbb{H} \to \operatorname{Im}\mathbb{H}, \qquad u \mapsto aua^{-1},$$

where $a$ runs through all quaternions $\neq 0$. (See HAMILTON, Quaternions: applications in geometry, in *Math. Papers* 3, 353–362, in particular formula (i') in the footnote on page 361; and CAYLEY: On certain results relating to quaternions, in *Math. Papers* 1, 123–126). CAYLEY himself assigns the priority to HAMILTON: "the discovery of the formula $q(ix + jy + kz)q^{-1} = ix' + jy' + kz'$, as expressing a rotation, was made by Sir W.R. HAMILTON some months previous to the date of this paper" (*Math. Papers* 1, p. 586).

In 1855 CAYLEY remarked in a paper which appeared in Vol. 50 of *Crelle's Journal* (p. 312; *Math. Papers* 2, p. 214), that every properly orthogonal mapping of $\mathbb{R}^4 = \mathbb{H}$ has the form

$$\mathbb{H} \to \mathbb{H}, \qquad x \mapsto \frac{axb}{|a|\,|b|},$$

where $a, b$ independently of each other run through all quaternions $\neq 0$. In the paragraphs which follow these theorems of HAMILTON and CAYLEY will be discussed in some detail. We shall, departing from the usual procedure, first deal with the situation in $\mathbb{R}^4 = \mathbb{H}$, and then obtain the perhaps more interesting case of $\mathbb{R}^3$ as a "gift" from the natural *embedding* of $\mathbb{R}^3 = \operatorname{Im}\mathbb{H}$ in $\mathbb{H}$.

**1. Orthogonal Groups.** Let $V$ denote a *finite dimensional inner product space*. A linear mapping $f: V \to V$ is said to be *orthogonal*, if

$$\langle f(x), f(y) \rangle = \langle x, y \rangle \quad \text{for} \quad x, y \in V;$$

this holds if and only if $f$ is *length-preserving*: $|f(x)| = |x|$ for all $x \in V$. Every orthogonal mapping is bijective, and its inverse mapping is likewise orthogonal. The orthogonal mappings of $V$ form a group $O(V)$ under composition; $O(V)$ is called *the orthogonal group of the inner product space $V$.*

Every endomorphism $f: V \to V$ has a determinant. The determinant has the value

$$\det f = \pm 1 \quad \text{when} \quad f \in O(V).$$

The subgroup $SO(V)$ of the *properly orthogonal mappings* is defined by

$$SO(V) = O^+(V) := \{f \in O(V): \det f = 1\};$$

the *coset of reflections* is given by

$$O^-(V) := \{f \in O(V): \det f = -1\},$$

and thus $O(V) = O^+(V) \cup O^-(V)$.

The groups $O(\mathbb{R}^n)$ and $SO(\mathbb{R}^n)$ of the Euclidean number space $\mathbb{R}^n$ are traditionally denoted by $O(n)$ and $SO(n)$ and are often identified with the matrix groups $\{A \in GL(n, \mathbb{R}): A^t A = E\}$ and $\{A \in GL(n, \mathbb{R}): A^t A = E$ and $\det A = 1\}$ respectively.

The mappings

$$s_a: V \to V, \quad x \mapsto x - 2\langle a, x \rangle a, \quad a \in V, \quad |a| = 1,$$

play a particularly important role. $S_a$ is always orthogonal, and represents *a reflection in the hyperplane* $\{x \in V: \langle a, x \rangle = 0\}$ *orthogonal to the line* $\mathbb{R}a$. We have

1) $s_a \in O^-(V)$, $s_a^2 = \text{id}$, $f \circ s_a \in O^+(V)$ for $f \in O^-(V)$.

2) $f \circ s_a = s_{f(a)} \circ f$ for $f \in O(V)$.

We state the following theorem taken from S. Lang, *An Introduction to Linear Algebra*, 2nd ed., Springer-Verlag.

**Generation Theorem for the Orthogonal Group.** *The group $O(V)$ is generated by its reflections. The mappings $f \in SO(V)$ are (just) the products of an even number $k$ of reflections, where $k \leq \dim V$.*

**2. The Group $O(\mathbb{H})$. CAYLEY's Theorem.** Every mapping

$$\mathbb{H} \to \mathbb{H}, \quad x \mapsto axb, \quad \mathbb{H} \to \mathbb{H}, \quad x \mapsto a\bar{x}b, \quad a, b \in S^3,$$

is orthogonal by virtue of the product rule 2.2(4). To show that these exhaust the orthogonal mappings of $\mathbb{H}$, we invoke the mappings $s_a: \mathbb{H} \to \mathbb{H}$.

It follows directly from the triple product identity 2.2(5) that

(1)                    $s_a(x) = -a\bar{x}a$   for all   $a \in S^3, x \in \mathbb{H}$;

in particular: $s_e(x) = -\bar{x}$. We denote by $p_a$ the mapping $x \mapsto axa$, $a \in S^3$. It follows from (1) that

(2)                    $s_a \circ s_b = p_a \circ p_{\bar{b}}$   for all   $a, b \in S^3$,

and in particular $p_a = s_a \circ s_e$.

We can now immediately deduce from (1), (2) and the generation theorem in 1 above the

**Generation Theorem for $O(\mathbb{H})$.** *Every orthogonal mapping $f \in SO(\mathbb{H})$ is a product of at most four mappings $p_a$, $a \in S^3$.*

*The group $O(\mathbb{H})$ is generated by the two mappings $x \mapsto axa$, $a \in S^3$, and $x \mapsto -\bar{x}$.*

*Example.* For the mapping $g: \mathbb{H} \to \mathbb{H}$, $x \to -x$, we have

$$g = p_i \circ p_j \circ p_k.$$                                                □

An immediate deduction from the generation theorem for $O(\mathbb{H})$ is the following result:

**Theorem (CAYLEY).** *To every orthogonal mapping $f: \mathbb{H} \to \mathbb{H}$ correspond two quaternions $a, b \in S^3$ with the following properties:*

a) $f(x) = axb$, *if* $f \in O^+(\mathbb{H})$.

b) $f(x) = a\bar{x}b$, *if* $f \in O^-(\mathbb{H})$.

**Proof.** a) When $f \in O^+(\mathbb{H})$ we have $f = p_{a_1} \circ \cdots \circ p_{a_4}$ with $a_1, \ldots, a_4 \in S^3$. If we put $a := a_1 a_2 a_3 a_4$, $b := a_4 a_3 a_2 a_1$, then $a, b \in S^3$ and $f(x) = axb$.

b) When $f \in O^-(\mathbb{H})$ we have $f \circ s_e \in O^+(\mathbb{H})$, hence $f(-\bar{x}) = f \circ s_e(x) = cxb$ with $b, c \in S^3$ by a). We thus see that $f(x) = a\bar{x}b$ with $a := -c$.   □

From this theorem of CAYLEY can be obtained the result already announced in 1.5.

**Theorem.** *Every $\mathbb{R}$-algebra automorphism $h: \mathbb{H} \to \mathbb{H}$ has the form $h(x) = axa^{-1}$, $a \in S^3$.*

**Proof.** By Theorem 2.4, $h \in O(\mathbb{H})$. As $h(e) = e$ it follows that

$$h(x) = axa^{-1}   \text{ or }   h(x) = a\bar{x}a^{-1}   \text{ with }   a \in S^3.$$

The second case is impossible since we should then have

$$h(xy) = a\overline{xy}a^{-1} = a\bar{y}a^{-1}a\bar{x}a^{-1} = h(y)h(x).$$                                □

**3. The Group $O(\mathrm{Im}\,\mathbb{H})$. HAMILTON's Theorem.** Every orthogonal mapping $\mathbb{H} \to \mathbb{H}$, $x \mapsto \pm ax\bar{a}$, $a \in S^3$, maps the subspace $\mathrm{Im}\,\mathbb{H} = \{u \in \mathbb{H} : \bar{u} = -u\}$ of all purely imaginary quaternions onto itself, since $\overline{au\bar{a}} = a\bar{u}\bar{a} = -au\bar{a}$, and thus induces an orthogonal mapping $\mathrm{Im}\,\mathbb{H} \to \mathrm{Im}\,\mathbb{H}$, $u \mapsto \pm au\bar{a}$. We assert that all orthogonal mappings of $\mathrm{Im}\,\mathbb{H}$ can be obtained in this way, for, since the space $\mathrm{Im}\,\mathbb{H}$ is orthogonal to the line $\mathbb{R}e$, every orthogonal mapping $f$ of $\mathrm{Im}\,\mathbb{H}$ can be extended *uniquely* to an orthogonal mapping $\hat{f} : \mathbb{H} \to \mathbb{H}$ by defining

$$\hat{f} := \mathrm{id} \text{ on } \mathbb{R}e, \quad \hat{f} := f \text{ on } \mathrm{Im}\,\mathbb{H}.$$

In matrix notation the matrix associated with $\hat{f}$ is $\begin{pmatrix} 1 & 0 \\ 0 & B \end{pmatrix}$, where $B$ is the $3 \times 3$ matrix associated with $f$. It is therefore clear that

$$\det \hat{f} = \det f, \text{ so that in particular } f \in O^+(\mathrm{Im}\,\mathbb{H}) \Leftrightarrow \hat{f} \in O^+(\mathbb{H}).$$

We can now easily derive

**HAMILTON's Theorem.** *To every orthogonal mapping* $f : \mathrm{Im}\,\mathbb{H} \to \mathrm{Im}\,\mathbb{H}$ *there corresponds a quaternion* $a \in S^3$ *with the following property*

a) $f(u) = au\bar{a}$, *if* $f \in O^+(\mathrm{Im}\,\mathbb{H})$.

b) $f(u) = -au\bar{a}$, *if* $f \in O^-(\mathrm{Im}\,\mathbb{H})$.

**Proof.** a) Suppose $f \in O^+(\mathrm{Im}\,\mathbb{H})$. Then $\hat{f} \in O^+(\mathbb{H})$, so that, by a) of Theorem 2, $\hat{f}(x) = axb$ with $a, b \in S^3$. From $\hat{f}(e) = e$ it follows that $ab = e$, or in other words $b = a^{-1} = \bar{a}$.

b) This clearly follows from a) since $f \in O^-(\mathrm{Im}\,\mathbb{H})$ implies $-f \in O^+(\mathrm{Im}\,\mathbb{H})$, as $\mathrm{Im}\,\mathbb{H}$ is of dimension 3. $\qquad\qquad\square$

**4. The Epimorphisms $S^3 \to SO(3)$ and $S^3 \times S^3 \to SO(4)$.** The theorems of HAMILTON and CAYLEY provide some important information about the classical groups $SO(3)$ and $SO(4)$. With every $a \in S^3$, and with every pair $(a, b) \in S^3 \times S^3$ we may associate the orthogonal mappings

$$\varphi(a) : \mathrm{Im}\,\mathbb{H} \to \mathrm{Im}\,\mathbb{H}, \quad u \mapsto au\bar{a}, \text{ and } \psi(a,b) : \mathbb{H} \to \mathbb{H}, \quad x \mapsto ax\bar{b}.$$

We consider the mappings $\varphi : S^3 \to O(\mathrm{Im}\,\mathbb{H})$, $\psi : S^3 \times S^3 \to O(\mathbb{H})$. Just as $S^3$ forms a compact non-abelian multiplicative group, so the Cartesian product $S^3 \times S^3$ forms a *compact non-abelian group* with respect to the composition $(a, b) \cdot (c, d) = (ac, bd)$.

**Theorem.** *The mappings* $\varphi : S^3 \to O(\mathrm{Im}\,\mathbb{H})$ *and* $\psi : S^3 \times S^3 \to O(\mathbb{H})$

*are group homomorphisms.*[5] *The kernel groups each have two elements:*
kernel $\varphi = \{\pm e\}$, kernel $\psi = \{\pm(e, e)\}$. *The image groups satisfy* $\varphi(S^3) = SO(\text{Im} \, \mathbb{H})$, $\psi(S^3 \times S^3) = SO(\mathbb{H})$.

**Proof.** That each is a homomorphism is a direct consequence of the definitions of the mappings concerned. For example $\psi((a, b) \cdot (c, d)) = \psi(a, b) \circ \psi(c, d)$ because

$$[\psi(a, b) \circ \psi(c, d)](x) = \psi(a, b)(cx\bar{d}) = acx\overline{db} = (ac)x(\overline{bd})$$
$$= \psi(ac, bd)(x), \quad x \in \mathbb{H}.$$

Suppose that $a \in$ kernel $\varphi$, so that $u = au\bar{a}$ for all $u \in \text{Im} \, \mathbb{H}$. By 1.5 such is the case if and only if $a \in \mathbb{R}e$. Since $|a| = 1$, it follows that $a = \pm e$, and hence kernel $\varphi = \{\pm e\}$. Suppose furthermore that $(a, b) \in$ kernel $\psi$, and thus $ax\bar{b} = x$ for all $x \in \mathbb{H}$. If $x := e$ then $a = b$ and thus $a \in$ kernel $\varphi$, that is $a = \pm e$ whence kernel $\psi = \{\pm(e, e)\}$.

Theorems 2 and 3 yield the non-trivial inclusion relations $\varphi(S^3) \supset SO(\text{Im} \, \mathbb{H})$, $\psi(S^3 \times S^3) \supset SO(\mathbb{H})$. In both cases we in fact have equality: this follows immediately on continuity grounds (by the usual argument based on determinants) or directly, as follows. If there were, for example, a $\psi(a, b) \in O^-(\mathbb{H})$, then by b) of Theorem 2 there would be elements $c, d \in S^3$, such that $ax\bar{b} = c\bar{x}d$ for all $x \in \mathbb{H}$. Thus we should always have $\bar{x} = pxq^{-1}$ with $p := c^{-1}a$, $q := db$. For $x := e$ it would follow that $p = q$ and for $x := p$, we should therefore have $\bar{p} = p$, and hence $p \in \mathbb{R}e$. This however leads to the absurdity $\bar{x} = x$. □

We see from the foregoing that there are natural group epimorphisms $S^3 \rightarrow SO(3)$, $S^3 \times S^3 \rightarrow SO(4)$, whose kernels each have 2 elements. As $S^3$, by 2.6 is isomorphic to $SU(2)$ there are also correspondingly epimorphisms $SU(2) \rightarrow SO(3)$, $SU(2) \times SU(2) \rightarrow SO(4)$, with kernels of 2 elements.

As $S^3$ is of dimension 3, and $S^3 \times S^3$ of dimension 6, the following consequences among others may be noted:
*The group* $SO(3)$ *is 3-dimensional, the group* $SO(4)$ *6-dimensional,* (and generally dim $SO(n) = \frac{1}{2}n(n - 1)$).
The sets $G := \psi(S^3 \times e)$, $G' := \psi(e \times S^3)$ are normal subgroups of $SO(4)$, which are isomorphic to the group $S^3$ under the isomorphisms $a \mapsto \psi(a, e)$, $b \mapsto \psi(e, b)$, respectively. While $G \cdot G' = SO(4)$, we have $G \cap G' = \pm\text{id}$ as is readily proved. We see in particular that:
*The group* $SO(4)$ *contains normal subgroups isomorphic to the group* $S^3$ *and is therefore not a "simple"* LIE *group.* On the other hand all groups

---

[5] As multiplication in $S^3$ and $S^3 \times S^3$ is *non-abelian*, one would no longer have a homomorphism if one associated with every $a \in S^3$, and with every $(a, b) \in S^3 \times S^3$, the isometric mappings $u \mapsto aua$ and $x \mapsto axb$, respectively. Note that $\bar{a} = a^{-1}$ for $a \in S^3$.

$SO(n)$, $n > 4$ are simple, that is to say they contain no nontrivial connected normal subgroups. The groups $SO(2n + 1)$ in fact have no proper normal subgroups $\neq \{e\}$ at all; the groups $SO(2n)$ have just the one nontrivial normal subgroup $\{(\pm e)\}$.

**5. Axis of Rotation and Angle of Rotation.** For $a \in S^3$ let $f_a : \operatorname{Im}\mathbb{H} \to \operatorname{Im}\mathbb{H}$ be defined by $f_a(u) := au\bar{a}$. Clearly $f_a \in O(\operatorname{Im}\mathbb{H})$, and $f_a$ is the identity if and only if $a = \pm e$ (note that $f_a = \varphi(a)$ by 4).

*If $f_a \neq \operatorname{id}$, it follows that $0 \neq a - \bar{a} \in \operatorname{Im}\mathbb{H}$ and $f_a(a - \bar{a}) = a - \bar{a}$; each point of the line generated by $a - \bar{a}$ is thus invariant under the mapping $f_a$.*

**Proof.** Since $a \neq \pm e$, it follows that $0 \neq a - \bar{a} \in \operatorname{Im}\mathbb{H}$ and also that $f_a(a - \bar{a}) = a(a - \bar{a})\bar{a} = a(a\bar{a}) - (a\bar{a})\bar{a} = a - \bar{a}$.                     □

To describe the mapping $f_a$ in a different way we use the following

**Lemma.** *Every quaternion $a \in S^3 \setminus \{\pm e\}$ has a unique representation in the form*

$$(1) \quad a = \cos\frac{1}{2}\omega \cdot e + \sin\frac{1}{2}\omega \cdot q \text{ with } q \in \operatorname{Im}\mathbb{H}, \ |q| = 1, \text{ and } 0 < \omega < 2\pi.$$

**Proof.** We write $a = \alpha e + \beta q$ with $q \in \operatorname{Im}\mathbb{H}$, $|q| = 1$ and $\beta > 0$. Since $\alpha^2 + \beta^2 = 1$ there is just one $\omega \in (0, 2\pi)$ such that $\alpha = \cos\frac{1}{2}\omega$, $\beta = \sin\frac{1}{2}\omega$.                     □

We shall now show that $f_a$ is a rotation about the axis $\mathbb{R}q$ through the angle $\omega$, or in other words that the plane in $\operatorname{Im}\mathbb{H}$ perpendicular to the line $\mathbb{R}q$ is rotated through the angle $\omega$. This and more is implicit in the following

**Theorem.** *If for any $a \in S^3 \setminus \{\pm e\}$ the quantities $\omega$ and $q$ are chosen to satisfy the equations (1) then*

$$f_a(u) = \cos\omega \cdot u + \sin\omega \cdot q \times u + (1 - \cos\omega)\langle q, u\rangle q \text{ for all } u \in \operatorname{Im}\mathbb{H}.$$

**Proof.** Using the abbreviations $\alpha := \cos\frac{1}{2}\omega$, $\beta := \sin\frac{1}{2}\omega$ we have

$$au\bar{a} = (\alpha e + \beta q)u(\alpha e - \beta q) = \alpha^2 u + \beta\alpha qu - \alpha\beta uq - \beta^2 quq.$$

From the definition of the vector product (see 1.4) we have $2q \times u = qu - uq$. As $\bar{u} = -u$ and $\langle q, q\rangle = 1$, it follows that $quq = u - 2\langle q, u\rangle q$ by the triple product identity (2.2(5)), and consequently

$$f_a(u) = (\alpha^2 - \beta^2)u + 2\alpha\beta q \times u + 2\beta^2\langle q, u\rangle q, \qquad u \in \operatorname{Im}\mathbb{H}.$$

From the definitions of $\alpha$, $\beta$ and the elementary formulae of trigonometry, it follows that $\alpha^2 - \beta^2 = \cos\omega$, $2\alpha\beta = \sin\omega$, $2\beta^2 = 1 - (\alpha^2 - \beta^2) = 1 - \cos\omega$.                                                                $\square$

**Corollary.** $f_a(q) = q$ *and* $\langle f_a(u), u \rangle = \cos\omega$ *for all* $u \in \operatorname{Im}\mathbb{H}$ *with* $|u| = 1$ *and* $\langle u, q \rangle = 0$.

We deduce from the foregoing results that $f_a$ is a rotation about the axis $\mathbb{R}q$ through the angle $\omega$. Incidentally it is easily shown that $\cos\omega = \operatorname{Re}(a^2)$. If $a$ is purely imaginary then $\omega = \pi$ and $f_a = -s_a$ is a rotation of 180° about the axis $\mathbb{R}a$.                                                                $\square$

*Remark.* As is well known every properly orthogonal mapping $\neq$ id of $\mathbb{R}^3$ is a rotation about a uniquely defined axis. Every $f \in SO(\operatorname{Im}\mathbb{H}) \setminus \{\mathrm{id}\}$ is therefore a rotation about an axis $\mathbb{R}q$, $q \in \operatorname{Im}\mathbb{H}$, $|q| = 1$, through an angle $\omega$, $0 < \omega < 2\pi$. If we now define $a \in S^3$ by (1), then $a = \pm e$ and $f_a \in SO(\operatorname{Im}\mathbb{H})$ is by the theorem a rotation through $\omega$ about the axis $\mathbb{R}q$. We have thus proved afresh statement a) of HAMILTON's theorem in 3, namely every $f \in SO(\operatorname{Im}\mathbb{H})$ has the form $f_a$ with $a \in S^3$.

**6. EULER's Parametric Representation of $SO(3)$.** The mapping

$$\mathbb{H} \setminus \{0\} \to SO(\operatorname{Im}\mathbb{H}), \quad a \mapsto h_a \quad \text{with} \quad h_a : \operatorname{Im}\mathbb{H} \to \operatorname{Im}\mathbb{H},$$

$$u \mapsto \frac{1}{|a|^2} a u \bar{a} = a u a^{-1},$$

is by Theorem 4, an *epimorphism* of the multiplicative group $\mathbb{H} \setminus \{0\}$ with $\mathbb{R}e \setminus \{0\}$ as kernel. If one sets $a := \kappa e + \lambda i + \mu j + \nu k$ and writes $u := xi + yj + zk$ as a column vector, we have

$$h_a(u) = A \begin{pmatrix} x \\ y \\ z \end{pmatrix}$$

where $A$ is a properly orthogonal matrix $A \in SO(3)$. This matrix is found by expressing $|a|^{-2} a u \bar{a}$ in terms of the basis $i, j, k$ of $\operatorname{Im}\mathbb{H}$. One obtains in this way the result discovered by EULER in 1770 (*Opera omnia* 6, Ser. 1, 287–315), the well-known

**Rational Parametric Representation of Orthogonal $3 \times 3$ Matrices.**
*For every quadruple* $(\kappa, \lambda, \mu, \nu) \in \mathbb{R}^4 \setminus \{0\}$ *the* $3 \times 3$ *matrix*

(1)
$$\frac{1}{\kappa^2 + \lambda^2 + \mu^2 + \nu^2}$$

$$\cdot \begin{pmatrix} \kappa^2 + \lambda^2 - \mu^2 - \nu^2 & -2\kappa\nu + 2\lambda\mu & 2\kappa\mu + 2\lambda\nu \\ 2\kappa\nu + 2\lambda\mu & \kappa^2 - \lambda^2 + \mu^2 - \nu^2 & -2\kappa\lambda + 2\mu\nu \\ -2\kappa\mu + 2\lambda\nu & 2\kappa\lambda + 2\mu\nu & \kappa^2 - \lambda^2 - \mu^2 + \nu^2 \end{pmatrix}$$

*is properly orthogonal, and all properly orthogonal $3 \times 3$ matrices can be expressed in this form.*

**Proof.** If we set $a := \kappa e + b \in (\mathbb{R} \times \operatorname{Im} \mathbb{H}) \backslash \{0\}$, then $a u \bar{a} = (\kappa e + b) u (\kappa e - b) = \kappa^2 u + 2 \kappa b \times u - b u b$. As $b u b = |b|^2 u - 2 \langle b, u \rangle b$ by the triple product identity since $\bar{u} = -u$, it is clear that $a u \bar{a} = (\kappa^2 - |b|^2) u + 2 \kappa b \times u + 2 \langle b, u \rangle b$. The representation (1) follows at once from this if $b := \lambda i + \mu j + \nu k$.    $\square$

The parametric representation for properly orthogonal $2 \times 2$ matrices given in 3.5.4 follows from (1), if we put $\mu = 0$, $\nu = 0$ (and write $-\lambda$ for $\lambda$) in the leading minor.

As the epimorphism $\mathbb{H} \backslash \{0\} \to SO(\operatorname{Im} \mathbb{H})$, $a \mapsto h_a$, has the group $\mathbb{R} e \backslash \{0\}$ as its kernel, the same matrix $A$ defined by (1) appertains to the two distinct quadruples $a, a' \in \mathbb{R}^4 \backslash \{0\}$ if and only if $a' = \alpha a$ with $\alpha \neq 0$, or in other words if and only if $a$ and $a'$ define the same point in the real projective 3-dimensional space $\mathbb{P}^3(\mathbb{R})$ with the homogeneous coordinates $\kappa$, $\lambda$, $\mu$, $\nu$. EULER's theorem can therefore also be expressed as follows:

*The mapping $\mathbb{P}^3(\mathbb{R}) \to SO(3)$ defined by (1) is bijective, and in particular $SO(3)$ is a rational manifold.*

This statement was generalized by CAYLEY in 1846 (*Math. Papers* **1**, 332–336):

*The group $SO(n)$ is an $\frac{1}{2} n(n-1)$-dimensional rational manifold. The $\frac{1}{2} n(n-1)$-dimensional real projective space is mapped birationally into $SO(n)$ by the CAYLEY mapping*

$$X \mapsto (\kappa E - X)^{-1} (\kappa E + X), \qquad X \in \operatorname{Mat}(n, \mathbb{R}),$$

*where $X$ is skew-symmetric.*

The case $n = 3$ of this CAYLEY representation is none other than the EULER parametric representation.

## REFERENCE

C.W. CURTIS, "The Four and Eight Square Problem and Division Algebras," pp. 100–125 of Studies in Modern Algebra. (A.A. Albert, ed.). Math. Assoc. of America 1963.

# 8

# The Isomorphism Theorems of Frobenius, Hopf and Gelfand–Mazur

*M. Koecher, R. Remmert*

## Introduction

**1.** In the second half of the nineteenth century, many other hypercomplex systems were discovered and investigated, in addition to that of the quaternions. Especially in England, this became almost an art and was held in high esteem. Shortly after the discovery of quaternions and *before* the introduction of matrices, John T. GRAVES and Arthur CAYLEY devised the non-associative division algebra of *octonions* (also called *octaves*). HAMILTON introduced, in his "Lectures on quaternions" of 1853, *biquaternions,* that is quaternions with complex coefficients, and noted that they do not form a division algebra. William Kingdon CLIFFORD (1845–1879) created in 1878, the associative algebras now called after him.

A flood of new hypercomplex systems now inundated the whole of algebra. The important question of how much freedom there really exists in this apparent profusion of examples, was one which moved only slowly into the foreground of interest. While GAUSS in 1831, had still been convinced at that time that no hypercomplex number systems existed for which the basic properties of the complex numbers would still hold (see 4.3.6), it was at first generally believed after the discovery of quaternions and octonions that new and interesting hypercomplex systems could now be everlastingly invented. It is nevertheless significant that HAMILTON was unable to prove that 3-dimensional, commutative and associative division algebras (that is to say fields) over ℝ do not exist. GRASSMANN as well had nothing to say on this point. In 1871 Benjamin PEIRCE (1809–1880) who was a Professor of Mathematics at Harvard, published an article entitled "Linear associative algebras" in which he gave a summary of all the then known algebras of this type (the article was reprinted in the *Amer. J. Math.* 4, 1881, 97–229).

**2.** An insight into the true situation that there are far fewer interesting ℝ-algebras than one might have expected, was first gained in the next generation of mathematicians. One of the first precise uniqueness theorems was proved in 1877 by Ferdinand Georg FROBENIUS, who was born in 1849 in Berlin, was a pupil of WEIERSTRASS, was appointed in 1875 Professor at the Zürich Polytechnic, and from 1892 onwards was a Professor at Berlin University. He promoted the development of abstract methods in algebra, and his theory of group representations was to find applications later in quantum theory after his death in 1917 in Charlottenburg. In his paper "Über lineare Substitutionen und bilineare Formen" [On linear substitutions and bilinear forms] published in *Crelle's Journal* (reproduced in *Ges. Abhandl.* 1, 343–405) he shows that there are only *three* isomorphically distinct real finite-dimensional associative division algebras, namely $\mathbb{R}$ itself, $\mathbb{C}$ and $\mathbb{H}$. This famous theorem which was proved independently in 1881 by the American mathematician Charles Sanders PEIRCE (1839–1914), the son of Benjamin PEIRCE in an Appendix to a work of his father (*Amer. J. Math.* 4), showed algebraists for the first time that there were limits to the construction processes which they had hitherto regarded as omnipotent. Had HAMILTON known of FROBENIUS's theorem, he would have been spared years of hard work in his fruitless search for three-dimensional associative division algebras.

The theorem of FROBENIUS is proved in the first two sections of this chapter. The central result is an Existence theorem for HAMILTONIAN triples, from which FROBENIUS's result follows. To avoid repetition later, we shall *not* assume from the outset that the algebras with which we shall be concerned are associative. Instead we shall in each case deliberately point out the (weaker) properties assumed, such as *power-associative*, or *alternative* or *quadratic*. This abstract point of view should not put off any reader nowadays, in the post-BOURBAKI era.

**3.** In the year 1940 Heinz HOPF posed the problem of specifying all finite-dimensional real *commutative* division algebras (dropping the requirement that they be associative). HOPF was a Swiss mathematician of German origin, born in 1894 in Grätschen (Silesia), who studied in Berlin, Heidelberg and Göttingen, where in 1925 he made the acquaintance of Paul ALEXANDROFF and Emmy NOETHER. In 1931 he succeeded Hermann WEYL at the Federal Technische Hochschule in Zürich and died in his adopted homeland at Zollikon in the Canton of Zürich in 1971. He did pioneering work on the topology of manifolds and their mappings as well as in differential geometry. A master of the true art of exposition, HOPF always gave the solutions to an individual problem, and at the same time created the method by which its difficulties could be overcome, in such a way as to bring out the main theme or guiding principle, and the deep underlying reason, so that further possibilities became clear.

The problem which HOPF had set himself, and which may appear at first sight to be somewhat contrived and artificial, leads to astonishing and unexpected insights. HOPF in his paper "Systeme symmetrischer Bilinearformen und euklidische Modelle der projektiven Räume" [Systems of symmetric bilinear forms and Euclidean models of projective spaces] which was to become famous (*Vierteljahreszeitschrift der Naturforschenden Gesellschaft in Zürich,* LXXXV, 1940, *Beibl.* No. 32, Festschrift Rudolf FEUTER: see also H. Hopf SELECTA, Springer-Verlag, 1964) showed that any real, commutative division algebra of finite dimension is *at most 2-dimensional.* The remarkable thing about HOPF's problem is that an algebraic question which can be formulated so simply, and which has such a simple answer, requires for its solution *nontrivial topological methods.* This is the first manifestation of the "topological thorn in the flesh of algebra" which many algebraists have found so painful to this very day.

**4.** In the year 1938 a Polish mathematician Stanislav MAZUR recognized that $\mathbb{R}$, $\mathbb{C}$ and $\mathbb{H}$ are the only BANACH division algebras, and in 1940 the Russian mathematician Izrail' Moiseevich GELFAND gave a proof by function theoretic methods.

MAZUR was born in 1905 in Lemberg (now Lvov), he studied in Lemberg and Paris, and was Professor at Lemberg in 1939, Lodz in 1946, and Warsaw in 1948. During 1946–1954 he was a member of the Sejm and died in 1981.

GELFAND was born in Odessa in 1913. In 1930 he worked as a porter at the Lenin Library in Moscow, and was a student in Moscow in 1932. He obtained his doctorate in 1939 and became president of the Moscow Mathematical Society in 1968. He was three times winner of the Lenin Prize. The GELFAND–MAZUR theorem can be placed into the framework of the ideas introduced by HOPF, and thus, as we shall show in §4, easily proved.

## §1. HAMILTONIAN TRIPLES IN ALTERNATIVE ALGEBRAS

The multiplicative behavior of HAMILTON's basis quaternions $i, j, k$ turns out to be of such importance for the general theory of algebras, that we shall find it useful to introduce a special definition.

In an algebra $\mathcal{A}$ with unit element $e$, we shall call three elements $u, v, w$ a *Hamiltonian triple* if the *nine* Hamiltonian conditions

$$u^2 = v^2 = w^2 = -e, \quad w = uv = -vu, \quad u = vw = -wv, \quad v = wu = -uw$$

are all satisfied.

The object of this preliminary section is to verify the existence of Hamiltonian triples in appropriately chosen algebras.

**1. The Purely Imaginary Elements of an Algebra.** In the $\mathbb{R}$-algebras $\mathbb{C}$ and $\mathbb{H}$, there is an *imaginary space* consisting of all elements $x \notin \mathbb{R}e \setminus \{0\}$, whose square is "real": $x^2 \in \mathbb{R}e$. We now introduce, for any algebra $\mathcal{A}$ with unit element $e$ the set

$$\operatorname{Im}\mathcal{A} := \{x \in \mathcal{A} : x^2 \in \mathbb{R}e \text{ and } x \notin \mathbb{R}e \setminus \{0\}\}$$

of *"purely imaginary"* elements. It is then trivial that

$$\mathbb{R}e \cap \operatorname{Im}\mathcal{A} = \{0\}, \quad u \in \operatorname{Im}\mathcal{A} \Rightarrow \alpha u \in \operatorname{Im}\mathcal{A} \text{ for every } \alpha \in \mathbb{R}.$$

However it is by no means obvious that $\operatorname{Im}\mathcal{A}$ is a subspace of $\mathcal{A}$ considered as a vector space. In other words, it does *not* automatically follows from $u, v \in \operatorname{Im}\mathcal{A}$, that $u + v \in \operatorname{Im}\mathcal{A}$ (see in this connection 2.1). Quite generally (and thus in particular for $\mathbb{H}$) the following proposition holds.

**Independence Lemma.** *If $u, v \in \operatorname{Im}\mathcal{A}$ are linearly independent, then so also are $e, u, v$.*

**Proof.** If this were not the case, one could assume without restriction that $v = \alpha e + \beta u$ with $\alpha, \beta \in \mathbb{R}$. It would then follow that $2\alpha\beta u = v^2 - \alpha^2 e - \beta^2 u^2 \in \mathbb{R}e$, and hence $\alpha\beta = 0$. Since $\alpha \neq 0$, because $u, v$ are linearly independent, and $\beta \neq 0$ because $v \in \operatorname{Im}\mathcal{A}$ is impossible, we should have a contradiction. $\qquad\square$

Since $uv + vu = (u + v)^2 - u^2 - v^2$ it is also clear that:

(1) $\qquad\qquad u, v, u + v \in \operatorname{Im}\mathcal{A} \Rightarrow uv + vu \in \mathbb{R}e.$

We now show that:

*If $\mathcal{A}$ has no zero divisors, then $u^2 = -\omega e$ with $\omega > 0$ for $u \in \operatorname{Im}\mathcal{A}$, $u \neq 0$.*

**Proof.** By hypothesis $u^2 = \alpha e$ for some $\alpha \in \mathbb{R}$. If $\alpha$ were $\geq 0$ we could write $\alpha = \beta^2$ with $\beta \in \mathbb{R}$, and we should have $(u - \beta e)(u + \beta e) = u^2 - \beta^2 e = u^2 - \alpha e = 0$. As $\mathcal{A}$ has no divisors of zero one of the two factors on the left must vanish and we should have $u \in \mathbb{R}e$ which is impossible. $\qquad\square$

In algebras without zero divisors one can therefore always (as in the case of $\mathbb{C}$ and $\mathbb{H}$) transform any purely imaginary element $u' \neq 0$, by scalar multiplication, into a normalized element $u = \gamma u'$, satisfying $u^2 = -e$.

**2. Hamiltonian Triple.** Every element of a Hamiltonian triple is purely imaginary. We can in fact assert more than this.

**Theorem.** *If $u, v, w$ is a Hamiltonian triple in $\mathcal{A}$, then*

1) *the mapping $f : \mathbb{H} \to \mathcal{A}$, $(\alpha, \beta, \gamma, \delta) \mapsto \alpha e + \beta u + \gamma v + \delta w$ is an algebra monomorphism,*

2) *$\mathbb{R}u + \mathbb{R}v + \mathbb{R}w \subset \operatorname{Im}\mathcal{A}$, and in particular $\operatorname{Im}\mathcal{A}$ contains a 3-dimensional vector subspace.*

**Proof.** 1) Since $f(e) = e$, $f(i) = u$, $f(j) = v$, $f(k) = w$, the mapping $f$ is an algebra homomorphism. To prove that $f$ is injective is equivalent to showing that $e, u, v, w \in \mathcal{A}$ are linearly dependent. Clearly $u, v$ are linearly independent because otherwise we should have $v \in \mathbb{R}u$ and also $uv = vu$, which would imply $w = -w$, and hence $w = 0$ in contradiction to $w^2 =$

$-e$. From the Independence Lemma 1 it follows that $e$, $u$, $v$ are linearly independent. If now $e$, $u$, $v$, $w$ were linearly independent there would be *uniquely* defined numbers $\rho, \sigma, \tau \in \mathbb{R}$ such that $w = uv = \rho e + \sigma u + \tau v$. Left-multiplication by $u$ gives $-v = \rho u - \sigma e + \tau w$ and the uniqueness of the representation means that $\tau$ has to satisfy the equation $\tau^2 = -1$ which would contradict $\tau \in \mathbb{R}$.

2) It is easily verified (by multiplying out) that $(\beta u + \gamma v + \delta w)^2 \in \mathbb{R}e$. $\qquad\qquad\qquad\qquad\qquad\qquad\qquad\qquad\qquad\qquad\qquad\qquad\qquad\quad\square$

An important preliminary stage in the construction of a Hamiltonian triple consists in finding for any given vector $p \in \operatorname{Im}\mathcal{A}$, a corresponding vector $q$ satisfying the HAMILTON condition $pq + qp = 0$. We first prove the:

**Lemma.** *Let $\mathcal{A}$ have no divisors of zero and let $U$ be a 2-dimensional vector subspace of $\operatorname{Im}\mathcal{A}$. Then to every $p \in U$, there exists a $q \in U \setminus \mathbb{R}p$, such that $pq + qp = 0$.*

**Proof.** We may assume $p \neq 0$, so that $p^2 = \alpha e$ with $\alpha \neq 0$. Choose $x \in U$ such that $p$ and $x$ are linearly independent, so that $px + xp = \beta e$ with $\beta \in \mathbb{R}$ (see 1). Then $q := x + \xi p$ with $\xi := -\beta(2\alpha)^{-1}$ has the required property. $\qquad\qquad\qquad\qquad\qquad\qquad\qquad\qquad\qquad\qquad\qquad\qquad\quad\square$

**3. Existence of Hamiltonian Triples in Alternative Algebras.** The next stage in the process is marked out by the lemma (in 2 above). If one has two vectors defined as in the lemma, we can immediately, in the case of an algebra without divisors of zero (by scalar multiplication, see the remark at the end of 1) find two vectors $u, v \in \operatorname{Im}\mathcal{A}$ with $u^2 = v^2 = -e$ and $uv = -vu$. Then $u, v$ and $w := uv$ are now candidates for a Hamiltonian triple. One cannot however "without further ado" show for example that the equation $vw = u$ holds; the removal of the brackets in the equation $v(uv) = -v(vu)$ is a step which would require justification. We can however make a virtue out of necessity and *postulate* this *weak associativity*.

*An algebra $\mathcal{A}$ is said to be alternative, if for all $x, y \in \mathcal{A}$*

1) $$x(xy) = x^2 y, \qquad (xy)y = xy^2.$$

Every associative algebra is alternative.[1] If $\mathcal{A}$ is alternative, then

2) $$(xy)x = x(yx) \quad \text{for all} \quad x, y \in \mathcal{A}.$$

To prove this we replace the element $y$ by $x + y$ in $(xy)y = xy^2$ and expand the left-hand side.

---

[1]The word alternative is used here as an adjective derived from the verb to alternate, and is intended to refer to the fact the *associator* $(xy)z - x(yz)$ alternates in sign when any two of its arguments are interchanged.

*Exercise.* Show that any two of the three identities $x(xy) = x^2y$, $(xy)y = xy^2$, $(xy)x = x(yx)$ together imply the third.

**Existence Theorem for Hamiltonian Triples.** *Let $A$ be an alternative algebra without divisors of zero and with unit element $e$, and let $U$ be a 2-dimensional vector subspace of $\operatorname{Im} A$. Then for every element $u \in U$ such that $u^2 = -e$, there is a $v \in U$ such that $u$, $v$, $uv$ form a Hamiltonian triple.*

**Proof.** The previous arguments show that there is a $v \in U$ with $v^2 = -e$ and $uv = -vu$. As $A$ is alternative the HAMILTON conditions are satisfied for $u$, $v$ and $w := uv$; namely, it follows that first $vw = v(uv) = -v(vu) = -v^2u = u$ and $wv = (uv)v = uv^2 = -u$. It can similarly be shown that $wu = v = -uw$. It remains to be shown that $w^2 = -e$. Since $vw^2 = (uw)w = uw = -v$, it follows that $v(w^2 + e) = 0$ and hence, since there are no divisors of zero, $w^2 = -e$.    □

**4. Alternative Algebras.** Alternative algebras acquire a special meaning through this last Existence theorem. We note two propositions which hold for such algebras and which will prove useful later on:

*Every alternative algebra $A$ is power associative.*

We have to verify that the exponentiation rule $x^m x^n = x^{m+n}$, $x \in A$, applies. This is easily done by induction using the rules in the definition. Thus for example if $x^{m-1}x^n = x^{m-1+n}$ is known to be true for all $n$, then $x^m x = (x^{m-1}x)x = x^{m-1}x^2 = x^{m+1}$. We leave the reader to fill in the details of this proof by induction.    □

The assumption made in the Existence theorem of paragraph 3 that $A$ has a unit element, is automatically satisfied for alternative algebras:

*Every alternative division algebra $A$ has a unit element.*

**Proof.** Choose $a \in A$, $a \neq 0$. Since $A$ is a division algebra there is an $e \in A$ with $ea = a$. We have $e \neq 0$, because $a \neq 0$. Furthermore $e(ea) = ea$ and therefore, since $A$ is alternative, $e^2a = ea$ which implies $(e^2 - e)a = 0$ and hence $e^2 = e$. It now follows that $e(ex - x) = e(ex) - ex = e^2x - ex = 0$, and therefore $ex = x$, for all $x \in A$. Similarly it can be seen that $xe = x$.    □

We may also mention, without proof, another interesting theorem:

**E. ARTIN's Theorem.** *An algebra $A$ is alternative if and only if any two of its elements $x, y \in A$ generate an associative subalgebra of $A$.*

The proof can be found on page 127 of a paper by ZORN, Abh. Math. Seminar Hamburg 8 (1931), 123–147.

## §2. FROBENIUS'S THEOREM

> Wir sind also zu dem Resultate gelangt,
> dass ausser den reellen Zahlen, den imaginären
> Zahlen und den Quaternionen keine andern complexen
> Zahlen in dem oben definirten Sinne existiren
> (G. FROBENIUS 1877).
>
> [We have thus arrived at the result that, apart from
> the real numbers, the imaginary numbers, and the
> quaternions, no other complex numbers in the sense
> defined above, exist.]

In order to be able to apply the Existence theorem for Hamiltonian triples, proved in 1.3 above, one first needs to have a 2-dimensional vector subspace of the imaginary space $\operatorname{Im}\mathcal{A}$. In this section we show that, in important cases which go well beyond $\mathbb{C}$ and $\mathbb{H}$, the set $\operatorname{Im}\mathcal{A}$ itself is a vector subspace of $\mathcal{A}$. In particular this is true for all so-called *quadratic algebras.* An algebra $\mathcal{A}$ with unit element $e$ is called a *quadratic* algebra if every element $x \in \mathcal{A}$ satisfies a quadratic equation $x^2 = \alpha e + \beta x$ with $\alpha, \beta \in \mathbb{R}$. The algebras $\mathbb{C}$ and $\mathbb{H}$ are quadratic (the latter by virtue of Theorem 6.1.2); and so also is the algebra $\operatorname{Mat}(2, \mathbb{R})$ of real $2 \times 2$ matrices. Quadratic algebras play an essential role in the general theory of algebras.[2] It turns out that every finite dimensional alternative division algebra is quadratic.

The main result of this section is the Quaternion lemma in 3 below, which leads immediately to the theorem of FROBENIUS.

**1. FROBENIUS's Lemma.** *If $\mathcal{A}$ is a quadratic algebra, then $\operatorname{Im}\mathcal{A}$ is a vector subspace of $\mathcal{A}$, and $\mathcal{A} = \mathbb{R}e \oplus \operatorname{Im}\mathcal{A}$.*

**Proof.** 1) We show that $u, v \in \operatorname{Im}\mathcal{A} \Rightarrow u + v \in \operatorname{Im}\mathcal{A}$. If $u, v$ are linearly dependent, say $v = \alpha u$, then $u+v = (1+\alpha)u \in \operatorname{Im}\mathcal{A}$ is obvious. Accordingly let $u, v$ be linearly independent. Since $\mathcal{A}$ is quadratic, the equations

$$(u + v)^2 = \alpha_1 e + \beta_1(u + v), \qquad (u - v)^2 = \alpha_2 e + \beta_2(u - v)$$

hold with $\alpha_1, \alpha_2, \beta_1, \beta_2 \in \mathbb{R}$. After multiplying out and adding, we obtain

$$(\beta_1 + \beta_2)u + (\beta_1 - \beta_2)v = 2u^2 + 2v^2 - (\alpha_1 + \alpha_2)e \in \mathbb{R}e$$

---

[2] The reader may want to become familiar with quadratic algebras by considering the following:

**Theorem.** *If $\mathcal{A}$ is a quadratic algebra, then for every $x \in \mathcal{A}$ the vector subspace $\mathbb{R}e + \mathbb{R}x$ is a commutative and associative $\mathbb{R}$-subalgebra of $\mathcal{A}$; in particular $\mathcal{A}$ is power associative.*

since $u, v \in \operatorname{Im} A$. It follows, from the Independence lemma 1.1 that $\beta_1 + \beta_2 = \beta_1 - \beta_2 = 0$, whence $\beta_1 = \beta_2 = 0$ and thus $(u + v)^2 = \alpha_1 e \in \mathbb{R} e$. As $u + v \notin \mathbb{R} e$ (by the Independence lemma), it follows that $u + v \in \operatorname{Im} A$.

2) Suppose $x \in A$, $x \notin \mathbb{R} e$ (but otherwise arbitrary). By hypothesis $x^2 = \alpha e + 2\beta x$ with $\alpha, \beta \in \mathbb{R}$. Hence $(x - \beta e)^2 = x^2 - 2\beta x + \beta^2 e = (\alpha + \beta^2)e$. Since $x - \beta e \notin \mathbb{R} e$, it follows that $x = \beta e + u$ with $u := x - \beta e \in \operatorname{Im} A$. We have thus shown that $A = \mathbb{R} e + \operatorname{Im} A$. Since $\mathbb{R} e \cap \operatorname{Im} A = \{0\}$, this implies $A = \mathbb{R} e \oplus \operatorname{Im} A$. $\qquad \square$

*Remark.* The device used in the first part of the proof, of taking the equations for $(u + v)^2$ and $(u - v)^2$ and adding them, is already to be found in FROBENIUS (*Ges. Abhandl.* 1, p. 403).

**2. Examples of Quadratic Algebras.** In power associative algebras, the *exponential law $x^m x^n = x^{m+n}$* (R.1) holds. An immediate generalization is the

**Substitution Law.** *Let $A$ be a power associative algebra with unit element $e$; for any polynomial $f = \alpha_0 + \alpha_1 X + \cdots + \alpha_n X^n \in \mathbb{R}[X]$ let $f(x)$ be defined by $f(x) := \alpha_0 e + \alpha_1 x + \cdots + \alpha_n x^n \in A$ (that is, by substituting $x \in A$ for $X$). Then*

$$(f \cdot g)(x) = f(x)g(x) \quad \text{for all polynomials } f, g \in \mathbb{R}[X] \quad \text{and all } x \in A.$$

The proof is simple and is taught in algebra. The Substitution law can also be stated in the following form:

*If $A$ is power associative and has a unit element then every element $x \in A$ defines an algebra homomorphism through the mapping $\mathbb{R}[X] \to A$, $f \mapsto f(x)$ (the so-called substitution homomorphism corresponding to $x$).*

The substitution law and the Fundamental theorem of algebra, quickly yield the

**Theorem.** *Every finite dimensional, alternative algebra $A$ is quadratic.*

**Proof.** By 1.4 $A$ is a power associative algebra with unit element $e$. Thus for every $x \in A$ the substitution homomorphism $\mathbb{R}[x] \to A$, $f \mapsto f(x)$, is defined; its kernel is an *ideal*, and in fact a *principal ideal*, since $\mathbb{R}[X]$ is a *principal ideal ring*. Since $\dim A < \infty$, this principal ideal is not the zero ideal and since $A$ has no divisors of zero, the kernel in question must be a *prime ideal*. There is therefore a monic prime polynomial $p \in \mathbb{R}[X]$ with $p(x) = 0$. Since every such polynomial has the form $X - \gamma$ or $X^2 - \beta X - \alpha$ (see 4.3.4), this proves the theorem. $\qquad \square$

**3. Quaternions Lemma.** *Every alternative quadratic real algebra $A$ without divisors of zero contains, in the case $\dim A \geq 3$, a Hamiltonian triple*

*and therefore subalgebras $B$ with $e \in B$ isomorphic to the quaternion algebra* $\mathbb{H}$.

**Proof.** By FROBENIUS's lemma Im$A$ is a vector subspace of $A$, and $\dim \text{Im} A \geq 2$, since $\dim A \geq 3$. By the Existence theorem 1.3 there are therefore Hamiltonian triples in $A$ and consequently also subalgebras $B$ in $A$ with $e \in B$ isomorphic to $\mathbb{H}$.                                  $\square$

In the associative case we derive the stronger

**4. Theorem of FROBENIUS (1877).** *Let $A \neq 0$, be an associative, quadratic real algebra without divisors of zero (for example, an associative, finite-dimensional, division algebra). Then there are three and only three possibilities:*

1) *$A$ is isomorphic to the field $\mathbb{R}$ of the real numbers.*

2) *$A$ is isomorphic to the field $\mathbb{C}$ of the complex numbers.*

3) *$A$ is isomorphic to the algebra $\mathbb{H}$ of the quaternions.*

**Proof.** Since $A \neq 0$, it follows that $\dim A \geq 1$ and that $A$ has a unit element $e$. If $\dim A = 1$, case 1) of R.4, the Repertory applies. If $\dim A = 2$, there is a $u \in A$ with $u^2 = -e$. Then $f : \mathbb{C} \to A, x + yi \mapsto xe + yu$ is an algebra homomorphism. Since $e, u$ are linearly independent, $f$ is injective and, since $\dim \mathbb{C} = \dim A$, bijective, that is, case 2) applies.

Suppose $\dim A \geq 3$. Then, by the Quaternions lemma a Hamiltonian triple $u, v, w \in \text{Im} A$ must exist with $w = uv$. Let $x \in \text{Im} A$ be chosen arbitrarily. By 1.1(1)

$$xu + ux = \alpha e, \quad xv + vx = \beta e, \quad xw + wx = \gamma e \quad \text{with} \quad \alpha, \beta, \gamma \in \mathbb{R}.$$

If we postmultiply the first equation by $v$, premultiply the second by $u$ and subtract, we obtain $xw - wx = \alpha v - \beta u$, since $A$ is associative. Therefore after combining with the third $2xw = \alpha v - \beta u + \gamma e$. Postmultiplication by $w$ gives $-2x = \alpha u + \beta v + \gamma w$ and we have thus shown that $\text{Im} A = \mathbb{R}u + \mathbb{R}v + \mathbb{R}w$ and therefore $A \cong \mathbb{H}$.

If $A$ is an associative, finite-dimensional division algebra, $A$ has no divisors of zero, and by Theorem 2 is quadratic.                              $\square$

*Note.* In 8.1.1 we shall, with the help of the natural scalar product in $A$, give a second (and simpler) proof of the fact that $A$ is isomorphic to $\mathbb{H}$ in the case when $\dim A \geq 3$.

Another elementary proof of FROBENIUS's theorem for the case of finite-dimensional division algebras will be found in R.S. PALAIS: *The classification of real division algebras, Am. Math. Monthly* **75**, 1968, 366–368.

FROBENIUS's theorem gave a decisive impetus to the problem of the classification of all finite dimensional associative algebras. D. HAPPEL in an article "*Klassifikationstheorie endlich-dimensionaler Algebren in der Zeit von 1880 bis 1920*" (*L'Enseignement Math.* **26**, 2e ser. 1980, 91–102) reports on subsequent developments in this theory.

## §3. Hopf's Theorem

> Auch wenn man die Gültigkeit des assoziativen
> Gesetzes der Multiplikation nicht ausdrücklich postuliert,
> ist der Körper der komplexen Zahlen der einzige
> kommutative Erweiterungskörper endlichen Grades
> über dem Körper der reellen Zahlen
> (H. Hopf 1940).

> [Even where the validity of the associative law of
> multiplication is not explicitly postulated, the field
> of the complex numbers still remains the only com-
> mutative extension field of finite degree over
> the field of the real numbers.]

In the previous sections we found, with the theorem of Frobenius, all finite-dimensional, real *associative* division algebras. We now turn our attention in this section to finite-dimensional real *commutative* division algebras, which no longer need necessarily be associative. We prove the theorem of Hopf that all algebras of this kind are at most 2-dimensional; that if they possess a unit element, then $\mathbb{C}$ is the only such algebra apart from $\mathbb{R}$, to within an isomorphism; and that therefore the associative law of multiplication is a consequence of the commutative law. The arguments leading to Hopf's theorem are *topological,* the fundamental theorem of algebra is not needed, but emerges as a by-product.

A central role in the investigations is played by the *quadratic mapping*

$$\mathcal{A} \to \mathcal{A}, \qquad x \mapsto x^2$$

defined for any algebra $\mathcal{A}$. Topological properties of this mapping are responsible for the validity of Hopf's theorem, and these are summarized in the Hopf lemma in 2. From this lemma follows not only Hopf's theorem, but also as we shall see in §4, the famous Gelfand–Mazur theorem in functional analysis.

In the present section (§3) we shall require the Implicit Function Theorem for differentiable mappings, and also use the fact that every space $\mathbb{R}^n \setminus \{0\}$ is *connected* for $n \geq 2$. However, the decisive step in the actual proof itself is a theorem belonging to the theory of coverings, which states that every connected covering of a space $\mathbb{R}^n \setminus \{0\}$, $n \geq 3$ is one-to-one (see 3.2).

**1. Topologization of Real Algebras.** Let $V$ be a (not necessarily finite-dimensional) real vector space. A mapping $V \to \mathbb{R}$, $x \mapsto |x|$, is called a *norm (function)*, if for all $x, y \in V$, $\alpha \in \mathbb{R}$, the following conditions are satisfied:

$$|x| > 0 \text{ for } x \neq 0, |\alpha x| = |\alpha| |x| \text{ and } |x + y| \leq |x| + |y|$$

(*the triangle inequality*). Every norm on a vector space $V$ induces through $(x, y) \mapsto |x - y|$, a metric on $V$, so that the usual topological concepts and expressions such as "convergent sequence, open, closed, compact, connected set, continuity and so on" become available for these so-called *normed* vector spaces. The mapping $V \to \mathbb{R}$, $x \mapsto |x|$ is continuous; and so are vector addition and scalar multiplication, that is, the two mappings

$$V \times V \to V, \quad (x, y) \mapsto x + y \quad \text{and} \quad \mathbb{R} \times V \to V, \quad (\alpha, x) \mapsto \alpha x.$$

The proof may be left to the reader. In finite-dimensional spaces one can work with any norm because they all lead to the same topology. In any normed space $V$, the *unit sphere* is defined by

$$S := \{x \in V : |x| = 1\};$$

$S$ is always *compact* for finite-dimensional spaces (HEINE–BOREL). Every Euclidean vector space $V$ with scalar product $\langle x, y \rangle$ has the norm

$$V \to \mathbb{R}, \quad x \mapsto |x| := +\sqrt{\langle x, x \rangle} \quad \text{(Euclidean length)}.$$

**Lemma.** *If $A = (V, \cdot)$ is a finite-dimensional real algebra and $x \mapsto |x|$ is a norm in $V$, then:*

(1) *the multiplication $A \times A \to A$, $(x, y) \mapsto xy$ is continuous;*

(2) *there exists a $\sigma \geq 0$, such that $|xy| \leq \sigma |x| |y|$ for all $x, y \in V$;*

(3) *if $A$ has no divisors of zero, then there is a $\rho > 0$, such that $|xy| \geq \rho |x| |y|$ for all $x, y \in A$.*

**Proof.** To prove (1): let $v_1, \ldots, v_n$ be a basis of $V$, and let $x := \xi_1 v_1 + \cdots + \xi_n v_n$ and $y = \eta_1 v_1 + \cdots + \eta_n v_n$. Then $xy$, being a sum of terms of the form $\xi_\mu \eta_\nu v_\mu u_\nu$ is clearly continuous. Each function $A \to \mathbb{R}, x \mapsto \xi_\nu$, is continuous since the function $A \to \mathbb{R}^n, x \mapsto (\xi_1, \ldots, \xi_n)$ is continuous.

To prove (2) and (3): since $|\alpha x| = |\alpha| |x|$ it suffices to show that there exist numbers $\sigma \geq \rho > 0$, such that $\rho \leq |xy| \leq \sigma$. For all $x, y \in A$ satisfying $|x| = |y| = 1$. But this is clearly true since the mapping $S \times S \to \mathbb{R}$, $(x, y) \mapsto |xy|$ is continuous by (1) and therefore, since $S \times S$ is compact, attains a maximum $\sigma \geq 0$ and a minimum $\rho$; in the case of a division algebra we necessarily have $\rho > 0$, since points $x, y \in S \subset A \setminus \{0\}$ satisfying $xy = 0$ cannot exist.  □

*Remark.* The norm can be chosen so that (2) holds with $\sigma = 1$. For if $\sigma > 0$, then $\|x\| := \sigma |x|$, $x \in V$ is a norm in $V$ with $\|xy\| \leq \|x\| \cdot \|y\|$ (proof left as an exercise).

As an application of (1) and (3) we prove the

**Theorem.** *If $A$ is finite-dimensional and has no divisors of zero, the set $\{x^2 : x \in A \setminus \{0\}\}$ of all non-zero squares is closed in $A \setminus \{0\}$.*

**Proof.** We have to show that if $(x_n)$ is a sequence in $A$ and $\lim x_n^2 = a$, then there is a $b \in A$ with $b^2 = a$. The convergent sequence $(x_n^2)$ is bounded. By (3) there is a $\rho > 0$ such that $|x_n^2| \geq \rho |x_n|^2$ for all $n$; and so the sequence $(x_n)$ is likewise bounded. By the BOLZANO–WEIERSTRASS theorem it has a convergent subsequence. If its limit is $b \in A$ then $b^2 = a$, by (1).    □

This theorem is needed in 3.3.

## 2. The Quadratic Mapping $A \to A$, $x \mapsto x^2$. HOPF's Lemma.

For every $\mathbb{R}$-algebra $A$, the quadratic mapping is well defined. We write $A^\times := A \setminus \{0\}$. If $A$ has no divisors of zero, then $x^2 \in A^\times$ whenever $x \in A^\times$ and we thus have a mapping

$$q : A^\times \to A^\times, \qquad x \mapsto x^2.$$

This mapping will play a predominant role in this and in the following section. We note straight away that

(1) *If $A$ is commutative and without divisors of zero, then every point of the image $q(A^\times)$ has exactly two inverse images in $A^\times$.*

**Proof.** Suppose $w \in q(A^\times)$, and $a^2 = w$ and $a \in A^\times$. Then we also have $(-a)^2 = w$ with $-a \in A^\times$. There are no other points $c \in A^\times$ with $c^2 = w$, because we should have then

$$0 = c^2 - a^2 = (c - a)(c + a) \quad \text{which implies } c = \pm a.$$

As $a \neq -a$, $q^{-1}(w)$ consists of exactly two points.    □

A mapping $f : X \to Y$ between topological spaces is said to be a *local homeomorphism* of $x \in X$ if there is an open neighborhood $U$ of $x$, whose image $f(U)$ is open in $Y$, and if the induced mapping $f \mid U : U \to f(U)$ is a homeomorphism. Of decisive importance in all that follows is now

**HOPF's Lemma.** *Let $A = (V, \cdot)$ be a commutative real algebra without divisors of zero, and having the two properties that:*

a) *there is a norm on $V$ such that the quadratic mapping $q : A^\times \to A^\times$ is a local homeomorphism at all points of $A^\times$; and*

b) *every element of $A$ is a square (existence of square roots in $A$).*

*Then $\mathcal{A}$ is 2-dimensional.*

The inequality $\dim \mathcal{A} \geq 2$ clearly holds, because otherwise $\mathcal{A}$ would be isomorphic to $\mathbb{R}$, by Theorem 4 in the repertory (preceding Chapter 7), and this cannot be so, since $-1$ is not a square in $\mathbb{R}$. The difficulty is to exclude the possibility $\dim \mathcal{A} > 2$. This is done by arguments from the theory of coverings.

A mapping $\pi: X \to Y$ between topological spaces is called a *covering*, if it has the following property.

(c) *Every point $y \in Y$ has an open neighborhood $W$, such that $\pi^{-1}(W)$ is the disjoint union of sets $U_j$, $j \in J$, open in $X$, and which are such that all the mappings $U_j \to W$ induced by $\pi$ are homeomorphisms.*

A covering $\pi: X \to Y$ is said to be *connected*, if the space $X$ is connected; it is a covering of degree $k$, $k \in \mathbb{N}$, $k \geq 1$, if every point of $Y$ has exactly $k$ different inverse images under the mapping $\pi$.

The proof of HOPF's Lemma is based on the following observation.

*The mapping $q: \mathcal{A}^\times \to \mathcal{A}^\times$ is a connected covering of degree 2.*

**Proof.** Since $q(\mathcal{A}^\times) = \mathcal{A}^\times$, by b) and since $\mathcal{A}^\times$ is connected because $\dim \mathcal{A} \geq 2$, we need only, in view of the remark (1), verify that the condition (c) is satisfied. Suppose $w \in \mathcal{A}^\times$ is fixed, and that $q^{-1}(w) = \{a, -a\}$. By a) there are open neighborhoods $\tilde{U}$, $\tilde{U}^-$ of $a$, $-a$ which are mapped homeomorphically onto open neighborhoods of $w$ by the mapping $q$. As $\mathcal{A}$ is a Hausdorff space, we may assume that $\tilde{U} \cap \tilde{U}^- = \emptyset$. Now

$$W := q(\tilde{U}) \cap q(\tilde{U}^-), \quad U := q^{-1}(W) \cap \tilde{U}, \quad U^- := q^{-1}(W) \cap \tilde{U}^-$$

are open neighborhoods of $w$, $a$, and $-a$ respectively, the induced mappings $U \to W$, $U^- \to W$ are homeomorphisms and $U \cap U^- = \emptyset$. However we also have $q^{-1}(W) = U \cup U^-$ because, of the two inverse image points of any given point in $W$, one lies in $U$ and the other in $U^-$.                           □

The proof of HOPF's Lemma can now be brought abruptly to a conclusion, with the observation that

(∗) *If $V$ is a normed $\mathbb{R}$-vector space with $\dim V \geq 3$, then every connected covering $\pi: X \to V \setminus \{0\}$ is of degree 1.*

It follows directly from this remark that in the situation described in HOPF's Lemma only the case $\dim \mathcal{A} \leq 2$ is possible.                           □

A few comments may be added in explanation of (∗). The assertion is a direct consequence of the following two facts:

(a) *In the case $\dim V \geq 3$, $V \setminus \{0\}$ is simply connected, that is, every closed path $\gamma$ in $V \setminus \{0\}$ is continuously contractible to a point, for example, to its initial point.*

(b) *If $\pi: X \to Y$ is a covering and if $X$, $y$ are path connected and Hausdorff and if in addition $Y$ is simply connected, then $\pi$ is of degree 1.*

To elucidate (a), consider first $V$, an $\mathbb{R}^n$ with $3 \leq n < \infty$. If $\mathbb{R}^+$ denotes the set of all positive real numbers and $S^{n-1}$ the Euclidean unit sphere in $\mathbb{R}^n$, the mapping $\mathbb{R}^n \setminus \{0\} \to \mathbb{R}^+ \times S^{n-1}$, $x \mapsto (|x|, x/|x|)$ is a homeomorphism. We therefore need to remember that all spheres $S^k$, $2 \leq k < \infty$, are simply connected. Let $\gamma$ be a closed path in $S^k$. Since $k > 1$ we can assume that there is a point $p \in S^k$, which does *not* lie on $\gamma$ (this can always be achieved by a suitable deformation of $\gamma$). On this "North Pole" $p$, the path $\gamma$ can now be continuously contracted to a point along the great circles on $S^k$ through $p$.

Now suppose $V$ to be infinite-dimensional and $\gamma$ a closed path in $V \setminus \{0\}$. We can first subdivide $\gamma$ into a finite number of parts each lying in a ball in $V \setminus \{0\}$. As balls are convex, these partial paths can be deformed inside their balls into line segments. We thus obtain a deformation of $\gamma$ in $V \setminus \{0\}$ into a closed polygonal path in $V \setminus \{0\}$. This however lies in a *finite-dimensional* subspace $\mathbb{R}^n$, $n \geq 3$ of $V$ and is thus contractible continuously to a point in $\mathbb{R}^n \setminus \{0\}$.

We now say a few words about (b). Let $x_1, x_2$ be points for which $\pi(x_1) = \pi(x_2)$, and let us choose a path $\hat{\gamma}$ in $X$ from $x_1$ to $x_2$. Then $\gamma := \pi \circ \hat{\gamma}$ is a closed path in $Y$ which, by hypothesis, is contractible in $Y$ to its initial point $\pi(x_1)$, by deformation through a continuous family of paths $\gamma_s$, $0 \leq s \leq 1$. Since $\pi: X \to Y$ is a covering, every path $\gamma_s$ can be lifted in one and only one way to a path $\hat{\gamma}_s$ above $\gamma_s$ with initial point $x_1$. Then $\hat{\gamma}_s$, $0 \leq s \leq 1$ is a continuous family of paths in $X$, with $\hat{\gamma}_0 = \hat{\gamma}$ and $\hat{\gamma}_1$ the path consisting of the point $x_1$ only. As all paths have the same end-point (Monodromy Theorem), it follows that $x_2 = $ end-point of $\hat{\gamma}_0 = $ end-point of $\hat{\gamma}_1 = x_1$. Consequently every set $\pi^{-1}(y)$, $y \in Y$ reduces to a single point.

**3. HOPF's Theorem.** In this subsection $\mathcal{A} = (V, \cdot)$ denotes a *finite-dimensional real division algebra*. We topologize by choosing a norm in $\mathcal{A}$ and study the mapping $q: \mathcal{A} \to \mathcal{A}$, $x \mapsto x^2$. We use the methods of the differential calculus which are familiar from a second calculus course. A mapping $f: V \to V$ is said to be *differentiable at the point* $v \in V$ if there is a *linear* mapping $f'(v): V \to V$ such that:

$$\lim_{h \to 0, h \neq 0} \frac{|f(v+h) - f(v) - f'(v)(h)|}{|h|} = 0;$$

the mapping $f'(v)$ is then *uniquely* defined, and is called the *differential* (or sometimes the *derivative*) of $f$ in $v$.

Each element $a \in \mathcal{A}$ defines, through multiplication on the *left* and on the *right* respectively, the two linear mappings

$$L_a: V \to V, \quad x \mapsto ax; \quad R_a: V \to V, \quad x \mapsto xa.$$

We now assert that:

(1) *the mapping $q: \mathcal{A} \to \mathcal{A}$, $x \mapsto x^2$ is differentiable at every point $a \in \mathcal{A}$, and $q'(a) = L_a + R_a$. If $\mathcal{A}$ is commutative, then $q'(a)$, $a \in \mathcal{A}^\times$ is always bijective.*

**Proof.** By 1(2) we have $|h^2| \leq \sigma|h|^2$. Since

$$q(a + h) - q(a) = (a + h)^2 - a^2 = ah + ha + h^2 = (L_a + R_a)h + h^2$$

it follows that

$$\frac{|q(a + h) - q(a) - (L_a + R_a)h|}{|h|} = \frac{|h^2|}{|h|} \leq \sigma|h|,$$

which tends to zero in the limit.

If $\mathcal{A}$ is commutative, $L_a = R_a$, then $q'(a)h = 2L_a h = 2ah$, $h \in V$. Since $\mathcal{A}$ has no divisors of zero, $q'(a)$ is injective for all $a$ and, because $\dim V < \infty$, is indeed bijective. $\qquad\qquad\square$

An everywhere differentiable mapping $f : V \to V$ induces the mapping $f' : V \to \operatorname{Hom}(V, V), v \mapsto f'(v)$. We call $f$ *continuously differentiable* if $f'$ is continuous. (Note that when $V$ is finite-dimensional, $\operatorname{Hom}(V, V)$ is likewise finite-dimensional, so that one can talk of continuity without running into any problems.) In differential calculus courses one proves the following as a special case of the theorem on implicit functions.

**Local Implicit Function Theorem (for Differentiable Functions).**
*Let $f: V \to V$ be continuously differentiable, and let $v \in V$ be a point such that the derivative $f'(v): V \to V$ is bijective. Then $f$ is a local homeomorphism at $v$.*

A corollary of this is

(2) *If $\mathcal{A}$ is commutative, then $q: \mathcal{A}^\times \to \mathcal{A}^\times$ is a local homeomorphism.*

**Proof.** The mapping $q$ induces, by (1) the mapping

$$q': V \to \operatorname{Hom}(V, V), \qquad v \mapsto 2L_v.$$

It is easily verified (by choosing bases and representing $q'$ by a matrix) that $q'$ is continuous. By virtue of (1) and the local implicit function theorem, $q$ is therefore a local homeomorphism at every point $a \in \mathcal{A}^\times$. $\qquad\square$

We can now complete in a few lines the proof of the famous

**HOPF's Theorem (1940).** *Every finite-dimensional real commutative division algebra $\mathcal{A} = (V, \cdot)$ is at most two-dimensional.*

**Proof.** By (2), the condition a) of HOPF's lemma is satisfied. Let $n :=\dim \mathcal{A} \geq 2$. Then $\mathcal{A}^\times$ is *connected*. The set $q(\mathcal{A}^\times)$ is, by Theorem 1, *closed* in $\mathcal{A}^{\times}$. But, in view of (2), it is also *open* in $\mathcal{A}^{\times}$. It follows therefore that $q(\mathcal{A}^{\times}) = \mathcal{A}^{\times}$, that is, $\mathcal{A}$ also satisfies the condition b) of HOPF's lemma, and consequently $n = 2$.    □

HOPF generalized his theorem considerably, immediately after its discovery in the year 1940, in his paper (*"Ein topologischer Beitrag zur reellen Algebra,"* in *Comment. Math. Helv.* **13**, 1940, 219–239, in particular p. 229) [A topological contribution to real algebra] where he was able to show, without any requirement of commutativity that:

*The dimension of a finite-dimensional real division algebra is necessarily a power of* 2.

The reader will find more details on this given in greater depth in Chapter 11.

**4. The Original Proof by HOPF.** Our proof of HOPF's theorem is an adaptation of his original proof. HOPF himself in 1940 dealt with the continuous mapping

$$g: \mathcal{A} \setminus \{0\} \to \mathcal{A}, \qquad x \mapsto \frac{x^2}{|x^2|}.$$

Each image vector is of unit length so that $\mathcal{A} \setminus \{0\}$ is mapped into the $(n-1)$-dimensional sphere

$$S^{n-1} := \{v \in V : |v| = 1\}, \qquad n := \dim \mathcal{A}.$$

Obviously $g(\alpha x) = g(x)$ for all $x \in \mathcal{A} \setminus \{0\}$, $\alpha \in \mathbb{R} \setminus \{0\}$ so that the mapping $g$ maps every straight line through 0 onto the same point. Now the real projective plane $\mathbb{P}^{n-1}$ is nothing more than the space of all straight lines in $V$ through 0, and so we have the famous "HOPF mapping"

$$h: \mathbb{P}^{n-1} \to S^{n-1}.$$

All this still applies for arbitrary division algebras $\mathcal{A}$; but in addition:

*If $\mathcal{A}$ is commutative then $h: \mathbb{P}^{n-1} \to S^{n-1}$ is injective.*

**Proof.** Let $\hat{x}, \hat{y} \in \mathbb{P}^{n-1}$ be points with $h(\hat{x}) = h(\hat{y})$. We represent $\hat{x}, \hat{y}$ by points $x, y \in V \setminus \{0\}$. The equation $h(\hat{x}) = h(\hat{y})$ then means, when we use the abbreviations $\xi := \sqrt{|x^2|}$, $\eta := \sqrt{|y^2|}$, $\alpha := \xi^{-1}\eta \in \mathbb{R}$

$$(\xi^{-1}x)^2 = (\eta^{-1}y)^2, \quad \text{that is} \quad y^2 = \alpha^2 x^2.$$

As $\mathcal{A}$ is commutative and without divisors of zero, it follows that

$$0 = y^2 - \alpha^2 x^2 = (y - \alpha x)(y + \alpha x), \quad \text{and hence} \quad y = \pm\alpha x, \text{ i.e. } \hat{y} = \hat{x}. \quad □$$

Thus corresponding to every $n$-dimensional, real, commutative division algebra, HOPF defined an associated *topological* mapping of the projective space $\mathbb{P}^{n-1}$ *into* the sphere $S^{n-1}$. He now argued as follows (see *Selecta*, p. 112) where we translate from the German original: "...as $\mathbb{P}^{n-1}$ and $S^{n-1}$ are closed manifolds of the same dimension $n-1$, $S^{n-1}$ must be identical to the image of $\mathbb{P}^{n-1}$, and the manifolds $S^{n-1}$ and $\mathbb{P}^{n-1}$ must therefore be homeomorphic. For $n-1 = 1$ this is indeed the case: the circle $S^1$ and the projective line are both (homeomorphic to) a simple closed line. If however $n-1 > 1$, the sphere $S^{n-1}$ is, in contrast to the case $n-1 = 1$, simply connected, whereas the projective space $\mathbb{P}^{n-1}$ is never simply connected, because the projective line can never be contracted into a point; the homeomorphism in question does not therefore exist when $n-1 > 1$."

Thus HOPF showed that $n-1 = 1$, that is, $n = 2$.                    □

To this day no "elementary" proof of HOPF's theorem is known. In 1954 the Dutch mathematician SPRINGER in a paper entitled "An algebraic proof of a theorem of H. HOPF" (*Indagationes Mathematicae* **16**, 33–35) gave a proof which uses results from algebraic geometry, amongst others the theorem of BEZOUT, instead of the argument of simple-connectedness.

**5. Description of All 2-Dimensional Algebras with Unit Element.**
Every 2-dimensional real algebra $\mathcal{A}$ with unit element $e$, has a basis $e$, $w$ with $w^2 = \omega e$, where $\omega = 0$ or $\omega = 1$ or $\omega = -1$. (The proof is left to the reader.) From this follows the

**Lemma.** *Every 2-dimensional real algebra $\mathcal{A}$ with unit element is both commutative and associative. There are the following three mutually exclusive possibilities:*

1) *$\mathcal{A}$ is isomorphic to the algebra $(\mathbb{R}^2, \cdot)$ of "dual numbers," that is $(1,0) \in \mathbb{R}^2$ is the unit element, and $\varepsilon := (0,1) \in \mathbb{R}^2$ satisfies the equation $\varepsilon^2 = 0$.*

2) *$\mathcal{A}$ is isomorphic to the direct sum $\mathbb{R} \oplus \mathbb{R}$, that is, for $a := (1,0)$, $b := (0,1) \in \mathbb{R}^2$ we have the relations $a^2 = a$, $b^2 = b$, $ab = 0$ (see R.2, 6).*

3) *$\mathcal{A}$ is isomorphic to the algebra $\mathbb{C}$.*

**Proof.** The three cases $\omega = 0$, $\omega = 1$, $\omega = -1$ lead to the cases 1), 2), 3) respectively. When $\omega = 1$, $u := \frac{1}{2}(e + w)$, $v := \frac{1}{2}(e - w)$ form a basis of $\mathcal{A}$ with $u^2 = u$, $v^2 = v$, $uv = 0$, and hence $\mathcal{A} \to \mathbb{R} \oplus \mathbb{R}$, $\alpha u + \beta v \mapsto \alpha a + \beta b$ is an isomorphism.                                                    □

From this lemma and HOPF's theorem follows at once (since $\mathcal{A}$ has divisors of zero in the first two cases of the lemma) the

**Corollary to HOPF's Theorem.** *Every finite-dimensional, real, commutative division algebra with unit element e is isomorphic to* $\mathbb{R}$ *or to* $\mathbb{C}$.

The fundamental theorem of algebra is implicitly contained in this statement, and has thus been proved anew.

*Exercise.* Find the fallacy in the following "direct proof" of the above corollary. If $n := \dim \mathcal{A} > 1$, there is a $j \in \mathcal{A}$ with $j^2 = -e$. Then $\mathcal{B} := \mathbb{R}e \oplus \mathbb{R}j$ is a subalgebra of $\mathcal{A}$, isomorphic to $\mathbb{C}$. For every $a \in \mathcal{A}$, the characteristic polynomial $\det(L_a - X \cdot \mathrm{id})$ of the left-multiplication $L_a : \mathcal{A} \to \mathcal{A}$, $x \mapsto ax$, has, by the Fundamental theorem of algebra, a zero $b \in \mathcal{B}$; that is, there is a $c \neq 0$ in $\mathcal{A}$ such that $(a - be)c = 0$. It follows that $a = be = b \in \mathcal{B}$, that is, $\mathcal{A} = \mathcal{B}$.

The assumption that $\mathcal{A}$ has a unit element is an essential part of the argument in the foregoing considerations. There are infinitely many non-isomorphic commutative 2-dimensional division algebras. For example we can derive one from $\mathbb{C}$ by defining multiplication of $w, z \in \mathbb{C}$ by $w \circ z := \overline{wz}$. The family of all these (non-isomorphic) algebras is two-dimensional and not connected.

*Exercise.* Show that every 2-dimensional alternative and commutative algebra is isomorphic to $\mathbb{C}$.

## §4. The Gelfand–Mazur Theorem

> Chaque domaine de rationalité du type $(B^*)$ est isomorphe au domaine de rationalité des nombres réels, des nombres complexes ou des quaternions (S. Mazur 1938).
>
> [Every domain of rationality of type $(B^*)$ is isomorphic to the domain of rationality of the real numbers, the complex numbers, or the quaternions.]

The theorem quoted above is nowadays known, mainly in the commutative case, as the theorem of Gelfand–Mazur. In functional analysis it is usually obtained from the fact that in a complex Banach algebra with unit element, every element has a non-empty spectrum. For this purpose one usually invokes Liouville's theorem (that a bounded entire holomorphic function is a constant) and thus in the final analysis the Cauchy theory of functions.

It does not appear to be generally known in the literature that the Gelfand–Mazur theorem is a simple corollary of Hopf's lemma. This will be explained in the following account and will again demonstrate the power of the quadratic mapping. If Hopf had known in 1940 of Mazur's note in the *Comptes Rendus* he would undoubtedly have taken the opportunity to prove the theorem then and there.

**1. BANACH Algebras.** An $\mathbb{R}$-algebra $\mathcal{A} = (V, \cdot)$ is said to be a *normed algebra*, if a norm $|\ |$ is defined on the vector space $V$, such that:

$$(1) \qquad\qquad |xy| \le |x|\,|y| \quad \text{for all} \quad x, y \in \mathcal{A}.$$

In that case the multiplication $\mathcal{A} \times \mathcal{A} \to \mathcal{A}$, $(x, y) \mapsto xy$ is continuous, (as well as the addition), as can immediately be seen from the inequality

$$|xy - x_0 y_0| \le |x - x_0|\,|y - y_0| + |x - x_0|\,|y_0| + |x_0|\,|y - y_0|.$$

If $\mathcal{A}$ has a unit element $e$, then $|e| \ge 1$.

An *associative* normed $\mathbb{R}$-algebra $\mathcal{A} = (V, \cdot)$ with norm $|\ |$ is called a real *Banach algebra*, if $V$ is a Banach space, that is, if every CAUCHY sequence in $V$ converges, where convergence is defined with respect to the metric $|x - y|$.

*Examples.* 1) The $\mathbb{R}$-algebras $\mathbb{R}$, $\mathbb{C}$, $\mathbb{H}$ with their natural norms are BANACH algebras.

2) The $\mathbb{R}$-algebra $\mathrm{Mat}(n, \mathbb{R})$ of all real $n \times n$ matrices, $1 \le n < \infty$, is a BANACH algebra with the norm

$$|A| := \sqrt{\mathrm{trace} A^t A} = \sqrt{\sum_{\mu,\nu=1}^{n} |a_{\mu\nu}|^2}, \quad \text{where} \quad A = (a_{\mu\nu}).$$

(Note that $|E| = \sqrt{n}$.)

3) The $\mathbb{R}$-algebra $\mathcal{C}[0,1]$ of all functions continuous in the closed interval $[0,1]$ is a BANACH algebra with respect to the maximum norm $|f| := \max_{0 \le x \le 1} |f(x)|$.

4) If $\mathcal{A} = (V, \cdot)$ is any finite-dimensional associative real algebra, there are always norms on $V$, such that $\mathcal{A}$ is a BANACH algebra. (This follows from the remark in 3.1, since finite-dimensional normed vector spaces are BANACH spaces.)

Normed $\mathbb{C}$-algebras, and complex BANACH algebras are defined almost word for word as above, and are not treated until §4.6 and §4.7.    □

In the following paragraphs $\mathcal{A}$ always denotes a real BANACH algebra with unit $e$. A power series

$$(*) \qquad \alpha_0 e + \alpha_1 X + \alpha_2 X^2 + \cdots + \alpha_n X^n + \cdots, \qquad \alpha_\nu \in \mathbb{R},$$

is said to be *absolutely convergent* at the point $a \in \mathcal{A}$, if $\sum |\alpha_\nu|\,|a|^\nu < \infty$. It then converges *absolutely and uniformly in the ball* $\{x \in A : |x| \le |a|\}$ *of radius* $|a|$ *around* 0 (*Abel's lemma*). Since all CAUCHY sequences converge in $\mathcal{A}$, this can be proved word for word as in the particular cases of $\mathcal{A} = \mathbb{R}$ and $\mathcal{A} = \mathbb{C}$. Uniformity of convergence implies that:

(2) *If the series* (∗) *converges absolutely at the point* $a \in \mathcal{A}$, $a \neq 0$, *then the function*

$$f : U \to \mathcal{A}, \qquad x \mapsto f(x) := \sum_0^\infty a_\nu x^\nu$$

*is continuous in the ball* $U$ *of radius* $|a|$ *with center* 0.

In the following argument, we shall make use of the *binomial series* $\sum_0^\infty \binom{1/2}{\nu} X^\nu$ with whose help we shall be able to extract square roots.

**2. The Binomial Series** with exponent $\alpha \in \mathbb{R}$ is defined by

$$b_\alpha(X) := \sum_0^\infty \binom{\alpha}{\nu} X^\nu = e + \alpha X + \frac{1}{2}\alpha(\alpha - 1)X^2 + \cdots,$$

$$\binom{\alpha}{\nu} := \frac{\alpha(\alpha - 1) \cdot \ldots \cdot (\alpha - \nu + 1)}{1 \cdot 2 \cdot \ldots \cdot \nu}.$$

As this series has a radius of convergence $\geq 1$ in $\mathbb{R}$, it converges absolutely and locally uniformly in the *unit ball* $\{x \in \mathcal{A} : |x| < 1\}$ and therefore defines a continuous mapping of this ball into $\mathcal{A}$.

In what follows a fundamental role is played by the *Addition theorem*

(1)     $b_\alpha(x)b_\beta(x) = b_{\alpha+\beta}(x)$ *for all* $x \in \mathcal{A}$, $|x| < 1$, *and all* $\alpha, \beta \in \mathbb{R}$.

**Proof.** The Multiplication theorem for absolutely convergent series holds for BANACH algebras as it does for $\mathbb{R}$. It follows therefore that, for all $x \in \mathcal{A}$ with $|x| < 1$, we have

$$\sum_0^\infty \binom{\alpha}{\mu} x^\mu \cdot \sum_0^\infty \binom{\beta}{\nu} x^\nu$$

$$= \sum_{n=0}^\infty \left[ \binom{\alpha}{0}\binom{\beta}{n} + \binom{\alpha}{1}\binom{\beta}{n-1} + \cdots + \binom{\alpha}{n}\binom{\beta}{0} \right] x^n.$$

The coefficient on the right is equal to $\binom{\alpha + \beta}{n}$,[3] and this proves (1).  □

---

[3]This so-called *Addition theorem for the binomial coefficients* is proved by induction on $n$. One can also however argue as follows: the binomial formula $b_n(x) = (1 + x)^n$ holds for all rational numbers $x$ and all natural numbers $n$. The Addition theorem for binomial coefficients must therefore be true for all $\alpha, \beta, n \in \mathbb{N}$, and hence for all $\alpha, \beta \in \mathbb{R}$, $n \in \mathbb{N}$ (Identity theorem for polynomials.)

We need (1) only for $\alpha = \beta = \frac{1}{2}$. Since $b_1(x) = e + x$ this equation is then equivalent to

$$(2) \qquad q(b(x)) = e + x \quad \text{for all} \quad x \in A \text{ with } |x| < 1,$$

where we have written $b(x)$ in place of $b_{1/2}(x)$ and where $q$ is the quadratic mapping $q: A \to A$, $x \mapsto x^2$.

**3. Local Inversion Theorem.** *Let $A$ be a commutative* BANACH *algebra without divisors of zero and with unit element, let $a \in A$ be invertible (that is, there is an $a^{-1} \in A$ with $aa^{-1} = e$). Then the quadratic mapping $q: A^\times \to A^\times$ is a local homeomorphism at $a$.*

**Proof.** The open ball $K \subset A^\times$ of radius $|a|$ and center $a$ is mapped injectively by $q$, for if $a + u$, $a + v$ are two distinct points satisfying $(a + u)^2 = (a + v)^2$, then $u + v = -2a$, so that $|u| < |a|$, $|v| < |a|$ cannot both hold.

As $a$ is invertible, $a^2$ is also invertible with inverse $a^{-2}$. In the open ball $L \subset A^\times$ of radius $1/|a^{-2}|$ around $a^2$, the function

$$\mu(x) := a \cdot b((x - a^2)a^{-2}) \quad \text{with} \quad \mu(a^2) = a$$

is thus well-defined and continuous. It follows directly from 2(2) that

$$(*) \qquad\qquad q \circ \mu = \text{id} \quad on \ L.$$

If therefore one chooses an open neighborhood $W \subset L$ of $a^2$ with $\mu(W) \subset C$, the open neighborhood $U := q^{-1}(W) \cap K$ is mapped onto $W$ continuously and injectively by $q$.

**Corollary.** *If, additionally, $A$ is a field, then $q: A^\times \to A^\times$ is a local homeomorphism (for now every element $a \in A^\times$ is invertible).*

*Remark.* The local inversion theorem can also be proved directly without power series if one is prepared to use the *general local inversion theorem for Banach spaces* which states:

*Let $f: V \to W$ be a continuous differentiable mapping between* BANACH *spaces, and let $v \in V$ be a point such that the linear mapping $f'(v): V \to W$ is an isomorphism. Then $f$ is a local homeomorphism at $v$.*

It can be verified—as in 3.3—that $q$ satisfies the hypotheses of this theorem: the derivative $q'(a): V \to V$, $x \mapsto 2ax$, is a homeomorphism since $a$ is invertible.

**4. The Multiplicative Group $A^\times$.** In every associative division algebra the set $A^\times$ of non-zero elements forms a group with respect to multiplication. The following can be said about the topology of this group.

*If $A$ is a* BANACH *division algebra of dimension $\geq 2$ there is no open subgroup of $A^\times$ other than $A^\times$ itself.*

**Proof.** Suppose there were such a group $G$. For every point $a \in A^\times \setminus G$ we have $aG \subset A^\times \setminus G$, because $ab = c$ with $b, c \in G$ implies $a = cb^{-1} \in G$. As $aG$ is open in $A^\times$ when $G$ is (the mapping $A^\times \to A^\times$, $x \mapsto ax$ is topological since $x \to a^{-1}x$ is continuous) it follows that all points of $A^\times \setminus G$ are interior points of $A^\times \setminus G$. Consequently $G$ is not only open, but also closed in $A^\times$. This means that $A^\times$ is not connected in contradiction to $\dim A \geq 2$.    □

We can now complete in a few lines the proof of

**5. The GELFAND–MAZUR Theorem.** *Every commutative* BANACH *division algebra $A$, is isomorphic to the field $\mathbb{R}$ or $\mathbb{C}$.*

**Proof.** We show that $A$, in the case $\dim A \geq 2$, has the properties a) and b) of HOPF's lemma 3.2. Corollary 3 asserts that a) holds. To verify b), that is, $q(A^\times) = A^\times$, we observe that $q(A^\times)$ is a *subgroup* of the group $A^\times$ which, by Corollary 3, is open in $A^\times$. Thus $q(A^\times) = A^\times$, by 4.    □

This theorem again contains the fundamental theorem of algebra, because every finite dimensional extension field of $\mathbb{R}$ is a commutative BANACH algebra (see 1.4).

*Remark.* The proof may also be carried out with the exponential mapping $x \mapsto \exp x := \sum x^\nu/\nu!$, which yields a group homomorphism $A \to A^\times$. This mapping is also a covering; in place of 2(2) we have:

(∗)  $\exp \log(e + x) = e + x$ for $|x| < 1$ where

$$\log(e + x) := \sum_1^\infty \frac{(-1)^{\nu-1}}{\nu} x^\nu.$$

This covering would be of degree one in the case $\dim A > 2$, that is, the groups $A$, $A^\times$ would be isomorphic, which is untrue. The proof of (∗) is incidentally somewhat troublesome, because we have to consider a power series whose terms are power series, and not as in 2 merely the product of two power series.

**Corollary.** *Let $A \neq 0$ be a commutative* BANACH *algebra with unit $e$, and let $\mathfrak{m}$ be a maximal ideal in $A$. Then the quotient algebra $A/\mathfrak{m}$ is isomorphic to $\mathbb{R}$ or to $\mathbb{C}$.*

**Proof.** The closure $\bar{\mathfrak{m}}$ of $\mathfrak{m}$ is an ideal in $A$ with $\mathfrak{m} \subset \bar{\mathfrak{m}}$. If $\mathfrak{m}$ were different from $\bar{\mathfrak{m}}$, we should have $\bar{\mathfrak{m}} = A$, that is $e \in \bar{\mathfrak{m}}$. There would then be an $a \in A$, such that $e - a \in \mathfrak{m}$ and $|a| < 1$. Now $e - a$ is invertible in $A$ (with inverse $e + a + a^2 + \ldots$) so that $e \in \mathfrak{m}$ and it would follow that $\mathfrak{m} = A$

which is absurd. Hence $\mathfrak{m}$ is a closed ideal in $\mathcal{A}$ and thus $\mathcal{A}/\mathfrak{m}$ (endowed with the residue class norm) is likewise a BANACH algebra. As $\mathcal{A}/\mathfrak{m}$ is also a commutative field, this proves the corollary.                                      □

**6. Structure of Normed Associative Division Algebras.** In proving the GELFAND–MAZUR theorem, the use of the "BANACH algebra" assumption was essential. From now on however we can dispense with the completeness of the algebra. This is made possible by the following embedding theorem:

*Every normed associative $\mathbb{R}$-algebra $\mathcal{A}$ with unit $e$ is an $\mathbb{R}$-subalgebra of a BANACH algebra $\bar{\mathcal{A}}$ with $e$ as unit element.*

We can, for example, choose for $\bar{\mathcal{A}}$, a completion of $\mathcal{A}$.                 □

There now follows, as an almost immediate consequence the

**GELFAND–MAZUR Theorem for Normed Algebras.** *Every normed commutative associative real division algebra $\mathcal{A} \neq 0$ is isomorphic to $\mathbb{R}$ or $\mathbb{C}$.*

**Proof.** Let $\bar{\mathcal{A}}$ be a BANACH algebra with $e$ as unit, of which $\mathcal{A}$ is a sub-algebra. We choose a maximal ideal $\mathfrak{m}$ in $\bar{\mathcal{A}}$ (Zorn's lemma). By Corollary 5, $\dim \bar{\mathcal{A}}/\mathfrak{m} \leq 2$. As $\mathcal{A} \cap \mathfrak{m} = \{0\}$, the residue class mapping $\bar{\mathcal{A}} \to \bar{\mathcal{A}}/\mathfrak{m}$ induces an $\mathbb{R}$-algebra monomorphism $\mathcal{A} \to \bar{\mathcal{A}}/\mathfrak{m}$. Hence $\dim \mathcal{A} \leq 2$.      □

**Corollary (OSTROWSKI 1918).** *Every valuated commutative and associative $\mathbb{R}$-algebra $\mathcal{A}$ with unit is isomorphic to $\mathbb{R}$ or $\mathbb{C}$.*

**Proof.** By hypothesis $|xy| = |x|\,|y|$ for all $x, y \in \mathcal{A}$, and so $\mathcal{A}$ is an integral domain. The valuation can be extended to a valuation of $K$, the quotient field of $\mathcal{A}$. The normed field $K$ is isomorphic to $\mathbb{R}$ or $\mathbb{C}$ by the theorem. As $\mathcal{A}$ is an $\mathbb{R}$-subalgebra of $K$, this proves the corollary.               □

*Remark.* The corollary becomes false if the condition "valuated" is replaced by the condition "normed and without divisors of zero." A counterexample is the polynomial algebra $\mathbb{R}[X]$ with the norm $|p| := \max_{0 \leq x \leq 1} |p(x)|$. The completion of this algebra is the BANACH algebra $\mathcal{C}([0,1])$ of all functions continuous in $[0,1]$ (WEIERSTRASS's approximation theorem), and this algebra has (infinitely) many divisors of zero.

**Structure Theorem (MAZUR 1938).** *Let $\mathcal{A}$ be a normed associative real division algebra. Then $\mathcal{A}$ is isomorphic either to $\mathbb{R}$ or to $\mathbb{C}$, or else to $\mathbb{H}$.*

**Proof.** If we can show that $\mathcal{A}$ is a quadratic algebra, the statement will follow from FROBENIUS's theorem. Suppose $x \in \mathcal{A}$. All non-zero elements of the commutative polynomial algebra $\mathbb{R}[x]$ are, by hypothesis, invertible in $\mathcal{A}$, so that $K := \{u/v : u, v \in \mathbb{R}[x],\ v \neq 0\}$ is a *normed commutative*, associative, real division algebra, and hence $\dim K \leq 2$ by the theorem. Consequently $x \in K$ satisfies a quadratic equation over $\mathbb{R}$.   □

**Consequence (GELFAND–MAZUR Theorem for $\mathbb{C}$-Algebras).** *Every normed associative complex division algebra $\mathcal{A}$ is isomorphic to $\mathbb{C}$.*

**Proof.** Since $\mathcal{A}$ is also an $\mathbb{R}$-algebra, $\dim \mathcal{A} < \infty$. It follows from the foonote on page 198 that $\mathcal{A} \cong \mathbb{C}$.

**7. The Spectrum.** If $\mathcal{A}$ is a normed associative $\mathbb{C}$-algebra with unit $e$, then the *spectrum* of an element $a \in \mathcal{A}$ is the set

$$\operatorname{Spec} a := \{\lambda \in \mathbb{C} : a - \lambda e \text{ is not invertible in } \mathcal{A}\}.$$

**Fundamental Lemma.** *Let $\mathcal{A}$ be a normed associative $\mathbb{C}$-algebra with unit $e$. Then $\operatorname{Spec} a \neq \emptyset$ for every $a \in \mathcal{A}$.*

**Proof.** Let $a \in \mathcal{A}$ be fixed. Every element $x \neq 0$ of the commutative polynomial algebra $\mathbb{C}[a]$ can (by the Fundamental theorem of algebra) be written as:

$$x = c(a - \lambda_1 e) \cdot \ldots \cdot (a - \lambda_n e), \qquad c, \lambda_1, \ldots, \lambda_n \in \mathbb{C}.$$

If $\operatorname{Spec} a$ were void, all the factors $a - \lambda_\nu e$ and hence all $x \neq 0$ in $\mathbb{C}[a]$ would be invertible in $\mathcal{A}$, and then $K := \{u/v : u, v \in \mathbb{C}[a],\ v \neq 0\}$ would be a normed commutative field. By the GELFAND–MAZUR theorem $K$ would be $\mathbb{C}e$, and $a - \lambda e$ would be zero for some $\lambda \in \mathbb{C}$, contradicting the assumption that $\operatorname{spec} a = \emptyset$.   □

*Remark.* Usually one proves the fundamental lemma for complex BANACH algebras by means of LIOUVILLE's theorem. The fundamental lemma, on the other hand, implies the GELFAND–MAZUR theorem for $\mathbb{C}$-algebras $\mathcal{A}$: if every $x \neq 0$ in $\mathcal{A}$ is a unit, then $x = \lambda e$ for $\lambda \in \operatorname{Spec} x$, that is $\mathcal{A} = \mathbb{C}e$.

A simple consequence of the fundamental lemma is:

*For every continuous (= bounded) endomorphism $\varphi : E \to E$ of a normed $\mathbb{C}$-vector space $E \neq 0$ there is a $\lambda \in \mathbb{C}$, such that $\varphi - \lambda$ id is not invertible.*

**Proof.** The set $\operatorname{End} E$ of all continuous endomorphisms of $\mathbb{E}$ is an associative $\mathbb{C}$-algebra with unit, with respect to composition. $\operatorname{End} E$ becomes a normed $\mathbb{C}$-algebra when the norm is defined by

$$\|u\| := \sup_{|x|=1} \{|u(x)|\} < \infty, \qquad u \in \operatorname{End} E.$$

For $\varphi \in \operatorname{End} E$ it now follows that $\operatorname{Spec} \varphi \neq \emptyset$.   □

In the statement which has just been proved is included the assertion
that every endomorphism $\varphi$ of a finite-dimensional $\mathbb{C}$-vector space $E \neq 0$,
has an eigenvalue $\lambda$ in $\mathbb{C}$ (see Theorem 4.3.4). In the case $\dim E < \infty$, we
have in fact

$$\operatorname{Spec} \varphi = \{\lambda \in \mathbb{C}: \quad \varphi - \lambda e : E \to E \text{ is not injective}\};$$

but in the case $\dim E = \infty$, $\varphi$ does not always have eigenvalues.

**8. Historical Remarks on the GELFAND–MAZUR Theorem.** The
starting point of the GELFAND–MAZUR theorem was the theorem of OS-
TROWSKI, proved in 1918, that every complete, commutative field with
Archimedean valuation is isomorphic to the field $\mathbb{R}$ or $\mathbb{C}$ (see Corollary 6
in this connection).

The original proof of OSTROWSKI is computational. In 1938 S. MAZUR
in a note in the *Comptes rendus* generalized this theorem and outlined a
proof; in 1941 I.M. GELFAND proved MAZUR's theorem with the help of
LIOUVILLE's theorem.

In 1952 E. WITT derived the GELFAND–MAZUR theorem in six lines: "In
the case $K \neq R$, $R(i)$ is of rank $[K : R] > 2$ and therefore the domain $(x \neq
0)$ is simply connected. The differential equation $x^{-1}dx = y$ then engenders
a global isomorphism between the multiplicative group $(x \neq 0)$ and the
additive group $(y)$. This however is impossible because the multiplicative
group contains the element 1 of order 2, whereas the additive group is of
characteristic zero." This is, of course, the proof mentioned in 5 based on
the exponential function.

REFERENCES

GELFAND, I.M.: Normierte Ringe, *Math. Sborn.* **9**, 3–23 (1941)

MAZUR, S.: Sur les anneaux linéaires, *C.R. Acad. Sci. Paris* **207**, 1025–
1027 (1938)

OSTROWSKI, A.: Über einige Lösungen der Funktionalgleichung $\varphi(x)\varphi(y) =
\varphi(xy)$, *Acta Math.* **41**, 271–284 (1918); *Coll. Math. Pap.* **2**, 322–335

WITT, E.: Über einen Satz von Ostrowski, *Arch. Math.* **3**, 334 (1952)

For more information see the survey article *Norm and Spectral Characterization in Banach Algebras* by V.A. BELFI and R.S. DORAN in L'Enseign.
Math., 2. Ser., 26 (1980), 103–130.

**9. Further Developments.** The GELFAND–MAZUR theorem has been
generalized in several directions. The starting point is the following

**Remark 1.** *Let $A$ be a BANACH algebra with unit $e$, and having the property
that for every $y \neq 0$ in $A$, a real number $M_y > 0$ can be found such that:*

(o) $$|x| \leq M_y |xy| \quad \text{for all} \quad x \in A.$$

*Then $A$ is a division algebra.*

**Proof.** We may assume that $\dim A > 1$. It is well known that the set $E$ of all invertible elements of $A$ is open in $A \setminus \{0\}$. If we can show that $E$ is also closed in $A \setminus \{0\}$, it will then follow that $E = A \setminus \{0\}$, because $A \setminus \{0\}$ is connected, since $\dim A > 1$.

Let $u_n \in E$ be a sequence with limit $u \neq 0$. Choose an $m > 0$, such that $|u - u_n| < \frac{1}{2} M_u^{-1}$ for all $n \geq m$. Then, since it is always true that

(*) $$u_n^{-1} u = u_n^{-1}(u - u_n) + e,$$

it follows, by (o), that for $n \geq m$

$$|u_n^{-1}| \leq M_u |u_n^{-1} u| \leq M_u |u_n^{-1}(u - u_n)| + M_u \leq \frac{1}{2}|u_n^{-1}| + M_u,$$

that is $|u_n^{-1}| \leq 2M_u$ for $n \geq m$. The sequence $u_n^{-1}$ is therefore bounded, and it follows from (*) that $\lim u_n^{-1} u = e$. As $E$ is open in $A \setminus \{0\}$, it follows that $u_n^{-1} u$ is invertible for large enough $n$, and hence $u$ itself is invertible, that is, $u \in E$. $\qquad\square$

We can now rapidly deduce the following result, if we invoke a classical theorem of BANACH.

**Theorem.** *A BANACH algebra $A$ with unit and without divisors of zero, and in which every principal ideal $Ay$, $y \in A$, is closed, is isomorphic to $\mathbb{R}$, $\mathbb{C}$, or $\mathbb{H}$.*

**Proof.** Let $y \neq 0$. The linear mapping $\varphi_y : A \to Ay$, $x \mapsto xy$, is bounded and bijective. As $Ay$ is, by hypothesis, a BANACH space, $\varphi_y^{-1}$ is bounded (BANACH). There is therefore an $M_y > 0$, such that $|x| = |\varphi_y^{-1}(xy)| \leq M_y |xy|$ for all $x \in A$. The statement now follows from Remark 1, by virtue of the structure theorem 6. $\qquad\square$

To apply the theorem, we need

**Remark 2.** *Let $A$ be a commutative BANACH algebra with unit $e$, and let $\mathfrak{a}$ be an ideal in $A$, whose topological closure $\bar{\mathfrak{a}}$ in $A$, can be finitely generated. Then $\mathfrak{a} = \bar{\mathfrak{a}}$.*

**Proof.** If $a_1, \ldots, a_n$ is a system of generators of $\bar{\mathfrak{a}}$, the mapping $\varphi : A^n \to \bar{\mathfrak{a}}$, $(x_1, \ldots, x_n) \mapsto \sum x_\nu a_\nu$ is linear, surjective and bounded. By BANACH's theorem $\varphi$ is therefore *open*. If $D(\varepsilon)$ denotes the ball of radius $\varepsilon > 0$ about $0$, then $\sum D(\varepsilon) a_\nu$ is a $0$-neighborhood in $\bar{\mathfrak{a}}$. As $\mathfrak{a}$ is dense in $\bar{\mathfrak{a}}$, it follows that

$$\bar{\mathfrak{a}} = \mathfrak{a} + \sum D(\varepsilon) a_\nu, \qquad \varepsilon > 0.$$

For every $\varepsilon > 0$ there are therefore elements $b_1, \ldots, b_n \in \mathfrak{a}$, $c_{\mu\nu} \in D(\varepsilon)$, such that

$$a_\nu = b_\nu + \sum_{\mu=1}^{n} c_{\mu\nu} a_\mu, \qquad 1 \le \nu \le n.$$

With $a := (a_1, \ldots, a_n)^t$, $b := (b_1, \ldots, b_n)^t$, $I := (\delta_{\mu\nu} e)$, $C := (c_{\mu\nu})$ it follows that

$$b = (I - C)a.$$

We now have $\det(I - C) = e - p$, where $p$ is a polynomial in the $c_{\mu\nu}$ without a constant term. For small $\varepsilon$, the element $e - p$ is therefore invertible, and thus also is the matrix $I - C$. It follows from $a = (I - C)^{-1} b$ that $a_1, \ldots, a_n \in \mathfrak{a}$, or in other words that $\bar{\mathfrak{a}} \subset \mathfrak{a}$. □

**Consequence.** *In commutative, Noetherian, BANACH algebras with unit element and without divisors of zero, all ideals are closed.*

We have thus obtained from the above proposition the

**Corollary.** *Every commutative, Noetherian, BANACH algebra with unit element and without divisors of zero is isomorphic to $\mathbb{R}$ or to $\mathbb{C}$.*

If we give up the condition regarding zero-divisors, the following generalization can be proved by purely algebraic arguments, which we omit:

**Theorem.** *Every commutative, Noetherian, BANACH algebra with unit element is finite-dimensional.*

In real or complex analysis there are therefore no function algebras, which possess on the one hand the NOETHER property, which is so convenient algebraically, and on the other hand the BANACH property which is so convenient analytically. The situation is better in $p$-adic analysis. The TATE algebras (which play such a fundamental role there) are commutative, valuated, Noetherian, BANACH algebras with unit element.

## REFERENCES

S. GRABINER: Finitely Generated, Notherian, and Artinian Banach Modules, Indiana Univ. Math. J. 26 (1977), 413–425.

H. GRAUERT and R. REMMERT: Analytische Stellenalgebren, *Grdl. Math. Wiss.* **176**, Springer, 1971, 52–56.

# 9

# Cayley Numbers or Alternative Division Algebras

*M. Koecher, R. Remmert*

> It is possible to form an analogous
> theory with seven imaginary roots of $(-1)$
> (A. CAYLEY 1845).

With the creation by HAMILTON of a "system of hypercomplex numbers" a process of rethinking began to take place. Mathematicians began to realize that, by abandoning the vague principle of permanence, it was possible to create "out of nothing" new number systems which were still further removed from the real and complex numbers than were the quaternions. In December 1843 for example, only two months after HAMILTON's discovery, GRAVES discovered the *eight-dimensional division algebra of octonions (octaves)* which—as HAMILTON observed—*is no longer associative*. GRAVES communicated his results about octonions to HAMILTON in a letter dated 4th January 1844, but they were not published until 1848 (*Note by Professor Sir* W.R. Hamilton, *respecting the researches of* John T. Graves, esq. *Trans. R. Irish Acad.*, 1848, *Science* 338–341). Octonions were rediscovered by CAYLEY in 1845 and published as an appendix in a work on elliptic functions (*Math. Papers* **1**, p. 127) and have since then been called CAYLEY numbers.

As the associative law does not hold in the CAYLEY algebra it is no longer possible, on principle, to draw on the resources of the matrix calculus to facilitate calculations, as was the case with quaternions. It is therefore inevitable that the derivation of the essential formulae, which are thoroughly familiar to us in the algebras of $\mathbb{C}$ and $\mathbb{H}$, should be more tedious. In the introductory §1 we have systematically gathered together the main identities that are valid for *alternative quadratic algebras* (*without divisors of zero*). In §2 the algebra of octonions will be explicitly constructed by a duplication process applied to the quaternion algebra $\mathbb{H}$.

There is a uniqueness theorem for octonions, analogous to FROBENIUS's Uniqueness theorem for quaternions. This theorem was discovered in 1933 by Max ZORN (well-known for his famous lemma) and published in his paper "Alternativkörper und quadratische Systeme" *Abh. Math. Sem. Hamburg* **9**, 395–402. We derive ZORN's theorem in §3.

## §1. Alternative Quadratic Algebras

*Every* real quadratic algebra $\mathcal{A}$ has the property $\mathcal{A} = \mathbb{R}e \oplus \text{Im}\,\mathcal{A}$, where the imaginary space $\text{Im}\,\mathcal{A}$ is a vector subspace of $\mathcal{A}$ (Frobenius's lemma, 8.2.1). There is therefore *just one* linear form

$$(1) \qquad \lambda \colon \mathcal{A} \to \mathbb{R} \quad \text{with} \quad \lambda(e) = 1 \quad \text{and} \quad \text{Ker}\,\lambda = \text{Im}\,\mathcal{A}$$

which we call the *linear form of the quadratic algebra*.

In the cases $\mathcal{A} = \mathbb{C}, \mathbb{H}$, the form $\lambda$ is the one introduced in 3.2.2 and 7.2.1 respectively and used to a considerable extent in those chapters under the name of the *real part* linear form Re. In the general case the notation Re is not normally used.

In the algebras $\mathbb{C}$ and $\mathbb{H}$ the conjugation mapping $x \mapsto \bar{x}$ proved to be very useful. This mapping can be defined in *any* quadratic $\mathbb{R}$-algebra, in terms independent of the basis, by

$$(2) \qquad {}^{-} \colon \mathcal{A} \to \mathcal{A}, \qquad x \mapsto \bar{x} := 2\lambda(x)e - x.$$

It is $\mathbb{R}$-linear, and

$$\bar{x} = \lambda(x)e - u \quad \text{for} \quad x = \lambda(x)e + u, \quad u \in \text{Im}\,\mathcal{A};$$

consequently $x \mapsto \bar{x}$ is, as in the case of $\mathbb{C}$, $\mathbb{H}$ a *reflection* of $\mathcal{A}$ in the line $\mathbb{R}e$. In particular

$$\bar{\bar{x}} = x \quad (\textit{Involution}), \qquad \lambda(\bar{x}) = \lambda(x)$$

and the *fixed point set* $\{x \in \mathcal{A} \colon \bar{x} = x\}$ is the $\mathbb{R}$-subalgebra $\mathbb{R}e$.  $\square$

One would therefore expect, to judge from the examples $\mathbb{C}$ and $\mathbb{H}$, that $\langle x, y \rangle := \lambda(x\bar{y})$ would give us a "natural" scalar product for $\mathcal{A}$. We shall see in this section, that this is indeed true for algebras *without divisors of zero*. Nevertheless we shall initially define $\langle x, y \rangle$ somewhat differently in §1.1 and derive the "desired equation" in §1.3 only under the additional assumption that $\mathcal{A}$ is *alternative*. Thus, as with $\mathbb{C}$ and $\mathbb{H}$, the important product rule $|xy| = |x|\,|y|$ will be valid for alternative algebras as well.

**1. Quadratic Algebras.** The identity $\langle x, y \rangle = 2\lambda(x)\lambda(y) - \lambda(xy)$ holds in $\mathbb{C}$ (trivially), and in $\mathbb{H}$. We shall make this equation, in a slightly modified form, serve as a base for the definition of a certain bilinear form in general quadratic algebras. We shall in fact prove the following

**Lemma.** *Let $\mathcal{A}$ be a quadratic algebra whose linear form is $\lambda$. Then*

$$\mathcal{A} \times \mathcal{A} \to \mathbb{R}, \quad (x, y) \mapsto \langle x, y \rangle := 2\lambda(x)\lambda(y) - \frac{1}{2}\lambda(xy + yx)$$

*is a symmetric bilinear form, and for all* $x, y \in \mathcal{A}$

(1) $$\langle x, x \rangle = 2\lambda(x)^2 - \lambda(x^2),$$

(2) $$\langle x, e \rangle = \lambda(x), \qquad \langle e, e \rangle = 1,$$

(3) $$x^2 = 2\lambda(x)x - \langle x, x \rangle e,$$

(4) $$xy + yx = 2\lambda(x)y + 2\lambda(y)x - 2\langle x, y \rangle e.$$

*If in addition, $\mathcal{A}$ has no divisors of zero, then $\langle x, x \rangle > 0$ for all $x \neq 0$.*

**Proof.** By definition $\langle x, y \rangle$ is a symmetric bilinear form for which (1) and (2) hold. To verify (3) we first note that $x - \lambda(x)e \in \text{kernel} \lambda = \text{Im} \mathcal{A}$. It follows from the definition of $\text{Im} \mathcal{A}$ (cf. 8.1.1) that $(x - \lambda(x)e)^2 = -\omega(x)e$ with $\omega(x) \in \mathbb{R}$, $x \in \mathcal{A}$. If we write this in the form $x^2 = 2\lambda(x)x - [\lambda(x)^2 + \omega(x)]e$ and apply $\lambda$ to it, we obtain $\langle x, x \rangle = \lambda(x)^2 + \omega(x)$ in view of (1), and this proves (3). When $\mathcal{A}$ has no divisors of zero, $\omega(x) \geq 0$ (see 8.1.1), and from this we deduce the last inequality $\langle x, x \rangle > 0$ for all $x \neq 0$, since $\lambda(x) = \omega(x) = 0$ is possible only when $x = 0$.

The equation (4) follows by linearization of (3) (putting $x + y$ instead of $x$). $\qquad\square$

The bilinear form $\langle x, y \rangle$ introduced in the last lemma will play a central role in what follows. We shall call it *the bilinear form of the quadratic algebra*. The reader should observe carefully that the equation (3) is a *universal* quadratic equation which holds for every $x$. In the original definition of quadratic algebras (see 8.2.E) the only requirement was that to every $x$ there should exist "somehow or other" elements $\alpha, \beta \in \mathbb{R}$, such that $x^2 = \beta x + \alpha e$. Now it has been shown that $\alpha$, $\beta$ can be chosen in a natural fashion by taking $\alpha = -\langle x, x \rangle$, $\beta = 2\lambda(x)$.

**Corollary.** *For all $x, y \in \text{Im} \mathcal{A}$*

(5) $$xy + yx = -2\langle x, y \rangle e.$$

*In particular $\langle x, y \rangle = 0$ is equivalent to $xy + yx = 0$.*

With the help of the identities (4) and (5) the product $xy$ can be expressed in terms of $yx$ and hence formulae can be simplified. Moreover, in the verification of identities applying to all elements of $\mathcal{A}$, one can often confine oneself to checking the identity for elements of $\text{Im} \mathcal{A}$.

We can now deduce from (5) by right-multiplication and left-multiplication respectively, and from (4) by replacing of $y$ by $yz$, the following identities, valid for $x, y, z \in \text{Im} \mathcal{A}$:

(A) $xy \cdot z + yx \cdot z = -2\langle x, y \rangle z,$

(B) $y \cdot xz + y \cdot zx = -2\langle x, z \rangle y,$

(C) $x \cdot yz + yz \cdot x = 2\lambda(yz)x - 2\langle x, yz \rangle\, e.$

Despite their simple derivation, these identities for alternative algebras are of crucial importance in §§2 and 4.

**2. Theorem on the Bilinear Form.** *The following identities hold for elements $x$, $y$, $z$ of an alternative quadratic algebra $\mathcal{A}$:*

(1)     $\lambda(xy) = \lambda(yx)$   *and hence*   $\langle x, y \rangle = 2\lambda(x)\lambda(y) - \lambda(xy),$

(2)     $\lambda(xyz) := \lambda(xy \cdot z) = \lambda(x \cdot yz)$   *(associativity of $\lambda$)*,

(3)     $\langle xy, z \rangle + \langle xz, y \rangle = 2\lambda(x)\langle y, z \rangle,$

(4)     $\langle xy, xy \rangle = \langle x, x \rangle\langle y, y \rangle$   *(product rule)*.

*If the bilinear form of $\mathcal{A}$ is positive definite, then $\mathcal{A}$ has no divisors of zero.*

A cyclic permutation of the elements in (2), yields, in combination with (1), the additional identity

(5)        $\lambda(xyz) = \lambda(xy \cdot z) = \lambda(yz \cdot x) = \lambda(zx \cdot y).$

In general however $\lambda(xy \cdot z)$ is different from $\lambda(yx \cdot z)$!

**Proof.** As multilinear identities such as (1), (2) or (3) remain unchanged if one adds scalar multiples of $e$ to arbitrary elements of $\mathcal{A}$, one can assume without restriction during the proof that $x, y, z \in \mathrm{Im}\,\mathcal{A}$, and hence that $\lambda(x) = \lambda(y) = \lambda(z) = 0$.

a) In 1(A) and 1(C) we put $z = x$, subtract and take account of 8.1.3(2), thus obtaining

$$0 = xy \cdot x - x \cdot yx = -2(\langle x, y \rangle + \lambda(xy))x + 2\langle x, yx \rangle e,$$

so that $\langle x, y \rangle + \lambda(xy) = 0$ and $\langle x, yx \rangle = 0$ for $x, y \in \mathrm{Im}\,\mathcal{A}$. Comparison with Lemma 1 and linearization yields the identities (1) and (3), respectively.

b) Since $\mathcal{A}$ is alternative, the relation $\lambda(x \cdot yx) = \lambda(xy \cdot x)$ follows from 8.1.3(2). If we replace $x$ by $x + z$ and compare the terms linear in $x, y, z$ we obtain

(*)        $\lambda(x \cdot yz) + \lambda(z \cdot yx) = \lambda(xy \cdot z) + \lambda(zy \cdot x)$

for $x, y, z \in \operatorname{Im} \mathcal{A}$. By 1(A) and 1(B) we have $\lambda(z \cdot yx) = -\lambda(x \cdot xy)$ and $\lambda(zy \cdot x) = -\lambda(yz \cdot x)$ respectively, so that (2) follows from (1) and (*).

c) We put $z = xy$ in (3) and apply the equations $x^2 y = x \cdot xy$ for $x, y \in \mathcal{A}$. It then follows that

$$\langle xy, xy \rangle = 2\lambda(x)\langle y, xy \rangle - \langle x^2 y, y \rangle = -\langle (x^2 - 2\lambda(x)x)y, y \rangle,$$

and 1(3) gives us (4).

d) From $xy = 0$ it follows that $\langle x, x \rangle = 0$ or $\langle y, y \rangle = 0$ by (4). In the positive definite case this implies $x = 0$ or $y = 0$. □

From (1) and Corollary (1) we now obtain the

**Corollary.** *The three following statements are equivalent for $x, y \in \operatorname{Im} \mathcal{A}$*

i) *$x$ and $y$ are orthogonal, that is, $\langle x, y \rangle = 0$,*

ii) *$xy + yx = 0$,*

iii) *$\lambda(xy) = 0$.*

*Remark 1.* With the help of these results we can obtain the following simple proof of FROBENIUS's theorem 8.2.4. Let $\mathcal{A}$ be an associative quadratic algebra without divisors of zero and let $u, v, w$ be a Hamiltonian triplet in $\operatorname{Im} \mathcal{A}$. For every $x \in \operatorname{Im} \mathcal{A}$, orthogonal to $u, v, w$ we then have $x = -xuvw = uxvw = -uvxw = uvwx = -x$, and hence $x = 0$.

*Remark 2.* For the quadratic algebra $\operatorname{Mat}(2, \mathbb{R})$ we have $X^2 - \operatorname{Trace} X \cdot X + \det X \cdot E = 0$ so that $2\lambda(X) = \operatorname{Trace} X$ and $\langle X, X \rangle = \det X$. Equation (4) is then the product rule for determinants. Of course $\operatorname{Mat}(2, \mathbb{R})$ has divisors of zero.

**3. Theorem on the Conjugation Mapping.** *The following three identities hold for all elements $x, y, z$ of an alternative quadratic algebra $\mathcal{A}$.*

(1) $\overline{xy} = \bar{y}\bar{x}$,

(2) $x(\bar{x}y) = \bar{x}(xy) = \langle x, x \rangle y$, *and in particular* $x\bar{x} = \bar{x}x = \langle x, x \rangle e$,

(3) $\langle x, y \rangle = \lambda(x\bar{y}) = \lambda(\bar{x}y)$, *and in particular* $\langle xy, z \rangle = \langle x, z\bar{y} \rangle = \langle y, \bar{x}z \rangle$.

**Proof.** Equation (1) follows from

$$\overline{xy} - \bar{y}\bar{x} = 2\lambda(xy)e - xy - [2\lambda(y)e - y][2\lambda(x)e - x]$$
$$= 2[\lambda(xy) - 2\lambda(x)\lambda(y)]e + 2\lambda(x)y + 2\lambda(y)x - (xy + yx),$$

because by 1(4) and 2(1) the right-hand side vanished. Equation (2) follows from 1(3):

$$x(\bar{x}y) = x[(2\lambda(x)e - x)y] = x(2\lambda(x)y - xy) = 2\lambda(x)xy - x^2 y = \langle x, x \rangle y;$$

and $\bar{x}(xy) = \langle x, x \rangle y$ is proved the same way.

To verify (3) we note that since $x = 2\lambda(x)e - \bar{x}$ we have $\lambda(xy) = 2\lambda(x)\lambda(y) - \lambda(\bar{x}y)$. Hence (3) follows using 2(1).

Now $\langle xy, z \rangle = \lambda(xy \cdot \bar{z}) = \lambda(x \cdot y\bar{z}) = \langle x, z\bar{y} \rangle$ by 2(2) and (1). Similarly $\langle y, \bar{x}z \rangle = \lambda(y \cdot \bar{z}x) = \lambda(y\bar{z} \cdot x) = \langle x, z\bar{y} \rangle$. $\qquad \square$

Along with (2) we also have the identity

$$(2') \qquad\qquad (x\bar{y})y = (xy)\bar{y} = \langle y, y \rangle x.$$

This can be proved either by conjugation of (2) or in the same way as (2).

**4. The Triple Product Identity.** *If $A$ is an alternative quadratic algebra, then the triple product identity*

$$(1) \qquad xy \cdot z + x \cdot yz = 2\lambda(yz)x - 2\lambda(xz)y + 2\lambda(xy)z + 2\lambda(xyz)e$$

*holds for all $x, y, z \in \operatorname{Im} A$. In particular, for $x, y \in A$*

$$(2) \qquad xyx := xy \cdot x = x \cdot yx = 2\lambda(xy)x - \langle x, x \rangle \bar{y}.$$

**Proof.** By linearizing $yx \cdot x = y \cdot x^2$ we first obtain

$$yx \cdot z + yz \cdot x = y \cdot xz + y \cdot zx$$

for $x, y, z \in A$. If one now forms the expression (A) $-$ (B) $+$ (C) from the identities in 1, the identity (1) now follows from 2(1).

Putting $z = x$ in (1) we immediately get $xyx = 2\lambda(xy)x - \lambda(x^2)y + \lambda(xyx)e$. By 1(3) we have $-\lambda(x^2) = \langle x, x \rangle e$, and after 2(1) and 2(2) we have $\lambda(xyx) = \lambda(x^2y) = -\langle x, x \rangle\lambda(y) = 0$ which proves (2) for $x, y \in \operatorname{Im} A$. The general case may be deduced from this by replacing $x$ by $x - \lambda(x)e$ and $y$ by $y - \lambda(y)e$. $\qquad \square$

**Corollary.** *If $x, y, z \in \operatorname{Im} A$ are pairwise orthogonal then $xy \cdot z = -zy \cdot x$.*

For by Corollary 2, we then have

$$xy \cdot z = 2\lambda(xyz)e - x \cdot yz = \overline{x \cdot yz} = \bar{z}\bar{y} \cdot \bar{x} = -zy \cdot x.$$

A subalgebra $B$ of $A$ which contains $e$, is conjugation invariant, in other words if $u$ belongs to $B$ so does $\bar{u} = 2\lambda(u)e - u$.

**Lemma.** *Let $A$ be an alternative quadratic algebra, $B$ a subalgebra of $A$ containing $e$, and $q$ a given element of $A$ with $\langle B, q \rangle = 0$. Then*
   a) $\langle B, Bq \rangle = 0$, *in particular $Bq \subset \operatorname{Im} A$ and $\lambda(B \cdot Bq) = 0$.*
   b) *For $u, v \in B$*

$$(3) \qquad\qquad u \cdot vq = vu \cdot q,$$

(4) $$uq \cdot v = u\bar{v} \cdot q, \quad \textit{in particular} \quad qv = \bar{v}q,$$

(5) $$uq \cdot vq = -\langle q, q \rangle \cdot \bar{v}u.$$

c) $\mathcal{B} + \mathcal{B}q := \{u + vq : u, v \in \mathcal{B}\}$ *is a subalgebra of* $\mathcal{A}$.

**Proof.** Since $e \in \mathcal{B}$, the algebra $\mathcal{B}$ is conjugation invariant.
a) By 3(3) we have $\langle u, vq \rangle = \langle \bar{v}u, q \rangle = 0$ for $u, v \in \mathcal{B}$.
We now apply Lemma 1.
b) To prove

(*) $$qv = \bar{v}q \quad \text{for} \quad v \in \mathcal{B}$$

we confine ourselves to $v \in \text{Im}\,\mathcal{B}$ and then apply Corollary 2. For $u \in \mathbb{R}e$ or $v \in \mathbb{R}e$, equations (3) and (4) are now trivial and (5) reduces to 8.1.3(1). It is therefore permissible to assume throughout that $u, v \in \text{Im}\,\mathcal{B}$.

To prove (3) we note that $u \cdot vq - vu \cdot q = u \cdot vq + uv \cdot q - (uv + vu)q = 2\lambda(uv)q + 2\lambda(uv \cdot q)e + 2\langle u, v \rangle q = 0$ by (1) and Corollary (2).

To prove (4) we can deduce from (3) by taking conjugates that $qv \cdot u = q \cdot uv$ apply (*) and obtain $-vq \cdot u = \overline{uv} \cdot q = vu \cdot q$, whence $vq \cdot u = v\bar{u} \cdot q$.

To prove (5) we note that $u, q, vq$ are pairwise orthogonal in $\text{Im}\,\mathcal{A}$ because of a) and of $\lambda(vq \cdot q) = \lambda(vq^2) = -\langle q, q \rangle \lambda(v) = 0$. If we now put $x = u$, $v = q$, $z = vq$ in the corollary, we obtain $uq \cdot vq = -(vq \cdot q)u = -\langle q, q \rangle \bar{v}u$.

Statement c) follows from b).                                           $\square$

*Remark.* The calculation leading to (1) can be performed for any elements of $\mathcal{A}$. We thus obtain for all $x, y, z$ of an alternative quadratic algebra the identity

(6) $$\begin{aligned} xy \cdot z + x \cdot yz = {} & 2\lambda(x)yz + 2\lambda(y)xz + 2\lambda(z)xy \\ & - 2\langle y, z \rangle x - 2\lambda(xz)y - 2\langle x, y \rangle z \\ & + (4\lambda(z)\langle x, y \rangle - 2\langle x, yz \rangle)e. \end{aligned}$$

This identity is, even for the associative algebras $\mathbb{H}$ and $\text{Mat}(2, \mathbb{R})$, little known. A proof for $2 \times 2$ matrices is implicit in a paper by H. HELLING (*Inv. Math.* **17**, 1972, 217–229).

## 5. The Euclidean Vector Space $\mathcal{A}$ and the Orthogonal Group $O(\mathcal{A})$.
In the results of 1 to 4 is included the

**Theorem.** *If* $\mathcal{A}$ *is a quadratic, alternative algebra without divisors of zero, then* $\mathcal{A}$ *is an inner product space with respect to the bilinear form* $\langle \, , \, \rangle$. *The product rule*

$$|xy| = |x||y| \quad \textit{holds for all} \quad x, y \in \mathcal{A}$$

*and in particular all the mappings* $p_a: \mathcal{A} \to \mathcal{A}$, $x \mapsto axa$, $a \in \mathcal{A}$, $|a| = 1$, *are isometries of* $\mathcal{A}$.

As a generalization of the generation theorem 7.3.2 for $O(\mathbb{H})$, we have the

**Generation Theorem for** $O(\mathcal{A})$. *Let* $\mathcal{A}$ *be a finite-dimensional, alternative, division algebra (and hence, in particular a quadratic algebra by 8.2.2). Then every proper isometry* $f \in O^+(\mathcal{A})$ *is a product of at most* $n := \dim \mathcal{A}$ *mappings* $p_a$.

*The full group* $O(\mathcal{A})$ *is generated by the mappings*

$$x \mapsto axa, \quad |a| = 1 \quad and \quad x \mapsto -\bar{x}.$$

**Proof** (analogous to that of Theorem 6.3.2). 1) Every $f \in O^+(\mathcal{A})$ is a product of an even number $k \leq n$ of reflections $s_a$. For any two reflections $s_a$, $s_b$ we have $s_a \circ s_b = p_a \circ p_b$ just as before, because $s_a(x) = -a\bar{x}a$.

2) for $f \in O^-(\mathcal{A})$ we have $f \circ s_e \in O^+(\mathcal{A})$, so that $O(\mathcal{A})$ is generated by the mappings $p_a$ and the mapping $s_e: x \mapsto -\bar{x}$ (note that $s_e = s_e^{-1}$).   □

*Warning.* Since the associative law is no longer available there is *no analogue of* CAYLEY's *theorem* in 7.3.2. Every mapping $x \mapsto a(xb)$ or $x \mapsto (ax)b$, $|a| = |b| = 1$, is, of course, orthogonal by the product rule, but $\mathcal{A}$ in the non-associative case has other such mappings, and so for example, the orthogonal mappings $x \mapsto [a(xb)]c$ for $|a| = |b| = |c| = 1$ cannot in general be written in the form $x \mapsto u(xv)$ or $x \mapsto (ux)v$.

## §2. EXISTENCE AND PROPERTIES OF OCTONIONS

As HAMILTON showed, the algebra $\mathbb{C}$ arises from the algebra $\mathbb{R}$ when, in the Cartesian product $\mathbb{R} \times \mathbb{R}$ of real number pairs, one introduces a new product defined by

$$(a_1, a_2)(b_1, b_2) = (a_1 b_1 - b_2 a_2, a_2 b_1 + b_2 a_1), \qquad a_1, a_2, b_1, b_2 \in \mathbb{R}.$$

By means of an analogous duplication process, the quaternion algebra $\mathbb{H}$ (defined to within isomorphism) can be derived from the algebra $\mathbb{C}$. One defines a multiplication in the product space $\mathbb{C} \times \mathbb{C}$ of complex numbers by

$$(a_1, a_2)(b_1, b_2) = (a_1 b_1 - \bar{b}_2 a_2, a_2 \bar{b}_1 + b_2 a_1), \quad \text{for} \quad a_1, a_2, b_1, b_2 \in \mathbb{C}.$$

This procedure is completely canonical: by means of the bijection

$$\mathbb{C} \times \mathbb{C} \to \mathcal{H}, \quad (a_1, a_2) \mapsto \begin{pmatrix} a_1 & a_2 \\ -\bar{a}_2 & \bar{a}_1 \end{pmatrix},$$

the matrix multiplication in $\mathcal{H}$ is carried over to $\mathbb{C} \times \mathbb{C}$ (see 7.1.2). We shall see below that this *duplication process* can again be carried out in $\mathbb{H}$, to give the new CAYLEY-algebra of octonions $\mathbb{O}$.

**1. Construction of the Quadratic Algebra $\mathbb{O}$ of Octonions.** Motivated by the considerations outlined in the introduction, we define a product $\mathbb{H} \times \mathbb{H}$ by

$$xy = (x_1, x_2)(y_1, y_2) := (x_1 y_1 - \bar{y}_2 x_2 \, , \, x_2 \bar{y}_1 + y_2 x_1),$$

where $x = (x_1, x_2)$, $y = (y_1, y_2)$ are any two elements of $\mathbb{H} \times \mathbb{H}$. It is easily verified that both the distributive laws hold, and thus $\mathbb{H} \times \mathbb{H}$ becomes an 8-*dimensional* $\mathbb{R}$-algebra, which we call the CAYLEY *algebra of octonions* and denote by $\mathbb{O}$. It must be emphasized that in the definition of the octonion product the order of the factors in the right-hand brackets is absolutely vital. If, for example, one were to write $x_1 y_2$ instead of $y_2 x_1$ in the second component, an uninteresting algebra would be obtained.

If we denote by $e'$ the unit element of $\mathbb{H}$, then $e := (e', 0)$ is the *unit element* of $\mathbb{O}$. We can also deduce directly that:

$\mathbb{O}$ *is a quadratic algebra: for every* $x = (x_1, x_2) \in \mathbb{O}$,

(1)              $$x^2 = 2\operatorname{Re}(x_1)x - (\langle x_1, x_1 \rangle + \langle x_2, x_2 \rangle)e.$$

**Proof.** It follows from the definition that $x^2 = (x_1^2 - \bar{x}_2 x_2, x_2 \bar{x}_1 + x_2 x_1)$. We know (see 7.2.1 and 7.2.2) that for quaternions $x_1^2 = 2\operatorname{Re}(x_1)x_1 - \langle x_1, x_1 \rangle e'$, $\bar{x}_2 x_2 = \langle x_2, x_2 \rangle e'$, $\bar{x}_1 + x_1 = 2\operatorname{Re}(x_1)e'$. Hence

$$x^2 = (2\operatorname{Re}(x_1)x_1 - (\langle x_1, x_1 \rangle + \langle x_2, x_2 \rangle)e', 2\operatorname{Re}(x_1)x_2). \qquad \square$$

**2. The Imaginary Space, Linear Form, Bilinear Form, and Conjugation of $\mathbb{O}$.** As $\mathbb{O}$ is a quadratic algebra, the imaginary space $\operatorname{Im}\mathbb{O}$, the linear form $\lambda$, the bilinear form $\langle x, y \rangle$ and the conjugation mapping $x \mapsto \bar{x}$, are all defined invariantly. The relationships with the corresponding constructs in $\mathbb{H}$, if octonions are written as quaternion-pairs $(x_1, x_2)$, $(y_1, y_2)$ are simple. It is easily checked that

(1)                           $$\lambda(x) = \operatorname{Re}(x_1),$$

(2)                     $$\langle x, y \rangle = \langle x_1, y_1 \rangle + \langle x_2, y_2 \rangle,$$

(3)                           $$\operatorname{Im}\mathbb{O} = \operatorname{Im}\mathbb{H} \times \mathbb{H},$$

(4)                           $$\bar{x} = (\bar{x}_1, -x_2).$$

As the bilinear form of $\mathbb{H}$ is positive definite, the relation (2) has the consequence that:

*The bilinear form of $\mathbb{O}$ is positive definite, that is, $\mathbb{O}$ is a Euclidean vector space.*

### 3. $\mathbb{O}$ as an Alternative Division Algebra.

As with quaternions, the following identities hold

$$(1) \qquad \overline{xy} = \bar{y}\bar{x}, \quad x\bar{x} = \bar{x}x = \langle x, x\rangle e, \quad x, y \in \mathbb{O},$$

$$(2) \qquad x(\bar{x}y) = \langle x, x\rangle y = (x\bar{x})y, \qquad x, y \in \mathbb{O}.$$

**Proof.** The statements (1) follow directly from the definition of octonion multiplication, since $\bar{x} = (\bar{x}_1, -x_2)$.

As regards (2) we have, with $x = (x_1, x_2)$, $y = (y_1, y_2)$

$$\bar{x}y = (\bar{x}_1, -x_2)(y_1, y_2) = (\bar{x}_1 y_1 + \bar{y}_2 x_2, -x_2\bar{y}_1 + y_2\bar{x}_1)$$

and consequently, since $\mathbb{H}$ is associative

$$x(\bar{x}y) = (x_1[\bar{x}_1 y_1 + \bar{y}_2 x_2] - [-y_1\bar{x}_2 + x_1\bar{y}_2]x_2, x_2[\bar{y}_1 x_1 + \bar{x}_2 y_2]$$
$$+ [-x_2\bar{y}_1 + y_2\bar{x}_1]x_1)$$
$$= (x_1\bar{x}_1 y_1 + y_1\bar{x}_2 x_2, x_2\bar{x}_2 y_2 + y_2\bar{x}_1 x_1) = (\langle x_1, x_1\rangle + \langle x_2, x_2\rangle)y,$$

which, by 2(2), is equivalent to the assertion made.          □

**Theorem.** *The algebra $\mathbb{O}$ is an alternative division algebra.*

**Proof.** With $\bar{x} = 2\lambda(x)e - x$ the identity (2) can be written as $x(2\lambda(x)y - xy) = (2\lambda(x)x - x^2)y$, from which it follows at once that $x(xy) = x^2 y$ for all $x, y \in \mathbb{O}$. Conjugation gives $(\bar{y}\bar{x})\bar{x} = \bar{y}\bar{x}^2$. Since $\bar{x}$, $\bar{y}$ run through *all* elements of $\mathbb{O}$ when $x, y$ do, it follows that $(yx)x = yx^2$ for all $x, y \in \mathbb{O}$. Hence $\mathbb{O}$ is alternative.

As the bilinear form of $\mathbb{O}$ is positive definite, $\mathbb{O}$ has no divisors of zero (see Theorem 1.2). As a finite dimensional algebra $\mathbb{O}$ is thus a division algebra (R.5).          □

By Theorem 1.3 the following identity holds for the algebra $\mathbb{O}$

$$\langle x, y\rangle = \lambda(x\bar{y}) = \lambda(\bar{x}y), \qquad x, y \in \mathbb{O},$$

which can of course also be verified directly.

The algebra $\mathbb{O}$ is, by FROBENIUS's theorem, non-associative. Thus, for example, if $e, i, j, k$ denote the standard basis of $\mathbb{H}$:

$$(0, e)[(0, i)(0, j)] = -(0, e)(k, 0) = (0, k),$$

$$[(0,e)(0,i)](0,j) = (i,0)(0,j) = -(0,k);$$

see also 6 on this point.

**4. The "Eight Squares" Theorem.** By 1.2(4),

**The product rule:** $\quad |xy| = |x|\,|y|$ *for* $x, y \in \mathbb{O}$

likewise holds for the alternative algebra $\mathbb{O}$.

This can of course also be proved directly, though somewhat tediously, from the definitions of $\mathbb{O}$. In view of the product rule for quaternions, if one writes $x = (x_1, x_2)$, $y = (y_1, y_2)$ and takes account of $xy = (x_1 y_1 - \bar{y}_2 x_2, x_2 \bar{y}_1 + y_2 x_1)$ as well as 2(2), one has to show that

$$(*) \quad |x_1 y_1 - \bar{y}_2 x_2|^2 + |x_2 \bar{y}_1 + y_2 x_1|^2 = (|x_1|^2 + |x_2|^2)(|y_1|^2 + |y_2|^2),$$

$x_1, x_2, y_1, y_2 \in \mathbb{H}$. This leads, after some work, to $\lambda(x_2 \bar{y}_1 \bar{x}_1 \bar{y}_2) = \lambda(x_1 y_1 \bar{x}_2 y_2)$ which is true by 7.2.1(8). The analogy between $(*)$ and GAUSS's identity in 7.2.3 should be noted.

From the product rule for octonions, follows an

**"Eight-squares theorem."** *For all* $p, q, r, s, t, u, v, w \in \mathbb{R}$ *and all* $P, Q, R,$ $S, T, U, V, W \in \mathbb{R}$ *the following formula holds:*

$$
\begin{aligned}
(P^2 + Q^2 + &\cdots + V^2 + W^2)(p^2 + q^2 + \cdots + v^2 + w^2) \\
= &\; (Pp - Qq - Rr - Ss - Tt - Uu - Vv - Ww)^2 \\
&+ (Pq + Qp + Rs - Sr + Tu - Ut - Vw + Wv)^2 \\
&+ (Pr - Qs + Rp + Sq + Tv + Uw - Vt - Wu)^2 \\
&+ (Ps + Qr - Rq + Sp + Tw - Uv + Vu - Wt)^2 \\
&+ (Pt - Qu - Rv - Sw + Tp + Uq + Vr + Ws)^2 \\
&+ (Pu + Qt - Rw + Sv - Tq + Up - Vs + Wr)^2 \\
&+ (Pv + Qw + Rt - Su - Tr + Us + Vp - Wq)^2 \\
&+ (Pw - Qv + Ru + St - Ts - Ur + Vq + Wp)^2.
\end{aligned}
$$

**Proof.** We apply the product rule to the two octonions

$$(Pe + Qi + Rj + Sk, Te + Ui + Vj + Wk),$$

$$(pe + qi + rj + sk, te + ui + vj + wk). \qquad \square$$

The "eight-squares theorem" was found by GRAVES in 1844 and by CAYLEY in 1845 with the help of his octonions. The theorem had, however, already been discovered in 1818 by C.F. DEGEN (*Adumbratio demonstrationis theorematis arithmetici maxime universalis*). DEGEN thought, wrongly, that the result could be extended to $2^n$ squares; GRAVES at first also believed in such an extension. Further historical information will be found in

L.E. DICKSON: *On quaternions and their generalization and the history of the eight square theorem, Ann. Math.* **20**, 1919, 155–171. In this paper are also given eight-squares formulae which go back to DEGEN.

## 5. The Equation $\mathbb{O} = \mathbb{H} \oplus \mathbb{H}p$.

For the algebras $\mathbb{C}$ and $\mathbb{H}$, we have the following representations

$$\mathbb{C} = \mathbb{R} \oplus \mathbb{R}i \qquad \text{and} \qquad \mathbb{H} = \mathbb{C} \oplus \mathbb{C}j$$

respectively, as direct sums of real vector spaces, if $\mathbb{R}$ in $\mathbb{C}$, is identified with the pairs $(\alpha, 0)$, $\alpha \in \mathbb{R}$, and $\mathbb{C}$ in $\mathbb{H}$ is identified with the quaternions $(\alpha, \beta, 0, 0)$, $\alpha, \beta \in \mathbb{R}$. In this way $\mathbb{R}$ is a subalgebra $\mathbb{C}$ containing the identity element of $\mathbb{C}$, and similarly $\mathbb{C}$ is a subalgebra of $\mathbb{H}$ containing the identity element of $\mathbb{H}$. The sum representations are orthogonal with respect to the natural scalar product in $\mathbb{C}$ and $\mathbb{H}$ respectively.

An analogous situation obtains for the algebra $\mathbb{O}$. In the first place it is clear that:

*The set $\{(u, 0) : u \in \mathbb{H}\}$ is a subalgebra of $\mathbb{O}$, isomorphic to the quaternion algebra $\mathbb{H}$, and containing the unit element $e$ of $\mathbb{O}$.*

We identify this subalgebra with $\mathbb{H}$ and the following multiplication rules then hold for all $u \in \mathbb{H}$ and all $(a_1, a_2) \in \mathbb{O}$:

$$u(a_1, a_2) = (ua_1, a_2 u), \qquad (a_1, a_2)u = (a_1 u, a_2 \bar{u}).$$

For $p := (0, e')$ it can be verified directly that

$$p^2 = -e, \quad (a_1, a_2) = a_1 + a_2 p \quad \text{for all} \quad (a_1, a_2) \in \mathbb{O},$$

and there follows easily the

**Theorem.** *Considered as a vector space, $\mathbb{O} = \mathbb{H} \oplus \mathbb{H}p$. This sum is orthogonal with respect to the Euclidean scalar product of $\mathbb{O}$. For all $u, v \in \mathbb{H}$ the following relations hold:*

(1)   $u(vp) = (vu)p$,

(2)   $(up)v = (u\bar{v})p$, *in particular* $pv = \bar{v}p$,

(3)   $(up)(vp) = -\bar{v}u$.

**Proof.** Since $(a_1, a_2) = a_1 e + a_2 p$ it is clear that $\mathbb{O} = \mathbb{H} + \mathbb{H}p$. As we always have $\langle a_1 e, a_2 p \rangle = \langle (a_1, 0), (0, a_2) \rangle = \langle a_1, 0 \rangle + \langle 0, a_2 \rangle = 0$, this representation is orthogonal and hence a direct sum. The rules (1)–(3) are easily checked by straightforward calculation, for example in the case of

(1): $u(vp) = u(0, v) = (0, vu) = (vu)p$,

(2): $(up)v = (0, u)v = (0, u\bar{v}) = (u\bar{v})p$,

(3): $(up)(vp) = (0, u)(0, v) = (-\bar{v}u, 0) = -\bar{v}u$. □

Equations of type (1)–(3) will play an important role in the next section.

**6. Multiplication Table for $\mathbb{O}$.** We know (R.6) that every multiplication in a vector space $V$ with basis $e_1, \ldots, e_n$ is completely determined by the $n^2$ individual products $e_\mu e_\nu$, $1 \leq \mu, \nu \leq n$. In the case $V := \mathbb{R}^8$ with the natural basis $e_1 := (1, 0, \ldots, 0), \ldots, e_8 := (0, \ldots, 0, 1)$ the table

|        | $e_2$   | $e_3$   | $e_4$   | $e_5$   | $e_6$   | $e_7$   | $e_8$   |
|--------|---------|---------|---------|---------|---------|---------|---------|
| $e_2$  | $-e_1$  | $e_4$   | $-e_3$  | $e_6$   | $-e_5$  | $-e_8$  | $e_7$   |
| $e_3$  | $-e_4$  | $-e_1$  | $e_2$   | $e_7$   | $e_8$   | $-e_5$  | $-e_6$  |
| $e_4$  | $e_3$   | $-e_2$  | $-e_1$  | $e_8$   | $-e_7$  | $e_6$   | $-e_5$  |
| $e_5$  | $-e_6$  | $-e_7$  | $-e_8$  | $-e_1$  | $e_2$   | $e_3$   | $e_4$   |
| $e_6$  | $e_5$   | $-e_8$  | $e_7$   | $-e_2$  | $-e_1$  | $-e_4$  | $e_3$   |
| $e_7$  | $e_8$   | $e_5$   | $-e_6$  | $-e_3$  | $e_4$   | $-e_1$  | $-e_2$  |
| $e_8$  | $-e_7$  | $e_6$   | $e_5$   | $-e_4$  | $-e_3$  | $e_2$   | $-e_1$  |

defines the octave multiplication, if $e_1$ is the unit element. We can immediately read from this table the non-associativity of $\mathbb{O}$, since for example $e_5(e_6 e_7) = e_8$, whereas $(e_5 e_6)e_7 = -e_8$. On the other hand, it would be extremely tedious to verify from this table that $\mathbb{O}$ is alternative.

# §3. UNIQUENESS OF THE CAYLEY ALGEBRA

> Ein größeres System kann nicht mehr alternativ sein (M. ZORN 1933).
>
> [A larger system can no longer be alternative.]

In this section $\mathcal{A}$ denotes an alternative quadratic algebra without divisors of zero. Our object is to prove ZORN's theorem which asserts that $\mathcal{A}$ is either associative or else isomorphic to the CAYLEY algebra.

As in 9.1.1 the bilinear form of $\mathcal{A}$ is denoted by $(x, y) \mapsto \langle x, y \rangle$, and by Lemma 9.1.1 it is positive definite.

**1. Duplication Theorem.** *Let $\mathcal{B}$ be a proper subalgebra of $\mathcal{A}$ containing $e$. Then:*

a) *$\mathcal{B}$ is associative.*

b) *There exists an element $q \in \text{Im}\,\mathcal{A}$ with $q^2 = -e$ and $\langle \mathcal{B}, q \rangle = 0$.*

c) *For every $q \in \operatorname{Im} A$ with $q^2 = -e$ and $\langle B, q \rangle = 0$, the set $B + Bq$ is a subalgebra of $A$ with $\dim(B + Bq) = 2 \dim B$, and the rules 4(3), 4(4) and 4(5) hold.*

**Proof.** b) By the FROBENIUS lemma 8.2.1, we have $A = \mathbb{R}e \oplus \operatorname{Im} A$ and $B = \mathbb{R}e \oplus \operatorname{Im} B$. Since $B \neq A$ there is a $q \in \operatorname{Im} A$ with $q \neq 0$ and $\langle B, q \rangle = 0$. Appropriate normalization of $q$ yields the required result.

c) Let $q \in \operatorname{Im} A$ be chosen such that $q^2 = -e$ and $\langle B, q \rangle = 0$. Lemma 1.4 can then be applied, so that by part c) of that lemma $B + Bq$ is a subalgebra of $A$, and by part a) the sum is direct and the mapping $v \mapsto vq$ of $B$ onto $B$ is injective, since $vq \cdot q = -v$. Furthermore

$$(*)\qquad\qquad\qquad \langle B, Bq \rangle = 0.$$

a) By 4(3) we have, in the first place $(uv \cdot w)q = w(uv \cdot q) = w(v \cdot uq)$ for $u, v, w \in B$. In view of $(*)$ we can now apply Lemma 1.4 for $uq$ instead of $q$. By 4(3) for $q$ and for $uq$, we thus have in the second place $(u \cdot vw)q = vw \cdot uq = w(v \cdot uq)$, and comparing the first terms of these two sets of equations we have $(uv \cdot w)q = (u \cdot vw)q$, and hence $uv \cdot w = u \cdot vw$.

**2. Uniqueness of the Cayley Algebra** (ZORN 1933). *Every alternative, quadratic, real, but non-associative algebra without divisors of zero is isomorphic to the CAYLEY algebra $\mathbb{O}$.*

**Proof.** By FROBENIUS's theorem 8.2.4, $\dim A > 4$, and by the Quaternion lemma 8.2.3 there is a subalgebra $B$ of $A$, and an algebra monomorphism $f: \mathbb{H} \to A$ with $f(\mathbb{H}) = B$. Since $A$ is non-associative, $B \neq A$. By part b) of the Duplication theorem 1, there is an element $q$ in $A$ such that $q^2 = -e$ and $\langle B, q \rangle = 0$. By part c) of the Duplication theorem 1, $B \oplus Bq$ is a subalgebra of $A$ containing $e$. On the other hand $\mathbb{O} = \mathbb{H}e \oplus \mathbb{H}p$, by Theorem 2.5. The mapping

$$(*)\qquad \mathbb{O} = \mathbb{H}e \oplus \mathbb{H}p \to B \oplus Bq, \quad ue + vp \mapsto f(u) + f(v)q,$$

is certainly bijective and $\mathbb{R}$-linear. As the rules (1)–(3) of Theorem 2.5 coincide with the rules (3)–(5) of Lemma 1.4, the mapping $(*)$ is in fact an algebra isomorphism.

If $B \oplus Bq$ were not equal to $A$, $B \oplus Bq$ would be associative by part a) of the Duplication theorem, in contradiction to 2.3, which says that $\mathbb{O}$ is non-associative. Consequently $A = B \oplus Bq \cong \mathbb{O}$. $\qquad\square$

We can now state this result in the form of a generalization of FROBENIUS's theorem 8.2.4.

**Structure Theorem.** *Every alternative, quadratic, real algebra without divisors of zero is isomorphic to $\mathbb{R}$, $\mathbb{C}$, $\mathbb{H}$ or $\mathbb{O}$.*

It should be remembered that, by Theorem 8.2.2, every finite dimensional alternative real division algebra is quadratic, and is thus likewise isomorphic to $\mathbb{R}$, $\mathbb{C}$, $\mathbb{H}$ or $\mathbb{O}$.

*Remark.* The structure theorem is itself capable of considerable generalization. The final result is associated with the names, among others, of BRUCK, KLEINFELD (*Proc. Am. Math. Soc.* **2**, 1951, 878–890), SHIRSHOV and SLATER (*Proc. Am. Math. Soc.* **19**, 1968, 712–715) and states that a simple alternative but non-associative algebra is a Cayley algebra over its center.

**3. Description of $\mathbb{O}$ by ZORN's Vector Matrices.** We introduced octonions in 2.1 as pairs $(x_1, x_2)$ of quaternions. Max ZORN in his classical work in 1933 gave a description of alternative algebras which comes closer to meeting the desire to facilitate explicit calculation. To bring out the motivation of ZORN's definition we start from the octonion product (see 2.1)

$$(1) \qquad xy = (x_1, x_2)(y_1, y_2) = (x_1 y_1 - \bar{y}_2 x_2, x_2 \bar{y}_1 + y_2 x_1).$$

With $x_k = \alpha_k e + u_k$, $y_k = \beta_k e + v_k$, where $\alpha_k, \beta_k \in \mathbb{R}$ and $u_k, v_k \in \operatorname{Im} \mathbb{H}$, this product $xy$, when we also bear in mind that $uv = -\langle u, v \rangle e + u \times v$, has the form

$$(2) \qquad ([\alpha_1 \beta_1 - \alpha_2 \beta_2 - \langle u_1, v_1 \rangle + \langle u_2, v_2 \rangle]e$$

$$+ \alpha_1 v_1 + \beta_1 u_1 + \alpha_2 v_2 - \beta_2 u_2 + u_1 \times v_1 - u_2 \times v_2,$$

$$[\alpha_2 \beta_1 + \alpha_1 \beta_2 + \langle u_2, v_1 \rangle - \langle u_1, v_2 \rangle]e$$

$$- \alpha_2 v_1 + \beta_2 u_1 + \alpha_1 v_2 + \beta_1 u_2 - u_2 \times v_1 - u_1 \times v_2).$$

We now regard $u_k$, $v_k$ as vectors of $\mathbb{R}^3$ and "complexify"

$$(3) \quad \alpha := \alpha_1 + i\alpha_2, \ \beta := \beta_1 + i\beta_2 \in \mathbb{C}; \ u := u_1 + iu_2, \ v := v_1 + iv_2 \in \mathbb{C}^3.$$

From now on a bar will be used only for complex conjugation; for vectors $w = (w_1, w_2, w_3)$, $z = (z_1, z_2, z_3) \in \mathbb{C}^3$ we write

$$(4) \ \langle w, z \rangle := \sum_1^3 w_\nu z_\nu, \quad w \times z =: (w_2 z_3 - w_3 z_2, w_3 z_1 - w_1 z_3, w_1 z_2 - w_2 z_1).$$

The expression (2) now takes the form

$$(5) \qquad (\operatorname{Re}([\alpha\beta - \langle \bar{u}, v \rangle]e + \bar{\alpha}v + \beta u + \bar{u} \times \bar{v}),$$

$$\operatorname{Im}([\alpha\beta - \langle \bar{u}, v \rangle]e + \bar{\alpha}v + \beta u + \bar{u} \times \bar{v})),$$

after using identities such as $\bar{u} \times \bar{v} = u_1 \times v_1 - u_2 \times v_2 + i(-u_2 \times v_1 - u_1 \times v_2)$. The formula (5) can be expressed particularly conveniently if one takes the eight-dimensional real vector space $\mathcal{L} := \mathbb{C}e \oplus \mathbb{C}^3$ with elements $\alpha e + u$, $\alpha \in \mathbb{C}$, $u \in \mathbb{C}^3$, and introduces the mapping

(6) $F: \mathbb{O} \to \mathcal{L}$, $\quad x = (x_1, x_1) = (\alpha_1 e + u_1, \alpha_2 e + u_2) \mapsto \alpha e + u = x_1 + ix_2$.

We can then, to summarize, say that:

*The mapping $F$ is an $\mathbb{R}$-vector space isomorphism; for any two octonions $x, y$ with $F(x) = \alpha e + u$, $F(y) = \beta e + v$ we have*

(7) $\qquad\qquad F(xy) = [\alpha\beta - \langle \bar{u}, v \rangle]e + [\bar{\alpha}v + \beta u + \bar{u} \times \bar{v}]$.

It is now clear how we should multiply in $\mathcal{L}$: we define

(8) $\qquad (\alpha e + u)(\beta e + v) := [\alpha\beta - \langle \bar{u}, v \rangle]e + [\bar{\alpha}v + \beta u + \bar{u} \times \bar{v}]$

and we know that in view of the foregoing:

*$\mathcal{L}$, with the multiplication defined by (8), is an $\mathbb{R}$-algebra; the mapping $F: \mathbb{O} \to \mathcal{L}$ is an $\mathbb{R}$-algebra isomorphism.*

Following ZORN's example we can also write the elements $\alpha e + u$ of $\mathcal{L}$ as vector matrices (that is, matrices with vector entries) $\begin{pmatrix} \alpha & u \\ -\bar{u} & \bar{\alpha} \end{pmatrix}$. Their product is then the "matrix product"

$$\begin{pmatrix} \alpha & u \\ -\bar{u} & \bar{\alpha} \end{pmatrix} \begin{pmatrix} \beta & v \\ -\bar{v} & \bar{\beta} \end{pmatrix} = \begin{pmatrix} \alpha\beta - \langle \bar{u}, v \rangle & \bar{\alpha}v + \beta u + \bar{u} \times \bar{v} \\ -\alpha\bar{v} - \bar{\beta}\bar{u} - u \times v & \alpha\bar{\beta} - \langle u, \bar{v} \rangle \end{pmatrix}.$$

*Remark.* One sometimes finds the octave product defined in the literature in a different way from that used here or in 2.1. Often an isomorphism is given, which is obtained by changing the product $(x, y) \mapsto xy$ into the "reversed" product $(x, y) \mapsto yx$. In view of the uniqueness theorem 3 all these different representations are of course isomorphic.

## ADDITIONAL READING

E. KLEINFELD: A characterization of the Cayley numbers, in *Studies in Modern Analysis* (A.A. Albert, editor), MAA (1963), pp. 126–143.

# 10

# Composition Algebras. Hurwitz's Theorem—Vector-Product Algebras

*M. Koecher, R. Remmert*

> Durch diesen Nachweis wird die alte Streitfrage, ob sich
> die bekannten Produktformeln für Summen von 2, 4 und 8
> Quadraten auf Summen von mehr als 8 Quadraten ausdehnen
> lassen, endgültig, und zwar in verneinendem Sinne
> entschieden (A. HURWITZ 1898).
>
> [By this proof, the long-debated question of whether
> the well-known product formulae for sums of 2, 4 and 8
> squares can be extended to more than 8 squares, has finally
> been answered in the negative.]

1. For multiplication in the algebras $\mathbb{R}$, $\mathbb{C}$, $\mathbb{H}$ and $\mathbb{O}$, the formula $|xy|^2 = |x|^2|y|^2$, holds, where $|\ |$ denotes the Euclidean length. If one expresses the vectors $x$, $y$ and $z := xy$ in terms of their coordinates with respect to an orthonormal basis, as $(\xi_\nu)$, $(\eta_\nu)$, and $(\zeta_\nu)$, respectively, then we obtain, in view of the bilinearity of the product $xy$ the

**Squares Theorem.** *In the four cases $n = 1, 2, 4, 8$ there are $n$ real (in fact rational integral) bilinear forms*

$$\zeta_\nu = \sum_{\lambda,\mu=1}^{n} \alpha_{\lambda\mu}^{(\nu)}\xi_\lambda\eta_\mu, \quad \alpha_{\lambda\mu}^{(\nu)} \in \mathbb{Z}, \quad \nu = 1, \ldots, n,$$

*such that, for all numbers $\xi_1, \ldots, \xi_n, \eta_1, \ldots, \eta_n \in \mathbb{R}$*

$$(*) \quad \zeta_1^2 + \zeta_2^2 + \cdots + \zeta_n^2 = (\xi_1^2 + \xi_2^2 + \cdots + \xi_n^2)(\eta_1^2 + \eta_2^2 + \cdots + \eta_n^2).$$

We have already discussed this theorem in detail in 3.3.4, 7.2.3 and 9.2.4.

The French mathematician LEGENDRE (1752–1833) was the first to give a proof of the impossibility of such an equation for $n = 3$. In his great work *Theorie des nombres* which appeared in Paris in 1830 he remarks on page 198 that although $3 = 1^2 + 1^2 + 1^2$ and $21 = 4^2 + 2^2 + 1^2$ their product $3 \cdot 21 = 63$ is not the sum of three squares of natural numbers, so that it follows that the squares theorem cannot possibly be valid for $n = 3$ with *rational* bilinear forms $\zeta_1, \zeta_2, \zeta_3$. "Had HAMILTON known of this remark of LEGENDRE he might perhaps have there and then abandoned his attempt to multiply triplets. Fortunately he had not read LEGENDRE: he was self-taught." So wrote van der WAERDEN in *"Hamiltons Entdeckung der Quaternionen"* (p. 14).

**2.** The question, which at once imposes itself, is for what values of $n \geq 1$ does the equation (∗)

$$\zeta_1^2 + \zeta_2^2 + \cdots + \zeta_n^2 = (\xi_1^2 + \xi_2^2 + \cdots + \xi_n^2)(\eta_1^2 + \eta_2^2 + \cdots + \eta_n^2)$$

have solutions in which $\zeta_1, \ldots, \zeta_n$ are suitably chosen bilinear forms in $\xi_1, \ldots, \xi_n$ and $\eta_1, \ldots, \eta_n$? The question was finally solved in 1898 by Adolf HURWITZ. HURWITZ was born in 1859 in Hildesheim, Germany, and was taught at the gymnasium by H.C.H. SCHUBERT, the father of "enumerative algebraic geometry." In 1877 he studied under KLEIN, WEIERSTRASS and KRONECKER and obtained his doctorate in 1881 in Leipzig. In 1882 he took his postdoctoral lecturing qualification in Göttingen, because those who had received their degrees via a Realgymnasium were not allowed to qualify as university lecturers in Leipzig. In 1884, at the age of 25, he became an "extraordinarius" (a sort of supernumerary lecturer of junior status, roughly equivalent to a reader in a British university, or associate professor in an American university) at Königsberg, where he became a friend of HILBERT and MINKOWSKI. In 1892 he declined the offer of becoming SCHWARZ's successor in Göttingen and took over as FROBENIUS's successor at the Federal Polytechnic in Zürich, where he died in 1919.

HURWITZ's main contributions were in the theory of functions and in particular the theory of modular functions, algebra and algebraic number theory. In his paper published in the *Nachrichten der k. Gesellschaft der Wissenschaften zu Göttingen:* Über die Komposition der quadratischen Formen von beliebig vielen Variablen, 1898, 309–316 (*Math. Werke* **2**, 565–571), he proved with the help of the matrix calculus, that the cases $n = 1, 2, 4, 8$ are the only ones for which the squares theorem holds.

In this chapter we shall derive HURWITZ's result from a structure theorem on composition algebras, which is itself based on the main theorem on alternative algebras.

**3.** In Euclidean $\mathbb{R}^3$, we have, as is well known, a *vector product*. To any two vectors $u, v \in \mathbb{R}^3$, corresponds the vector $u \times v \in \mathbb{R}^3$, of length $|u|\,|v| \sin \measuredangle$ $(u, v)$, which is perpendicular to $u$ and $v$, and directed in such a way that the three vectors $u, v, u \times v$ form a "right-handed screw." By means of the mapping $\mathbb{R}^3 \times \mathbb{R}^3 \to \mathbb{R}^3$, $(u, v) \mapsto u \times v$, the vector space $\mathbb{R}^3$ becomes a *vector product algebra*. Such algebras will be studied systematically in §3. As an application of the HURWITZ structure theorem for composition algebras we shall show that there are vector-product algebras of dimensions 1, 3, and 7 only, and that two such algebras of the same dimension are always isometrically isomorphic.

## §1. COMPOSITION ALGEBRAS

Let $V$ be a real vector space and $(x, y) \mapsto \langle x, y \rangle$ a scalar product on $V$. An algebra $\mathcal{A} = (V, \cdot) \neq \{0\}$ is called a *composition algebra* (with respect to the given scalar product), if multiplication in $\mathcal{A}$ is isometric, that is to say, if

$$|xy| = |x|\,|y| \quad \text{for all} \quad x, y \in V \qquad \text{(product rule)}.$$

*A composition algebra cannot have divisors of zero.* The algebras $\mathbb{R}$, $\mathbb{C}$, $\mathbb{H}$, $\mathbb{O}$ are composition algebras with unit element. The object of this section is to show that these four algebras are the only non-isomorphic composition algebras of finite dimension with a unit.

**1. Historical Remarks on the Theory of Composition.** To explain the choice of the word "composition" in this context, we shall briefly outline the historical origins of the concept. In the famous *Disquisitiones arithmeticae*, the masterpiece which GAUSS wrote as a young man and which was published in 1801 (*Werke* 1), there is a section, beginning at Article 153, devoted to a systematic study of the arithmetic of *binary* quadratic forms, that is to say polynomials of the form $f(\xi_1, \xi_2) = a\xi_1^2 + 2b\xi_1\xi_2 + c\xi_2^2$ whose coefficients are rational integers. Since the days of FERMAT mathematicians had been interested in questions of the representation of integers by such forms, or in other words in the question of whether the equation $f(\xi_1, \xi_2) = n$ where $n \in \mathbb{Z}$ is given, possesses solutions in integers. This arithmetical problem is significantly more difficult than those which were considered in the arithmetic of the Ancient Greeks (*Euclid*, Book 9), and the first general results were obtained by LAGRANGE.

In the context of his researches into the problem of the representability of natural numbers by quadratic forms, GAUSS introduced the concept of the *composition of quadratic forms* (*Disq. Arith.* Art. 235 *et seq.*) If $f, g, h$ are any three given binary quadratic forms with coefficients $a, b, c$; $a', b', c'$; $A, B, C$ respectively, then he said that $h$ is *composed* of $f$ and $g$ (or is the result of the *composition* of $f$ and $g$), if the equation

$$(*) \quad A\zeta_1^2 + 2B\zeta_1\zeta_2 + C\zeta_2^2 = (a\xi_1^2 + 2b\xi_1\xi_2 + c\xi_2^2)(a'\eta_1^2 + 2b'\eta_1\eta_2 + c'\eta_2^2),$$

holds identically for all $\xi_1, \xi_2$, and all $\eta_1, \eta_2$ where $\zeta_1$ and $\zeta_2$ are suitably chosen bilinear forms in $\xi_1, \xi_2$ and $\eta_1, \eta_2$ with *integer* coefficients. GAUSS's theory of composition is one of the culminating points attained in the *Disquisitiones*. It is now known that this theory is essentially equivalent to the theory of ideals in quadratic number fields (for details, see SCHARLAU and OPOLKA: *From Fermat to Minkowski*, Springer-Verlag, 1985, p. 88ff.).

One of the main results of the Gaussian theory (to simplify greatly) is that the "equivalence-classes" of integral quadratic forms of a given *discriminant* $d := b^2 - ac$ form a finite Abelian group (the so-called class group). GAUSS, in effect, proves the group properties without knowing the group concept. The theory is purely arithmetic; if one allows the coefficients to be any real numbers and restricts one's self to positive definite forms (that is, $a$, $c$, $-d$ positive) then (*) is transformed by a suitable change of variables into the two-squares formula $(\zeta_1'^2 + \zeta_2'^2) = (u^2 + v^2)(x^2 + y^2)$, which is solved by $\zeta_1' = ux - vy$, $\zeta_2' = uy + vx$ (Two-squares theorem 3.3.4).

Following in GAUSS's footsteps, general compositions of quadratic forms in $n$ variables were considered. Some interesting problems also arose even when the restrictive arithmetic requirement that the coefficients must be integers was dropped. HURWITZ begins his paper on the squares theorem, to which he gave, quite deliberately, the title "Über die Komposition der quadratischen Formen ..." [On the composition of quadratic forms] with the following words: "In the domain of quadratic forms in $n$ variables, a theory of composition exists, if for any three quadratic forms $\varphi$, $\psi$, $\chi$ of non-vanishing determinant the equation

$$(1) \qquad \varphi(x_1, x_2, \ldots, x_n)\psi(y_1, y_2, \ldots, y_n) = \chi(z_1, z_2, \ldots, z_n)$$

can be satisfied by replacing the variables $z_1, z_2, \ldots, z_n$ by suitably chosen bilinear functions of the variables $x_1, x_2, \ldots, x_n$ and $y_1, \ldots y_n$. As a quadratic form can be expressed as a sum of squares by a suitable linear transformation of the variables,[1] one can consider, without loss of generality, in place of the equation (1), the following equation:

$$(2) \qquad (x_1^2 + x_2^2 + \cdots + x_n^2)(y_1^2 + y_2^2 + \cdots + y_n^2) = z_1^2 + z_2^2 + \cdots + z_n^2.$$

In view of this the question as to whether a composition theory exists for quadratic forms with $n$ variables is essentially equivalent to this other question, as to whether the equation (2) can be satisfied by suitably chosen bilinear functions $z_1, \ldots, z_n$ of the $2n$ independent variables $x_1, \ldots, x_n$, $y_1, \ldots, y_n$."

We note a simple criterion for the existence of composition theories.

*The identity*

$$(*) \qquad \begin{aligned}(\xi_1^2 + \cdots + \xi_n^2)(\eta_1^2 + \cdots + \eta_n^2) = \phi_1(\xi_1, \ldots, \xi_n, \eta_1, \ldots, \eta_n)^2 + \cdots \\ + \phi_n(\xi_1, \ldots, \xi_n, \eta_1, \ldots, \eta_n)^2\end{aligned}$$

*holds for $n$ bilinear forms $\phi_\nu(x, y)$, $1 \leq \nu \leq n$, if and only if $(\mathbb{R}^n, \cdot)$ with $x \cdot y := (\phi_1(x, y), \ldots, \phi_n(x, y))$ is a composition algebra.*

**Proof.** Any $n$ bilinear forms $\phi_1, \ldots, \phi_n$ make $\mathbb{R}^n$, as already explained, into an algebra. The product rule $|xy| = |x||y|$ holds if and only if (*) is satisfied.                                                                                      □

---

[1] For Hurwitz quadratic forms here are always positive definite.

As every $n$-dimensional Euclidean vector space $V$ is isometrically isomorphic to the number space $\mathbb{R}^n$ of $n$-tuples $x = (\xi_1, \ldots, \xi_n)$, $y = (\eta_1, \ldots, \eta_n)$ with its canonical scalar product $\langle x, y \rangle = \sum_1^n \xi_\nu \eta_\nu$, it also follows that:

*There is a theory of composition for real quadratic forms in $n$ variables, if and only if there is an $n$-dimensional composition algebra.*

**2. Examples.** The algebras $\mathbb{R}$, $\mathbb{C}$, $\mathbb{H}$, $\mathbb{O}$ give the classical composition theories for $n = 1, 2, 4, 8$. For $\mathbb{C}$ this is the identity

$$(1) \qquad (x_1^2 + x_2^2)(y_1^2 + y_2^2) = (x_1 y_1 - x_2 y_2)^2 + (x_1 y_2 + x_2 y_1)^2$$

but we shall not repeat the corresponding identities for $\mathbb{H}$ and $\mathbb{O}$ which were given in 7.2.3 and 9.2.4.

Every one-dimensional composition algebra is isomorphic to $\mathbb{R}$ and in particular has a unit element (see R.4). It is easy to give examples of composition algebras of dimension 2, 4 or 8 which have no unit element. Suppose first that $n = 2$. We define on $\mathbb{R}^2$ three different multiplications with the help of the ordinary complex product $wz$ in $\mathbb{R}^2 = \mathbb{C}$. We set

$$w \mathbin{\underset{1}{\square}} z := \bar{w}z, \qquad w \mathbin{\underset{2}{\square}} z := w\bar{z}, \qquad w \mathbin{\underset{3}{\square}} z = \overline{wz}.$$

It is easily verified that:

$\mathcal{A}_\nu := (\mathbb{R}^2, \underset{\nu}{\square})$, $1 \le \nu \le 3$ *is a non-alternative composition algebra without unit element, only $\mathcal{A}_3$, being commutative. Their associated composition formulae are:*

$$(2) \qquad (x_1^2 + x_2^2)(y_1^2 + y_2^2) = (x_1 y_1 + x_2 y_2)^2 + (x_1 y_2 - x_2 y_1)^2,$$

$$(3) \qquad (x_1^2 + x_2^2)(y_1^2 + y_2^2) = (x_1 y_1 + x_2 y_2)^2 + (-x_1 y_2 + x_2 y_1)^2,$$

$$(4) \qquad (x_1^2 + x_2^2)(y_1^2 + y_2^2) = (x_1 y_1 - x_2 y_2)^2 + (-x_1 y_2 - x_2 y_1)^2.$$

The equations (1), (4) differ from one another only inessentially in sign, and the same is true of the pair (2), (3). It can easily be shown that:

*Of the algebras $\mathbb{C}$, $\mathcal{A}_1$, $\mathcal{A}_2$, $\mathcal{A}_3$ no two are isomorphic, and every other two-dimensional composition algebra $\mathcal{A}$ is isometrically isomorphic to one of these four algebras.*

Now suppose $n = 4$. By taking any two quaternions $a, b \in \mathbb{H}$ of unit length, we can define a multiplication by adopting any of the four definitions

$$x \mathbin{\square} y := axyb, \qquad x \mathbin{\square} y := a\bar{x}yb, \qquad x \mathbin{\square} y := ax\bar{y}b, \qquad x \mathbin{\square} y := a\overline{xy}b,$$

where the expressions on the right denote the ordinary quaternion products. In this way we obtain infinitely many non-isomorphic composition algebras $(\mathbb{R}^4, \square)$ without unit element. The same method can also be used to construct infinitely many non-isomorphic composition algebras $(\mathbb{R}^8, \square)$ when $n = 8$.

**3. Composition Algebras with Unit Element.** In this paragraph $\mathcal{A} = (V, \cdot)$ denotes a real, but not necessarily finite-dimensional composition algebra. We use the *product rule* in the (squared) form:

$$(*) \qquad\qquad \langle xy, xy \rangle = \langle x, x \rangle \langle y, y \rangle, \qquad x, y \in V.$$

To appreciate the fairly drastic consequences of this condition, we shall apply the *linearization process* twice. We write $x + x'$ in place of $x$ in $(*)$ and obtain after a straightforward reduction

$$\langle xy, xy \rangle + 2 \langle xy, x'y \rangle + \langle x'y, x'y \rangle = (\langle x, x \rangle + 2 \langle x, x' \rangle + \langle x', x' \rangle) \langle y, y \rangle.$$

It now follows from $(*)$ that

$$(0) \qquad\qquad \langle xy, x'y \rangle = \langle x, x' \rangle \langle y, y \rangle.$$

If we now write $y + y'$ in place of $y$, we obtain

$$\langle x'y, xy \rangle + \langle x'y', xy \rangle + \langle x'y, xy' \rangle + \langle x'y', xy' \rangle$$
$$= \langle x, x' \rangle (\langle y, y \rangle + 2 \langle y, y' \rangle + \langle y', y' \rangle).$$

By $(0)$ the first and last terms on the left are equal to the corresponding terms on the right and hence:

$$(1) \quad \langle xy, x'y' \rangle + \langle xy', x'y \rangle = 2 \langle x, x' \rangle \langle y, y' \rangle \quad \text{for all} \quad x, x', y, y' \in V.$$

If we now put $x' := z$, $y' := y$ and then $x' := x$, $y' := z$, we obtain (respectively)

$$(2) \qquad\qquad \langle xy, zy \rangle = \langle x, z \rangle \langle y, y \rangle \quad \text{for all} \quad x, y, z \in V.$$

$$(3) \qquad\qquad \langle xy, xz \rangle = \langle x, x \rangle \langle y, z \rangle \quad \text{for all} \quad x, y, z \in V.$$

On the other hand if we put $x' := z$, $y' := e$ or $x' := e$, $y' := z$ then $(1)$ gives

$$(4) \ \langle xy, z \rangle + \langle x, zy \rangle = 2 \langle y, e \rangle \langle x, z \rangle \quad \text{and} \quad \langle xy, z \rangle + \langle xz, y \rangle = 2 \langle x, e \rangle \langle y, z \rangle.$$

After these preliminaries we are now in a position to demonstrate the fundamental

**Theorem.** *Every real composition algebra $\mathcal{A}$ with unit element $e$ is quadratic and alternative.*

**Proof.** In the first equation of (4) we write $xy$ for $x$ and in the second equation $xz$ for $z$. In view of (2) and (3) respectively we obtain

$$\langle xy \cdot y, z \rangle + \langle y, y \rangle \langle x, z \rangle = 2 \langle y, e \rangle \langle xy, z \rangle,$$
$$\langle x \cdot xz, y \rangle + \langle x, x \rangle \langle y, z \rangle = 2 \langle x, e \rangle \langle y, xz \rangle.$$

From these formulae for real numbers we obtain, since the scalar product $\langle x, y \rangle$ is *non-singular*,[2] identities valid for all elements $x, y \in \mathcal{A}$

(5) $$\qquad\qquad xy \cdot y = 2\langle y, e \rangle xy - \langle y, y \rangle x,$$

(6) $$\qquad\qquad x \cdot xy = 2\langle x, e \rangle xy - \langle x, x \rangle y.$$

If we put $y := e$ in the last equation we obtain

(7) $$\qquad x^2 = 2\langle x, e \rangle x - \langle x, x \rangle e \quad \text{for all} \quad x \in \mathcal{A},$$

whence $\mathcal{A}$ is quadratic. Right-multiplication of (7) by $y$ yields $x^2 \cdot y = x \cdot xy$ by (6); left-multiplication of $y^2 = 2\langle y, e \rangle y - \langle y, y \rangle e$ by $x$ leads, by (5), to $x \cdot y^2 = xy \cdot y$. Consequently $\mathcal{A}$ is also alternative. $\qquad\square$

*Analysis of the proof.* The derivation of the equations (1)–(3) uses *only* the symmetry of the bilinear form $\langle x, y \rangle$ and the fact that $\mathbb{R}$ is a field of characteristic $\neq 2$. The derivation of (5), (6) needs the non-singularity of $\langle x, y \rangle$. We have therefore in reality proved the more general result:

*Let $K$ be a commutative field of characteristic $\neq 2$, and let $\mathcal{A} = (V, \cdot) \neq 0$ be a $K$-algebra with unit element. Let $\langle x, y \rangle$ be a nonsingular $K$-bilinear form on $V$, such that $\langle xy, xy \rangle = \langle x, x \rangle \langle y, y \rangle$ for all $x, y \in V$. Then $\mathcal{A}$ is quadratic and alternative.*

**4. Structure Theorem for Composition Algebras with Unit Element.** Composition algebras have no divisors of zero. In view of Theorem 3 and the structure theorem of 9.3.2, we can therefore state the

**Structure Theorem.** *Let $\mathcal{A}$ be a composition algebra with unit element. Then $\mathcal{A}$ is isometrically isomorphic to one of the four algebras $\mathbb{R}$, $\mathbb{C}$, $\mathbb{H}$ and $\mathbb{O}$.*

This theorem can be generalized to arbitrary ground fields (even with characteristic 2). We refer the reader to the article by KAPLANSKY: *Infinite-dimensional quadratic forms admitting composition*, in *Proc. Am. Math. Soc.* 4, 1954, 956–960.

---

[2] A bilinear form $\langle x, y \rangle$ is said to be *non-singular*, if $\langle w, v \rangle = 0$ for all $v \in V$ implies $w = 0$. Positive definite bilinear forms are non-singular. The equation (5) is obtained from the identity $\langle xy \cdot y - 2\langle y, e \rangle xy + \langle y, y \rangle x, z \rangle = 0$ which holds for all $z \in V$. The equation (6) follows similarly when one finally substitutes $y$ for $z$.

*Historical Note.* The dimension of a composition algebra with unit element is, by the structure theorem, 1, 2, 4 or 8. This weaker statement can be derived in a direct fashion, and in this connection we should mention a note, published in 1959, on *Multiplication in n dimensions,* in *Nord. Mat. Tidskr.* **7**, 111–116 by ÖGMUNDSSON. As regards the author of this note, HELGASON writes to us: "He was a farmer on Snaefellsnes (West Iceland) and had a very limited mathematical training, probably on the level of an American High School. He found the quaternions on his own."

Another proof of this theorem which works consistently with bases, avoids alternative algebras, and does not use the fundamental theorem of algebra either, is to be found in BOS: *Multiplikation in euklidischen Räumen,* in *Jber. Deutsch. Math.-Verein.* **73**, 1971, 53–59.

The existence of a unit element in $\mathcal{A}$ is essential to the validity of the Structure theorem, as the examples of the algebras $\mathcal{A}_1, \mathcal{A}_2, \mathcal{A}_3$ in paragraph 2 show.

## §2. MUTATION OF COMPOSITION ALGEBRAS

In the light of 1.1 the statement that there exists a theory of composition for forms in $n$ variables is equivalent to the assertion that there exists an $n$-dimensional composition algebra. Such algebras do not need to have a unit element, and as we saw in §1.2, they exist in bewildering profusion. The structure theorem 1.4 appears therefore at first sight to be of no great help in solving HURWITZ's problem. And yet in reality the problem has already been essentially solved. There is, in fact, a simple process which enables us to go *from any arbitrary composition algebra* $(V, \cdot)$ *to a composition algebra* $(V, \square)$ *with unit element.* We shall begin by describing a general method of changing the multiplication in an arbitrary algebra.

**1. Mutation of Algebras.** *Let* $(V, \cdot)$ *be a* $K$-*algebra, and let* $f: V \to V$, $g: V \to V$ *be two* $K$-*linear mappings. Let the multiplication* $\square$ *be defined by*

$$x \,\square\, y := f(x)g(y) \quad \text{for all} \quad x, y \in V.$$

*Then* $(V, \square)$ *is a* $K$-*algebra.*

**Proof.** The distributive laws for $\square$ follow from the distributive laws for $\cdot$ and the linearity of $f$, $g$. $\square$

Every element $a$ of an algebra $\mathcal{A} = (V, \cdot)$ defines by right- and left-multiplication two $K$-linear mappings

$$L_a: V \to V, \quad x \mapsto ax; \qquad R_a: V \to V, \quad x \mapsto xa.$$

If *both the mappings* $L_a$ and $R_a$ are bijective, then there exists by the preceding argument the algebra $\mathcal{A}(a) := (V, \square)$ with the product

$$x \,\square\, y := R_a^{-1}(x) \cdot L_a^{-1}(y),$$

and we call $\mathcal{A}(a)$ the *mutation of $\mathcal{A}$ with respect to a*. The mappings $R_a^{-1}$, $L_a^{-1}$ serve as a substitute in $\mathcal{A}$ for the inverse $a^{-1}$ of $a$ which does not in general exist. If $a^{-1}$ does exist, then $x \mathbin{\Box} y = (xa^{-1})(a^{-1}y)$. In every case

$$(1) \qquad\qquad xa \mathbin{\Box} ay = xy.$$

If $\mathcal{A}$ has a unit element, then $\mathcal{A}(e)$ exists and $\mathcal{A}e = \mathcal{A}$. In general however mutations are entirely distinct from the original algebra (see §2.2 on this point).

**Existence Criterion for Mutations.** *If $\mathcal{A}$ is finite-dimensional, then the mutation $\mathcal{A}(a)$ exists for every element $a \in \mathcal{A}$ that is not a divisor of zero. In particular the mutation $\mathcal{A}(a)$ exists for every element $a \in \mathcal{A} \setminus \{0\}$ in a finite dimensional composition algebra.*

**Proof.** When $a$ is not a divisor of zero, the two mappings $L_a$, $R_a$ are injective and thus, if $\mathcal{A}$ is finite-dimensional, bijective as well. $\quad\Box$

In the next paragraph we shall need the two following propositions on mutations.

1) *Every mutation $\mathcal{A}(a)$ of a finite-dimensional algebra has $a^2$ as unit element.*

2) *A mutation $\mathcal{A}(a)$ of a composition algebra $\mathcal{A}$ is itself a composition algebra if $|a| = 1$.*

**Proof.** By (1), we have $a^2 \mathbin{\Box} ax = ax$ and $xa \mathbin{\Box} a^2 = xa$. Since however the mappings $x \mapsto ax$ and $x \mapsto xa$ are bijective, it follows that $a^2 \mathbin{\Box} x = x = x \mathbin{\Box} a^2$, $x \in \mathcal{A}$. This proves the first statement. As for the second, it follows from (1) that $|xa|\,|ay| = |xy| = |xa \mathbin{\Box} ay|$ and hence $|u|\,|v| = |u \mathbin{\Box} v|$ for all $u, v \in \mathcal{A}$. $\quad\Box$

**2. Mutation Theorem for Finite-Dimensional Composition Algebras.** *Every finite-dimensional composition algebra $\mathcal{A}$ possesses a mutation $\mathcal{A}(a)$ with $|a| = 1$, such that $\mathcal{A}(a)$ is isometrically isomorphic to one of the four algebras $\mathbb{R}$, $\mathbb{C}$, $\mathbb{H}$, $\mathbb{O}$; in particular, therefore, $\dim \mathcal{A} = 1, 2, 4$ or $8$.*

**Proof.** As $\mathcal{A} \neq \{0\}$ there are elements $a \in \mathcal{A}$ with $|a| = 1$. By 1.1) and 1.2), the algebra $\mathcal{A}(a)$ is a composition algebra with unit element. The last statement now follows from the Structure theorem 1.4. $\quad\Box$

All the composition algebras mentioned in 1.2 such as for example the algebras $\mathcal{A}_\nu = (\mathbb{R}^2, \underset{\nu}{\Box})$, $1 \leq \nu \leq 3$, fall under the mutation theorem. Direct verification gives immediately:

*Every mutation $\mathcal{A}_1(1)$, $\mathcal{A}_2(1)$, $\mathcal{A}_3(1)$, where $1 := (1, 0)$ denotes the "complex unity" is isometrically isomorphic to the algebra $\mathbb{C}$.*

This proposition is especially instructive because it illustrates how multiplication can be altered by mutation: *algebras which are neither commutative nor alternative can become commutative and alternative after mutation.*

In contrast to this may be noted:

*If $A = (V, \cdot)$ is a finite-dimensional associative division algebra, then*

$$A(a) = (V, \Box) \quad \text{with} \quad x \Box y = xa^{-2}y \quad \text{for every} \quad a \in A \setminus \{0\};$$

*and the mapping $f: A(a) \to A$, $x \mapsto a^{-2}x$, is an algebra isomorphism.*

**Proof.** By Lemma R.5 there exists $a^{-1} \in A$ for all $a \neq 0$, so that, by 1.(1) $x \Box y = (xa^{-1})(a^{-1}y)$. As $A$ is associative, it follows that $x \Box y = xa^{-2}y$. The mapping $f$ is linear and bijective, and furthermore $f(x \Box y) = a^{-2}(xa^{-2}y) = (a^{-2}x)(a^{-2}y) = f(x)f(y)$. $\qquad\Box$

**3. HURWITZ's Theorem (1898).** *Let $n \geq 1$ be a natural number and let $\zeta_1, \ldots, \zeta_n$ be real bilinear forms in the real variables $\xi_1, \ldots, \xi_n$, and $\eta_1, \ldots, \eta_n$, such that*

$$\zeta_1^2 + \zeta_2^2 + \cdots + \zeta_n^2 = (\xi_1^2 + \xi_2^2 + \cdots + \xi_n^2)(\eta_1^2 + \eta_2^2 + \cdots + \eta_n^2).$$

*Then $n = 1, 2, 4$ or $8$.*

**Proof.** We use the mutation theorem 2 and the criterion 1.1. $\qquad\Box$

At the end of his classical work HURWITZ formulated a generalization of the composition problem:

Let $m \geq 1$, $n \geq 1$ be given natural numbers. Determine the largest natural number $p$ for which the equation

$$\zeta_1^2 + \cdots + \zeta_m^2 = (\xi_1^2 + \xi_2^2 + \cdots + \xi_p^2)(\eta_1^2 + \eta_2^2 + \cdots + \eta_n^2)$$

is solvable by bilinear forms $\zeta_1, \ldots, \zeta_m$ in $\xi_1, \ldots, \xi_p$, and $\eta_1, \ldots, \eta_n$.

In 1923 this problem was solved completely for the case $m = n$ in HURWITZ's paper: *Über die Komposition der quadratischer Formen, Math. Ann.* **88**, 1–25, (published after his death, and reproduced in *Math. Werke* **2**, 641–666). RADON in 1922 solved the problem by a different method in his note: *Lineare Scharen orthogonaler Matrizen, Abh. Math. Sem. Hamburg* **1**, 1–14. In his formulation the solution is given by the following proposition.

**The HURWITZ–RADON Theorem (1923).** *Let $n = u2^{4\alpha+\beta}$, where $1 \leq u$ odd and $0 \leq \alpha$, $0 \leq \beta \leq 3$. Then the following two statements are equivalent:*

   i) *There are $n$ real bilinear forms $\zeta_1, \ldots, \zeta_n$ in $\xi_1, \ldots, \xi_p$, and $\eta_1, \ldots, \eta_n$ such that $\zeta_1^2 + \zeta_2^2 + \cdots + \zeta_n^2 = (\xi_1^2 + \xi_2^2 + \cdots + \xi_p^2)(\eta_1^2 + \eta_2^2 + \cdots + \eta_n^2)$.*

ii) $p \le 8\alpha + 2^\beta$.

It is trivial that $p \le n$. An elementary argument shows that $p = n$ if and only if $n = 1, 2, 4$ or $8$, as it must be in accordance with the original HURWITZ theorem. The cases $n := 4k$, $p := 4$ and $n := 8k$, $p := 8$ are realized by the $\mathbb{R}$-vector spaces $\mathbb{H}^k$ and $\mathbb{O}^k$, respectively, with the norm defined by

$$|x|^2 := |x_1|^2 + \cdots + |x_k|^2, \quad \text{where} \quad x := (x_1, \ldots, x_k).$$

We put $qx := (qx_1, \ldots, qx_n)$ for $q \in \mathbb{H}$, $x \in \mathbb{H}^k$ or $q \in \mathbb{O}$, $x \in \mathbb{O}^k$, as the case may be and verify that $|qx|^2 = |q|^2|x|^2$. As $\dim_{\mathbb{R}}\mathbb{H}^k = 4k$ and $\dim_{\mathbb{R}}\mathbb{O}^k = 8k$, statement i) of the HURWITZ–RADON theorem clearly holds in both cases.

The methods of proof used by HURWITZ and RADON were devised for the purpose. In 1943 B. ECKMANN in a paper: *Gruppentheoretischer Beweis des Satzes von* HURWITZ–RADON *über die Komposition der quadratischer Formen, Comm. Math. Helv.* **15**, 358-366, gave a proof based on the ideas of the theory of group representations into which the theorem can be placed. Today, one knows (from the theorem of ADAMS) that, after translation into the language of vector fields on spheres, the HURWITZ–RADON number $p-1$ can be interpreted as an upper bound for the number of independent vector fields on the $(n - 1)$-sphere (cf. the next chapter).

## §3. VECTOR-PRODUCT ALGEBRAS

Anyone who has ever successfully worked with the vector product in geometrical investigations in $\mathbb{R}^3$ (see 6.1.4) is bound to wonder whether a product with analogous properties exists in spaces of other dimensions. The frequently voiced opinion that this is possible only in $\mathbb{R}^3$ because it is only in this space that there are only two choices for a product vector $u \times v$ perpendicular to each of the two component vectors $u$, $v$ is certainly not a conclusive argument. We shall see that our question has a close connection with composition algebras. We shall show, among other results, that a vector product also exists in $\mathbb{R}^7$, but in no other spaces of dimension $n > 1$.

**1. The Concept of a Vector-Product Algebra.** Let $W$ be an Euclidean vector space with a Euclidean scalar product $(u, v) \mapsto \langle u, v \rangle$, and let $|u| := \sqrt{\langle u, u \rangle}$ be the Euclidean length of the vector $u \in W$.

**Lemma.** *Let* $\mathcal{W} = (W, \times)$ *be an* $\mathbb{R}$-*algebra, such that*

(1)        $u \times v = -v \times u$ *for all* $u, v \in W$ *(anticommutativity)*,

(2)        $\langle u \times v, w \rangle = \langle u, v \times w \rangle$ *for all* $u, v, w \in W$ *(interchange rule)*.

*Then the following three statements are equivalent:*

i) *If $|u| = |v| = 1$ and $\langle u, v \rangle = 0$ where $u, v \in W$ then $|u \times v| = 1$.*

ii) $|u \times v|^2 = |u|^2|v|^2 - \langle u, v \rangle^2$ *for all $u, v \in W$.*[3]

iii) $u \times (u \times v) = \langle u, v \rangle u - |u|^2 v$ *for all $u, v \in W$.*

In view of (1), it follows that, in particular

(3) $$u \times u = 0 \quad \text{for all} \quad u \in W.$$

**Proof.** i) $\Rightarrow$ ii): In view of (3) it will be sufficient to prove ii) for linearly independent vectors $u, v \in W$ satisfying $|u| = |v| = 1$. For such vectors $\lambda := |v - \langle u, v \rangle u| \neq 0$ and $w := (v - \langle u, v \rangle u)/\lambda$ satisfies $|w| = 1$ and $\langle u, w \rangle = 0$. Consequently, by i) we therefore have $|u \times w| = 1$, so that

$$\lambda^2 = |u \times (v - \langle u, v \rangle u)|^2 = |u \times v|^2, \quad \text{as} \quad u \times u = 0 \quad \text{by (3)}.$$

On the other hand since

$$\lambda^2 = |v - \langle u, v \rangle u|^2 = |v|^2 - 2\langle u, v \rangle^2 + \langle u, v \rangle^2 |u|^2 = 1 - \langle u, v \rangle^2,$$

the statement ii) follows from $|u| = |v| = 1$.

ii) $\Rightarrow$ i): This is obvious.

ii) $\Leftrightarrow$ iii): By linearization (with $v + w$ in place of $v$) the statement ii) is clearly equivalent to

$$\langle u \times v, u \times w \rangle = |u|^2 \langle v, w \rangle - \langle u, v \rangle \langle u, w \rangle.$$

As, by (1) and (2), $\langle u \times (u \times v), w \rangle = -\langle u \times v, u \times w \rangle$ the above can be written as

$$\langle u \times (u \times v), w \rangle = \langle q, w \rangle \quad \text{with} \quad q := \langle u, v \rangle u - |u|^2 v.$$

As this identity holds for all $w \in W$, it follows that ii) is equivalent to iii). □

We have already met the identities i) – iii) in 7.1.4 for the *vector product* in the imaginary space of $\mathbb{H}$. From now on we shall call an algebra $W = (W, \times) \neq \{0\}$ having the properties specified in the lemma, a *vector-product algebra* and $(u, v) \mapsto \langle u, v \rangle$ the associated scalar product.

*The Euclidean vector space $\mathbb{R}^1$, with the scalar product $(u, v) \mapsto uv$, is a vector-product algebra with respect to the trivial multiplication $(u, v) \mapsto 0$.*

Furthermore it follows immediately from 7.1.4 that:

---

[3]The expression $|u|^2|v|^2 - \langle u, v \rangle^2$ is the GRAM determinant, $\det \begin{pmatrix} \langle u, u \rangle, \langle u, v \rangle \\ \langle v, u \rangle, \langle v, v \rangle \end{pmatrix}$.

*The Euclidean vector space* $\operatorname{Im} \mathbb{H}$—*with the canonical scalar product—is a three-dimensional vector-product algebra with respect to the multiplication* $(u, v) \mapsto \frac{1}{2}(uv - vu)$. *If one identifies* $\operatorname{Im} \mathbb{H}$ *with* $\mathbb{R}^3$, *then* $u \times v$ *is simply the well-known vector product in* $\mathbb{R}^3$.

*Exercise.* Let $W$ be an Euclidean vector space, and $\mathcal{W} = (W, \times)$ an $\mathbb{R}$-algebra, such that for all $u, v \in W$:

   a) $u \times v$ is perpendicular to $u$, that is $\langle u, u \times v \rangle = 0$,

   b) $|u \times v|^2 = |u|^2 |v|^2 - \langle u, v \rangle^2$.

Show that $\mathcal{W}$ is a vector-product algebra.

**2. Construction of Vector-Product Algebras.** Let $\mathcal{A}$ be a composition algebra with unit element $e$, scalar product $(x, y) \mapsto \langle x, y \rangle$, norm $|x| := \sqrt{\langle x, x \rangle}$, and

$$(1) \qquad \operatorname{Im} \mathcal{A} := \{u \in \mathcal{A} : \langle e, u \rangle = 0\}.$$

By Theorem 1.3, $\mathcal{A}$ is an alternative quadratic algebra, whose imaginary space is in fact given by (1). However knowledge of this relationship with alternative quadratic algebras is not actually required for the construction of vector-product algebras.

   It follows directly from 1.3(4) that

$$(2) \qquad \langle uv, w \rangle + \langle u, wv \rangle = 0 = \langle uv, w \rangle + \langle v, uw \rangle$$

for $u, v, w \in \operatorname{Im} \mathcal{A}$ (cf. 9.1.3(3)). By 1.3(7), $u^2 = -\langle u, u \rangle e$ for $u \in \operatorname{Im} \mathcal{A}$, and hence

$$(3) \qquad uv + vu = -2 \langle u, v \rangle e \quad \text{for} \quad u, v \in \operatorname{Im} \mathcal{A} \quad \text{(see 9.1.1(5))}.$$

**Theorem.** *If* $\mathcal{A}$ *is a composition algebra with unit element* $e$ *and scalar product* $(x, y) \mapsto \langle x, y \rangle$, *then* $\operatorname{Im} \mathcal{A}$ *is a vector-product algebra with respect to the product*

$$(4) \qquad u \times v := \frac{1}{2}(uv - vu) = uv + \langle u, v \rangle e$$

*with associated scalar product* $(u, v) \mapsto \langle u, v \rangle$.

**Proof.** By (3), the second equality sign in (4) is valid, because (2) is equivalent to $2\langle u \times v, e \rangle = \langle uv - vu, e \rangle = 0$, and hence $u \times v \in \operatorname{Im} \mathcal{A}$. Thus $u \times v$ and $e$ in (4) are orthogonal, and so $|uv|^2 = |u \times v|^2 + \langle u, v \rangle^2$. The remaining property ii) of Lemma 1 now follows from the product rule for $\mathcal{A}$. □

We have therefore found a new vector-product algebra, apart from the one in $\mathbb{R}^3$, namely:

*The imaginary space of the* CAYLEY *algebra, together with the product defined by* (4) *is a 7-dimensional vector-product algebra* (see 9.2.1–9.2.3).

With this result however as we shall now prove, all vector-product algebras are now known.

**3. Specification of all Vector-Product Algebras.** If $W = (W, \times)$ is a real algebra, $(u, v) \mapsto \langle u, v \rangle$ a scalar product of $W$ and $e$ an element not belonging to $W$, one can define in the vector space $\mathbb{R} \oplus W$ a scalar product by

$$(1) \qquad \langle \alpha e + u, \ \beta e + v \rangle := \alpha\beta + \langle u, v \rangle,$$

and a product by

$$(2) \qquad (\alpha e + u) \cdot (\beta e + v) := (\alpha\beta - \langle u, v \rangle)e + \alpha v + \beta u + u \times v.$$

The resulting algebra in the Euclidean vector space $\mathbb{R}e \oplus W$ will be denoted by $(\mathcal{W}, \langle \cdot, \cdot \rangle, e)$; it has $e$ as unit element.

**Lemma.** *If* $\mathcal{W} = (W, \times)$ *is a vector-product algebra with associated scalar product* $(u, v) \mapsto \langle u, v \rangle$, *then* $(\mathcal{W}, \langle \cdot, \cdot \rangle, e)$ *is a composition algebra with unit element.*

**Proof.** In the first place, we have, by (1) and (2)

$$|(\alpha e + u) \cdot (\beta e + v)|^2 = (\alpha\beta - \langle u, v \rangle)^2 + |\alpha v + \beta u + u \times v|^2.$$

However by 1(2) and 1(3), $\langle u, u \times v \rangle = \langle v, u \times v \rangle = 0$, so that the right-hand side becomes

$$(\alpha\beta - \langle u, v \rangle)^2 + |\alpha v + \beta u|^2 + |u \times v|^2$$
$$= \alpha^2\beta^2 + \langle u, v \rangle^2 + \alpha^2|v|^2 + \beta^2|u|^2 + |u \times v|^2.$$

By the property ii) of Lemma 1 this immediately becomes

$$(\alpha^2 + |u|^2)(\beta^2 + |v|^2) = |\alpha e + u|^2 \cdot |\beta e + v|^2. \qquad \square$$

Since the embedding of $W$ in $\mathbb{R}e \oplus W$, as well as the projection of the subspace $W$ of $\mathbb{R}e \oplus W$ onto $W$ are isometric mappings, the structure theorem of 1.4 gives the

**Isometry Theorem for Vector-Product Algebras.** *To within an isometric isomorphism, the three imaginary spaces*

$$\begin{array}{ll} \operatorname{Im}\mathbb{C} & (\textit{the null algebra on } \mathbb{R}) \\ \operatorname{Im}\mathbb{H} & (\textit{the vector-product space in } \mathbb{R}^3) \\ \operatorname{Im}\mathbb{O} & \end{array}$$

*together with the product* $(u,v) \mapsto \frac{1}{2}(uv - vu)$ *are the only vector-product algebras. In particular therefore there are no infinite-dimensional vector-product algebras.*

The vector-product algebra $\operatorname{Im} \mathbb{H}$ is a LIE algebra, for by 7.1.4 the JACOBI identity

$$(3) \qquad u \times (v \times w) + v \times (w \times u) + w \times (u \times v) = 0$$

holds for this algebra.

On the other hand $\operatorname{Im} \mathbb{O}$ is not a LIE algebra, since for example

$$p \times (i \times j) + i \times (j \times p) + j \times (p \times i) = 3(p \times k) \neq 0.$$

The subalgebras of vector-product algebras are themselves vector-product algebras, so that it follows at once from the isomorphism theorem that:

*Every proper subalgebra* $\neq \{0\}$ *of* $\operatorname{Im} \mathbb{O}$ *is a* LIE *algebra of dimension 1 or 3.*

Of course the one-dimensional subalgebras of $\operatorname{Im} \mathbb{O}$ have the null product.

**4. MALCEV-Algebras.** Parallel to alternative algebras considered as a generalization of associatve algebras, there exists a generalization of LIE algebras, the so-called MALCEV algebras. An algebra $W = (W, \times)$ is called a MALCEV algebra, if it is anticommutative, so that $u \times v = -v \times u$, and if it also satisfies the MALCEV identity

$$(1) \quad m(u,v,w) := (u \times v) \times (u \times w) + u \times [(u \times v) \times w] - u \times [u \times (v \times w)]$$

$$+ v \times [u \times (u \times w)] = 0$$

for all $u, v, w \in W$.

**Proposition.** *Every* LIE *algebra is a* MALCEV *algebra.*

**Proof.** The JACOBI-identity 3(2) implies

$$v \times [u \times (u \times w)] + u \times [(u \times w) \times v] + (u \times w) \times (v \times u) = 0,$$

and hence

$$m(u,v,w) = u \times [(u \times v) \times w - u \times (v \times w) - (u \times w) \times v] = 0. \qquad \square$$

Our object is now to prove the

**Theorem.** $\operatorname{Im} \mathbb{O}$ *together with the product* $(u,v) \mapsto \frac{1}{2}(uv - vu)$ *is a* MALCEV *algebra.*

**Proof.** We linearize the identity iii),

$$(2) \qquad\qquad u \times (u \times w) = \langle u, w \rangle u - |u|^2 w,$$

of Lemma 1 and obtain

$$u \times (v \times w) + v \times (u \times w) = \langle u, w \rangle v + \langle v, w \rangle u - 2\langle u, v \rangle w.$$

We now substitute $u \times v$ for $v$ and note that $\langle u, u \times v \rangle = 0$, thus obtaining

$$(3) \quad u \times ((u \times v) \times w) + (u \times v) \times (u \times w) = \langle u, w \rangle u \times v + \langle u \times v, w \rangle u.$$

From (3) and (2) we now have

$$\begin{aligned} m(u, v, w) &= \langle u, w \rangle u \times v + \langle u \times v, w \rangle u - \langle u, v \times w \rangle u \\ &\quad + |u|^2 v \times w + v \times [\langle u, w \rangle u - |u|^2 w] = 0. \qquad \square \end{aligned}$$

*Note.* In connection with so-called "analytical loops" MALCEV (1909–1967) in the year 1955 was probably the first to consider anticommutative algebras satisfying the condition (1) and to observe that such algebras could be constructed from alternative algebras (*Mat. Sbornik* **78**, 1955, 569–578). It was shown by SAGLE (*Pacific J. Math.* **12**, 1962, 1057–1078) that, subject to a certain additional condition, Im $\mathbb{O}$ is the only simple proper MALCEV algebra over the field $\mathbb{C}$. The additional condition was shown to be superfluous by LOOS (*Pacific J. Math.* **18**, 1966, 553–562). A systematic account of the theory of MALCEV algebras is given in the book by HYO CHUH MYUNG, *Malcev-admissible algebras*, Birkhäuser, 1986.

**5. Historical Remarks.** The problem of determining all spaces with a vector product does not appear to have been treated in the classical literature. It was first discussed in 1942 by ECKMANN and completely solved. We know of only a few places in the literature which deal with the subject. Among these are:

ECKMANN B. *Stetige Lösungen linearer Gleichungssysteme, Comm. Math. Helv.* **15**, 1942/43, 318–339 (particularly p. 338–339).

ECKMANN B. *Continuous solutions of linear equations—some exceptional dimensions in topology, Battelle Rencontres 1967*, 516–526, W.A. Benjamin, 1968.

MASSEY W.S. *Cross products of vectors in higher dimensional Euclidean spaces, Am. Math. Monthly* **90**, 1983, 697–701.

WALSH B. *The scarcity of cross products on Euclidean spaces, Am. Math. Monthly* **74**, 1967, 188–194.

ZVENGROWSKI P. *A 3-fold vector product in* $\mathbb{R}^8$. *Comm. Math. Helv.* **40**, 1965/66, 149–152.

The most elegant treatment is given by MASSEY. WALSH, who uses extremely simple arguments, is not mentioned in MASSEY's work.

# 11

# Division Algebras and Topology

*F. Hirzebruch*

The preceding chapters examined the division algebras of the real numbers, the complex numbers, the quaternions and the octonions. These are of dimension 1, 2, 4 and 8, respectively. So far no algebraist has been able to show that every division algebra has to be of one of these four dimensions, though this surprising fact can be proved by topological methods. HOPF was able to prove in 1940 [7], that the dimension of a division algebra must be a power of 2. His proof, which used the homology groups of projective spaces, will be given in §1. In the year 1958, KERVAIRE and MILNOR independently of one another proved that the power of 2 must be equal to 1, 2, 4 or 8 [9]. They used for this purpose the periodicity theorem of BOTT on the homotopy groups of unitary and orthogonal groups. The periodicity theorem had led to the development of $K$-theory ([4], [3]), a new cohomology theory with whose help many of the classical problems of topology, which had resisted the ordinary homology and cohomology theory, could be solved. We shall describe in §2 a proof of the $(1, 2, 4, 8)$-Theorem, which is based on $K$-theory.

## §1. THE DIMENSION OF A DIVISION ALGEBRA IS A POWER OF 2

Following HOPF we shall prove a theorem on continuous odd mappings of spheres, which yields the required theorem on division algebras as a corollary. As the homology of projective spaces will be used, a brief introduction to homology theory (see DOLD: *Lectures on algebraic topology*, Berlin, etc: Springer, 1980, 2nd edn.) will be included in §1.2.

Projective spaces are examples of manifolds. These are topological spaces which, in the neighborhood of any one of their points, admit of $n$ real coordinates, $n$ being the dimension of the manifold. These coordinates can also be regarded as a homeomorphism of the neighborhood onto an open subset of the Euclidean space $\mathbb{R}^n$. (Strictly speaking the topology should satisfy the "Hausdorff" condition and have an enumerable basis; furthermore we shall be considering only differentiable manifolds, that is, those in which the different coordinate systems are related to one another by coordinate transformations which are differentiable arbitrarily often.) We should also make it clear that we restrict ourselves to connected and compact manifolds only. This assumption is of importance in §1.2.

## 1. Odd Mappings and HOPF's Theorem.

If $A$ is a division algebra of dimension $n$, one can choose a vector space isomorphism of $A$ onto $\mathbb{R}^n$ and transfer the multiplication defined in $A$ over to $\mathbb{R}^n$:

$$\mathbb{R}^n \times \mathbb{R}^n \to \mathbb{R}^n, \qquad (x, y) \mapsto z = x \cdot y.$$

The vector $z = (\zeta_1, \dots, \zeta_n)$ depends on $x = (\xi_1, \dots, \xi_n)$ and $y = (\eta_1, \dots, \eta_n)$ and in fact $\zeta_i$ is a bilinear form in the $\xi_r$ and $\eta_s$. In $\mathbb{R}^n$ the usual Euclidean length is defined by $\|x\| = \xi_1^2 + \xi_2^2 + \cdots + \xi_n^2$, and the $(n-1)$-dimensional sphere $S^{n-1}$ of vectors of length 1 can be introduced. The above multiplication mapping $\mathbb{R}^n \times \mathbb{R}^n \to \mathbb{R}^n$ can be restricted to $S^{n-1} \times S^{n-1}$ and the restriction denoted by $f$. As the algebra has no divisors of zero, $f$ never assumes the value $0 \in \mathbb{R}^n$ and hence the mapping $g = f/\|f\|$ is well-defined

$$g: S^{n-1} \times S^{n-1} \to S^{n-1}.$$

HOPF uses for his proof only the fact that the continuous mapping $g$ is *odd*, that is to say that

$$g(-x, y) = g(x, -y) = -g(x, y) \quad \text{for} \quad x, y \in S^{n-1}.$$

(The mapping $z \mapsto -z$ associates with every point on the sphere its antipodal point.)

**Theorem.** *If there exists a continuous odd mapping of $S^{n-1} \times S^{n-1}$ into $S^{n-1}$, then $n$ is a power of 2.*

**Corollary.** *The dimension of a division algebra over $\mathbb{R}$ is a power of 2.*

The theorem is considerably more general than its corollary. The odd mapping could for example be given by real algebraic forms which are homogeneous of odd degree in the $\xi_r$ and also homogeneous of odd degree in the $\eta_s$. HOPF also turns to another generalization. He discusses continuous odd mappings of the type $S^{p-1} \times S^{n-1} \to S^{m-1}$. It does not appear to be known, even today, for what values of $p$, $n$, $m$ such mappings exist. If the composition problem of HURWITZ mentioned in 10.2.3 is solvable, then there exists an odd mapping $S^{p-1} \times S^{n-1} \to S^{m-1}$. We shall confine ourselves here to the case $p = n = m$.

To prove the above theorem of HOPF, we still have quite a long way to go, because we have to make use of the homology of real projective spaces, and before we can do that we shall first need to give a short introduction to homology theory in the next paragraph.

The real projective space $\mathbb{P}^{n-1}$ is the $(n-1)$-dimensional manifold which results from the sphere $S^{n-1}$ when we identify every point with its antipodal point. We shall denote this identification by the mapping $\alpha: S^{n-1} \to$

$\mathbb{P}^{n-1}$. Every $k$-dimensional linear subspace of $\mathbb{R}^n$ defines an embedding of the sphere $S^{k-1}$ in $S^{n-1}$ and hence (as an image under the mapping $\alpha$) a $(k-1)$-dimensional projective subspace of $\mathbb{P}^{n-1}$, which can frequently be denoted simply by $\mathbb{P}^{k-1}$. For $k = 2$ we obtain the great circles on $S^{n-1}$, whose images under $\alpha$ are the projective lines in $\mathbb{P}^{n-1}$.

An odd mapping $g: S^{n-1} \times S^{n-1} \to S^{n-1}$ induces a mapping

$$G: \mathbb{P}^{n-1} \times \mathbb{P}^{n-1} \to \mathbb{P}^{n-1},$$

to which Hopf applies homology theory. (The homology of spheres is too trivial to get any results!) But now it is time to turn to homology theory itself.

**2. Homology and Cohomology with Coefficients in $F_2$.** Let $X$ be a topological space. Two points $P$ and $Q$ in $X$ are said to be homologous if they can be joined to each other by a path in $X$. The set $S_0 X$ of homology classes of points is therefore equivalent to the set of (pathwise) connected components of $X$. The "zeroth" or zero-dimensional homology group $H_0(X)$ with coefficients in $F_2$, the finite field with two elements, is the $F_2$-vector space of all formal linear combinations of the elements of $S_0 X$ with coefficients in $F_2$. If $X$ is path connected then $H_0(X) \cong F_2$.

To define the $q$-dimensional homology group $H_q(X)$ for $q > 0$, one needs to consider $q$-dimensional structures (cycles), instead of points, to construct formal linear combinations of them with coefficients in $F_2$, and to operate with these combinations modulo a certain equivalence relation known as a "homology." We cannot go into the full details here, but the following salient points must be mentioned.

a) Every closed path $w$ in $X$ represents a homology class $|w| \in H_1(X)$.

b) Every $q$-dimensional submanifold $M$ of an $n$-dimensional manifold $X$ represents a homology class $|M| \in H_q(X)$. If $q = 0$, then we come back to the homology classes of points mentioned above. Moreover, $H_n(X) \cong F_2$, and $|X|$ is the nonzero element of $H_n(X)$.

In general, not all homology classes are represented by a) and b), but for spheres $S^n$ ($n > 0$) and the projective spaces $\mathbb{P}^n$, this does apply. (Note, incidentally, that the dimension index $n - 1$ in paragraph 1 has now been replaced by $n$.)

The homology groups $H_0(S^n)$ and $H_n(S^n)$ both have rank 1 as $F_2$-vector spaces. Otherwise $H_q(S^n) = 0$. The nonzero elements are $|P| \in H_0(S^n)$, where $P$ is an arbitrary point, and $|S^n| \in H_n(S^n)$. For the projective spaces $\mathbb{P}^n$, the situation is as follows. The homology groups $H_q(\mathbb{P}^n)$ are of rank 1 for $0 \le q \le n$ (that is to say $H_q(\mathbb{P}^n) \cong F_2$). Otherwise $H_q(\mathbb{P}^n) = 0$. All $q$-dimensional projective subspaces $\mathbb{P}^q \subset \mathbb{P}^n$ ($0 \le q \le n$) are homologous to one another, and in fact $|\mathbb{P}^q|$ is the nonzero element of $H_q(\mathbb{P}^n)$.

We continue with our general description of homology. Any continuous mapping $f: X \to Y$ between the topological spaces $X$, $Y$ *induces a homomorphism* $f_*: H_q(X) \to H_q(Y)$. The definition of $f_*$ for homology classes of points and homology classes of paths runs as follows. For a point $P \in X$, $f_*|P| = |f(P)|$. For a closed path $w$ in $X$, $f \cdot w$ is a closed path in $Y$ and $f_*|w| = |f \cdot w|$.

The transformation from $f$ to $f_*$ is compatible with the operation of composition, or, in other words, if $f: X \to Y$ and $g: Y \to Z$ then $(g \cdot f)_* = g_* \cdot f_*$. Thus $(H_q, f_*)$ gives us a "covariant functor." Hopf's proof requires cohomology as well as homology. The $q$th *cohomology group* of $X$ with coefficients in $F_2$ is the vector space, dual to $H_q(X)$

$$H^q(X) = \text{Hom}(H_q(X), F_2),$$

whose elements are the linear mappings $u: H_q(X) \to F_2$. If $x \in H_q(X)$ then the value of $u$ on $x$ is denoted by $\langle u, x \rangle \in F_2$. Given a continuous mapping $f: X \to Y$, to the homomorphism $f_*: H_q(X) \to H_q(Y)$ corresponds the dual homorphism $f^*$ defined by

$$f^*: H^q(Y) \to H^q(X), \quad f^*(u) = u \cdot f_*, \quad \text{so that } \langle f^*(u), x \rangle = \langle u, f_* x \rangle.$$

In the transformation from $f$ to $f^*$, direction becomes reversed, and $(g \cdot f)^* = f^* \cdot g^*$.

So far cohomology contributes nothing essentially new. However, one can now define a *product*, that is, a bilinear mapping

$$H^p(X) \times H^q(X) \to H^{p+q}(X), \quad (u, v) \to u \cdot v,$$

which makes the direct sum $H^*(X) = \bigoplus_{p \geq 0} H^p(X)$ into an associative and commutative ring (a graded $F_2$-algebra). How this is done cannot be detailed here. The homomorphism $f^*$ is compatible with the product, $f^*(u \cdot v) = f^*(u) \cdot f^*(v)$, and is therefore a ring homomorphism

$$f^*: H^*(Y) \to H^*(X).$$

The rank of $H_p(X)$ as an $F_2$-vector space is denoted by $b_p(X)$ (the $p$th Betti number). If the Betti numbers are finite, as will always be the case for us, $b_p(X)$ is also equal to the rank of $H^p(X)$. For an $n$-dimensional manifold $X$, $b_p(X) = b_{n-p}(X)$. This is the *Poincaré duality theorem* (1895), which today can be expressed in the following form. There is a canonical isomorphism

$$\pi: H_{n-p}(X) \xrightarrow{\cong} H^p(X),$$

which makes it possible to give a *geometrical interpretation* of the cohomology product in manifolds: if $M$ and $N$ are submanifolds of $X$ of codimension $p$ and $q$, which lie transversely to one another, so that their intersection is a submanifold of codimension $p + q$, then

$$\pi(|M|) \cdot \pi(|N|) = \pi(|M \cap N|).$$

Thus the product in $H^*(X)$ corresponds to the intersection operation. The direct sum $H_*(X) = \bigoplus_{p \geq 0} H_p(X)$ is likewise a ring, in view of the Poincaré isomorphism for manifolds. The multiplication

$$H_{n-p}(X) \times H_{n-q}(X) \rightarrow H_{n-(p+q)}(X)$$

defines what is known as the intersection product. If $X$ is connected, $H_0(X) \cong F_2$, and then for $x \in H_{n-p}(X)$, $y \in H_p(X)$ the intersection product $x \cdot y \in H_0(X)$ is to be regarded as an element of $F_2$ (the intersection number), and we have

$$\langle \pi(x), y \rangle = x \cdot y.$$

The intersection ring for manifolds was known long before the cohomology ring of an arbitrary topological space (see §1.4). The cohomology ring of the projective space $\mathbb{P}^n$ can now be easily determined. The intersection of $q$ projective subspaces of dimension $n-1$ in general position is a projective subspace $\mathbb{P}^{n-q}$ of dimension $n-q$ ($0 \leq q \leq n$). If we denote $\pi(|\mathbb{P}^{n-1}|)$ by $u$, then $u$ is the nonzero element of $H^1(\mathbb{P}^n)$ and $u^q = \pi(|\mathbb{P}^{n-q}|)$ is the nonzero element of $H^q(\mathbb{P}^n)$. Thus $H^*(\mathbb{P}^n)$ is the polynomial ring over $F_2$ in $u$ with the relation $u^{n+1} = 0$, which results from $H^q(\mathbb{P}^n) = 0$ for $q > n$.

The cohomology ring of the Cartesian product $\mathbb{P}^n \times \mathbb{P}^n$ is found equally easily. The homology classes $|\mathbb{P}^r \times \mathbb{P}^s|$ with $r + s = q$ and $0 \leq r \leq n$, $0 \leq s \leq n$ form a basis for $H_q(\mathbb{P}^n \times \mathbb{P}^n)$. In view of the intersection product $|\mathbb{P}^r \times \mathbb{P}^s| \cdot |\mathbb{P}^k \times \mathbb{P}^l| = |\mathbb{P}^r \cap \mathbb{P}^k \times \mathbb{P}^s \cap \mathbb{P}^l|$ the cohomology ring is the polynomial ring over $F_2$ with indeterminates $u$ and $v$ modulo the relations $u^{n+1} = 0$, $v^{n+1} = 0$, where $u$, $v$ come, via the Poincaré isomorphism of $\mathbb{P}^n \times \mathbb{P}^n$, from the homology classes $|\mathbb{P}^{n-1} \times \mathbb{P}^n|$ and $|\mathbb{P}^n \times \mathbb{P}^{n-1}|$, respectively, which form a basis of $H_{2n-1}(\mathbb{P}^n \times \mathbb{P}^n)$. By forming the intersection numbers one sees that

$$\langle u, |\mathbb{P}^1 \times \text{point}| \rangle = 1, \qquad \langle u, |\text{point} \times \mathbb{P}^1| \rangle = 0,$$

and correspondingly for $v$. This will be important for the next section.

### 3. Proof of HOPF's Theorem. In §1.1 the mapping

$$G: \mathbb{P}^{n-1} \times \mathbb{P}^{n-1} \rightarrow \mathbb{P}^{n-1}$$

was considered. The first homology of $\mathbb{P}^{n-1} \times \mathbb{P}^{n-1}$ has $|\mathbb{P}^1 \times \text{point}|$ and $|\text{point} \times \mathbb{P}^1|$ as basis, where $\mathbb{P}^1$ is any one-dimensional projective subspace of $\mathbb{P}^{n-1}$ (arising from a great circle by identifying antipodes). We now assert that

$$G_*(\mathbb{P}^1 \times \text{point}|) = G_*(|\text{point} \times \mathbb{P}^1|) = |\mathbb{P}^1|.$$

Why is $G_*(|\mathbb{P}^1 \times \text{point}|)$ the nonzero element of $H_1(\mathbb{P}^{n-1})$? We use the following criterion for the homology class $|w| \in H_1(\mathbb{P}^{n-1})$ of a closed path in $\mathbb{P}^{n-1}$. There is a path $\tilde{w}$ in $S^{n-1}$ that under the antipodal identification

$\alpha: S^{n-1} \to \mathbb{P}^{n-1}$ maps into the path $w$ (traversed *once*). The path $\tilde{w}$ is either closed, or else it joins two antipodal points. In the first case $|w| = 0$, in the second case $|w| \neq 0$.

Now $\mathbb{P}^1 \times$ point is a closed path $w \times$ point in $\mathbb{P}^{n-1} \times \mathbb{P}^{n-1}$, and the path $\tilde{w}$ is "half a great circle" and thus joins two antipodal points. As the mapping $g: S^{n-1} \times S^{n-1} \to S^{n-1}$ inducing $G$ satisfies the equation $g(-x, y) = -g(x, y)$, the image path $g \cdot (\tilde{w} \times$ point$)$ in $S^{n-1}$ also joins two antipodal points. Under the antipodal identification this path is transformed into $G \cdot (w \times$ point$)$, which represents the homology class $G_*(|\mathbb{P}^1 \times$ point$|)$ which is therefore, as asserted, nonzero.

The cohomology ring of $\mathbb{P}^{n-1}$ can be written as a polynomial ring over $F_2$ in the indeterminate $t$, with the relation $t^n = 0$, while the cohomology ring of $\mathbb{P}^{n-1} \times \mathbb{P}^{n-1}$ is the polynomial ring in $u, v$ with the relations $u^n = 0$, $v^n = 0$ (see §1.2). We assert that

$$G^*(t) = u + v.$$

To prove this we note that

$$\langle G^*(t), |\text{point} \times \mathbb{P}^1| \rangle = \langle t, G_* |\text{point} \times \mathbb{P}^1| \rangle = \langle t, |\mathbb{P}^1| \rangle = 1 \in F_2$$

and that a corresponding relation holds for $|\mathbb{P}^1 \times \text{point}|$. On the other hand, we also have

$$\langle u + v, |\text{point} \times \mathbb{P}^1| \rangle = \langle u + v, |\mathbb{P}^1 \times \text{point}| \rangle = 1,$$

(see the last formula in §1.2). A cohomology class is however determined by its value on the homology classes.

Now we come to the real proof of HOPF's theorem [7], which, once homology theory is known, is impressively short. From $t^n = 0$ follows:

$$0 = G^*(t^n) = (G^*(t))^n = (u + v)^n.$$

Now

$$0 = (u + v)^n = \sum_{k=1}^{n-1} \binom{n}{k} u^k v^{n-k}$$

(because of the relations $u^n = 0$, $v^n = 0$). Thus the binomial coefficients must all be even ($1 \leq k < n$) and hence $n$ must be a power of 2.

## 4. Historical Remarks on Homology and Cohomology Theory.
The development of algebraic topology began with POINCARÉ (1854–1912). However, no homology groups $H_q(X)$ are yet to be found in his writings, but only the Betti numbers $b_q(X)$. (POINCARÉ used the integers as his coefficient domain. The coefficients $F_2$ used by us have the advantage that questions of sign and orientation can be ignored.) In those days one spoke of combinatorial analysis situs rather than algebraic topology. The group

theoretical formulation of homology is due to Emmy NOETHER (1882–1935). The intersection product in manifolds was known a decade or two before the cohomology ring. POINCARÉ already used intersection numbers for his duality theorem. The ring homomorphism $f^*: H^*(Y) \to H^*(X)$ for arbitrary topological spaces $X$, $Y$ and continuous mappings $f: X \to Y$, was foreshadowed by the inverse homomorphism $\phi: H_*(N) \to H_*(M)$ introduced by HOPF in 1928–1930, and defined for manifolds $M$, $N$ and a continuous mapping $f: M \to N$. SAMELSON [12] writes about this:

"A continuous mapping $f: M \to N$ induces a mapping of the intersection rings, which is linear, but unfortunately, *not* in general multiplicative. This led HOPF to set himself the task of finding out whether *something* multiplicative could be assigned to $f$. Now one knows that in set theory a mapping $f$ preserves the sum though not the intersection (analogously to the fact just mentioned) whereas the operation $f^{-1}$ (= full inverse image) is both additive and multiplicative. HOPF, perhaps prompted by this analogy, defined for every homology class of $N$ an "inverse image" as the correctly interpreted full inverse image in $M$. (Roughly speaking, for any cycle $z$ in $N$, one intersects the cycle $M \times z$ (in $M \times N$) with the graph of $f$ and projects the result into $M$.) The construction yields a ring (algebra-) homomorphism $\phi$ of the ring of $N$ into that of $M$, which is connected with $f$ by the formula $f_*(\phi(z) \cdot x) = z \cdot f_*(x)$, analogous to the formula $f(f^{-1}(A) \cap B) = A \cap f(B)$ in set theory. Incidentally the construction does not require that $M$ and $N$ should have the same dimension; $\phi$ increases the dimension by $\dim M - \dim N$ ...."

The construction of the inverse homomorphism, which HOPF had derived with a sure feel from the analogy in set theory, turned out later to have been the "first appearance of cohomology." Nowadays one interprets $\phi$ as the composition of (a) Poincaré duality in $N$ (from homology to cohomology), (b) the cohomology mapping $f^*$ induced by $f$, and (c) Poincaré duality in $M$ (from cohomology to homology). The cohomology mapping is multiplicative for arbitrary spaces, and Poincaré duality maps the intersection (in homology) to the cup product (in cohomology). (The intersection is often *defined* in this way.) It was not until a few years later, though, that cohomology was discovered in 1935 (Alexander, Kolmogorov, Whitney)."

About the "first appearance of cohomology" HOPF had this to say in the talk which he gave in 1966 entitled "Einige persönliche Erinnerungen aus der Vorgeschichte der heutigen Topologie" [Some personal recollections from the prehistory of present-day topology] [8]:

"The year 1935 was especially significant in the development of topology for several reasons. In September the first International Conference on Topology took place in Moscow. The presentations given at this conference, completely independently of one another, by Alexander, Gordon and Kolmogorov, may be regarded as the beginning of cohomology theory (though Lefschetz with his pseudocycles of 1930 played the role of a precursor).

What surprised me—and probably many other topologists—at that time were not so much the cohomology groups—these are, after all, nothing more than the character groups of the homology groups—as the fact that, in arbitrary complexes and more general spaces, a multiplication could be defined, and hence a cohomology ring, which generalized the intersection ring in manifolds. We had until then thought this to be possible only in manifolds, thanks to their locally Euclidean nature."

## 5. STIEFEL's Characteristic Homology Classes.

In addition to the foregoing considerations, we need to bring the question of the dimension of division algebras into relation with some other topological problems.

Let $M$ be a manifold of dimension $n$. The tangent space $T_x M$ is well defined for every point $x \in M$. It is an $n$-dimensional vector space. A vector field $v$ in $M$ is a function which assigns a vector $v(x) \in T_x M$ to every $x \in M$. Of course $v$ is assumed to be continuous. When does there exist a vector field $v$ which vanishes nowhere? The answer is given in a famous theorem of HOPF, dating from 1926 (POINCARÉ, BROUWER and HADAMARD were precursors). See also MILNOR's book [10].

*A vector field $v$ without zeros exists if and only if the Euler–Poincaré characteristic of $M$ vanishes.*

The Euler–Poincaré characteristic $\chi(M)$ is the alternating sum of the Betti numbers (§1.2), so that $\chi(M) = \sum_{i=0}^{n}(-1)^i b_i(M)$. For the sphere $S^n$ ($n \geq 1$), $\chi(S^n) = 2$ for even $n$ and $\chi(S^n) = 0$ for odd $n$. For the projective spaces $\chi(\mathbb{P}^n) = 1$ for even $n$ and $\chi(\mathbb{P}^n) = 0$ for odd $n$. Thus when $n$ is odd, there exist nonvanishing vector fields on $S^n$ and on $\mathbb{P}^n$, whereas these do not exist when $n$ is even.

By a $k$-field we mean a $k$-tuple $v_1, \ldots, v_k$ of vector fields on $M$, such that the vectors $v_1(x), \ldots, v_k(x)$ at each point $x \in M$ are linearly independent. The largest $k$ for which a $k$-field exists will be called Span($M$). Clearly $0 \leq \text{Span}(M) \leq n = \dim M$.

If Span($M$) $= n$, then the manifold is said to be parallelizable (one also says that there is a trivialization). This terminology arises from the fact that one can then regard two vectors at different points $x$ and $y$ of $M$ as being parallel if they have the same coefficients with respect to the bases $v_1(x), \ldots, v_n(x)$ of $T_x(M)$ and $v_1(y), \ldots, v_n(y)$ of $T_y M$. The space of all tangent vectors, that is, $\bigcup_{x \in M} T_x M$, can then be mapped bijectively onto $M \times \mathbb{R}^n$.

It is a difficult problem to determine Span($M$) for any given manifold $M$, for example for the spheres and projective spaces. Obviously Span($S^n$) $\geq$ Span($\mathbb{P}^n$), since a vector field on $\mathbb{P}^n$ is nothing but a vector field on $S^n$ which is transformed into itself by the antipodal mapping.

It is now known that Span($S^n$) = Span($\mathbb{P}^n$), and its precise value is also known. A few remarks on this topic are given in §3.

There is a fairly close connection between division algebras and related algebraic structures and the existence of vector fields on spheres and projective spaces (see §3). The simplest example is the following

**Theorem.** *If a division algebra of dimension $n$ over $\mathbb{R}$ exists, then the projective space $\mathbb{P}^{n-1}$ and the sphere $S^{n-1}$ are parallelizable.*

**Proof.** As in §1.1 we consider the multiplication $\mathbb{R}^n \times \mathbb{R}^n \to \mathbb{R}^n$ $((x,y) \mapsto z = x \cdot y)$. Let $e_1, \ldots, e_n$ be the standard basis vectors of $\mathbb{R}^n$ and let $y \in S^{n-1}$. Then the vectors $e_1 \cdot y, \ldots, e_n \cdot y$ are linearly independent. If we orthonormalize them we obtain $n$ vectors $w_1(y), \ldots, w_n(y)$ with $w_1(y) = e_1 \cdot y / \|e_1 \cdot y\|$. The vectors $w_2(y), \ldots, w_n(y)$ are tangential to $S^{n-1}$ at the point $w_1(y)$. As $y \mapsto w_1(y)$ is a bijective mapping of $S^{n-1}$ onto itself, we have found an $(n-1)$-field on $S^{n-1}$ which obviously remains unaltered by the antipodal transformation.

HOPF (who from 1931 onwards was Professor at the Eidgenössische Technische Hochschule in Zürich) proposed the following problem to his first pupil, STIEFEL:

*In what manifolds $M$ of dimension $n$ does an $m$-field exist?* In other words: *when is $\mathrm{Span}(M) \geq m$?*

In his dissertation [14], STIEFEL developed the theory of characteristic homology classes. He allowed $m$-fields with singularities. A singular point of an $m$-field $v_1, \ldots, v_m$ is a point $x \in M$ at which $v_1(x), \ldots, v_m(x)$ are linearly dependent. STIEFEL showed that there is always an $m$-field whose set of singularities is $(m-1)$-dimensional and which can be regarded as an $(m-1)$-dimensional cycle for the homology with coefficients in $F_2$.

**The Main Result.** *The homology class $s_{m-1} \in H_{m-1}(M)$ $(m = 1, \ldots, n)$ of the set of singularities is independent of the choice of the $m$-field.*

(We have simplified STIEFEL's theory; for certain $m$, STIEFEL uses homology with integer coefficients, but this can be reduced modulo 2 so that our $s_{m-1}$ is obtained.)

The $s_{m-1}$ are the characteristic homology classes of STIEFEL. In the homology of $M$ certain elements are therefore distinguished by the properties of the tangent bundle $\bigcup_{x \in M} T_x(M)$. We have

$$\mathrm{Span}(M) \geq m \Rightarrow s_0 = 0, \; s_1 = 0, \ldots, s_{m-1} = 0.$$

By HOPF's theorem on vector fields, $s_0 = 0$ if and only if $\chi(M)$ is even. The calculation of the STIEFEL classes thus leads to statements about $\mathrm{Span}(M)$ which, for example, for $M = \mathbb{P}^n$ have a good chance of producing successful results, because the homology of $\mathbb{P}^n$ is not trivial. With $M = S^n$, however, they are doomed to failure.

In a later paper [15] STIEFEL calculated the characteristic homology classes of $\mathbb{P}^n$ by the construction of special $m$-fields with singularities. This work was submitted for publication in the *Commentarii* on the same day as HOPF's paper [7]. STIEFEL's result can be given here in the form of an almost word for word quotation:

> *The characteristic homology class $s_{m-1}$ of $\mathbb{P}^n$ is the null class or the class which contains $\mathbb{P}^{m-1}$, according to whether $\binom{n+1}{m}$ is even or odd.*

(Note that, for $m = 1$, $\binom{n+1}{m} = n + 1 = \chi(\mathbb{P}^n) \bmod 2$.)

What conclusions can now be drawn from $\mathrm{Span}(\mathbb{P}^{n-1}) \geq m - 1$? Answer:

$$\mathrm{Span}(\mathbb{P}^{n-1}) \geq m - 1 \Rightarrow \binom{n}{k} \quad \text{is even for} \quad 0 < k < m.$$

In particular the parallelizability of $\mathbb{P}^{n-1}$ implies that $\binom{n}{k}$ is even for $0 < k < n$, and that $n$ must therefore be a power of 2. We have thus given a new proof, based on the above theorem, of the proposition that a division algebra of dimension $n$ can exist only when $n$ is a power of 2.

## §2. THE DIMENSION OF A DIVISION ALGEBRA IS 1, 2, 4 OR 8

The following eight subsections contain a proof that there can be division algebras only in the dimensions 1, 2, 4 or 8. The proof uses the methods of algebraic topology, the cohomology theory discussed in §1.2 of this chapter, as well as the theory of vector space bundles and the theory of characteristic classes, to be introduced in §2.3 and §2.4 of this section (see [11] for a more detailed exposition). The decisive element in the proof is the BOTT periodicity theorem. It will be stated in §2.6 without any hint of the proof. All proofs of the (1,2,4,8)-theorem make use of BOTT periodicity. The proof to be described here stems from ATIYAH and HIRZEBRUCH [5]. The first proofs were, as already mentioned at the beginning of this chapter, found independently of one another by KERVAIRE and MILNOR in 1958 shortly after the appearance of the Periodicity theorem.

**1. The mod 2 Invariants $\alpha(f)$.** Given a continuous mapping $\phi: S^{n-1} \to S^{n-1}$ we say that $y$ is a regular value, if every $x$ with $\phi(x) = y$ has a neighborhood which is mapped homeomorphically by $\phi$ onto a neighborhood of $y$. In this case the number $\#\phi^{-1}(y)$ is finite. Its parity $\#\phi^{-1}(y)$ modulo 2 does not depend on the choice of $y$. It is called the "degree mod 2" of $\phi$. A homeomorphism has degree $\neq 0$. SARD's theorem asserts that every $C^\infty$-mapping $\phi$ (that is, every mapping which is differentiable arbitrarily many times) always has regular values and hence possesses a mapping degree. If $\phi$ is merely continuous, it can be approximated by $C^\infty$-mappings. All sufficiently good approximations have the same degree, so that even with continuous mappings $\phi: S^{n-1} \to S^{n-1}$ one can talk of the degree mod 2 of the mapping (see MILNOR [10]). One can use the same method to assign an integer as the degree of the mapping, which by reduction modulo 2 yields the degree mod 2. (The points in $\phi^{-1}(y)$ have to be counted with multiplicity $-1$ or $+1$, depending on whether orientation is changed or not.)

Let $GL(n)$ be the topological group of $n \times n$ invertible matrices. If one considers the $n$ columns, one can also regard $GL(n)$ as the set of bases of $\mathbb{R}^n$. We now consider continuous mappings

$$f: S^{n-1} \to GL(n).$$

By choosing a fixed vector $v \in S^{n-1}$ we can define the continuous mapping

$$\phi: S^{n-1} \to S^{n-1}, \qquad \phi(x) = \frac{f(x) \cdot v}{\|f(x) \cdot v\|}.$$

Its degree mod 2 does not depend on the choice of $v$. It is called the mod 2 invariant $\alpha(f)$ of $f$. We now have the following deep result

**Theorem.** *If the mod 2 invariant of a continuous mapping $f: S^{n-1} \to GL(n)$ differs from 0, then $n = 1, 2, 4$ or 8.*

In the following subsection this result will be applied to division algebras. From the next subsection on we shall describe the methods by which this theorem is proved.

**2. Parallelizability of Spheres and Division Algebras.** Suppose that the sphere $S^{n-1}$ is parallelizable. Then, for every vector $x \in S^{n-1}$, there are $n - 1$ linearly independent vectors $w_2(x), \ldots, w_n(x)$, perpendicular to $x$ and depending continuously on $x$. The $n$ "columns" $x, w_2(x), \ldots, w_n(x)$ form an element $f(x) \in GL(n)$. If $v$ is the vector $(1, 0, \ldots, 0)$ of $\mathbb{R}^n$, then $f(x)v = x$ and consequently $\alpha(f) = 1$. If one assumes the theorem of §2.1, then it follows that:

*The sphere $S^{n-1}$ is parallelizable only for $n = 1, 2$ 4, 8; and hence a division algebra exists only in the dimensions 1,2,4,8 at most (see the theorem in §1.5).*

(Strictly speaking, the case $n = 1$ should be excluded because it leads to additional, though trivial, considerations.)

If one starts directly from the division algebra (with multiplication $\mathbb{R}^n \times \mathbb{R}^n \to \mathbb{R}^n$, $(x, y) \to x \cdot y$), then one defines

$$f: S^{n-1} \to GL(n)$$

by $f(x) \cdot v = x \cdot v$ for $x \in S^{n-1}$ and $v \in \mathbb{R}^n$ and one then has $\alpha(f) = 1$.

**3. Vector Bundles.** The theorem of §2.1 will be reformulated at the end of §2.4 as a statement about characteristic classes of vector bundles. By an $n$-dimensional vector bundle over the topological space $X$ is meant another topological space $E$ together with a continuous mapping (bundle projection) $p: E \to X$, such that each fibre $E_x = p^{-1}(x)$ ($x \in X$) is an $n$-dimensional real vector space. A further requirement is that $E$ be locally trivial in the following sense: for every point in $X$ there exists a neighborhood $U$ and $n$ cross-sections $v_1, \ldots, v_n: U \to E$ (that is, continuous mappings with $p \cdot v_i = \mathrm{id}$), such that for every $x \in U$, the $n$ vectors $v_1(x), \ldots, v_n(x)$ form a basis of $E_x$. Vector bundles belong to the fundamental concepts of differential topology and differential geometry. Perhaps the most important example is the tangent bundle $E = TM$ of an $n$-dimensional manifold $M$, which is constituted by forming the union of all tangent spaces, $TM = \bigcup_{x \in M} T_x M$ (see §1.5).

In the present case, however, we are interested in other bundles, namely, the $m$-dimensional bundles $E_f$ over $S^n$ that are derived from a mapping $f : S^{n-1} \to GL(m)$ by the following gluing process: $S^n$ is divided into its upper and lower hemispheres

$$S^n = H^+ \cup H^-, \qquad H^+ \cap H^- = S^{n-1} \qquad \text{(the equator)}$$

and we form the trivial bundles $H^+ \times \mathbb{R}^m$ and $H^- \times \mathbb{R}^m$ then join them together along the equator $S^{n-1}$ by identifying each point $(x, v) \in S^{n-1} \times \mathbb{R}^m$ with $(x, f(x) \cdot v) \in S^{n-1} \times \mathbb{R}^m$. The identification space (quotient space) so obtained is denoted by $E_f$. The bundle projection $p : E_f \to S^n$ arises from the projections $H^+ \times \mathbb{R}^m \to H^+$ and $H^- \times \mathbb{R}^m \to H^-$ on the first factor. Every bundle over $S^n$ can be obtained in this way. If $f : S^{n-1} \to GL(n)$ is derived from a division algebra as in §2.2, the bundle $E_f$, obtained by this gluing process, is called the HOPF *bundle* of the algebra. (In the case of $n = 1$ it represents the well-known MÖBIUS strip.)

**4. WHITNEY's Characteristic Cohomology Classes.** In defining the $n$-dimensional vector bundle $E$ over $X$ it was required that locally there were always $n$ linearly independent cross sections. WHITNEY who was the founder of the theory of bundles, concerned himself with the problem of what obstacles might be encountered in trying to find $k$ global (that is, defined over the whole of $X$) everywhere linearly independent cross-sections for an $n$-dimensional bundle. He succeeded in describing such obstacles in cohomological terms. Let $H^i(X)$ denote the $i$th cohomology of $X$, with coefficients in the two-element field $F_2$ as already introduced in §1.2. Then WHITNEY defines the so-called characteristic cohomology classes

$$w_i(E) \in H^i(X), \qquad i = 1, \ldots, n.$$

We shall not present this definition here. Reference [11] may be recommended as an appropriate textbook.

We may however mention that $w_i(E) = 0$ for $i > n - k$, if $k$ everywhere linearly independent global cross-sections exist. If $w_{n-k+1}(E) \neq 0$, then such a $k$-dimensional cross-section does not exist.

We discussed in §1.5, the STIEFEL classes $s_0, s_1, \ldots, s_{n-1}$ of the tangent bundle $TM$ of an $n$-dimensional manifold $M$, which are homology classes $s_{k-1} \in H_{k-1}(M)$. Under the Poincaré duality (§1.2), $s_{k-1}$ goes over into the Whitney class $w_{n-k+1}$ of the tangent bundle. ($s_{k-1}$ and $w_{n-k+1}$ respectively are the first obstacle to a $k$-cross-section.)

We now return to the $n$-dimensional bundles $E_f$ over $S^n$, which arose from the mappings $f : S^{n-1} \to GL(n)$ by the gluing process. As $H^i(S^n) = 0$ for $i \neq 0$, $i \neq n$, the only class of interest is $w_n(E_f) \in H^n(S^n) \cong F_2$. Now $w_n(E_f) = \alpha(f)$, the mod 2 invariant introduced in §2.1. The reformulated version of the theorem at the end of §2.1 therefore runs as follows:

**Theorem.** *If there is an $n$-dimensional vector bundle $E$ over $S^n$ with $w_n(E) \neq 0$, then $n = 1, 2, 4$ or $8$.*

The existence of the HOPF bundles corresponding to the division algebras of the real numbers, the complex numbers, the quaternions and the octonions, proves that such bundles $E$ do in fact exist in these four dimensions.

In order to prove this reformulated theorem one needs to make use of the survey of all possible vector bundles over the spheres $S^n$, which is given by the periodicity theorem of BOTT. We shall formulate this theorem with the help of the ring $KO(X)$ which we introduce in the next paragraph.

**5. The Ring of Vector Bundles.** It is well known that, given two vector spaces $E$ and $F$ one can produce new vector spaces by forming their direct sum $E \oplus F$ or their tensor product $E \otimes F$. The same applies to vector bundles $E$ and $F$ over $X$. We can form the direct-sum bundle $E \oplus F$, so that for the fibres on $x \in X$, we have $(E \oplus F)_x = E_x \oplus F_x$. Similarly we can define the tensor product $E \otimes F$, with $(E \otimes F)_x = E_x \otimes F_x$.

We now consider the set $N(X)$ of isomorphism classes of vector bundles over $X$. In this set the compositions $\oplus$ and $\otimes$ are defined and satisfy, like ordinary addition and multiplication in $\mathbb{N}$, the associative, commutative and distributive laws, and one has elements which are neutral for $\oplus$ and $\otimes$. This suggests that $N(X)$ should be made into a ring, just as $\mathbb{N}$ is extended to $\mathbb{Z}$. We form $N(X) \times N(X)$ and define the following equivalence relation

$$(a, b) \sim (c, d) \Leftrightarrow \text{there is an } f \text{ such that } a \oplus d \oplus f = c \oplus b \oplus f.$$

(We have to use $f$ because the cancellation rule does not hold in $N(X)$ as it does in $\mathbb{N}$.) The set of equivalence classes is denoted by $KO(X)$.

The compositions $\oplus$ and $\otimes$ go over into $+$ and $\cdot$ in $KO(X)$ thus turning $KO(X)$ into a commutative ring with unit element. The natural mapping $N(X) \rightarrow KO(X)$ is not injective, because the cancellation law does not hold in $N(X)$.

The correspondence $X \rightarrow KO(X)$ behaves like a cohomology theory: if $f: Y \rightarrow X$ is a continuous mapping, any $n$-dimensional vector space bundle $E$ over $X$ (bundle projection $p: E \rightarrow X$) is lifted to the following vector space bundle (likewise of dimension $n$) $f^*E$ over $Y$:

$$f^*E = \{(y, v) \in Y \times E : f(y) = p(v)\}.$$

In this way a mapping $f^*: N(X) \rightarrow N(Y)$ is induced, which is compatible with $\oplus$ and $\otimes$, and consequently we obtain a ring homomorphism

$$f^!: KO(X) \rightarrow KO(Y)$$

with $(f \cdot g)^! = g^! \cdot f^!$ for $Z \xrightarrow{g} Y \xrightarrow{f} X$.

**6. Bott Periodicity.** If we assign to every vector space bundle its fibre dimension, we obtain an epimorphism

$$\varepsilon: KO(X) \rightarrow \mathbb{Z}.$$

(For this purpose $X$ is assumed to be connected.) The kernel of $\varepsilon$ is denoted by $\widetilde{KO}(X)$.

**Bott's Periodicity Theorem.** *The following relations hold*

$$\widetilde{KO}(S^1) = \widetilde{KO}(S^2) = \mathbb{Z}/2, \qquad \widetilde{KO}(S^3) = 0, \qquad \widetilde{KO}(S^4) = \mathbb{Z},$$

$$\widetilde{KO}(S^5) = \widetilde{KO}(S^6) = \widetilde{KO}(S^7) = 0, \qquad \widetilde{KO}(S^8) = \mathbb{Z}.$$

*In the dimensions $n = 1$, 2, 4 and 8 the generating elements are represented by the Hopf bundles associated with the division algebras of the real numbers, complex numbers, quaternions and octonions respectively (less the $n$-dimensional trivial bundle).*

*For all $n$, $\widetilde{KO}(S^n) \cong \widetilde{KO}(S^{n+8})$.*

(All isomorphisms are merely additive and are not to be understood as ring isomorphisms.)

As regards the proof of this theorem, which was published by BOTT in 1957 in another form (see §2.9), no details can be given here. What we shall need is a description of the isomorphism $\widetilde{KO}(S^n) \cong \widetilde{KO}(S^{n+8})$.

For any two spheres $S^n$ and $S^m$ we form the Cartesian product $S^n \times S^m$. In addition we choose a base point on each, $x_0$ on $S^n$ and $y_0$ on $S^m$. Then the "axial cross" $S^n \vee S^m = \{x_0\} \times S^m \cup S^n \times \{y_0\}$ lies in $S^n \times S^m$. We collapse it to a point. Then $S^n \times S^m$ becomes $S^{n+m}$, and we obtain the mappings $S^n \vee S^m \overset{i}{\to} S^n \times S^m \overset{p}{\to} S^{n+m}$. This yields an exact sequence. (We omit the proof.)

$$0 \to \widetilde{KO}(S^{n+m}) \overset{p^!}{\to} \widetilde{KO}(S^n \times S^m) \overset{i^!}{\to} \widetilde{KO}(S^n \vee S^m) \to 0,$$

that is, $p^!$ is injective, $i^!$ is surjective, and the kernel of $i^! = \text{Image of } p^!$. Let $\pi_1, \pi_2$ be the projections of $S^n \times S^m$ on the two factors. Given $a \in \widetilde{KO}(S^n)$ and $b \in \widetilde{KO}(S^m)$, we form

$$a \cdot b = \pi_1^! a \cdot \pi_2^! b \in KO(S^n \times S^m).$$

As $i^!(a \cdot b) = 0$, $a \cdot b$ is the image under $p^!$ of precisely one element in $\widetilde{KO}(S^{n+m})$, which we shall also denote by $a \cdot b$.

*The Bott isomorphism $\widetilde{KO}(S^n) \cong \widetilde{KO}(S^{n+8})$ is described by $a \mapsto a \cdot (\rho_8 - 8)$ where $\rho_8$ denotes the Hopf bundle corresponding to the octonions and 8 denotes the 8-dimensional trivial bundle over $S^8$.*

The survey of the possible vector bundles over the spheres which we mentioned earlier as being needed for the proof of the theorem in §2.4 has now been achieved. The only thing lacking is a method for calculating the characteristic classes.

*Remark.* Our formulation of the Bott periodicity theorem will be found, in essentials, in:

BOTT R. *Lectures on $K(X)$*, New York: W.A. Benjamin, 1969, on page 73 but without proofs.

A detailed proof within the framework of $K$-theory is given in the textbook:

KAROUBI M. *K-theory. An Introduction*, Berlin, etc.: Springer, 1978.

The reader will have a certain amount of difficulty, however, in extracting the results used here from KAROUBI's formulation.

Much simpler is the $K$-theory for complex vector bundles; see

ATIYAH M. *K-theory*, New York: W.A. Benjamin, 1967.

In the appendix to this book (*On K-theory and reality*) there is a concise exposition of how one can arrive at $KO$ by suitable modifications.

**7. Characteristic Classes of Direct Sums and Tensor Products.** In §2.4 following WHITNEY, we assigned to an $n$-dimensional bundle $E$ over $X$ the classes $w_i(E) \in H^i(X)$ for $i = 1, \ldots, n$. It is convenient to supplement these by $w_0(E) =$ unit element $\in H^0(X)$ and $w_i(E) = 0$ for $i > n$ and to combine all these classes in the total Stiefel–Whitney class

$$w(E) = 1 + w_1(E) + \cdots + w_n(E) = \sum_{i=0}^{\infty} w_i(E) \in H^*(X).$$

Whitney showed (1941) that, for the direct sum

$$w(E \oplus F) = w(E) \cdot w(F),$$

or, written out,

$$w_i(E \oplus F) = \sum_{r+s=i} w_r(E) \cdot w_s(F).$$

We now restrict ourselves to spaces $X$ where $H^i(X) = 0$ for almost all $i$. All elements $a \in H^*(X)$ whose 0-dimensional component equals 1 then form a multiplicative group $G(X)$. The Whitney sum formula means that:

*The total Stiefel–Whitney class defines a homomorphism of the additive group $KO(X)$ into the multiplicative group $G(X)$*

$$w: KO(X) \to G(X).$$

The Stiefel–Whitney classes are furthermore compatible with the lifting of the bundles: $w_i(f^*E) = f^*w_i(E)$. The diagram

$$KO(X) \xrightarrow{f^!} KO(Y)$$

$$w \downarrow \qquad\qquad \downarrow w$$

$$G(X) \xrightarrow{f^*} G(Y)$$

is therefore commutative (for a continuous mapping $f: Y \to X$).

The vector bundles behave contravariantly under continuous mappings. How lucky it was that Whitney introduced his classes as cohomology classes thereby ensuring contravariance. For arbitrary vector bundles, however, there was really no other possibility open to him. Making definitions in mathematics is not just an arbitrary game.

Now for the characteristic classes of tensor products. In the case of one-dimensional bundles $E$, $F$ this is simple

$$w_1(E \otimes F) = w_1(E) + w_1(F).$$

If $E = E_1 \oplus \cdots \oplus E_m$ and $F = F_1 \oplus \cdots \oplus F_n$ are direct sums of one-dimensional bundles, then $E \otimes F = \bigoplus(E_i \otimes F_j)$, where the summation is over all $i, j$ such that $1 \leq i \leq m$, $1 \leq j \leq n$. By the Whitney sum formula, we then have $w(E \otimes F) = \prod_{i,j}(1 + w_1(E_i) + w_1(F_j))$.

For arbitrary vector bundles $E$, $F$ of dimensions $m$ and $n$, this result remains true, in the following sense:

Consider the polynomial $\prod_{i,j}(1 + x_i + y_j)$ with coefficients in $F_2$. As it is symmetric in the $x_i$ and $y_j$, it can be expressed as a polynomial in the elementary symmetric functions $\sigma_1, \ldots, \sigma_m$ of the $x_i$ and $\tau_1, \ldots, \tau_n$ of the $y_j$:

$$\prod_{i,j}(1 + x_i + y_j) = P(\sigma_1, \ldots, \sigma_m, \tau_1, \ldots, \tau_n).$$

We then have

$$w(E \otimes F) = P(w_1(E), \ldots, w_m(E), w_1(F), \ldots, w_n(F)).$$

**8. End of the Proof.** We remind the reader that the main result of §2.2 depended on the theorem of §2.1 and that the latter was reformulated in §2.4. With the following proof of this last theorem we shall therefore have achieved our objective.

If an $n$-dimensional vector bundle $E$ over $X$ is replaced by $E - n \in \widetilde{KO}(X)$ the Stiefel–Whitney class remains unchanged in view of the Whitney formula. To prove the theorem it therefore suffices to show that $w(c) = 1$ (that is, $w_n(c) = 0$) for all $c \in \widetilde{KO}(S^n)$ when $n \neq 1, 2, 4$ or $8$. The periods given by the Bott theorem show that this is true for $n = 3, 5, 6$ and $7$. If, now, $n > 9$, we write $n = m + 8$ and, because of the periodicity we can write $c = a(\rho_8 - 8)$, where $a \in \widetilde{KO}(S^m)$. By the definition of $\widetilde{KO}(S^m)$, $a$ can be represented in the form $a = E - F$, where $E$, $F$ are equidimensional bundles over $S^m$, so that the equation

$$c = (E - F) \cdot (\rho_8 - 8) = E \cdot \rho_8 \ - \ F \cdot \rho_8 \ - \ 8 \cdot E \ + \ 8 \cdot F,$$

holds in $KO(S^m \times S^8)$, and hence by the Whitney formula

$$w(c) = w(E \cdot \rho_8) \cdot w(F \cdot \rho_8)^{-1} \cdot w(8 \cdot E)^{-1} \cdot w(8 \cdot F).$$

It can now be shown that each of the four factors has the value 1. Consider, for example, $w(E \cdot \rho_8)$ where $E \cdot \rho_8 \in KO(S^m \times S^8)$ is the element determined by $\pi_1^* E \otimes \pi_2^* \rho_8$. We use the following

**Lemma.** *Let $\xi$ and $\eta$ be even-dimensional vector bundles over $X$ with $w(\xi) = 1 + w_r(\xi)$ and $w(\eta) = 1 + w_s(\eta)$, $s$ even and $w_r(\xi)^2 = w_s(\eta)^2 = 0$. Then $w(\xi \otimes \eta) = 1$.*

The lemma follows from the expression of the Stiefel–Whitney classes of a tensor product in terms of symmetric polynomials with coefficients in $F_2$ described at the end of §2.7.

We apply this lemma to $X = S^m \times S^8$, $\xi = \pi_1^* E$ and $\eta = \pi_2^* \rho_8$. One can assume, without further ado, that $E$ and $F$ are of even dimensions because if not one can add the trivial one-dimensional bundle to each of them without altering $a = E - F$. We thus obtain $w(\pi_1^* E \otimes \pi_2^* \rho_8) = 1$.

**9. Historical Remarks.** The first textbook on fibre bundles is due to N. STEENROD [13] who, in the preface to his work, wrote:

"The recognition of the domain of mathematics called fibre bundles took place in the period 1935–1940. The first general definitions were given by H. Whitney. His work and that of H. Hopf and E. Stiefel demonstrated the importance of the subject for the applications of topology to differential geometry. Since then, some seventy odd papers dealing with bundles have appeared. The subject has attracted general interest, for it contains some of the finest applications of topology to other fields, and gives promise of many more. It also marks a return of algebraic topology to its origin; and after many years of introspective development, a revitalization of the subject from its roots in the study of classical manifolds."

HOPF reports in [8] that at the Moscow conference in 1935 he gave a talk on STIEFEL's theory, and continues, writing: "After I had presented all this in Moscow, H. Whitney pointed out in the discussion that a large part of it was contained in his recent note on 'Sphere spaces' (*Proc. Nat. Acad. Sci.* **21**, 1935). He was quite correct, but Stiefel and I had not known of this note. In any case, it is entirely right that the characteristic classes should now mostly be called "Stiefel–Whitney" classes. I find that in Whitney, everything is treated somewhat more generally than in Stiefel, whereas Stiefel's interest is more directed towards particular problems, which do not occur in Whitney's work."

WHITNEY's theory is indeed more general. He defines the characteristic classes for an arbitrary vector bundle over a base space $X$ and not just for the tangent bundle of a manifold. He had to use cohomology. It is only for manifolds that one can make do with homology alone.

It took a long time before one could really work with the Stiefel–Whitney classes. We refrain from giving any detailed references to the literature on the historical development of the subject, but would refer the interested reader to the textbook by MILNOR and STASHEFF [11].

HOPF [8] would have regarded the subject matter of §1 of this chapter as belonging to the prehistory of topology, but that of §2 together with cohomology, vector bundles, the detailed theory of characteristic classes, Bott periodicity, and $K$-theory as part of the modern era. Bott originally formulated his theorem in the language of homotopy groups, and proved it by the methods of differential geometry. It was first announced in *Proc. Nat. Acad. Sci. USA* **43**, 1957, 933–935, and a detailed exposition is given in "The stable homotopy of the classical groups" *Ann. Math.* **70**, 1959, 313–337. See also J. MILNOR, *Morse theory*, Princeton University Press, 1963. An essential tool was the theory of MORSE.

In 1958, GROTHENDIECK in the context of algebraic geometry, introduced with the help of algebraic vector bundles, a ring $K(X)$ for an algebraic variety $X$ and used it for his generalized version of the RIEMANN–ROCH–HIRZEBRUCH theorem. His ring of vector bundles behaves contravariantly, like the cohomology ring $H^*(X)$ of a topological space. (GROTHENDIECK chose from the letters near $H$, and picked out $K$.) Following GROTHENDIECK's lead, the ring $K(X)$ for topological spaces $X$ was then introduced with the help of topological vector bundles whose fibres are *complex* vector spaces ([4], [3]). If one takes real vector spaces, one arrives at $KO(X)$, the $O$ being a reminder of the role of the orthogonal group in real vector spaces. In order to make $K$ and $KO$ into a complete cohomology theory, one needs the Bott periodicity result, which is simpler for $K$ than for $KO$. In fact $\tilde{K}(S^n) = \mathbb{Z}$ for $n$ even and $\tilde{K}(S^n) = 0$ for $n$ odd.

## §3. ADDITIONAL REMARKS

Naturally, the main object of this chapter was to indicate how the (1,2,4,8)-theorem for division algebras can be proved by topological methods, but at the same time we have made a little excursion taking us from the "pre-history" (the thirties and forties) up to the beginning of the sixties. This account would however be incomplete if we were to leave out any mention of the Hopf invariant (HOPF [6]). We also take another brief look at vector fields on spheres (see §2.5).

**1. Definition of the HOPF Invariant** (see [12]). Let $F: S^{2n-1} \to S^n$ ($n \geq 2$) be a continuous mapping. After deformation we can assume that $F$ has derivatives of all orders. The inverse image $F^{-1}(x)$ of a point $x \in S^n$ is in general an $(n-1)$-dimensional submanifold of $S^{2n-1}$, which bounds an $n$-dimensional manifold $M$; $M$ is then mapped by $F$ onto $S^n$ with a certain mapping degree $\gamma_F$ (see §2.1). The *integer* $\gamma_F$ is called the Hopf invariant of $F$. Under this definition $\gamma_F$ is also the intersection number of $F^{-1}(y)$ ($y \neq x$, $y$ in general position) with $M$, or also the linking number (or looping coefficient) of $F^{-1}(x)$ in relation to $F^{-1}(y)$. An orientation argument shows that $\gamma_F$ vanishes when $n$ is odd. The number $\gamma_F$ depends only on the homotopy class of $F$, and is a homomorphism of the homotopy group $\pi_{2n-1}(S^n)$ to the integers.

**2. The HOPF Construction** (see [12]). HOPF proposed the following problem. *For a fixed even $n$, determine the additive subgroup consisting of those integers that occur as invariants $\gamma$ for a continuous mapping $F: S^{2n-1} \to S^n$.*

With the help of the Hopf construction, we can construct, from a given mapping $g: S^{n-1} \times S^{n-1} \to S^{n-1}$ a mapping $F: S^{2n-1} \to S^n$ as follows:

The sphere $S^{2n-1}$ can be topologically described as the boundary of $E^n \times E^n$ where $E^n$ is the $n$-dimensional ball. Consequently

$$S^{2n-1} = \partial(E^n \times E^n) = S^{n-1} \times E^n \cup E^n \times S^{n-1},$$

so that this sphere is divided into the two products $S^{n-1} \times E^n$ and $E^n \times S^{n-1}$ with $S^{n-1} \times S^{n-1}$ as the common boundary.

The sphere $S^n$ is divided by $S^{n-1}$ into two hemispheres $H^+$ and $H^-$. We extend $g$ in the obvious way into a mapping of $E^n \times S^{n-1}$ into $H^+$ and of $S^{n-1} \times E^n$ into $H^-$. The mapping $F: S^{2n-1} \to S^n$ obtained in this way, is the Hopf construction associated with $g$. Following Hopf $\gamma_F = c_1 \cdot c_2$, if $g$ has the bidegree $(c_1, c_2)$. (Here $c_1$ is the degree with which $S^{n-1} \times$ point is mapped to $S^{n-1}$, and $c_2$ is defined analogously.)

*If $g: S^{n-1} \times S^{n-1} \to S^{n-1}$ is odd (§1.1) then $\gamma_F$ is odd.*

Every function $f: S^{n-1} \to GL(n)$ (see §2.1) defines a mapping $g: S^{n-1} \times S^{n-1} \to S^{n-1}$ by

$$g(x, v) = \frac{f(x)v}{\|f(x)v\|}$$

of bidegree $(c, 1)$ and by means of the Hopf construction a mapping $F: S^{2n-1} \to S^n$ with $\gamma_F = c$. Here $c$ is even or odd according to whether $w_n(E_f) = 0$ or $w_n(E_f) \neq 0$ (see §2.4). If one takes, for even $n$, the glueing function $f$ of the tangent bundle of $S^n$, then $\gamma_F = 2$. If one takes for $f$ the functions $S^{n-1} \to GL(n)$ derived from the division algebras (see §2.1), then $\gamma_F = 1$.

In answer to Hopf's problem therefore we have the following result:

*All integers occur as Hopf invariants for $n = 2, 4, 8$. For the other even $n$, at least all even integers occur as Hopf invariants.*

### 3. ADAMS's Theorem on the HOPF Invariants.

ADAMS [1] showed that *mappings $f : S^{2n-1} \to S^n$ with odd $\gamma_f$ exist only for $n = 2, 4, 8$.* He used the so-called secondary cohomology operations. Meanwhile a proof based on $K$-theory (ADAMS and ATIYAH, 1966, see [3]) has appeared, which is very simple, once $K$-theory has been fully developed.

### 4. Summary.

*Suppose $n \geq 2$.* The results of §2 and §3 have shown that the following "mathematical objects" exist only for $n = 2, 4, 8$.

Division algebras of dimension $n$,
Odd mappings $S^{n-1} \times S^{n-1} \to S^{n-1}$,
Parallelization of $\mathbb{P}^{n-1}$,
Parallelization of $S^{n-1}$,
Vector bundles $E$ over $S^n$ with Stiefel–Whitney class $w_n(E) \neq 0$,
Mappings $f: S^{2n-1} \to S^n$ with odd Hopf invariant.

Starting from a division algebra of dimension $n$, one can, as we have seen, construct the other mathematical objects very simply. From any of the objects listed here, one can fairly simply (using the Hopf construction) obtain a mapping with an odd Hopf invariant. In this sense, the ADAMS theorem on the nonexistence of mappings with an odd Hopf invariant is the most general result; it implies the nonexistence of the other objects.

### 5. ADAMS's Theorem About Vector Fields on Spheres.

We begin by referring to the HURWITZ–RADON theorem (§2.3 of Chapter 10). If the statement i) is satisfied, one can easily construct $p-1$ linearly independent tangent vector fields on $\mathbb{P}^{n-1}$. This is fully analogous to the theorem in §1.5. Consequently $\mathrm{Span}(S^{n-1}) \geq \mathrm{Span}(\mathbb{P}^{n-1}) \geq 8\alpha + 2^\beta - 1$, when $n = u \cdot 2^{4\alpha+\beta}$, $1 \leq u$ is odd, $0 \leq \alpha$, $0 \leq \beta \leq 3$. ADAMS [2] showed that $\mathrm{Span}(S^{n-1}) = \mathrm{Span}(\mathbb{P}^{n-1}) = 8\alpha + 2^\beta - 1$. This involved much highly sophisticated $K$-theory. ADAMS's theorem is a marvelous generalization of the theorem that only the spheres $S^1, S^3, S^7$ are parallelizable (for example, $\mathrm{Span}(S^{15}) = 8$), and thus also a generalization of our (1,2,4,8)-theorem.

The theorem of ADAMS had many precursors. The inequality

$$\mathrm{Span}(\mathbb{P}^{n-1}) \leq \max\left\{ k > 0 \,\middle|\, \binom{n}{k} \text{ even} \right\}$$

is one such (see 11.1.5). Together with the formula $\mathrm{Span}(\mathbb{P}^{n-1}) \geq 8\alpha + 2^\beta - 1$ this immediately gives us, for example, $\mathrm{Span}(\mathbb{P}^{4m+1}) = 1$ and $\mathrm{Span}(\mathbb{P}^{8m+3}) = 3$.

ECKMANN and WHITEHEAD had already proved in the forties that moreover $\mathrm{Span}(S^{4m+1}) = 1$ and $\mathrm{Span}(S^{8m+3}) = 3$.

## REFERENCES

[1] ADAMS, J.F.: On the non-existence of elements of Hopf invariant one. *Ann. of Math.* **72**, 20–104 (1960)

[2] ADAMS, J.F.: Vector fields on spheres. *Ann. of Math.* **75**, 603–632 (1962)

[3] ATIYAH, M.F.: *K-Theory*. W.A. Benjamin, Inc., New York, Amsterdam 1967

[4] ATIYAH, M.F., HIRZEBRUCH, F.: Vector bundles and homogeneous spaces. Proc. of Symposia in Pure Mathematics, Vol. 3, pp. 7–38. Am. Math. Soc. 1961

[5] ATIYAH, M.F., HIRZEBRUCH, F.: Bott periodicity and the parallelizability of the spheres. *Proc. Cambridge Phil. Soc.* **57**, 223–226 (1961)

[6] HOPF, H.: Über die Abbildungen von Sphären auf Sphären niedrigerer Dimension. *Fundamenta Math.* **25**, 427–440 (1935)

[7] HOPF, H.: Ein topologischer Beitrag zur reellen Algebra. *Comm. Math. Helvetici* **13**, 219–239 (1940/41)

[8] HOPF, H.: Einige persönliche Erinnerungen aus der Vorgeschichte der heutigen Topologie. Colloque de Topologie, Centre Belge de Recherches Mathématiques 1966, pp. 9–20

[9] MILNOR, J.: Some consequences of a theorem of Bott. *Ann. of Math.* **68**, 444–449 (1958)

[10] MILNOR, J.: *Topology from the differentiable viewpoint.* The University Press of Virginia, 1965

[11] MILNOR, J., STASHEFF, J.: Characteristic classes. *Ann. of Math. Studies* **76**, Princeton University Press 1974

[12] SAMELSON, H.: Heinz Hopf zum Gedenken. II. Zum wissenschaftlichen Werk von Heinz Hopf. Jber. Deutsch. Math., Verein **78**, 126–146 (1976)

[13] STEENROD, N.: *The topology of fibre bundles.* Princeton University Press 1951

[14] STIEFEL, E.: Richtungsfelder und Fernparallelismus in $n$-dimensionalen Mannigfaltigkeiten. *Comm. Math. Helvetici* **8**, 305–353 (1935/36)

[15] STIEFEL, E.: Über Richtungsfelder in den projektiven Räumen und einen Satz aus der reellen Algebra. *Comm. Math. Helvetici* **13**, 201–218 (1940/41)

# Part C

# Infinitesimals, Games, and Sets

# 12

# Nonstandard Analysis

*A. Prestel*

## §1. INTRODUCTION

In this chapter, our objective will be to extend the field $\mathbb{R}$ of real numbers to a field $^*\mathbb{R}$ in which there are both infinitely small and infinitely large "numbers." In particular we shall find that it is possible in $^*\mathbb{R}$, to define precisely the Leibniz differentials $dx$, $dy$ and to establish a connection between the differential coefficient $dy/dx$ and the derivative $f'(x)$ of a function $y = f(x)$ at the point $x$.

Calculations involving infinitely small quantities such as $dx$ were made as a matter of course in the mathematics and physics of earlier centuries, even if their legitimacy did not always go unchallenged. (Some insight into this use of infinitesimals and the criticisms raised against such use can be gained, for example, from the book by Edwards [1].) It was not until the advent of "epsilon techniques," and the creation by WEIERSTRASS of a firm foundation for analysis based on the concept of the limit, that infinitesimal magnitudes were banished from mathematics. To be more accurate, they were banned from use in exact proofs. For heuristic purposes in mathematics and physics they maintained their rightful place as before.

In the "epsilon technique" the differential quotient $dy/dx$ of a function $y = f(x)$ is defined as the limit of the quotient of the difference

$$\frac{f(x+h) - f(x)}{h}$$

as $h$ tends to zero, whenever this limit exists. Its value is then denoted by $f'(x)$. Although the notation $dy/dx$ is still commonly used, it is always emphasized that the quantities $dy$ and $dx$ by themselves are meaningless. Of course this is true enough if one has in mind only the real numbers: there is no real number, say, lying between 0 and all the strictly positive real numbers.

If therefore one wishes to work with infinitely small quantities, one obviously has to take these from a domain larger than $\mathbb{R}$. Mathematicians of earlier centuries worked with such quantities, as though it were the most natural thing in the world, and just as with real numbers, but they were always quite clear about the distinction. The fact that they did not bother about the construction of such an enlarged domain need not surprise us. It was not usual at that time to consider consistency problems of this kind.

It was quite enough for most mathematicians that such quantities existed in their mathematical intuition and that their use led to correct results. Those among the mathematicians and philosophers who declined to have any dealings with them probably did so because they felt that there was a contradiction in treating these quantities like real numbers and yet refusing to recognize them as "finite."

In the year 1960 Abraham ROBINSON constructed an extension $^*\mathbb{R}$ of the field $\mathbb{R}$ of real numbers, in which quantities exist, which are infinitely small, but whose properties in general are nevertheless indistinguishable in many respects from those of the numbers of $\mathbb{R}$. ROBINSON used for this purpose a construction which had been applied for the first time by SKOLEM to obtain an extension of the natural numbers, which provided an alternative model of a system satisfying the Peano axioms [6]. In this context the Peano axioms were formulated in the so-called first-order predicate logic, which represents a certain restriction, in comparison with the usual set-theory formulation (see Chapter 1, §2). Such models were called "nonstandard" models of the Peano system of axioms. The construction method used, which essentially represents the present-day ultra power method, can be applied to any structure and always leads to an extension which, within a certain framework—the 1st order logic—possesses the same properties as those of the original structure from which one started. ROBINSON applied this method to the field $\mathbb{R}$ and obtained an extension field $^*\mathbb{R}$, whose elements he called nonstandard numbers. The term *nonstandard analysis* was used to cover mathematical operations carried out within $^*\mathbb{R}$. A detailed exposition of this method and its applications is given in ROBINSON's book [5].

We now propose to mention briefly a few of the properties of $^*\mathbb{R}$, to explain how the differential $dy$ of a function $y = f(x)$ can be defined in this domain, and how the quotient $dy/dx$ is connected with the limit $f'(x)$ of the quotient of the differences.

The field $^*\mathbb{R}$ is an ordered overfield of $\mathbb{R}$, in which there are elements $a$ with the property that $r < a$ for all $r \in \mathbb{R}$; in other words elements $a$ which are infinitely large. Clearly any element $1/a$ is then infinitely small; it lies between 0 and all positive $\varepsilon \in \mathbb{R}$. The elements of the set

$$\mathcal{D} = \{x \in {}^*\mathbb{R} : |x| \leq r \text{ for an } r \in \mathbb{R}\}$$

are said to be *finite;* the elements of

$$\mathcal{M} = \{x \in {}^*\mathbb{R} : |x| \leq \varepsilon \text{ for all } \varepsilon \in \mathbb{R}^+\}$$

are said to be *infinitely small.* Here $\mathbb{R}^+$ denotes the set of positive real numbers. Elements $x, y \in^* \mathbb{R}$ are said to be *neighboring elements* or *neighbors,* if $x - y \in \mathcal{M}$, and we then write $x \approx y$. Every finite number $x$ is a neighbor of just one number $r \in \mathbb{R}$. We write $r = st(x)$, and call $r$ the *standard part* of $x$. To avoid any possible confusion with the real part of a complex number, one deliberately refrains from using the much more suggestive term "real part." An essential property of $^*\mathbb{R}$ is that every real function $y = f(x)$ can be "canonically" extended to a function $^*f$ in $^*\mathbb{R}$; this is to be taken in the sense that such properties as can be expressed in the language of the first order predicate calculus, continue to hold. Using this extension, we can define the value of $^*f(x + dx)$ in $^*\mathbb{R}$ for every nonzero element $dx$ in $\mathcal{M}$ whenever $x$ lies in the domain of definition of $f$. The difference

$$df = \; ^*f(x + dx) - f(x)$$

is called the differential of the function $f$. As both $df$ and $dx$ are elements of $^*\mathbb{R}$, one can obviously form their quotient $df/dx$ (for $dx \neq 0$). This differential quotient is thus an element of $^*\mathbb{R}$. If $f$ is differentiable at the point $x$, that is to say, if $\lim_{h \to 0}(f(x + h) - f(x))/h = f'(x)$ exists, then it can be shown that

$$\frac{df}{dx} \approx f'(x)$$

that is, $f'(x)$ is the standard part of $df/dx$. Note that the differential quotient $df/dx$, unlike the limit of the quotient of the differences, always exists. It does not necessarily have to be finite however and even where it happens to be finite for all $dx$, its standard part need not necessarily be independent of the choice of $dx$. If however this is the case, then the limit of the quotient of the differences does exist and is equal to the standard part of $df/dx$.

Before we prove all this and more in the sections which follow, we should now like to indicate a path which, after laying down a few very natural requirements, and starting out from the kind of statements made by, say, LEIBNIZ and L'HOSPITAL on infinitesimals and the way in which they handled them, leads almost inevitably to the domain $^*\mathbb{R}$ used by ROBINSON.

As already mentioned, mathematicians of earlier times were quite clear in their own minds that quantities such as $dx$ or $f(x + dx)$ could not simply be real numbers. At the beginning of his textbook *Analyse des infiniments petits* (Paris, 1696) the Marquis de l'Hospital gave the following definitions:

"*Definition I.* Variable quantities are those which continually increase or decrease. And constant quantities are those which always stay the same, while others change ..."

"*Definition II.* The infinitely small part by which variable quantities continually increase or decrease, is called the differential of this quantity."

A quantity such as $dx$ is therefore something variable, something which can vary "with time." LEIBNIZ says in a letter written to the French Professor Pierre VARIGNON in Paris, in 1702, among other things:

"It should however be borne in mind that the incomparably small quantities, even taken in their popular sense, are by no means constant and definite, and that rather, since one can assume them to be as small as one wishes, they play the same role in geometrical considerations as do the infinitely small in the strict sense of the term."

By the words "incomparably small quantities" Leibniz means here the
differentials introduced by him.

These quotations suggest, in our opinion, that the quantity $dx$ should be
regarded as a (variable) function—let us say a function of time—which in
the long run assumes ever smaller values. On the other hand, the number 2
say, can be thought of as a function of time which constantly has the value
2. Viewed in this way, our "quantities" would therefore all be mappings of
a time axis $T$ onto the set $\mathbb{R}$. In this context it is not particularly important
(as one can convince one's self later on) whether the time runs continuously
or not. We could in fact have taken $\mathbb{R}^+$ for $T$ but we have decided on technical grounds, and for typographical convenience, on $T = \mathbb{N}' = \{1, 2, \ldots\}$.
Accordingly our quantities will be sequences of real numbers. The mapping
$t \to 1/t$, gives us for example the sequence

$$\left(1, \frac{1}{2}, \frac{1}{3}, \ldots, \frac{1}{t}, \frac{1}{t+1}, \ldots\right).$$

The terms of this sequence assume smaller and smaller values in the course
of time. We are therefore entitled to regard it as infinitely small in comparison with the constant sequences

$$(\varepsilon, \varepsilon, \ldots, \varepsilon, \ldots)$$

which represents the real numbers $\varepsilon$. We would like the sequence of quotients

$$\left(\frac{f(x + (1/t)) - f(x)}{1/t}\right)_{t \in \mathbb{N}'}$$

to represent the differential quotients for $x \in \mathbb{R}$. The sequence of quotients
itself, be it emphasized, and not its limit as $t$ tends to infinity, which may
or may not exist. However, here we are faced with a difficulty. Whereas for
sequences

$$(a_1, a_2, \ldots) \quad \text{and} \quad (b_1, b_2, \ldots)$$

a formal addition, subtraction, and multiplication can easily be defined by
the obvious canonical termwise definition for sequences

$$(a_1 \pm b_1, a_2 \pm b_2, \ldots) \quad \text{and} \quad (a_1 \cdot b_1, a_2 \cdot b_2, \ldots)$$

this procedure fails for division. In other words the set $R$ of all sequences
indexed by $\mathbb{N}'$ and with terms belonging to $\mathbb{R}$, constitutes a ring with respect
to the operations defined as above, but is not a field. To arrive at a field—
and this will be our *one and only* requirement—we shall have to enlarge
somewhat our concept of "quantity." If two sequences are indistinguishable
from one another from some point in time onwards, then we would like to
regard them as being equal, because in such cases they differ only trivially.
If we agree to regard sequences as equal only in such cases, then this is
tantamount to forming the residue classes of the ring $R$ with respect to the

ideal $D$, consisting of those sequences $(a^{(t)})_{t \in \mathbb{N}'}$, for which all but a finite number of the terms are equal to 0. This residue class however still does not lead to a field. We can only obtain a field by choosing a maximal ideal $M$ over $D$, (which means $M \supset D$), and taking as our new domain $^*\mathbb{R}$ the residue class field $R/M$. It is well-known (see 2.3.4) that $R/M$ is a field when $M$ is a maximal ideal. Our quantities are therefore the congruence classes of the sequences $(a^{(t)})_{t \in \mathbb{N}'}$ with respect to the ideal $M$. In other words two sequences are regarded as representing the same "quantity" if and only if their difference lies in $M$, that is, they differ only by an element of $M$.

We shall show in the following sections not only that the field $^*\mathbb{R} = R/M$ defined in this way contains infinitely small and infinitely large elements, but also that all functions mapping $\mathbb{R}$ into $^*\mathbb{R}$ can be extended canonically, and that $\mathbb{R}$ and $^*\mathbb{R}$ have in common all those properties which can be formulated within a certain definite axiomatic framework. We stress once again that all this follows merely from the single requirement that $R/M$ should be a field. It is immaterial which maximal ideal we choose provided that it contains $D$. We shall go into this point in rather more detail in the epilogue.

## §2. The Nonstandard Number Domain $^*\mathbb{R}$

**1. Construction of $\mathbb{R}$.** As was already established in the introduction, $R$ is the ring of sequences $a = (a^{(n)})_{n \in \mathbb{N}'}$ of real numbers, with addition, subtraction and multiplication each defined componentwise. Furthermore. $D$ is the ideal in $R$, which comprises just those sequences $(a^{(n)})_{n \in \mathbb{N}'}$ for which $a^{(n)}$ is almost always zero (that is, for which all but a finite number of the terms $a^{(n)}$ are zero). Finally $M$ is a maximal ideal having $D$ as a subset (that is, $M \supset D$). The existence of such an ideal is guaranteed by Zorn's lemma (see 14.3.2).

The ring $R$ contains a canonical, isomorphic image of the field $\mathbb{R}$ of real numbers. This canonical embedding is given by

$$r \mapsto (r, r, r, \ldots)$$

for $r \in \mathbb{R}$. We shall identify $\mathbb{R}$ with its image, that is to say, we shall regard the constant sequences as real numbers. For sequences $a, b \in \mathbb{R}$, we define

$$a \equiv b \bmod M :\Leftrightarrow a - b \in M.$$

This is an equivalence relation on $R$. We denote the set of equivalence classes by $^*\mathbb{R}$, so that

$$^*\mathbb{R} = R/M.$$

The operations $+$, $-$, and $\cdot$ carry over in the usual way from $R$ to the quotient domain $R/M$, which thereby becomes a ring. The maximality of

$M$ ensures that $R/M$ is indeed a field. As every nonzero constant sequence in $R$ is invertible the subfield $\mathbb{R}$ of $R$ is not affected by the operation of forming the residue class modulo $M$, or in other words an isomorphic image of $\mathbb{R}$ is still to be found in $R/M$. Here again we shall identify $\mathbb{R}$ with its isomorphic image. Finally therefore we have obtained $^{*}\mathbb{R} = R/M$ as an overfield of $\mathbb{R}$ (and as we shall see later, a proper one).

We next wish to show that every function $f\colon \mathbb{R}^m \to \mathbb{R}$ can be extended to a function $^{*}f\colon {}^{*}\mathbb{R}^m \to {}^{*}\mathbb{R}$ which retains all the properties which are expressible within the framework of the first-order logic (we shall explain in §3 precisely what this means). We first define, for any given function $f\colon \mathbb{R}^m \to \mathbb{R}$ a componentwise extension $\bar{f}$ on $R^m$ by:

$$\bar{f}(a_1,\ldots,a_m) = (f(a_1^{(1)},\ldots,a_m^{(1)}), f(a_1^{(2)},\ldots,a_m^{(2)}),\ldots),$$

where $a_i = (a_i^{(n)})_{n\in\mathbb{N}'}$ for $1 \le i \le m$ are sequences belonging to $R$. We then set

$$^{*}f(a_1,\ldots,a_m) \equiv \bar{f}(a_1,\ldots,a_m) \bmod M,$$

where the sequences $a_1,\ldots,a_m$ are representatives of certain residue classes modulo $M$. It remains to be shown that this definition is independent of the choice of these representatives. Suppose therefore that

$$a_1 \equiv b_1,\ldots,a_m \equiv b_m \bmod M.$$

We then have to show that

$$\bar{f}(a_1,\ldots,a_m) \equiv \bar{f}(b_1,\ldots,b_m) \bmod M.$$

The proof of this rather general statement is not immediately obvious, as it would be, say, in the case of addition where $a_1 - b_1$, $a_2 - b_2 \in M$, naturally imply $(a_1 + a_2) - (b_1 + b_2) \in M$. The proof applies to any arbitrary ideal $M$, the maximality of $M$ not being an essential requirement. We shall first give the proof for the ideal $D$, which will thus point the way. Obviously $a \equiv b \bmod D$ simply means that $a^{(n)} = b^{(n)}$ for almost all $n$ (that is, for all $n$ save for at most a finite number of possible exceptions). But we also have, in our case,

$$a_1^{(n)} = b_1^{(n)} \quad \text{and} \quad \cdots \quad \text{and} \quad a_m^{(n)} = b_m^{(n)}$$

for almost all $n \in \mathbb{N}'$. Thus, for almost all $n$

$$f(a_1^{(n)},\ldots,a_m^{(n)}) = f(b_1^{(n)},\ldots,b_m^{(n)}),$$

which naturally implies $\bar{f}(a_1,\ldots,a_m) \equiv \bar{f}(b_1,\ldots,b_m) \bmod D$.

For an arbitrary ideal $M$, we now proceed as follows: for $a \in R$ we define

$$Z(a) = \{n \in \mathbb{N}' : a^{(n)} = 0\}.$$

Next we form the set

$$U = U_M = \{Z(a): a \in M\}.$$

$U$ has the following properties:

(0) $\emptyset \notin U$,

(i) $\mathbb{N}' \in U$,

(ii) $Z_1, Z_2 \in U \Rightarrow Z_1 \cap Z_2 \in U$,

(iii) $Z \in U, Z \subset A \subset \mathbb{N}' \Rightarrow A \in U$,

that is, to say $U$ is a *filter* on $\mathbb{N}'$. If $M$ is a maximal ideal, then $U_M$ is an *ultrafilter*, that is, it has the additional property

(iv) $A \subset \mathbb{N}' \Rightarrow A \in U$ or $\mathbb{N}' \setminus A \in U$.

If furthermore $M$ includes $D$, then $U$ is a *non-trivial* ultra filter, that is, we also have

(v) $A \subset \mathbb{N}', |\mathbb{N}' \setminus A| < \infty \Rightarrow A \in U$.

The verification of all these properties is very simple, as we exemplify by carrying it out for (iv). We choose a sequence $a$ of zeros and ones, such that $Z(a) = A$ holds, and assume $A \notin U$. In particular therefore $a \notin M$. As $M$ is maximal, there are elements $b \in M$ and $c \in R$ with $1 = b + ac$. It follows at once that $Z(b) = Z(1 - ac) \subset \mathbb{N}' \setminus A$. Since $b \in M$ we have $Z(b) \in U$ and hence, by (iii) we also have $\mathbb{N}' \setminus A \in U$.

Now, for $a \in R$

$$a \in M \Leftrightarrow Z(a) \in U$$

holds generally. We need only to prove that $Z(a) \in U$ implies $a \in M$, because the truth of the reverse implication follows from the definition of $U$. Since $Z(a) \in U$ there is a $b \in M$ satisfying $Z(a) = Z(b)$, or in other words $a$ and $b$ have the same 0-components. We define a sequence $c = (c^{(n)})_{n \in \mathbb{N}'}$ by

$$c^{(n)} = \begin{cases} a^{(n)}/b^{(n)} & \text{for } n \notin Z(b), \\ 0 & \text{otherwise.} \end{cases}$$

Then obviously $a = bc \in M$.

If we now observe that $Z(a - b) = \{n: a^{(n)} = b^{(n)}\}$, we get for $a, b \in R$

(vi) $a \equiv b \bmod M \Leftrightarrow \{n: a^{(n)} = b^{(n)}\} \in U_M$.

This important relation now enables us to complete the still outstanding independence proof and at the same time will serve as a guide to what follows.

We return therefore to the definition of $^*f$. By hypothesis $a_i \equiv b_i \bmod M$ for $1 \leq i \leq m$, that is, we have $\{n: a_i^{(n)} = b_i^{(n)}\} \in U$ for $1 \leq i \leq m$, whence also

$$\bigcap_{i=1}^{m} \{n: a_i^{(n)} = b_i^{(n)}\} = \{n: a_1^{(n)} = b_1^{(n)}, \ldots, a_m^{(n)} = b_m^{(n)}\} \in U.$$

Since

$$\{n: a_1^{(n)} = b_1^{(n)}, \ldots, a_m^{(n)} = b_m^{(n)}\} \subset \{n: f(a_1^{(n)}, \ldots a_m^{(n)}) = f(b_1^{(n)}, \ldots, b_m^{(n)})\}$$

it follows from this, in conjunction with (iii) that

$$\{n: f(a_1^{(n)}, \ldots, a_m^{(n)}) = f(b_1^{(n)}, \ldots, b_m^{(n)})\} \in U.$$

With (vi) this is however equivalent to the statement which had to be proved. We thus know how we can extend real functions on $R/M$, for any ideal $M$. Which of their properties carry over from $\mathbb{R}$ to $^*\mathbb{R}$, is a matter that we shall analyze precisely in §3. As one can easily see after thinking about it, the relation

$$^*(f \circ g) = {}^*f \circ {}^*g$$

holds for the extensions of two functions $f$, $g$ and their composition. We shall make use of this property without specially drawing attention to it.

The fact that we have been dealing only with functions defined on the whole set of $\mathbb{R}^m$ need not trouble us, because it is a trivial matter to extend the domain of definition to the whole of $\mathbb{R}^m$ for any function initially defined only on a subset of $\mathbb{R}^m$.

**2. Properties of $^*\mathbb{R}$.** We shall from now on again assume that $M$ is a maximal ideal over $D$. By doing so we shall on the one hand ensure that $^*\mathbb{R} = R/M$ is a field. On the other hand, the ordering of the real numbers, defined by the relation $\leq$, can be extended canonically to an ordering of $^*\mathbb{R}$. Let us frame our definition of the extension of the ordering relation $\leq$, which we shall continue to denote by the same symbol $\leq$, on the relation (vi). For $a, b \in R$ we set

$$a \leq b \bmod M :\Leftrightarrow \{n: a^{(n)} \leq b^{(n)}\} \in U_M.$$

Like (vi) this asserts that the relation $a \leq b$ modulo $M$ holds if the corresponding property holds for "very many" components. It remains to be shown that this definition is likewise independent of the particular representatives chosen. Suppose therefore that $a \equiv a_1$ and $b \equiv b_1 \bmod M$. Then

$$\{n: a^{(n)} \leq b^{(n)}\} \cap \{n: a^{(n)} = a_1^{(n)}\} \cap \{n: b^{(n)} = b_1^{(n)}\} \subset \{n: a_1^{(n)} \leq b_1^{(n)}\}.$$

By hypothesis and by (ii), the left-hand side is in $U$; and hence by (iii) so also is the right, which is what had to be proved. It now follows immediately, with the help of the properties (iii) and (iv) of $U$, that for any $a, b \in R$:

$$\{n: a^{(n)} \leq b^{(n)}\} \in U \quad \text{or} \quad \{n: b^{(n)} \leq a^{(n)}\} \in U,$$

that is, that at least one of the two relations $a \leq b \bmod M$ or $b \leq a \bmod M$ must hold. The other properties of an ordering, namely

$$a \leq a,$$
$$a \leq b, \quad b \leq a \Rightarrow a = b,$$
$$a \leq b, \quad b \leq c \Rightarrow a \leq c,$$
$$a \leq b \qquad\qquad \Rightarrow a + c \leq b + c,$$
$$0 \leq a, \quad 0 \leq b \Rightarrow 0 \leq a \cdot b,$$

all follow immediately by using the filter properties of $U_M$.

Until now we have made no use of the assumption $D \subset M$. This will be used for the first time in proving that *ℝ has an element which exceeds all real numbers. In fact the relation

$$r \leq \omega \bmod M,$$

holds for all $r \in \mathbb{R}$, if we set

$$\omega = (1, 2, 3, \ldots, n, n+1, \ldots).$$

This is clearly true since by (v) $\{n: r^{(n)} \leq \omega^{(n)}\} = \{n: r \leq n\} \in U_M$.

As already indicated in the introduction, we would now like to define the ring of the finite elements of *ℝ. To do this we use the extension *| | of the absolute value of the real numbers. Just as in ℝ, so also in *ℝ, we have

$$*|a| = \begin{cases} a, & \text{if } 0 \leq a, \\ -a, & \text{if } a \leq 0. \end{cases}$$

If here $a \in R$ is a representative, then strictly speaking we should interpret both the equation and the inequality in the modulo $M$ sense. Since however we have chosen a fixed $M$ once and for all, we shall in future omit the qualification "mod $M$," at least when this entails no risk of confusion. We shall also often omit the asterisk in denoting the extended form of the function, particularly when the function already has a definite name or notation associated with it, such as | |.

The property of the extended version of | | which has just been indicated above, follows by virtue of the general transference principle to be discussed in §3, from the corresponding property of the absolute value of real numbers. Nevertheless, it can also be deduced immediately from the following considerations. If for a sequence $a = (a^{(n)})_{n \in \mathbb{N}}$ we have $0 \leq a$, then this means that $\{n: 0 \leq a^{(n)}\} \in U$. This of course implies that the set

$\{n: |a^{(n)}| = a^{(n)}\}$ is also an element of $U$, and hence it follows by (vi) that $|a| = a \bmod M$. A similar argument holds good for $a \leq 0$.

If we now define

$$\mathcal{D} = \{a \in {}^*\mathbb{R}: |a| \leq r \text{ for an } r \in \mathbb{R}\},$$

we see at once that $\mathcal{D}$ is a proper convex subring of ${}^*\mathbb{R}$. By the convexity of $\mathcal{D}$ is meant the property that

$$0 \leq b \leq a \in \mathcal{D} \Leftrightarrow b \in \mathcal{D}.$$

In addition, we define

$$\mathcal{M} = \{a \in {}^*\mathbb{R}: |a| \leq \varepsilon \text{ for all } \varepsilon \in \mathbb{R}^+\}.$$

One can see at once that $\mathcal{M}$ is a convex ideal in $\mathcal{D}$, that is to say, that

$$a, b \in \mathcal{M} \Rightarrow a + b \in \mathcal{M},$$
$$a \in \mathcal{M}, b \in \mathcal{D} \Rightarrow a \cdot b \in \mathcal{M},$$
$$0 \leq b \leq a \in \mathcal{M} \Rightarrow b \in \mathcal{M}.$$

$\mathcal{M}$ contains nonzero elements because since $n \leq \omega$ it obviously follows that $0 < 1/\omega \leq 1/n$ for all $n \in \mathbb{N}'$, so that $1/\omega \in \mathcal{M}$. The elements of $\mathcal{M}$ are said to be *infinitely small* or *infinitesimal* quantities. The elements of $\mathcal{D}$ are called *finite* quantities. All other elements of ${}^*\mathbb{R}$ are said to be *infinite*. For $a, b \in {}^*\mathbb{R}$, we use the notation

$$a \approx b \Leftrightarrow a - b \in \mathcal{M},$$

to indicate that $a$ and $b$ differ from each other by an infinitesimal amount. We therefore say that $a$ and $b$ are neighboring quantities. Obviously $\approx$ is an equivalence relation on ${}^*\mathbb{R}$. We now prove the important

**Theorem.** *Every finite quantity $a \in {}^*\mathbb{R}$ is a neighbor of just one real number $r$, and $r$ then is called the standard part $\mathrm{st}(a)$ of $a$.*

**Proof.** To prove *existence* we consider the sets $X_a = \{r \in \mathbb{R}: r \leq a\}$ and $Y_a = \{s \in \mathbb{R}: a \leq s\}$. Obviously $X_a, Y_a$ define a cut in $\mathbb{R}$, and because of the completeness of $\mathbb{R}$ (see Chapter 2, §2.2) there is a $t \in \mathbb{R}$ such that $r \leq t \leq s$ for $r \in X_a$, $s \in Y_a$. We at once deduce that $|t - a| \leq \varepsilon$ for all $\varepsilon \in \mathbb{R}^+$. Hence $a \approx t \in \mathbb{R}$.

To prove *uniqueness*, let us assume that $t_1 \approx a \approx t_2$ for $t_1, t_2 \in \mathbb{R}$. Then $t_1 \approx t_2$, that is, $|t_1 - t_2| < \varepsilon$ for all $\varepsilon \in \mathbb{R}^+$; but this is possible only if $t_1 - t_2 = 0$.                                                                                $\square$

From this theorem we see, in particular, that the mapping $\mathrm{st}: \mathcal{D} \to \mathbb{R}$ is an order preserving ring homomorphism whose kernel is $\mathcal{M}$ and that st | $\mathbb{R}$ = id. Thus we know in particular that all the field operations can be validly performed with finite quantities, provided that we replace = by $\approx$ and divide by an $a \in \mathcal{D}$, only if $a \not\approx 0$.

It should also be noted that corresponding to every subfield $K$ of $\mathbb{R}$ a field *$K$ can be constructed in the same way as *$\mathbb{R}$ was constructed from $\mathbb{R}$. The above theorem can therefore be sharpened to "In *$K$ every finite quantity possesses a standard part if and only if $K = \mathbb{R}$." This theorem therefore expresses the completeness (with respect to cuts) of $\mathbb{R}$; the calculation with standard parts replaces the explicit application of the completeness of the real number continuum in classical analysis.

We now turn to the *continuity* of a function $f: \mathbb{R} \to \mathbb{R}$ at the point $x \in \mathbb{R}$. In the usual modern definition this is equivalent to asserting that to every $\varepsilon \in \mathbb{R}^+$ corresponds a $\delta \in \mathbb{R}^+$ such that for all $h \in \mathbb{R}$,

$$|h| \leq \delta \Rightarrow |f(x+h) - f(x)| \leq \varepsilon.$$

As we have seen previously, the function $f$ can be extended to a function *$f$: *$\mathbb{R} \to$ *$\mathbb{R}$ in such a way that for $a = (a^{(n)})_{n \in \mathbb{N}'}$,

$$*f(a) \equiv (f(a^{(n)}))_{n \in \mathbb{N}'} \bmod M.$$

We now consider an $h \in$ *$\mathbb{R}$, with $|h| \leq \delta$, so that $\{n : |h^{(n)}| \leq \delta\} \in U$. Since $\{n : |h^{(n)}| \leq \delta\} \subset \{n : |f(x + h^{(n)}) - f(x)| \leq \varepsilon\}$ this latter set is also in $U$, or in other words $|*f(x + h) - *f(x)| \leq \varepsilon$. If, in particular $h \in \mathcal{M}$, then the statement that $|h| \leq \delta$ for all $\delta \in \mathbb{R}^+$ is true. Consequently $|*f(x + h) - *f(x)| \leq \varepsilon$ for all $\varepsilon \in \mathbb{R}^+$ is also true, and hence

$$*f(x + h) \approx *f(x).$$

We have therefore proved the 'only if' part of the following

**Theorem.** *The function $f: \mathbb{R} \to \mathbb{R}$ is continuous at the point $x \in \mathbb{R}$ if, and only if* $*f(x + h) \approx f(x)$ *for all $h \approx 0$.*

**Proof of the "if" part of the theorem.** We have to prove that $f$ is continuous. Suppose $f$ were not continuous at $x$. Then there would be an $\varepsilon \in \mathbb{R}^+$ such that, for every $n \in \mathbb{N}'$ an $h^{(n)} \in \mathbb{R}$ exists satisfying the two inequalities $|h^{(n)}| \leq \frac{1}{n}$ and $|f(x + h^{(n)}) - f(x)| \geq \varepsilon$. If we now set $h = (h^{(n)})_{n \in \mathbb{N}}$, then obviously $|h| \leq 1/\omega$. Since $1/\omega \in \mathcal{M}$ it follows that $h \in \mathcal{M}$, or in other words $h \approx 0$. On the other hand, because of the way in which $h^{(n)}$ was chosen, $\{n : |f(x + h^{(n)}) - f(x)| \geq \varepsilon\} \in U$, that is $|*f(x + h) - f(x)| \geq \varepsilon$. This however contradicts the hypothesis that $*f(x + h) \approx f(x)$ for $h \approx 0$ and thus proves the "if" part of the theorem. $\square$

## §3. Features Common to ℝ and *ℝ

The proof of the last theorem and (even more clearly) the verification of the
characteristic property of the absolute value show that certain properties
of functions on ℝ are transmitted via their components to the extensions
of these functions to *ℝ. In this section we shall try to formulate a general
principle governing the transference of properties from ℝ to *ℝ. It is clear
from the outset that not every property can be carried over—after all ℝ and
*ℝ are different entities. Nevertheless we shall attempt to find the largest
possible domain of transferable properties. We shall construct this domain
inductively, starting from very simple properties and proceeding to more
and more complex properties by means of a specifically designed procedure.

In order to carry out such a program we first need to consider how we can
describe general properties. One possibility is the following: we introduce
an artificial language—a formal language—in which we can describe the
properties in which we are interested. The inductive procedure which we
have just mentioned will then operate by an induction on these "descriptions," for example, on the length of the description. The formal language
to which we have referred will naturally be very similar to the everyday
language of mathematics. This is anyhow advisable because it makes it
easier to read so that the correct interpretation is immediately suggested.

Let us consider an example. One of the properties which ℝ and *ℝ have in
common is the commutativity of addition. In the formal language which we
yet have to introduce, this property is described by the formal expression

$$\forall x \, \forall y \quad x + y = y + x.$$

This is a row of symbols, consisting of eleven individual symbols. From the
notational viewpoint, the symbols have been chosen so that a momentary
glance at the row of symbols is enough to make one think at once of the right
interpretation. There is really only one interpretation left open, namely that
of $\forall x$. This depends on whether we wish to interpret the above formula in
ℝ or in *ℝ. In the first case $\forall x$ should be interpreted as meaning "for all
$a \in ℝ$," but in the second "for all $a \in$ *ℝ." We shall therefore also have to
define a relation $\models$ between ℝ and formulae, or between *ℝ and formulae,
which expresses the validity of a formula in ℝ, or in *ℝ respectively.

We shall use for the *basic symbols* (the individual symbols) of the language now to be defined, the known symbols

$$\neg \quad \wedge \quad \exists \quad = \quad \leq \quad ) \quad , \quad ($$

as well as the symbols for variables

$$v_0 \quad v_1 \quad v_2 \quad \cdots .$$

We also use for each function $f: ℝ^m \to ℝ$ a symbol with which we can
denote $f$ or *$f$, respectively. For convenience we shall use $f$ itself as the

symbol for this purpose. It should be noted therefore that the interpretation of $f$ is simply $f$ when interpreted as a function on $\mathbb{R}$, but $*f$ when interpreted as a function on $*\mathbb{R}$. Lastly, we shall introduce, for every $a \in *\mathbb{R}$ a symbol $\underline{a}$ (the *name* of $a$). The interpretation of $\underline{a}$ is of course $a$.

From these basic symbols we can now build up entities called *terms*, defined as follows:

(1) variables and $\underline{a}$ for $a \in *\mathbb{R}$ are terms;

(2) if $t_1, \ldots, t_m$ are terms and $f$ is a function of $m$ arguments, then $f(t_1, \ldots, t_m)$ is a term.

If a term $t$ contains no variables it is said to be a *constant*. The interpretation of a constant (term) in $\mathbb{R}$ or in $*\mathbb{R}$ is obvious and is the same in both cases, provided the only constants $\underline{a}$ present are from $\mathbb{R}$.

Next we define *formulae*

(1) If $t_1$ and $t_2$ are terms, then $t_1 = t_2$ and $t_1 \leq t_2$ are *formulae* (the so-called primitive formulae).

(2) If $\varphi_1$ and $\varphi_2$ are formulae and $v$ is a variable, then $\neg\varphi_1$, $(\varphi_1 \wedge \varphi_2)$, $\exists v \varphi_1$ are formulae.

We shall call a formula with no free variables a *statement*. In this connection a variable $v$ which occurs as a free variable in a formula $\varphi$ is no longer a free variable in the formula $\exists v \varphi$. This can be put rather more precisely as follows: for any term $t$ let $F(t)$ be the (finite) set of all variables occurring in $t$. We now define recursively

$$F(t_1 = t_2) = F(t_1 \leq t_2) = F(t_1) \cup F(t_2),$$
$$F(\neg\varphi) = F(\varphi),$$
$$F(\varphi_1 \wedge \varphi_2) = F(\varphi_1) \cup F(\varphi_2),$$
$$F(\exists v \varphi) = F(\varphi) \setminus \{v\}.$$

and $\varphi$ is accordingly a statement in the case where $F(\varphi) = \emptyset$.

We now define what is meant by the *validity* in $\mathbb{R}$ or in $*\mathbb{R}$ as the case may be, of a statement $\alpha$. In other words we define the relation $\mathbb{R} \models \alpha$ (read as "the statement $\alpha$ holds in $\mathbb{R}$") or $*\mathbb{R} \models \alpha$ (the statement $\alpha$ holds in $*\mathbb{R}$). The first case makes sense only if there are no constants $\underline{a}$ with $a \in *\mathbb{R} \setminus \mathbb{R}$. In this case we call $\alpha$ an $\mathbb{R}$-statement. As far as primitive formulae are concerned, their validity in $\mathbb{R}$ or in $*\mathbb{R}$ is immediately apparent. Suppose now that the validity of $\alpha_1$ and $\alpha_2$ in $*\mathbb{R}$ has already been defined. We then put

$$*\mathbb{R} \models \neg\alpha_1 \quad \Leftrightarrow *\mathbb{R} \not\models \alpha_1,$$
$$*\mathbb{R} \models (\alpha_1 \wedge \alpha_2) \Leftrightarrow [*\mathbb{R} \models \alpha_1 \text{ and } *\mathbb{R} \models \alpha_2],$$
$$*\mathbb{R} \models \exists v \varphi \quad \Leftrightarrow \text{there is an } a \in *\mathbb{R} \text{ with } *\mathbb{R} \models \varphi(\underline{a}).$$

Here $\varphi(\underline{a})$ means the result of substituting $\underline{a}$ for $v$ in the formula $\varphi$. Another quantification $\exists v\psi$ which may possibly occur in $\varphi$ is of course not allowed to be replaced by $\exists \underline{a}\psi(\underline{a})$ because this is no longer a formula under the above construction. In the last case of the definition the hypothesis that $\exists v\varphi$ is a statement naturally ensures the same property for $\varphi(\underline{a})$. The inductive structure of the definition of the validity of $\mathbb{R}$-statements in $\mathbb{R}$ is analogous to that of the above definition. For example

$$\mathbb{R} \models \exists v\varphi \Leftrightarrow \text{ there is an } a \in \mathbb{R} \text{ with } \mathbb{R} \models \varphi(\underline{a}).$$

We are now finally in a position to formulate and to prove the general principle of transference.

**General Transfer Principle.** *Let $\alpha$ be a statement in which, at most the constants $\underline{a_1}, \ldots, \underline{a_m}$ appear, so that $\alpha = \alpha(\underline{a_1}, \ldots, \underline{a_m})$. Then*

$$^*\mathbb{R} \models \alpha(\underline{a_1}, \ldots, \underline{a_m}) \Leftrightarrow \{n : \mathbb{R} \models \alpha(\underline{a_1^{(n)}}, \ldots, \underline{a_m^{(n)}})\} \in U_M.$$

**Proof.** Suppose first that $\alpha$ is a primitive statement, and therefore $t_1 = t_2$ or $t_1 \leq t_2$. Then (vi) and the definition of $\leq$ in $^*\mathbb{R}$ provide us with just the equivalence asserted.

We now deduce the general result by induction on the structure of the statement $\alpha$. If $\alpha$ is of the form $\neg\alpha_1$ and we assume that the above equivalence holds for $\alpha_1$, then we can argue as follows:

$$^*\mathbb{R} \models \neg\alpha_1(\underline{a_1}, \ldots) \Leftrightarrow {}^*\mathbb{R} \not\models \alpha_1(\underline{a_1}, \ldots)$$
$$\Leftrightarrow \{n : \mathbb{R} \models \alpha_1(\underline{a_1^{(n)}}, \ldots)\} \notin U$$
$$\Leftrightarrow \{n : \mathbb{R} \not\models \alpha_1(\underline{a_1^{(n)}}, \ldots)\} \in U$$
$$\Leftrightarrow \{n : \mathbb{R} \models \neg\alpha_1(\underline{a_1^{(n)}}, \ldots)\} \in U.$$

This is clear since for any ultrafilter $U$ on $\mathbb{N}'$

$$A \notin U \Leftrightarrow \mathbb{N}' \setminus A \in U.$$

If $\alpha$ is of the form $(\alpha_1 \wedge \alpha_2)$, then we argue as follows:

$$^*\mathbb{R} \models (\alpha_1(\underline{a_1}, \ldots) \wedge \alpha_2(\underline{a_1}, \ldots)) \Leftrightarrow [^*\mathbb{R} \models \alpha_1(\underline{a_1}, \ldots) \text{ and } {}^*\mathbb{R} \models \alpha_2(\underline{a_1}, \ldots)]$$
$$\Leftrightarrow [\{n : \mathbb{R} \models \alpha_1(\underline{a_1^{(n)}}, \ldots)\} \in U \text{ and } \ldots]$$
$$\Leftrightarrow \{n : \mathbb{R} \models \alpha_1((\underline{a_1^{(n)}}, \ldots)$$
$$\wedge \alpha_2(\underline{a_1^{(n)}}, \ldots))\} \in U_M.$$

This again is right because

$$A \cap B \in U \Leftrightarrow A \in U \text{ and } B \in U.$$

Finally, if $\alpha$ is of the form $\exists v\varphi$, then obviously

$$^*\mathbb{R} \models \exists v\varphi(\underline{a}_1, \ldots) \Leftrightarrow \text{there is an } a \in {}^*\mathbb{R} \text{ with } {}^*\mathbb{R} \models \varphi(\underline{a}, \underline{a}_1, \ldots)$$
$$\Leftrightarrow \text{there is an } a \in {}^*\mathbb{R} \text{ with}$$
$$\{n : \mathbb{R} \models \varphi(\underline{a}^{(n)}, a_1^{(n)}, \ldots)\} \in U$$
$$\Rightarrow \{n : \mathbb{R} \models \exists v\varphi(\underline{a_1^{(n)}}, \ldots)\} \in U.$$

This follows from the inclusion

$$\{n : \mathbb{R} \models \varphi(\underline{a}^{(n)}, \ldots)\} \subset \{n : \mathbb{R} \models \exists v\varphi(\ldots)\}.$$

There remains the converse of the last implication to be proved. Suppose therefore that $\{n : \exists v\varphi(a_1^{(n)}, \ldots)\} \in U$. We define a sequence $a = (a^{(n)})_{n \in \mathbb{N}'}$ as follows. Let $a^{(n)}$ be an $r \in \mathbb{R}$, for which $\mathbb{R} \models \varphi(\underline{r}, a_1^{(n)}, \ldots)$ holds, provided that such an $r$ exists at all. Otherwise, we put $a^{(n)} = 0$. Obviously with this sequence $a$

$$\{n : \mathbb{R} \models v\varphi(\underline{a_1^{(n)}}, \ldots)\} \subset \{n : \mathbb{R} \models \varphi(\underline{a}^{(n)}, a_1^{(n)}, \ldots)\}.$$

If the first set is in $U_M$, then so is the second. This proves the required converse. $\qquad\square$

As a corollary we obtain the

**Transfer Principle.** *Let $\alpha$ be an $\mathbb{R}$-statement. Then $\alpha$ holds in $\mathbb{R}$ if and only if it holds in $^*\mathbb{R}$.*

**Proof.** Since for numbers $r \in \mathbb{R}$, all the component terms $r^{(n)}$ are equal to $r$, it is clear that for $a_i \in \mathbb{R}$

$$\{n : \mathbb{R} \models \alpha(a_1^{(n)}, \ldots, a_m^{(n)})\} = \mathbb{N}' \text{ or } \emptyset,$$

depending on whether $\alpha(a_1^{(n)}, \ldots, a_m^{(n)})$ is true or untrue in $\mathbb{R}$. Since $\mathbb{N}' \in U$ and $\emptyset \notin U$ it therefore follows that

$$\mathbb{R} \models \alpha \Leftrightarrow \{n \mid \mathbb{R} \models \alpha(a_1^{(n)}, \ldots)\} \in U$$
$$\Leftrightarrow {}^*\mathbb{R} \models \alpha. \qquad\square$$

From now on we shall simply apply this transfer principle. We can forget all about the way in which $^*\mathbb{R}$ was originally defined. This will actually turn out to be very useful on most occasions, since fiddling about with indices can be rather confusing. By simply using the transfer principle on its own we can to a certain extent treat the elements of $\mathbb{R}$ and $^*\mathbb{R}$ on an equal footing as "atoms" or primitive elements. If however we work with the particular construction of $^*\mathbb{R}$, then we have to deal on the one hand with real numbers, and on the other, with equivalence classes of sequences of real numbers.

The only difficulty which arises in applying the transfer principle, and it is one which should not be underestimated, is that involved in formalizing the properties to be transferred. It needs a certain amount of practice before one develops a feel for it. This is the price one has to pay for the convenience of working with infinitesimals.

In formalizing the properties which are to carry over one will naturally try to improve the intelligibility of the formal language used by employing abbreviations which make it easier to read. For example one uses

$$
\begin{array}{lll}
(\varphi \lor \psi) & \text{for} & \neg(\neg\varphi \land \neg\psi), \\
(\varphi \to \psi) & \text{for} & \neg(\varphi \land \neg\psi), \\
(\varphi \leftrightarrow \psi) & \text{for} & (\varphi \to \psi) \land (\psi \to \varphi), \\
\forall v \varphi & \text{for} & \neg\exists v \,\neg\varphi.
\end{array}
$$

We should like to conclude this section by giving another proof of the theorem of §2 on continuity, this time using the transfer principle alone. At the same time, with a view to a later application, we shall present a slight generalization. We shall prove the

**Limit Theorem.** *If $x_0, b \in \mathbb{R}$ and $g$ is a function from $\mathbb{R}$ to $\mathbb{R}$ then the statement $\lim_{0 \neq h \to 0} g(x_0 + h) = b$ is equivalent to $^*g(x_0 + h) \approx b$ for all $h \approx 0$ with $h \neq 0$.*

**Proof.** If the limit statement holds for $g$, then for every $\varepsilon \in \mathbb{R}^+$ there exists a $\delta \in \mathbb{R}^+$ such that

$$
\mathbb{R} \models \forall h (0 < |h| < \underline{\delta} \to |g(\underline{x_0} + h) - \underline{b}| < \underline{\varepsilon}).
$$

By the transfer principle, we deduce from this that

$$
^*\mathbb{R} \models \forall h (0 < |h| < \underline{\delta} \to |g(\underline{x_0} + h) - \underline{b}| < \underline{\varepsilon}).
$$

Hence, since $|h| < \delta$ we obtain $|^*g(x_0 + h) - b| < \varepsilon$ for all $h \approx 0$ with $h \neq 0$. As this applies for every $\varepsilon \in \mathbb{R}^+$, we have $^*g(x_0 + h) \approx b$. Conversely, let us assume that $^*g(x_0 + h) \approx b$ for $h \approx 0$ with $h \neq 0$. Suppose that $h_0 \approx 0$ where $h_0$ is fixed and $0 < h_0$. Then for $0 < |h| < h_0$ we have $^*g(x_0+h) \approx b$, and thus in particular, if we think of $\delta = h_0$ we have

$$
^*\mathbb{R} \models \exists \delta (0 < \delta \land \forall h (0 < |h| < \delta \to |g(\underline{x_0} + h) - \underline{b}| < \underline{\varepsilon}).
$$

Here $\varepsilon \in \mathbb{R}^+$ can be any positive number, and the transfer principle then gives us

$$
\mathbb{R} \models \exists \delta (0 < \delta \land \forall h (0 < |h| < \delta \to |g(\underline{x_0} + h) - \underline{b}| < \underline{\varepsilon}).
$$

This however is precisely what the limit statement asserts.                         □

This proof shows clearly the convenience of applying the transfer principle and its advantage over the use of sequences constructed specially for the purpose. This is not really surprising because the constructed sequences have been incorporated in the proof of the transfer principle.

## §4. DIFFERENTIAL AND INTEGRAL CALCULUS

**1. Differentiation.** We now introduce, for a given function $f: \mathbb{R} \to \mathbb{R}$ its *differential* $df(x)$ at the point $x \in \mathbb{R}$. To this end we fix an $h \approx 0$ with $h \neq 0$ and set

$$df(x) = {}^*f(x + h) - f(x).$$

In the case of the identity function $f(x) = x$, we obtain, in particular, $dx = x + h - x = h$. From now on therefore we shall always use $dx$ instead of $h \approx 0$ with $h \neq 0$. We thus obtain, for the differential of $f$ at the point $x$

$$df(x) = {}^*f(x + dx) - f(x).$$

Note however that this differential depends on the choice of the quantity $dx \in \mathcal{M} \setminus \{0\}$.

The *differential quotient* $df(x)/dx$ can be formed for *every* function $f: \mathbb{R} \to \mathbb{R}$; it is a definite element of ${}^*\mathbb{R}$. The connection with the derivative of the function $f$ at the point $x$ is described by the following

**Theorem.** *If, for a given function $f: \mathbb{R} \to \mathbb{R}$, the limit of the quotient of the differences at the point $x \in \mathbb{R}$ exists and has the value $f'(x)$, then $df(x)/dx \approx f'(x)$ for all $dx \in \mathcal{M} \setminus \{0\}$, and the converse is also true.*

**Proof.** If, in the limit theorem of §3, we put

$$g(h) = \frac{f(x + h) - f(x)}{h}$$

and $x_0 = 0$, then we obtain the result that the statement about the limit is equivalent to the assertion that

$$\frac{df(x)}{dx} = \frac{{}^*f(x + dx) - f(x)}{dx} = {}^*g(dx) \approx f'(x)$$

for all $dx \in \mathcal{M} \setminus \{0\}$. $\qquad\square$

The usual rules on the differentiation of functions can now easily be obtained:

(1) *If $f$ is differentiable at the point $x$, then $f$ is continuous there.*

Since $df(x)/dx \approx f'(x)$ it follows that ${}^*f(x + dx) - f(x) = df(x) \approx f'(x)dx \approx 0$. Hence ${}^*f(x + dx) \approx f(x)$ for all $dx$. This is the continuity of $f$ at $x$.

(2) *If f and g are differentiable at x, then so are $(f+g)$ and $(f \cdot g)$ and*

$$(f+g)'(x) = f'(x) + g'(x), \quad (f \cdot g)'(x) = (f' \cdot g)(x) + (f \cdot g')(x).$$

We shall carry out the proof of the rule for the case of multiplication:

$$\begin{aligned}
d(f \cdot g)(x) &= {}^*(f \cdot g)(x + dx) - (f \cdot g)(x) \\
&= {}^*f(x + dx) \cdot {}^*g(x + dx) - f(x)g(x) \\
&= (df(x) + f(x)) \cdot (dg(x) + g(x)) - f(x)g(x) \\
&= df(x)dg(x) + f(x)dg(x) + g(x)df(x).
\end{aligned}$$

Division by $dx$ then yields

$$\begin{aligned}
\frac{d(f \cdot g)(x)}{dx} &= f(x) \cdot \frac{dg(x)}{dx} + g(x) \cdot \frac{df(x)}{dx} + df(x) \cdot \frac{dg(x)}{dx} \\
&\approx f(x)g'(x) + g(x)f'(x).
\end{aligned}$$

This last line is a consequence of the hypothesis of the differentiability of $f$ and $g$ at the point $x$ which, by (1), implies in particular that $df(x) \approx 0$. We have thus shown that

$$(f \cdot g)'(x) \approx f(x) \cdot g'(x) + g(x) \cdot f'(x).$$

As however both sides are elements of $\mathbb{R}$, they must be equal.

(3) *If f is differentiable at the point x and $f(x) \neq 0$, then $1/f$ is differentiable at x and $(1/f)'(x) = -f'(x)/f(x)^2$.*

Since $f(x) \approx {}^*f(x+dx)$ then naturally ${}^*f(x+dx) \neq 0$ as well. We therefore have

$$d\frac{1}{f}(x) = \frac{1}{{}^*f(x + dx)} - \frac{1}{f(x)} = \frac{f(x) - {}^*f(x + dx)}{f(x) \cdot {}^*f(x + dx)}.$$

It follows from this that

$$\frac{d\frac{1}{f}(x)}{dx} = \frac{-\frac{df(x)}{dx}}{f(x)^* f(x + dx)} \approx \frac{-f'(x)}{f(x)^* f(x + dx)} \approx \frac{-f'(x)}{f(x)^2}$$

which, as in (2) proves the assertion.

(4) *If f is differentiable at the point x and g is differentiable at the point $f(x)$ then $g \circ f$ is also differentiable at the point x and $(g \circ f)'(x) = g'(f(x)) \cdot f'(x)$.*

In the case where $df(x) \neq 0$ we obtain, with

$$\begin{aligned}
d(g \circ f)(x) &= {}^*g({}^*f(x + dx)) - g(f(x)) \\
&= {}^*g(f(x) + df(x)) - g(f(x)) \\
&= dg(f(x))
\end{aligned}$$

a differential of $g$ at the point $f(x)$, formed with $h = df(x)$. Division by $dx$ then yields

$$\frac{d(g \circ f)(x)}{dx} = \frac{dg(f(x))}{df(x)} \cdot \frac{df(x)}{dx}.$$

By taking the standard parts we obtain

$$\text{st}\left(\frac{d(g \circ f)(x)}{dx}\right) = g'(f(x)) \cdot f'(x).$$

This equation also holds when $df(x) = 0$. Indeed, it follows from this that, on the one hand, $f'(x) = \text{st}(df(x)/dx) = 0$ and on the other hand that $d(g \circ f)(x) = {}^*g(f(x)) - g(f(x)) = 0$. Therefore, as the above equation holds for all $dx \in \mathcal{M} \setminus \{0\}$, $(g \circ f)$ is also differentiable in $x$ and has the derivative stated.                                                                □

It will certainly not have escaped the attention of the observant reader that, in the course of the proof which has just been given, we made implicit use of the transfer principle in a few places, for example, in asserting that

$$^*(f \cdot g)(y) = {}^* f(y) \cdot {}^* g(y).$$

Indeed, since the statement

$$\forall v (f \cdot g)(v) = f(v) \cdot g(v)$$

is true in $\mathbb{R}$, it must also be true in $^*\mathbb{R}$, that is to say, it holds for the extensions of the three functions $f$, $g$ and $(f \cdot g)$.

**2. Integration.** In this last section we should like to sketch how the integral of a function $f$, continuous in the closed interval $[a, b]$ can be described as a sum of the areas of rectangles of infinitely small width. Understandably enough the "sum" involved cannot be a finite sum, that is, the summation cannot run from 0 to $n$, where $n \in \mathbb{N}$. This is clear because a finite sum of elements of $\mathcal{M}$ must itself be in $\mathcal{M}$. We shall instead take as the upper limit of summation an infinitely large "natural number." We still however have to define what is meant by the word "sum" in this case.

First, as regards the infinitely large "natural" numbers. The characteristic function $\chi$ of $\mathbb{N}$, defined for $x \in \mathbb{R}$ by

$$\chi(x) = \begin{cases} 1, & \text{if } x \in \mathbb{N} \\ 0, & \text{otherwise} \end{cases}$$

has, like every other function, an extension $^*\chi$ defined on $^*\mathbb{R}$. This extension obviously retains (thanks to the transfer principle) the property of being a 0, 1-function. We now define

$$^*\mathbb{N} = \{a \in {}^*\mathbb{R} : {}^*\chi(a) = 1\}.$$

Since $\{n: \chi(\omega^{(n)}) = 1\} = \mathbb{N}' \in U$ it follows at once from the transfer principle that $\omega \in {}^*\mathbb{N}$. There are therefore infinitely large natural numbers in ${}^*\mathbb{R}$. It is easy to satisfy one's self that $\mathbb{N}$ is a subset of ${}^*\mathbb{N}$ and that the new elements of ${}^*\mathbb{N}$ are larger than any of the elements of $\mathbb{N}$.

Before coming to integration, we shall use infinitely small subintervals to prove the following well-known lemma.

**Lemma.** *If a function $f$ is continuous in the closed interval $[a, b]$, it assumes its maximum (and its minimum) value within that interval.*

**Proof.** If $n \in \mathbb{N}$ and we set $a_i = a + ((b - a)/n)(i - 1)$ for $1 \le i \le n + 1$, then naturally, among the finitely many values of $f(a_i)$ with $1 \le i \le n + 1$ there is a maximal one (that is, one which is not smaller than any of the others). That this is true for all $n \in \mathbb{N}$, can be expressed in the form of a valid statement in $\mathbb{R}$ with the parameters $\underline{a}$ and $\underline{b}$ (an exercise which will provide good practice for the reader), and it therefore holds good for all elements of ${}^*\mathbb{N}$, for example, for $\omega$. Now suppose that ${}^*f(a_j)$ is maximal among the values ${}^*f(a_i)$ for $1 \le i \le \omega + 1$. Suppose also that $x = \mathrm{st}(a_j) \in [a, b]$. To every real number $y \in [a, b]$ corresponds an $i \le \omega$ with $a_i \le y \le a_{i+1}$. This again follows from the transfer principle, because the statement is true for every $n \in \mathbb{N}$ in place of $\omega$. Since $a_{i+1} - a_i = (b - a)/\omega \approx 0$ we obtain in particular the result that $y = \mathrm{st}(a_i)$. Owing to the continuity of $f$ it follows therefore that $f(y) \approx {}^*f(a_i) \le {}^*f(a_j) \approx f(x)$. This at once implies $f(y) \le f(x)$. $\qquad\square$

Now back to the integral. Let $f$ be a function from $\mathbb{R}$ to $\mathbb{R}$ and suppose $a, b \in \mathbb{R}$ with $a < b$. To every $h \in \mathbb{R}^+$ with $h \le b - a$ corresponds an $n \in \mathbb{N}$ such that $nh \le b - a < (n + 1)h$. The function

$$S_f(a, b, h) = \sum_{i=1}^{n} f(a_i) \cdot h + f(a_{n+1})(b - a_{n+1}),$$

where we have here set $a_i = a + (i - 1)h$, is the sum of the areas of rectangles of width $h$ or $(b - a_{n+1})$ and height $f(a_i)$. For infinitely small $h$, such a sum should adequately describe the area under the curve $y = f(x)$ between the abscissa $a$ and $b$, and this is in fact correct.

Like every other real function, $S_f$ can also be extended to ${}^*\mathbb{R}$ (and indeed in such a way as to retain its properties expressible by statements of the type considered in §3). If for every positive $h \in \mathcal{M}$ the value ${}^*S_f(a, b, h)$ is finite and has the same standard part $c$, then we call $c$ the *integral* of the function from $a$ to $b$. It can be established without much trouble that $c$ is equal to the Riemann integral of $f$ over the closed interval $[a, b]$, whenever this integral, as normally defined, exists.

We shall now prove (at least in part) the following theorem.

**Theorem.** *If the function $f$ is continuous in the real interval $[a, b]$, then its integral from $a$ to $b$ exists.*

**Proof.** We first show that $^*S_f(a, b, h)$ is finite for $h \in \mathcal{M} \setminus \{0\}$. As $f$ is continuous on $[a, b]$ in $\mathbb{R}$, $|f|$ is bounded in view of the lemma proved earlier. For $h \in \mathbb{R}^+$ with $h \leq b - a$ we therefore have

$$|S_f(a, b, h)| \leq \sum_{i=1}^{n} |f(a_i)| \cdot |h| + |f(a_{n+1})|(b - a_{n+1}) \leq (b - a)c,$$

where $c \in \mathbb{R}^+$ is an upper bound for $|f|$ on $[a, b]$. We thus have

$$\mathbb{R} \models \forall h (0 < h \leq (\underline{b} - \underline{a}) \rightarrow |S_f(\underline{a}, \underline{b}, h)| \leq (\underline{b} - \underline{a})\underline{c}).$$

The same statement holds in $^*\mathbb{R}$. Consequently $^*S_f(a, b, h)$ is finite for every positive $h \in \mathcal{M}$.

The proof that it does not depend on $h$ is somewhat harder, and we leave it to the interested reader. □

Again suppose $f$ to be continuous in $[a, b]$. Then $f$ is also continuous in $[a, x]$ for every $x \in [a, b]$. We write

$$I(a, x) = \text{st}(^*S_f(a, x, h)),$$

where $h$ is an arbitrary positive element of $\mathcal{M}$. We now wish to demonstrate the additivity of the integral, that is, we wish to show that for $\varepsilon \in \mathbb{R}^+$ and $x + \varepsilon \in [a, b]$

$$I(a, x) + I(x, x + \varepsilon) = I(a, x + \varepsilon).$$

This follows at once from the relation

$$^*S_f(a, x, h) + {}^*S_f(x, x + \varepsilon, h) = {}^*S_f(a, x + \varepsilon, h),$$

where, because of the independence of the standard part, a suitable positive $h \in \mathcal{M}$ can be chosen. In fact this relation holds for $h = (x - a)/\omega$, because the relation

$$S_f(a, x, h) + S_f(x, x + \varepsilon, h) = S_f(a, x + \varepsilon, h)$$

holds in $\mathbb{R}$ for $h = (x - a)/n$ and every sufficiently large $n \in \mathbb{N}$.

**Main Theorem.** *If $f$ is a function continuous in the interval $[a, b]$ then $F(x) = I(a, x)$ is an antiderivative of $f$, that is, $F'(x) = f(x)$ for $x \in (a, b)$.*

**Proof.** We have to show that

$$\frac{^*F(x + dx) - F(x)}{dx} \approx f(x)$$

for $dx \in \mathcal{M} \setminus \{0\}$. Because of the additivity of $I$, which naturally carries over to $^*\mathbb{R}$, this means

$$\frac{^*I(x, x + dx)}{dx} \approx f(x),$$

where, for $dx < 0$, we naturally interpret $^*I(x, x + dx)$ as $-^*I(x + dx, x)$. Now this can be shown as follows: by the transfer principle $^*f$ has a maximum $c_1$ and a minimum $c_2$ in $^*\mathbb{R}$ within the interval $[x, x + dx]$ and for positive $dx$

$$c_2 dx \leq {}^*I(x, x + dx) \leq c_1 dx.$$

Hence

$$f(x_2) = c_2 \leq \frac{{}^*I(x, x + dx)}{dx} \leq c_1 = f(x_1),$$

where $x_1, x_2$ are appropriate elements of the interval $[x, x + dx]$ in $^*\mathbb{R}$. Owing to the continuity of $f$ however, $f(x_1) \approx f(x) \approx f(x_2)$. This proves the theorem.                                                                                       □

## EPILOGUE

We should now like to go briefly into three particular points in connection with our presentation of the introduction of the nonstandard domain $^*\mathbb{R}$: these are, the uniqueness of $^*\mathbb{R}$, extensions of context, and lastly other approaches.

**Uniqueness of $^*\mathbb{R}$.** If one assumes the truth of the continuum hypothesis $2^{\aleph_0} = \aleph_1$ then it follows from some general theorems of model theory (see, for example, [4], Chapter 5, Corollary 23.6) that the ordered field $^*\mathbb{R}$ is uniquely determined to within an isomorphism. This means that $^*\mathbb{R}$ does not depend on the particular choice of the maximal ideal $M$ in $R$, as long as it has $D$ as a subset. If this condition is satisfied the resulting ultrafilter is nontrivial (called "free" in [4]). If on the other hand a maximal ideal $M$ is chosen such that $D \not\subset M$, then $U_M$ becomes a so-called principal ultra-filter which implies that $R/M \simeq \mathbb{R}$.

The independence (to within an isomorphism) from the choice of the maximal ideal $M \supset D$ no longer holds for the extensions of all real functions on $^*\mathbb{R}$.

**Extension of the Context.** The principle proved in §3 allows certain properties to be carried over from $\mathbb{R}$ to $^*\mathbb{R}$. The context (or universe of discourse) within which this transference can take place is determined by the formal language which was discussed there. This context can to a certain extent, be chosen arbitrarily. In this particular case we have chosen it to be as simple as possible. It can be widened considerably. If this is done a difficulty which has so far remained unnoticed can, and as a rule will, arise. We would like to explain this briefly.

To every subset $A$ of $\mathbb{R}$ corresponds (as with $^*\mathbb{N}$) an extension $^*A$ in $^*\mathbb{R}$. Not every subset of $^*\mathbb{R}$ however is of this form. Furthermore if one extends the formal language in such a way that quantification over all subsets of $\mathbb{R}$ becomes possible, then the quantification interpreted in $^*\mathbb{R}$ no longer runs over *all* subsets, but only over the so-called "internal" subsets of $^*\mathbb{R}$. Thus, for example, $^*\mathbb{N}$ is such an internal subset, but not $\mathbb{N}$. One can easily see this, if one formalizes the following statement in the extended context

> "every subset that contains 0, and $x + 1$ whenever it contains $x$, 'exceeds' every element."

In $\mathbb{R}$ this statement is true in an obvious sense. Interpreted in $^*\mathbb{R}$ it cannot validly refer to all subsets because $\mathbb{N}$ certainly satisfies the hypotheses, but does not "exceed" every element of $^*\mathbb{R}$.

**Other Approaches.** In the approach to nonstandard analysis adopted here, we have constructed the domain $^*\mathbb{R}$ from the already available domain $\mathbb{R}$. This approach corresponds to the construction of the real numbers from the rational numbers by means of sequences. Another possibility—analogous to the axiomatic introduction of real numbers (which then contain the rational numbers as subsets)—is to introduce $^*\mathbb{R}$ axiomatically as well, so that $\mathbb{R}$ is then a special or distinguished subset. This approach is found for example in the book *Elementary Calculus* by H.J. Keisler [2]. While in Keisler the axiomatics are specially tailored to $^*\mathbb{R}$, those chosen by E. Nelson in [3] are far more general and are based on set theory.

## REFERENCES

[1] EDWARDS, C.H., JR.: The historical development of the calculus. Springer-Verlag, New York-Heidelberg-Berlin 1979

[2] KEISLER, H.J.: Elementary calculus. Prindle, Weber & Schmidt, Inc., Boston 1976

[3] NELSON, E.: Internal set theory: a new approach to non-standard analysis. Bull. of the Amer. Math. Soc. **83**, 1165–1198 (1977)

[4] POTTHOFF, K.: Einführung in die Modelltheorie und ihre Anwendungen. Wiss. Buchges., Darmstadt 1981

[5] ROBINSON, A.: Non-standard analysis. North-Holland Publ. Comp., Amsterdam, London 1966

[6] SKOLEM, TH.: Über die Nichcharakterisierbarkeit der Zahlreihe mittels endlich oder abzählbar unendlich vieler Aussagen mit ausschließlich Zahlvariablen. Fund. Math. **23**, 150–161 (1934)

# 13

# Numbers and Games

*H. Hermes*

## §1. INTRODUCTION

This penultimate chapter will be devoted to presenting a new method by which the real numbers can be introduced. This method was published in the seventies by the English mathematician JOHN CONWAY. In contrast to the previous chapters we shall not be giving a systematic exposition of the subject matter. Our aim instead in the passages which follow will primarily be to explain the ideas on which the Conway Theory is based. The technical details of its implementation will be found in CONWAY's book [1], in [2] and—in a popularized version—in [5].

**1. The Traditional Construction of the Real Numbers.** We shall confine ourselves here to pointing out a few of the characteristic features (a detailed exposition is given in Chapters 1 and 2). The basis is set theory. The real numbers are constructed by a step-by-step procedure. There are several variants but they are not fundamentally different. One of these variants leads to the goal in three steps.

The natural numbers are introduced in the first step. By means of the von Neumann construction the number 0 is identified with the empty set $\emptyset$, and the number $n + 1$ with the set $n \cup \{n\}$ (see 14.1.3 and 14.2.1).

In the second step (which in Chapter 1 is subdivided into two) the rational numbers are regarded as classes of number triples (ordered triples of natural numbers); for example, to the class $-\frac{2}{3}$ belongs the triple $(13, 17, 6)$, since $-\frac{2}{3} = (13 - 17)/6$. The rational numbers constitute an ordered field.

The third step leads from the rational numbers to the real numbers. A real number is a *Dedekind cut* [3] (see Chapter 2.2), that is, an (ordered) pair $(x_1, x_2)$, where $x_1$ and $x_2$ are sets of rational numbers. It is usual (in German) to call $x_1$ the upper and $x_2$ the lower class, but we shall here follow Conway and speak of the *left class* $x_1$ and the *right class* $x_2$ of the cut $(x_1, x_2)$.

DEDEKIND lays down four requirements for a cut $(x_1, x_2)$:

(D1) *Every rational number lies in precisely one of the two classes $x_1, x_2$.*

(D2) *$x_1$ and $x_2$ are (each) non-empty.*

(D3) *Every element of $x_1$ is smaller than every element of $x_2$.*

(D4) $x_1$ *has no largest element.*

In this three-stage construction of the real numbers the arithmetic operations have to be defined three times, and at each stage the system of numbers so far defined has to be isomorphically embedded in the system defined by the next stage.

**2. The CONWAY Method.** Here again the starting point is set theory. The real numbers are obtained in a single step. To achieve this CONWAY makes use of two ideas. The first involves a suitable generalization of DEDEKIND cuts. However, it is not at first sight at all clear how to define an order relation for these generalized cuts. This is where CONWAY's second idea comes in. He saw that the generalized cuts could be regarded as defining a two-person game, and that the theory of such games provided a key to a definition of order.

One of the great advantages of CONWAY's method is that it avoids a step-by-step construction of the real numbers and hence the tedious repetitions associated with this approach. It could also be considered to be a further advantage that it links the number concept with the game concept. Games belong to the oldest experiences of mankind as well as to the earliest experiences of every individual (see [4]). Any link of this kind is of value to a science like mathematics with its tendency to ever greater abstraction.

Of course we do not in any way assert, or even suggest, that CONWAY's method will supersede the traditional construction of the real numbers. Indeed one cannot hide the fact that this method, besides having the advantages which have just been indicated, also has its adverse side. Among the disadvantages is the often tedious verification of the validity of the arithmetic rules of calculation. Furthermore, the primary product of the CONWAY process is not just the real set of numbers, but an ordered field which includes the real numbers as a proper subfield. The "nonstandard numbers" which are the elements of this larger field are either infinitely large or infinitely small or infinitely close to a real number (see Chapter 12). If one wishes to arrive at the real numbers one has to separate them out from the other elements of the CONWAY field.

**3. Synopsis.** In §2, we shall discuss the DEDEKIND postulates (D1–D4) with CONWAY's proposed generalization in view and we shall introduce the concept of a CONWAY game. CONWAY games may be regarded as a particular type of game, and we shall define the relevant concept here in §3. A few fundamental theorems about such games will be proved in §4, based on the idea of a winning strategy. In §5, it will be shown that the games concerned constitute (modulo an equivalence relation =) a partially ordered group. We end our account of the game-theory part in §6 by proving the result that CONWAY games can be regarded as "standard forms" of games (of the type being considered here).

CONWAY's two basic postulates (C1), (C2) will be formulated in §7. They are motivated by the considerations in §2 and make use of the partial ordering relation introduced in connection with games. Finally, §8 contains the definition of the arithmetic operations for the ordered field of CONWAY numbers and ends with a brief summary of CONWAY's results.

A *warning* here may not be out of place. Although the basic ideas of CON-WAY's theory are very simple and illuminating, their precise implementation—which will mostly be waived here—proves to be quite troublesome (see say 8.2) or at least non-trivial (for example, the proof of the closure property of $K_0$ mentioned in 8.3).

## §2. CONWAY GAMES

> Poesis doctrinae tamquam somnium (Francis BACON)
>
> [The poetry of learning, a kind of dream ...]

CONWAY's first idea, as already remarked in 1.2, consists in generalizing DEDEKIND cuts. We propose to examine DEDEKIND's postulates (D1)–(D4) in greater detail in order to extract from them what is important for this generalization. We shall then eventually come to the definition of CONWAY games. This definition may be regarded as preparing the stage for the CONWAY postulates (C1), (C2) corresponding to (D1)–(D4) (see §7).

**1. Discussion of the DEDEKIND Postulates.** The purpose of (D4) is to prevent a real number $r$ from being represented by the two different pairs of sets

(the set of rationals $\leq r$, the set of rationals $> r$)

(the set of rationals $< r$, the set of rationals $\geq r$).

If one is prepared to allow a real number to be represented by different pairs of sets, then (D4) becomes superfluous.

(D2) forbids, for example, the pair of sets

(the set of all rational numbers, the empty set).

A "number" given by this set would, intuitively speaking, be a positive infinite number. It would even be the only such number which is in conflict with the axioms of an ordered field. Such a conflict could possibly be avoided if, by generalizing DEDEKIND's construction, one could produce an infinity of infinitely large positive numbers. Not so very long ago such numbers were banned in mathematics, after a period of critical examination of the foundations had created a "horror infiniti." Nowadays however we have ceased to be frightened by "infinite" objects (see also Chapter 12).

(D1) implies that the right class, say of a DEDEKIND cut, is uniquely defined by the left class. Logically, therefore, it would be simpler always to operate with the left class alone and to abandon DEDEKIND's "poetic" concept of a real number as a *pair* of sets. CONWAY, however, takes DEDEKIND's "poem" very much to heart: for him it is the left and right classes of the set-pair which generate the number, not the reverse. CONWAY therefore has to throw overboard (D1). For example, it becomes possible to define the real number 0 by the set-pair (the set of rational numbers $-1/2^n$, the set of rational numbers $1/2^n$) among an infinity of others. Moreover the set-pair $(\{0\}, \{1\})$ also defines, for him, a "number" (see 8.2).

(D3) remains as the last postulate. This requirement ensures, for DEDEKIND, that the real numbers form a fully ordered domain. CONWAY also makes a corresponding demand. If we modify the formulation of (D3) by using $\leq$ instead of $<$ we obtain the following version to which we shall come back later:

(D3′)  no element of the right-hand class is less than or equal to an element of the left-hand class.

## 2. CONWAY's Modification of the DEDEKIND Postulates. CON-

WAY, like DEDEKIND regards numbers as pairs of sets $(x, y)$. However, while DEDEKIND allows as elements of the sets $x$ and $y$ only rational numbers— which have previously been constructed—CONWAY, in forming a number $(x, y)$, allows the elements of $x$ and $y$ to be any numbers whatsoever which are capable of having already been constructed "earlier" by this method.

The formation of pairs is, however, (as with DEDEKIND) limited by the restriction (D3′). Here a problem arises to which we have already alluded in 1.2: with DEDEKIND (D3), or (D3′), is meaningful because an order is already defined for the rational numbers. For CONWAY's intended generalization to make sense it has to be assumed that an order relation $\leq$ between the elements of $x$ and $y$ has already been defined.

This suggests that until one can visualize what such a definition might be, one should initially abandon the limitation on pair formation imposed by (D3′) and investigate what sets can be formed when one ignores this restriction. One would then naturally expect that other objects besides numbers could be produced. We shall see in §3 that all the objects constructible in this way can be thought of informally as *games*. In anticipation of the definitions and explanations given later we shall therefore call the objects constructible by the CONWAY pair formation process *Conway games*.

The theory of games provides us, in a suggestive manner, with a partial order relation $\leq$ between games (§6). This partial order is then finally used in §7 to formulate the restriction (D3′).

## 3. CONWAY Games. In accordance with the explanation in §2 we shall introduce the concept of a CONWAY game by the postulate

(CG) *If $x$ and $y$ are Conway games, then the (ordered) pair $(x, y)$ is a Conway game.*

(CG) defines a CONWAY game inductively. With the help of the standard techniques of set theory one could turn (CG) into an explicit definition, but it is more convenient to work with the inductive definition.

Inductive definitions are well-known, for example, in elementary number theory where, for example, addition can be defined inductively by the two requirements that $x + 0 = x$ and $x + S(y) = S(x + y)$ (see 1.2.3).

A few *examples* should illustrate how (CG) can be applied.

Considering that to construct a CONWAY game $(x, y)$, the elements of $x$ and $y$ need already to have been constructed as CONWAY games, one might think that it would be quite impossible to construct any CONWAY games at all with (CG). However, this would be a wrong conclusion because if $x$ and $y$ are both empty, then $x$ and $y$ are sets of CONWAY games (that is, every member of $x$ and every member of $y$ is a CONWAY game). Accordingly, by (CG) the pair $(\emptyset, \emptyset)$ is a CONWAY game. It will be shown later that this game can be identified with the number 0:

$$(1) \qquad\qquad 0 = (\emptyset, \emptyset).$$

Since 0 is a CONWAY game, $\{0\}$ is a set of CONWAY games. Accordingly one can now obtain, with the help of (CG), the CONWAY games $(\{0\}, \emptyset)$, $(\emptyset, \{0\})$ and $(\{0\}, \{0\})$. In particular it can be seen that the following sets are CONWAY games:

$$(2) \qquad \begin{array}{cc} 1 = (\{0\}, \emptyset), & 2 = (\{0, 1\}, \emptyset), \\ \cdots\cdots\cdots\cdots & \cdots\cdots \quad \cdots\cdots\cdots\cdots \\ n + 1 = (\{0, \ldots, n\}, \emptyset), & \omega = (\{0, 1, 2, \ldots\}, \emptyset) \end{array}$$

and one recognizes that the process by which VON NEUMANN constructed the ordinal numbers (see 14.1.3) also yields CONWAY games, so that

(3) *All ordinal numbers are Conway games.*

In order to show that a set $z$ is a CONWAY game, there is only the one postulate (CG) available. Thus $z$ has to be a pair $(x, y)$ in which $x$ and $y$ are each sets of CONWAY games. We shall call the elements of $x$ the **left elements of $z$**, and the elements of $y$ the **right elements of $z$**. We therefore have

(4) *Every Conway game is a pair of sets. The left and right elements of a Conway game are themselves Conway games.*

## §3. Games

We shall be concerned with a special class of games between two persons. To this class belong many well-known games and (which is of particular interest here) all CONWAY games. Later (§6) we shall even be able to show that to every game of the class considered here we can assign an "equivalent" CONWAY game.

When we talk of "games" from now on, we generally mean games of the particular class considered here.

**1. The Concept of a Game.** We consider *games* played between two persons, $L$ the *left player* and $R$ the *right player*. Before each *play* it is agreed which player is to begin, and thereafter the players play alternately (each player making a move). A *move* leads from one *position* to another. There is a *set $S$ of positions*, one of which is distinguished as *the initial position* $s_0$. Two binary (two-place) *game relations* $\to_L$ and $\to_R$ exist between the positions. If the play has reached a position $s$, in which it is, for example, $L$'s turn to *move*, then a move by $L$ consists in changing the position from $s$ to a position $s'$ where $s \to_L s'$. If there is no such $s'$ (that is, $L$ has no legal move), then $L$ has (by the agreed rules of the game) *lost* the play (and $R$ has *won*). The same goes for $R$.

We define

(1)        $s \to s'$ *if and only if* $s \to_L s'$ *or* $s \to_R s'$,

and lay down as a requirement of a game the

**Finiteness Condition.** *There exists no infinite sequence of positions* $s_0$, $s_1, s_2, \ldots$ *such that* $s_0 \to s_1 \to s_2 \to \cdots$.

It follows that every game must end after a finite number of moves and that one of the two players must win. Henceforth there can be no draws. There can also be no plays in which there is a return to the initial position.

A game is defined by the set $S$ of positions, the initial position $s_0$, and the two relations $\to_L$ and $\to_R$ (the permissible moves for $L$ and $R$), so that it can be identified with the tuple $(S, s_0, \to_L, \to_R)$.

**2. Examples of Games.** One can easily convince one's self that the following examples fall within the scope of the games defined in the first paragraph.

(a) NIM, for example in the following version. The initial position $s_0$ is any prescribed $m$-tuple $(N_1, \ldots, N_m)$ of natural numbers. The positions are the $m$-tuples $(n_1, \ldots, n_m)$ with $n_i \leq N_i$ $(i = 1, \ldots, m)$. Between any two positions $(n_1, \ldots, n_m)$ and $(n'_1, \ldots, n'_m)$ the relations $\to_L$ and $\to_R$, hold, provided that $n_i = n'_i$ for all $i$ except for one index $i_0$, where $n'_{i_0} < n_{i_0}$. (Thus the player who has the move is obliged to remove something from one of the remaining "heaps.")

(b) The following DOMINO type game. The initial position $s_0$ is a finite set of squares on a checkered plane. The positions are the subsets of $s_0$. The relations $s \to_L s'$ resp. $s \to_R s'$, hold if $s'$ can be derived from $s$ by removing two vertically, resp. horizontally adjacent squares. (This game can be played in practice by covering the squares with dominoes.)

(c) *Conway games.* Every CONWAY game $x$ can be regarded as a game. The initial position $s_0$ is identified with $x$. The positions are, apart from the initial position, the left and right elements of $x$, then their left and right elements, and so on. All positions are therefore, by (2.4), themselves CONWAY games. The relations $s \to_L s'$ resp. $s \to_R s'$ hold if and only if $s'$ is a left resp. right element of $s$. The finiteness condition is satisfied, because if there were an infinite sequence of positions such that $s_0 \to s_1 \to s_2 \to \ldots$ there would be an infinite sequence of sets $s_0, s_1, s_2, \ldots$ such that each set would be a left or right element of the preceding set. By the axiom of foundation of the theory of sets, such a sequence cannot exist (see 14.2.2).

**3. An Induction Principle for Games.** Let $x$ be a given game $(S, s_0, \to_L, \to_R)$. To every $s_0'$ with $s_0 \to s_0'$ we can assign a game $x' = (S', s_0', \to_L', \to_R')$ as follows: let $s \in S'$ if and only if there is a chain $s_0' \to \ldots \to s$ (including the case where $s = s_0'$). For $s, s' \in S'$, let $s \to_L' s'$ if and only if $s \to_L s'$. Let $s \to_R' s'$ be defined analogously.

Every game $x'$ defined in this way will be called a *predecessor (game)* of $x$. More specifically, we may talk of a game as being a *left predecessor* or *right predecessor*, according as to whether $s_0 \to_L s_0'$ or $s_0 \to_R s_0'$.

The induction principle for games is concerned with a property $P$ defined for games. We write $Px$ to mean that the game $x$ has the property $P$.

**Induction Principle for Games.** *If the statement $Px$ follows as a necessary consequence of the induction hypothesis that $P$ applies to every predecessor game $x'$ or $x$, then every game $x$ has the property $P$.*

**Proof.** Suppose that the induction statement $Px$ follows from the induction hypothesis, but that there is nevertheless a game $x_0$ which does not have the property $P$. Then there would be a predecessor game $x_0'$ which did not have the property, and $x_0'$ would have a predecessor $x_0''$ without the property $P$, and so on. The initial positions $s_0, s_0', s_0'', \ldots$ of the games $x_0, x_0', x_0'', \ldots$ would therefore satisfy the relations

$$s_0 \to s_0' \to s_0'' \to \cdots$$

in contradiction to the finiteness condition for games, and this proves the falsity of the original assumption. □

## §4. On the Theory of Games

It signifies nothing to play well if you lose. (Proverb)

We shall show that in every game (as defined above), either the player $L$ or the player $R$ or the player who begins (the "first" player) or the one who does not (the "second" player) can force a win. The idea of a winning strategy plays a decisive role here. In particular this idea can be used to define "positive" and "negative" games. We can do this in such a way that the natural numbers, which we have already learned in §2 to be Conway games, are all positive in this sense.

**1. Winning Strategies.** The concept of a *strategy* is one of the fundamental ideas of the theory of games. Suppose that in an actual game played according to the rules of the game $x$ the player $A$ is the one who has to move (so that $A$ may be either $L$ or $R$). If there is no possible move open to $A$ then the game is over and $A$ has lost, but if $A$ has any options at all then in general there will be several legitimate moves which he can make. A *strategy $\sigma$ for $A$ in $x$* in that case prescribes unambiguously the move to be played by $A$.

The move prescribed by a given strategy can depend on the course of the play up to that point. (It would be possible to restrict the definition of a strategy so that it depended only on the position reached, but we shall not use this simpler concept because we should have to prove more in this case.)

We say that *a player $A$ in playing a game $x$ plays a play with the strategy $\sigma$* if $\sigma$ is a strategy for $A$ in $x$ and each move made by $A$ is the one prescribed by $\sigma$.

In defining a *winning strategy* we distinguish between the player using it, and the player who has the first move.

$\sigma$ is called a *winning strategy for $L$ in the game $x$ in the case where $R$ begins if* (and only if) $\sigma$ is a strategy for $L$ in $x$ such that $L$ wins every play in which $R$ begins, if $L$ adopts the strategy $\sigma$.

We shall write $LxR$ to denote that $L$ has a winning strategy in game $x$, when $R$ begins. We define $LxL$, $RxL$ and $RxR$ analogously.

Later on we shall use the two following lemmas:

(1) *Let $x'$ be a game which is a right predecessor of $x$, then $Rx'L$ implies $RxR$.*

(2) *If $Lx'L$ for every game $x'$ that is a right predecessor of the game $x$, then $LxR$.*

**Proof of (1).** Let $\sigma'$ be a winning strategy for $R$ in $x'$, when $L$ begins. A winning strategy $\sigma$ for $R$ in $x$, when $R$ begins, can be devised as follows:

$R$'s first move is to bring about the initial position in $x'$; thereafter $R$ plays according to the strategy $\sigma'$. By hypothesis this strategy guarantees him a win.

**Proof of (2).** A winning strategy for $L$ in $x$, when $R$ begins, is to allow $R$ to make any move (if there is no legitimate move available to $R$ then $L$ wins immediately). This initial move leads to a game $x'$, a right predecessor of $x$, in which $L$ has the move. Now $L$ can use a winning strategy $\sigma'$ which, by the hypothesis $Lx'L$, must exist. □

Note that, for reasons of symmetry, each of the statements (1) and (2) (and the later statements using this terminology) has a valid *dual* obtained by interchanging $L$ and $R$ and replacing the word "right" by "left" and vice versa.

If in a particular game the player $R$, say, has the first move, then the players $R$ and $L$ cannot both have winning strategies. The following proposition shows that at least one of the two players must have a winning strategy. For every game $x$ the statement is valid:

(3) *($LxR$ or $RxR$) and ($LxL$ or $RxL$).*

The **proof** uses the induction principle for games. The first bracketed statement may be proved as follows: if there exists a game $x'$, which is a right predecessor of $x$ and which satisfies $Rx'L$, then $RxR$ by (1), and the statement is true. If not, $Rx'L$ is false for every right predecessor $x'$ of $x$, and hence by the induction hypothesis $Lx'L$, so that $LxR$ by (2) and again the bracketed statement is true. The second bracketed statement in (3) is proved in the same way with the help of the propositions dual to (1) and (2). □

**2. Positive and Negative Games.** If at the beginning of a play the player $R$ has no move, then $LxR$ is trivially true. This applies to all the CONWAY games named in 2.3(2). All these numbers are positive (in the sense of $\geq 0$). These examples provide a motive for introducing a property "$0 \leq$" defined by the following

**Definition.** $0 \leq x$ *if and only if $LxR$.*

Dual to this we introduce a property "negative," abbreviated to "$\leq 0$" by the

**Definition.** $x \leq 0$ *if and only if $RxL$.*

With the help of this definition and of (3) the statements (1) and (2) can be reformulated. We follow CONWAY here and use $x^L$, $x^R$ as variables for *the left and right predecessor games of $x$.* We thus obtain:

(1') *If an $x^R \leq 0$, then $0 \leq x$ is false.*

(2′) *If for all $x^R$, $x^R \leq 0$ is false, then $0 \leq x$.*

By combining these two statements we obtain joint inductive characterizations of "$0 \leq$" and "$\leq 0$", namely

(4) $0 \leq x$ *if and only if, for all $x^R$,* not $x^R \leq 0$,

and the dual of this

(5) $x \leq 0$ *if and only if, for all $x^L$* not $0 \leq x^L$.

**3. A Classification of Games.** Equivalence of games. By applying the distributive law of the Boolean *and* operation to (3) we have, for every game $x$:

$$(LxR \text{ and } LxL) \text{ or } (LxR \text{ and } RxL) \text{ or } (RxR \text{ and } LxL)$$
$$\text{or } (RxR \text{ and } RxL).$$

If the first bracketed statement holds for $x$ then $L$ has a winning strategy for any play in which $L$ begins and also a winning strategy for any play in which $R$ begins. We shall say that such a game belongs to the *class* **L**. Similarly a game belongs to the *class* **R** if the last bracketed statement holds.

If the third bracketed statement applies then the player who *begins* has a winning strategy, that is the *first* player. We shall say that a game of this kind belongs to the class **F**.

Lastly if the second bracketed statement holds, the *second* player who *does not begin*, has a winning strategy and we assign such a game to the class **S**.

Clearly no game can belong to two different classes, so that

(6) *Every game falls into one of the mutually exclusive classes* **L**, **R**, **F**, **S**.

**Definition.** Games which fall into the same class are said to be *of equal value*.

*Examples.* The domino games with the initial positions ⊟ , ⊡ , ⊞ , ⊡ ⊢ belong to the classes **L**, **R**, **F**, **S** respectively as is easily verified. Let $D_n$ be a domino game whose initial position is an $n \times n$ *square*.

$D_0$, $D_1$, $D_5$ belong to **S** and $D_2$, $D_3$, $D_4$ to **F**. The CONWAY game $0 = (\emptyset, \emptyset)$ defined in §2(1) belongs to **S** because neither player in the initial position can make a legitimate move.

One could naturally also define the above-mentioned classes with the help of the relations $\leq 0$ and $\geq 0$. Thus

(a) $x \in$ **S**, if and only if $x \leq 0$ and $0 \leq x$,

(b) $x \in \mathbf{L}$, *if and only if* $0 \le x$ *and* $x \not\le 0$,

(c) $x \in \mathbf{R}$ *if and only if* $x \le 0$ *and* $0 \not\le x$,

(d) $x \in \mathbf{F}$ *if and only if* $x \not\le 0$ *and* $0 \not\le x$.

In particular $0 \in \mathbf{S}$, and thus in the sense of the definitions for $0 \le x$ and $x \le 0$, we have

(7) $$0 \le 0.$$

If we define $0 < x$ as meaning $0 \le x$ and $x \not\le 0$, we see that $\mathbf{L}$ contains all and only the strictly positive games, and in the same way $\mathbf{R}$ contains all and only the strictly negative games.

## §5. A Partially Ordered Group of Equivalent Games

In the preceding paragraph we introduced the concepts of "positive" and "negative" games. Instead of writing $x$ *is positive,* we also wrote: $x$ *has the property* "$0 \le$" or "$0 \le$"$x$ or more shortly $0 \le x$; in the same way for $x$ *is negative* we wrote $x$ has the property "$\le 0$",$x$ "$\le 0$", or $x \le 0$. The notations $0 \le x$ and $x \ge 0$ suggest that $x$ can be compared with $0$, although the property is not explicitly mentioned in the definitions.

In this paragraph we shall introduce a *binary* relation $\le$ between games and show that "$0 \le$"$x$ if and only if $0 \le x$ and that $x$"$\le 0$" if and only if $x \le 0$.

We shall also define two operations $-x$ and $x + y$. We then interpret $x \le y$ as "$0 \le$" $y - x$, where $y - x$ is as usual an abbreviation for $y + (-x)$.

The $\le$ relation (and this applies to $-$ and $+$ as well) is a contribution from the theory of games to Conway's theory of numbers. It is the relation which we lacked in §2.

The relationship which exists between two games $x$ and $y$, when $x \le y$ and $y \le x$ both hold is an equivalence relation compatible with respect to $\le$, $-$ and $+$. The corresponding congruence classes constitute the elements of a partially ordered Abelian group whose zero element is $\mathbf{S}$.

**1. The Negative of a Game.** The negative of a game

$$x = (S, s_0, \to_L, \to_R)$$

may be defined as the game

$$-x = (S, s_0, \to_R, \to_L),$$

that is the game derived from $x$ by transposing the game-relations for $R$ and $L$. Clearly we have

(1) $$-(-x) = x, \qquad -0 = 0$$

where (see §2,3(1)) the CONWAY game 0 must be interpreted as the game with the one and only position $(\emptyset, \emptyset)$ in which neither player has a legitimate move.

(2) *If* $0 \le x$, *then* $-x \le 0$ (*and conversely*).

**Proof.** We have to show that $R(-x)L$, if $LxR$. This follows from the remark that a winning strategy for $L$ in $x$, when $R$ has the first move, is also a winning strategy for $R$ in $-x$ when $L$ has the first move. $\qquad \square$

**2. The Sum of Two Games.** First an example: $x_1$ could be a game of NIM and $x_2$ a game of dominoes. Then $x_1 + x_2$ would mean the game in which $x_1$ and $x_2$ are played *simultaneously,* on the understanding that each opponent when it is his turn to move has the option of making a move either in $x_1$ or in $x_2$ (but not both).

The general definition is: if

$$x_i = (S_i, s_{0i}, \rightarrow_{Li}, \rightarrow_{Ri}), \qquad (i = 1, 2)$$

then

$$x_1 + x_2 = (S, s_0, \rightarrow_L, \rightarrow_R),$$

where $S = S_1 \times S_2$ is the set of pairs of positions of the games $x_1$, $x_2$; $s_0$ is the pair $(s_{01}, s_{02})$ and

$$(s_1, s_2) \rightarrow_L (s_1', s_2')$$

*if and only if*

$$(s_1 \rightarrow_{L1} s_1' \text{ and } s_2 = s_2') \quad \text{or} \quad (s_1 = s_1' \text{ and } s_2 \rightarrow_{L2} s_2').$$

(The relation $\rightarrow_R$ is defined similarly.) It is clear that:

(3) $$-(x + y) = -x - y \qquad (= -x + (-y)).$$

Furthermore

(4)  a) $0 \le x - x$ *and* $x - x \le 0$.

  b) *If* $0 \le x$ *and* $0 \le y$, *then* $0 \le x + y$.

  c) *If* $0 \le x + y$ *and* $y \le 0$, *then* $0 \le x$.

**Proof.** a) $0 \le x - x$ means that $L(x - x)R$. If $R$ begins, $L$ can win the game $x - x$, if $L$ copies the move played by $R$ in the other component. The second assertion follows by duality.

b) Let $LxR$ and $LyR$. We have to show that $L(x + y)R$. We obtain a winning strategy for $L$ in the game $x + y$, when $R$ begins, by adopting the rule that $L$ responds to every move of $R$ by one in the same component game as that chosen by $R$, and by making the move required by the winning strategy for that component game.

c) We show that $0 \neq x$ and $y \leq 0$ together imply $0 \not\leq x + y$. By (4.3) it will suffice to prove that

$$if \; RxR \; and \; RyL, \; then \; R(x+y)R.$$

$R$ begins by making a move in the component $x$ where $R$ has a winning strategy. Thereafter he always makes his move in the game in which his opponent has chosen to move, and in accordance with the winning strategy which exists for $R$ in that game.                                                  □

**3. Isomorphic Games.** Isomorphism for games can be defined in the usual way.

It is easily seen that the game $x + y$ is isomorphic to $y + x$ and that $(x + y) + z$ is isomorphic to $x + (y + z)$.

If $y$ is isomorphic to $x$ and $LxR$, then clearly $LyR$, and so on.

*Example.* The domino game with initial position ⊞ ⊟ is isomorphic to the sum of the domino games with initial positions ⊡ and ⊟ .

**4. A Partial Ordering of Games**

**Definition.** $x \leq y$ *if and only if* $0 \leq y - x$ (where naturally the "$0 \leq$" on the right means the property "$0 \leq$" introduced in §4).

We wish to show that $0 \leq y$ if and only if "$0 \leq$" $y$ (see the opening comment). (The proof that $x \leq 0$ if and only if $x$"$\leq 0$" is proved similarly.) We have to prove that, for the property $0 \leq$ we have:

$$0 \leq y - 0 \quad if \; and \; only \; if \quad 0 \leq y.$$

If $0 \leq y$, then $0 \leq y - 0$ follows from $0 \leq 0$ (4.7), $-0 = 0$ (1) and (4b).
    If $0 \leq y - 0$ then $0 \leq y$ follows from $0 \leq 0$, $-0 = 0$ and (4c).          □

$\leq$ *is a partial order relation. Reflexivity* comes from (4a), and it only remains to prove *transitivity.* Suppose $x \leq y$ and $y \leq z$, then

$$0 \leq y - x \quad and \quad 0 \leq z - y$$
$$0 \leq (z - y) + (y - x) \quad (4b)$$
$$0 \leq (z - x) + (y - y) \quad (isomorphism)$$
$$0 \leq z - x \quad \quad by \; (4a) \; and \; (4c)$$

and hence $x \leq z$.

(5)    a) *If $x \leq y$, then $-y \leq -x$.*
       b) *If $x \leq y$, then $x + z \leq y + z$.*

**Proof.** a) Let $x \leq y$. Then $0 \leq y-x, 0 \leq -x-(-y)$ (isomorphy), $-y \leq -x$.

b) Let $x \leq y$. Then $0 \leq y-x, 0 \leq (y-x)+(z-z)$ (4b), $0 \leq (y+z)-(x+z)$ (isomorphy) which proves the assertion                                       $\square$

(6)                          $No\ x^R \leq x\ \ and\ no\ \ x \leq x^L$

(see 4.2 for the definitions of $x^R$ and $x^L$). We shall prove the the first statement (the second is its dual). $x^R - x^R$ is a right predecessor game of $x - x^R$. By (4a) we have $R(x^R - x^R)L$. It now follows from §4(1) that $R(x - x^R)R$ and hence that $L(x - x^R)R$ and $x^R \leq x$ *are both false.*   $\square$

In §4 we characterized the property "$\leq 0$" inductively. There is a corresponding inductive characterization for the binary relation $\leq$.

**Theorem.** $x \leq y$ *if and only if* (a) *no* $y^R \leq x$ *and* (b) *no* $y \leq x^L$.

**Proof.** Suppose $x \leq y$. To prove (a) we note that $x \leq y$ and $y^R \leq x$, would together imply $y^R \leq y$ by transitivity, and this would contradict (6). The statement (b) is proven similarly.

Suppose that we never have $y^R \leq x$ and never have $y \leq x^L$, but that $x \leq y$ is false. Then we would have $R(y - x)R$. $R$ thus has a winning strategy for the game $y - x$, when $R$ begins. There are two conceivable cases to be considered for $R$'s first move:

(i) $R$ makes a move in the component $y$. This move yields a $y^R$, and $R(y^R - x)L$, so that $L(x - y^R)R$, or in other words $y^R \leq x$ contrary to hypothesis.

(ii) $R$ makes a move in the component $-x$. This move yields a right predecessor of $-x$, and thus a left predecessor $x^L$ of $x$. This implies $R(y - x^L)L$ and thus $L(x^L - y)R$ or in other words, $y \leq x^L$ contrary to hypothesis.                                                                     $\square$

**5. Equality of Games.** In the foregoing we have shown that $\leq$ has all the properties which characterize the binary relation expressed by "$x \leq y$ *and* $y \leq x$" as an equivalence relation compatible with $\leq$, $-$ and $+$. We now follow CONWAY's terminology and call two games *equal* $(=)$ when this relation holds between them. *It should be noted that until now we have always understood equality to mean logical identity.* To avoid confusion we shall from now on use the symbol $\equiv$ to denote the latter. Accordingly we now adopt the following definition of equality.

**Definition.** $x = y$ *means* $(x \leq y$ *and* $y \leq x)$.

We shall spare ourselves the details of the construction of the equivalence classes corresponding to this definition of equality and of the extension of the definition $\leq$, $+$ and $-$ to these classes and content ourselves with formulating the result.

**Theorem.** *The classes of equal games constitute a partially ordered Abelian group with respect to $\leq$, $-$, $+$ whose zero element is* **S**.

*Equal games are of course also of equal value* in the sense defined in 4.3 (of having the same value). Each of the classes **S**, **L**, **R**, **F** thus splits up into classes of equal games. All games of the class **S** are equal to one another, but the other classes split up into more than one class of equal games (indeed into an infinite number of such classes). For example, the two domino games, $x$ with initial position $\boxed{\phantom{0}\phantom{0}}$ , and $y$ with the initial position $\boxed{\phantom{0}\phantom{0}}$ $\boxed{\phantom{0}\phantom{0}}$ are obviously both in $R$, but they are not equal. In fact $x$ is isomorphic to a $y^R$. For such a $y^R$ we have trivially $y^R \leq x$. We can therefore deduce by the theorem in §4, that $x \leq y$ is false, and hence $x \neq y$.

## §6. GAMES AND CONWAY GAMES

We saw in 3.2 that every CONWAY game $c$ can be regarded as a game. More precisely, we have shown how given a CONWAY game $c$ we can define a corresponding game $c_{\mathbf{G}}$. We now propose to show that *conversely* to every game $x$ can be assigned a CONWAY game $x_{\mathbf{C}}$ and the game $x_{\mathbf{CG}}$ corresponding to this CONWAY game $x_{\mathbf{C}}$, is a game equal to $x$, where the word "equal" is to be understood in the sense defined in the last paragraph.

One could denote $x_{\mathbf{C}}$ as *the normal form of $x$*. CONWAY bases his theory from the outset on normal forms. This has the advantage of greater mathematical simplicity though at the cost of intuitive appeal.

The two mappings $c \mapsto c_{\mathbf{G}}$ and $x \mapsto x_{\mathbf{C}}$ enables the relations $\leq$ and $=$ and the operations $+$ and $-$, defined initially for games, to be carried over to CONWAY games.

**1. The Fundamental Mappings.** We begin by repeating the definition of $c_{\mathbf{S}}$ which, in principle has already been given in 3.2 namely:

$$(1) \qquad\qquad c_{\mathbf{S}} \equiv (S_c, c, \underset{c}{\rightarrow} L, \underset{c}{\rightarrow} R),$$

where the positions $c_{\mathbf{G}}$ are, apart from the initial position $c$, the left and right elements of $c$, their left and right elements, and so on indefinitely. The move $s \underset{c}{\rightarrow} L\, s'$ is valid if and only if $s$, $s'$ are positions and $s'$ is a left element of $s$; $\underset{c}{\rightarrow} R$ is defined analogously.

We introduced $x^L$, $x^R$ in 4.2 as variables for the left and right predecessors of a game $x$. We shall similarly use $c^L$, $c^R$ as variables for the *left and right elements of a Conway game* (see 2.3). It is easily verified that

$$(2) \qquad \textit{The } c_{\mathbf{G}}^L \textit{ coincide with the } c^L{}_{\mathbf{G}} \textit{ and the } c_{\mathbf{G}}^R \textit{ with the } c^R{}_{\mathbf{G}}.$$

We now wish to assign a CONWAY game $x_C$ to each game $x$. We define this correspondence inductively on the assumption that $z_C$ has already been defined for all predecessors $z$ of the game $x$. (One can justify this procedure with help of the inductive principle for games in 3.3.) Accordingly we define:

$$(3) \qquad\qquad x_C \equiv (\text{set of } x^L{}_C, \text{ set of } x^R{}_C).$$

By induction over games one can see at once that $x_C$ is a CONWAY game; and it is immediately apparent from (6.3) that

(4)   *The $x_C{}^L$ coincide with the $x^L{}_C$ and similarly the $x_C{}^R$ with the $x^R{}_C$.*

(5)   *For every Conway game $c_{GC} \equiv c$.*

To *prove* this we need a principle of induction for CONWAY games analogous to the one for games, and which can be proved in an analogous fashion.

*Induction Principle for Conway Games.* If from the *induction assumption* that $Px'$ holds for every left or right element $x'$ of an arbitrary CONWAY game $x$ the *induction consequence* $Px$ follows, then every CONWAY game $x$ has the property $P$.

We deduce from this that:

$$
\begin{aligned}
c_{GC} &\equiv (\text{set of } c_G{}^L{}_C, \text{ set of } c_G{}^R{}_C) \qquad (3)\\
&\equiv (\text{the set of } c^L{}_{GC}, \text{ set of } c^R{}_{GC}) \qquad (2)\\
&\equiv (\text{set of } c^L, \text{ set of } c^R) \qquad (\text{induction hypothesis})\\
&\equiv c.
\end{aligned}
$$

(6)   $\qquad\qquad x = x_{CG}$ *for every game $x$.*

We prove that $x \le x_{CG}$ (the proof that $x_{CG} \le x$ is similar). We use the inductive characterization of the relation $\le$ given in 5.4, together with the (2), (4) and the induction hypothesis.

$$
\begin{aligned}
x \le x_{CG} \quad &\text{if and only if no } x_{CG}{}^R \le x \text{ and } x_{CG} \le \text{ no } x^L,\\
&\text{if and only if no } x^R{}_{CG} \le x \text{ and } x_{CG} \le \text{ no } x^L{}_{CG},\\
&\text{if and only if no } x^R \le x \quad \text{ and } x_{CG} \le \text{ no } x^L{}_{CG},
\end{aligned}
$$

and the last conjunction holds by §5(6).                                              □

## 2. Extending to CONWAY Games the Definitions of the Relations and Operations Defined for Games.

We begin by defining the relation $\le$ between CONWAY games $c, c'$:

(7)   $\qquad\qquad c \le c' \quad means \quad c_G \le c'_G.$

As with the definition of equality for games we write $c = c'$ if $c \leq c'$ and $c' \leq c$ both hold.

The extension of the operations $-$ and $+$, defined for games, to CONWAY games is achieved canonically by means of the two following definitions:

(8) $$-c \equiv (-c_G)_C,$$

(9) $$c_1 + c_2 \equiv (c_{1G} + c_{2G})_C.$$

One can also characterize the relation $\leq$ and the operations inductively $-$, $+$ by:

(7I) $c \leq c'$ *when and only when,*

    (a) *we never have $c'^R \leq c$, and*

    (b) *we never have $c' \leq c^L$.*

(8I) $-c \equiv$ (set of the $-(c^R)$, set of the $-(c^L)$).

(9I) $c_1 + c_2 \equiv$ (set of the $(c_1^L + c_2) \cup$ set of the $(c_1 + c_2^L)$,

        (set of the $(c_1^R) + c_2) \cup$ set of the $(c_1 + c_2^R)$).

(7I) follows at once with the help of (2) from the inductive characterization of the $\leq$ relation between games.

We prove (8I). By (8) and (3) we have

$$-c \equiv (\text{the set of the } (-c_G)^L_C, \text{ set of the } (-c_G)^R_C).$$

We deduce from (8) using (6) that $(-c)_G = -c_G$ and thus

$$-c \equiv (\text{the set of the } (-c)^L_{G\,C}, \text{ set of the } (-c)^R_{G\,C})$$

and from this, together with (2) and (8)

$$-c \equiv (\text{set of the } (-c)^L, \text{ set of the } (-c)^R).$$

We therefore have (8I), if we take into account the fact that the sets of the $(-c)^L$ and of the $-(c^R)$ have the same elements and that similarly the set of the $(-c)^R$ has the same elements as the set of the $-(c^L)$. □

From the result in 5.5 we can now easily deduce

**Theorem.** *The classes of equal Conway games form a partially ordered Abelian group with respect to $\leq$, $-$ and $+$.*

**3. Examples.** We shall determine by way of example the CONWAY games corresponding to one or two of the games of dominoes discussed at the end of 4.2. Since $D_0$ and $D_1$ have no predecessor, $D_{0C} \equiv D_{1C} \equiv (\emptyset, \emptyset) \equiv 0$. The domino game with initial position ⊟ has $D_0$ as its left predecessor, but has no right predecessor. The CONWAY game corresponding to this domino game is therefore $(\{0\}, \emptyset) \equiv 1$. In the same way we see that the domino games with the initial positions ⊞ , and ⊞⊟, correspond respectively to the CONWAY games $(\emptyset, \{0\}) \equiv -1$ and $(\{-1\}, \{1\})$. The domino game with initial position ⊞ which has $D_1$ as its only left and only right predecessor, corresponds to the CONWAY game $(\{0\}, \{0\})$.

## §7. CONWAY NUMBERS

In §2 we discussed the DEDEKIND postulates (D1) to (D4). In the intended generalization, apart from the basic concept of regarding a number as a pair of sets, whose elements were numbers that had already been constructed, only the postulate D3 (or the version (D3'), see 2.1) was to have been retained. This led to the problem of how the relation $\leq$ should be defined. This problem has now been solved. The CONWAY numbers are, from the way in which they are constructed, in any case CONWAY games, and we have already introduced in 6.2 a partial ordering relation for CONWAY games, motivated by game theory. We are therefore now in a position to formulate the two CONWAY postulates (C1) and (C2). (C1) generalizes the DEDEKIND postulate (D3') and (C2) contains the inductive characterization of $\leq$ (see 6.2).

**1. The CONWAY Postulates (C1) and (C2).** The CONWAY numbers, which from now on we shall simply call numbers, will be introduced by the two following postulates. We shall use $z^L$ and $z^R$, as in 6.1, as variables to denote the left and right elements of a pair of sets.

(C1) *If $z = (x, y)$, where $x$ and $y$ are both sets of numbers, and $z^R \leq z^L$ is never true, then $z$ is a number.*

(C2) *For numbers $x, y$ the statement $x \leq y$ is equivalent to the combined statement that $y^R \leq x$ is never true and that $y \leq x^L$ is never true.*

CONWAY develops his theory entirely on the basis of these two axioms— apart of course from the definition of the arithmetic operations (see 8.1). Thus we have to derive all the properties of $\leq$ from these postulates and we shall have no need to refer back to the game theory definition in §5.

We find ourselves here in a situation analogous to that in §2 where we defined CONWAY games with the help of the postulate (CG).

From (C1) it follows that:

(1) Every number is a pair of sets. The left and right elements of a number are themselves numbers. Every number is a CONWAY game.

If $x$ is a set of numbers, then $(x, \emptyset)$ and $(\emptyset, x)$ are numbers because the restrictive condition in (C1) is trivially satisfied. In particular therefore (see 2.3) it follows that:

(2) All ordinal numbers are numbers.

As we shall repeatedly be giving inductive proofs, we shall formulate an induction principle for numbers which corresponds to the one for CONWAY games in 6.2 and the one for games in (3.4) and which is most simply proved in the same way. In addition to formulating an induction principle *for a property*, we shall also formulate one for *a relation*.

**Induction Principle for Numbers** (for a *property P*). If from the *induction hypothesis* that $Px'$ holds for every left or right element $x'$ of a number $x$, the *induction conclusion* follows that $Px$ holds for every such number $x$, then every number has the property $P$.

**Induction Principle for Numbers** (for a *relation R*).
*Induction conclusion: $Rx_1, \ldots, x_n$,*
*Induction hypothesis: $Rx'_1, \ldots, x'_n$* for every n-tuple $x'_1, \ldots, x'_n$, where, for every $i$, $x'_i$ is equal to $x_i$ or is a left or right element of $x_i$ and where, for at least one $i$, $x'_i$ is a left or right element of $x_i$.
If (for all $x_1, \ldots, x_n$) the above conclusion follows from the above hypothesis, then $Rx_1, \ldots, x_n$ is true for all numbers $x_1, \ldots, x_n$.

**2. Elementary Properties of the Order Relation.** We first show using the induction principle that $\leq$ is *reflexive*. At the same time we prove two further statements:
*For every number x*

(3)
$$\begin{cases} \text{(a)} & x^R \not\leq x \quad \text{for every } x^R, \\ \text{(b)} & x \not\leq x^L \quad \text{for every } x^L, \\ \text{(c)} & x \leq x. \end{cases}$$

**Proof.** As to (a) (the proof of (b) is analogous), if there were an $x^R \leq x$, then by (C2) in particular $z \leq x^R$ is false when $z$ is any right element of $x$. Now $x^R$ is a right element of $x$, so that we should have $x^R \not\leq x^R$, contrary to part (c) of the induction hypothesis.

As to (c), if $x \not\leq x$ then by §6 (7I) there is an $x^R \leq x$ or $x \leq$ an $x^L$ in contradiction to (a) or (b) respectively. □

As with CONWAY games and games we now introduce an equivalence relation $=$ for numbers by the following

**Definition.** $x = y$ means $x \leq y$ and $y \leq x$.

It follows from (3) that:

(4) *For every number x, $x = x$.*

We will now show that $\leq$ is *transitive:*

(5) *For all numbers $x, y, z$ we have: if $x \leq y$ and $y \leq z$, then $x \leq z$.*

(Of course we already know from the earlier paragraphs that this applies when the relation $\leq$ is defined by the game theory definition. What we are concerned with here is to deduce it from the CONWAY postulates.) We use the induction principle for the ternary relation $R$, defined by

$$Rxyz \text{ holds if and only if } (x \leq y \text{ and } y \leq z \text{ imply } x \leq z)$$
$$\text{and} \quad (y \leq z \text{ and } z \leq x \text{ imply } y \leq x)$$
$$\text{and} \quad (z \leq x \text{ and } x \leq y \text{ imply } z \leq y).$$

We have to show that the induction conclusion $Rxyz$ is a consequence of the induction hypothesis. On grounds of symmetry it suffices to show that $x \leq z$, if $x \leq y$ and $y \leq z$ follows from the induction hypothesis. Accordingly, suppose $x \leq y$ and $y \leq z$. If $x \not\leq z$ were true, there would be by (C2) a $z^R$ with $z^R \leq x$ and an $x^L$ with $z \leq x^L$. We confine ourselves to the first case (the second can be dealt with in a similar way). It follows from $z^R \leq x$ and $x \leq y$ by the induction hypothesis (in particular the *third* term of the conjunction defining $Rxyz$) that $z^R \leq y$. It now follows from $z^R \leq y$ and $y \leq z$ and the *first* term of this conjunction that $z^R \leq z$. This contradicts (3). □

In this proof of reflexivity and transitivity we have not made use of the fact that there is a restriction in (C1) on the formation of pairs of sets. This restriction will however be essential in what follows.

We define $x < y$ in the usual way by $x \leq y$ and $y \not\leq x$ (or equivalently by $x \leq y$ and $x \neq y$) and assert:

(6) *For every number $x$, $x^L < x$ and $x < x^R$.*

(Note that the corresponding statement *for Conway games is false* for it would then follow that $x^L < x^R$ would always hold, whereas there are of course CONWAY games $x$ and $z$ such that $z$ is both a left element and a right element of $x$.)

**Inductive Proof for $x^L < x$.** We have already shown in (3) that $x \not\leq x^L$. It will therefore be sufficient to prove that $x^L \leq x$. If $x^L \not\leq x$ were true, there would, by (C2), be an $x^R$ with $x^R \leq x^L$ or an $x^{LL}$ with $x \leq x^{LL}$.

$x^R \leq x^L$ contradicts (C1).

If $x \leq x^{LL}$ were true, then by the induction hypothesis $x^{LL} < x^L$ and by the transtivity of $\leq$ we should also have $x \leq x^L$ in contradiction to (C3). □

We now intend to show that the numbers are totally ordered by the relation $\leq$. (This does not hold for games in general. We have in fact already indicated in 4.3 an example of a game $x$ of the class **F**, and for this game $x \not\leq 0$ and $0 \not\leq x$.)

(7) *For any numbers $x$, $y$, either $x \leq y$ or $y \leq x$.*

**Proof.** We assume that $y \not\leq x$ and have to show that $x \leq y$. It follows from $y \not\leq x$, by (C2), that there is an $x^R \leq y$ or that $x \leq y^L$.

$$x \leq x^R \ (6) \text{ and } x^R \leq y \quad \text{imply} \quad x \leq y,$$

$$x \leq y^L \text{ and } y^L \leq y \ (6) \quad \text{imply} \quad x \leq y. \qquad \square$$

**3. Examples.** We have seen that all ordinals are numbers. If the ordinals are constructed in succession (see 2.3 where the first ordinals are defined) one sees at once that each ordinal is different from (that is, $\not\equiv$ to) any of its predecessors. However, more than this is true because every ordinal number is also unequal to any of its predecessors. We shall content ourselves with proving this for the natural numbers $n$. To do this it will be sufficient to show that $n < n + 1$ always holds.

(a) $n \leq n + 1$: we use (C2): ($a_1$) $(n + 1)^R \leq n$ can never be true, because there is no right element of $n + 1$. ($a_2$) If $n + 1$ were $\leq$ an $n^L$, then by the definition of $n + 1$ such an $n^L$ would also be an $(n + 1)^L$ and $n + 1$ would be $\leq$ an $(n + 1)^L$ contrary to (3).

(b) $n + 1 \not\leq n$; in view of (C2) it is enough to show that $n \leq$ an $(n+1)^L$; but $n$ is an $(n + 1)^L$ and $n \leq n$ by (3).

## §8. THE FIELD OF CONWAY NUMBERS

In the preceding paragraphs we introduced the CONWAY numbers, together with the order relation $\leq$, and the equivalence relation $=$. We shall now give the definitions for the arithmetic operations and a few examples (more will be found in [1], [2] and [5]), and outline the properties of the field of CONWAY numbers.

**1. The Arithmetic Operations for Numbers.** These are defined inductively. As regards $-$ and $+$ it will be recalled that we have already defined such operations for CONWAY games in 6.2. We take over these inductive definitions (8I) and (9I) and recast them in the form of two postulates (C$-$) and (C$+$) for numbers:

(C$-$) *for every number $x$, let*

$$-x \equiv (set \ of \ all \ -x^R, \ set \ of \ all \ -x^L).$$

(C$+$) *for any two numbers $x, y$ let*

$$x + y \equiv (set \ of \ all \ (x^L + y) \cup set \ of \ all \ (x + y^L),$$
$$set \ of \ all \ (x^R + y) \cup set \ of \ all \ (x + y^R)).$$

It can be shown that the operations $-$ and $+$ never lead outside the domain of numbers, and that the relation of equality defined in 7.2 is a congruence relation for these operations.

As regards multiplication, there appears to be no model ready to hand in the domain of games and CONWAY games. After some difficulty (see [1]) CONWAY succeeded in finding the following inductive definition (C*) of multiplication formulated on the analogy of (C$-$) and (C$+$).

(C*) *For any two numbers* $x, y$ *let*

$$x * y \equiv (set\ of\ all\ x^L y + xy^L - x^L y^L) \cup set\ of\ all\ (x^R y + xy^R - x^R y^R),$$
$$set\ of\ all\ (x^L y + xy^R - x^L y^R) \cup set\ of\ all\ (x^R y + xy^L - x^R y^L)).$$

Multiplication does not lead outside the domain of numbers, and the equality defined in 7.2 is a congruence relation for this operation.

CONWAY shows that the set of all numbers modulo equality constitute an ordered field with respect to $\leq$, $-$, $+$, $*$.

**2. Examples.** The following examples are intended to illustrate the definitions in §1. We shall show by induction that $x + 0 \equiv x$ and $x + y \equiv y + x$, that $x + -x = 0$ (not $\equiv$), $1 + 1 = 2$ and that $\frac{1}{2} + \frac{1}{2} = 1$, for $\frac{1}{2} \equiv (\{0\}, \{1\})$. We use the abbreviation $S(\ldots)$ to denote the set of elements $(\ldots)$.

(a)                              $x + 0 \equiv x$    (and similarly $0 + x \equiv x$)

$$x + 0 \equiv (S(x^L + 0) \cup S(x + 0^L), S(x^R + 0) + S(x + 0^R))\quad \text{(C+)}$$
$$\equiv (S(x^L + 0), S(x^R + 0))$$
$$\equiv (S(x^L), S(x^R))\quad \text{(induction hypothesis)}$$
$$\equiv x.$$

(b)                              $x + y \equiv y + x.$

By induction over $y$:

$$x + y \equiv (S(x^L + y) \cup S(x + y^L), S(x^R + y) \cup S(x + y^R))\quad \text{(C+)}$$
$$\equiv (S(x + y^L) \cup S(x^L + y), S(x + y^R) \cup S(x^R + y))$$
$$\equiv S(y^L + x) \cup S(y + x^L), S(y^R + x) \cup S(y + x^R))$$
$$\text{(induction hypothesis)}$$
$$\equiv y + x\quad \text{(C+)}.$$

(c)                              $x + -x = 0.$

(Here the sign $=$ cannot be replaced by $\equiv$, as can be seen for example by putting $x \equiv 1$.)

We use the definition of $=$ in 7.2 and restrict ourselves to show that $x + -x \leq 0$. It is clear that no $0^R \leq x + -x$, because there is no $0^R$. If there were a $z \equiv (x + -x)^L$ with $0 \leq z$, then we should have, by (C+), $z \equiv x^L + -x$ or $z \equiv x + (-x)^L$. *In the first case* 0 would be $\leq x^L + -x$, and hence by (C2), $(x^L + -x)^R$ would never be $\leq 0$; but, by (C+) and (C−), $x^L + -x^L$ is such an $(x^L + -x)^R$ and by virtue of the induction hypothesis $x^L + -x^L \leq 0$. *In the second case* there would be an $x^R$ with $z \equiv x + -x^R$, and 0 would be $\leq x + -x^R$. Consequently there would, by (C2), be no $(x + -x^R)^R \leq 0$; and this would contradict the induction hypothesis that there is an $x^R + -x^R \leq 0$.                    □

(d)                    $1 + 1 = 2.$

In paragraph 2.2 we defined $1 \equiv (\{0\}, \emptyset)$, $2 \equiv (\{0, 1\}, \emptyset)$. It follows that

$$1 + 1 \equiv (S(1^L + 1) \cup S(1 + 1^L), S(1^R + 1) \cup S(1 + 1^R)) \quad \text{(C+)}$$
$$\equiv (\{0 + 1\} \cup \{1 + 0\}, \emptyset)$$
$$\equiv (\{1\}, \emptyset) \quad \text{(a)}.$$

(d$_1$) $(\{1\}, \emptyset) \leq (\{0, 1\}, \emptyset)$. As there can be no $\emptyset^R$, it will be sufficient to show that $2 \leq$ no $(\{1\}, \emptyset)^L$, or in other words that $2 \not\leq 1$. This follows from $1 < 2$ (see 7.3).

(d$_2$) $(\{0, 1\}, \emptyset) \leq (\{1\}, \emptyset)$. It suffices to show that $(\{1\}, \emptyset) \leq$ no $(\{0, 1\}, \emptyset)$, that is, that $(\{1\}, \emptyset) \not\leq 0$ and $(\{1\}, \emptyset) \not\leq 1$. If $(\{1\}, \emptyset) \leq 0$ were true then 0 would be $\leq$ no $(\{1\}, \emptyset)^L$ contrary to $0 \leq 1$.

(e) We define

$$\frac{1}{2} \equiv (\{0\}, \{1\}).$$

$\frac{1}{2}$ is a number by (C1) since $1 \not\leq 0$. The notation will be justified by proving that $\frac{1}{2} + \frac{1}{2} = 1$.

(f) $0 \leq \frac{1}{2}$.

It suffices to show that no $\left(\frac{1}{2}\right)^R \leq 0$ and this is true because $1 \not\leq 0$.

(g) $1 \not\leq \frac{1}{2}$.

This follows from the fact that 1 is a right element of $\frac{1}{2}$ and $1 \leq 1$.

(h) $1 + \frac{1}{2} \equiv (\{\frac{1}{2}, 1\}, \{1 + 1\})$.

This results from (C+) with (a) and (b).

(i)

$$\frac{1}{2} + \frac{1}{2} = 1$$

$$\frac{1}{2} + \frac{1}{2} \equiv (\{0 + \frac{1}{2}\} \cup \{\frac{1}{2} + 0\}, \{1 + \frac{1}{2}\} \cup \{\frac{1}{2} + 1\}) \qquad \text{(C+)}$$

$$\equiv (\{\frac{1}{2}\}, \{1 + \frac{1}{2}\}) \qquad \text{(a), (b).}$$

It only remains therefore to prove $(i_1)$ and $(i_2)$.

$(i_1)$ $1 \leq (\{\frac{1}{2}\}, \{1 + \frac{1}{2}\})$. To do this we first prove $(i_{11})$ $1 + \frac{1}{2} \not\leq 1$; this follows from $1 \leq 1$ because $1$ by (h), is a $(1 + \frac{1}{2})^L$. We then prove $(i_{12})$, $(\{\frac{1}{2}\}, \{1 + \frac{1}{2}\}) \not\leq 0$, which results from $0 \leq \frac{1}{2}$ (f).

$(i_2)$ $(\{\frac{1}{2}, \{1 + \frac{1}{2}\}) \leq 1$. To prove this it suffices to show that $1 \leq$ no $(\{\frac{1}{2}\}, \{1 + \frac{1}{2}\})^L$, or in other words that $1 \not\leq \frac{1}{2}$, which was done in (g).    □

**3. Properties of the Field of Numbers.** The totality of all numbers forms a proper class, and is not therefore a set (see Chapter 14). This already follows from the fact that every ordinal is a CONWAY number and the ordinals do not form a set (see 14.2.4).

We have already mentioned that the class of all numbers constitutes an ordered field $K_0$ with respect to the operations defined in §1. $K_0$ is real, closed, and is uniquely defined (to within an isomorphism) by the property of being a universal embedding ordered field. By this is meant the following: To every ordered subfield $K_1$ of $K_0$ such that $K_1$ is a set, and to every extension $K_2$ of $K_1$ such that $K_2$ is both a set and an ordered field, there is a subfield, $K_2'$, of $K_0$ which is isomorphic to $K_2$ with respect to the field operations and ordering relation and where the isomorphism on $K_1$ is the identity. It follows in particular that every ordered field is embeddable in $K_0$. All the "nonstandard models" considered in Chapter 12 belong here.

If numbers are constructed by means of (C1), but with *finite* sets only being allowed, then one obtains just the *dyadic* numbers, that is, the numbers of the form $\pm m/2^n$, where $m, n$ are natural numbers. For every *real* number $x$, let $x_1$ and $x_2$ denote the set of dyadic numbers $< x$ and $> x$ respectively. Then $x = (x_1, x_2)$. A number $x$ is a real number if and only if there exists a natural number $n$ with $-n < x < n$ and

$$x = (\text{the set of all numbers } x - 1/2^k, \text{ set of all numbers } x + 1/2^k).$$

For ordinal numbers the operations defined by (C+) and (C*) yield the so-called natural sum and natural product respectively.

There are infinite numbers, for example, the number $\omega$, and there are also therefore infinitely small numbers, for example, $1/\omega$.

## REFERENCES

[1] CONWAY, J.H.: On Numbers and Games. Academic Press 1976, $^3$1979

[2] BERLEKAMP, E.R., CONWAY, J.H., GUY, R.K.: Winning Ways for your Mathematical Plays. Academic Press, I, II 1982

[3] DEDEKIND, R.: Stetigkeit und irrationale Zahlen. Vieweg, 1872, $^7$1965

[4] HUIZINGA, J.: Homo Ludens. Rowohlt 1956

[5] KNUTH, D.E.: Surreal Numbers. Addison-Wesley 1974

# 14

# Set Theory and Mathematics

*H.-D. Ebbinghaus*

> Gesetzt, es gebe eine große nützliche mathematische Wahr-
> heit, auf die der Erfinder durch einen offenbaren Trugschluß
> gekommen wäre; – wenn es dergleichen nicht gibt, so könnte
> es doch dergleichen geben – leugnete ich darum diese Wahr-
> heit, entsagte ich dann, mich dieser Wahrheit zu bedienen?
> (LESSING, Theologische Streitschriften)

**Introduction.** On the 7th of December 1873, the theory of sets left behind
forever its age of innocence, for on that day Georg CANTOR proved that the
set of real numbers is uncountable, or in other words that the real numbers
cannot be enumerated in the form $r_0, r_1, r_2, \ldots$ [2, p. 115 et seq.]. He thus
laid down, at a time when the idea of the existence of the *actual infinite*
in mathematics was still a matter of controversy, a foundation stone in the
*theory of infinite cardinalities.*

In 1878 he showed that the linear continuum of the real numbers could
be mapped bijectively onto continua of higher dimensions, onto a plane or
space,... and so on, so that these various continua of different dimensions
have the same power or cardinality [2, p. 119 et seq.]. With this unexpected
result, he provided a motive and impulse for the development of dimension
theory. Afterwards he investigated the formation of the set $H(A)$ of accu-
mulation points of a set $A$ of real numbers, and by defining the sets

$$A^{(0)} := A, \quad A^{(1)} := H(A), \ldots, A^{(n+1)} := H(A^{(n)}), \ldots,$$

$$A^{(\infty)} := \bigcap_{n \in \mathbb{N}} A^{(n)}, \quad A^{(\infty+1)} := H(A^{(\infty)}), \ldots$$

was able to continue the process of forming new sets beyond the finite into
the transfinite, and by so doing to create the *theory of transfinite ordinals*
[2, p. 145 et seq.]. The occasion which gave rise to these researches was a
paper [2, p. 92 et seq.] on a uniqueness theorem for trigonometrical series,
a fact which caused ZERMELO [2, p. 102] to remark that the birthplace of
CANTOR's set theory could be found in the theory of trigonometric series.

Long before CANTOR's epoch-making work, however, the idea of a set
and of infinity had already been the subject of much deep thought and
perspicacious speculation. Thus in the height of the Middle Ages discussions about the actual infinite had led to the comparison of infinite sets by

means of one-to-one correspondences. ALBERT OF SAXONY (circa 1320–1390) for example, proves in his *Questiones subtilissime in libros de caelo et mundo* that a beam of infinite length has the same volume as infinite three-dimensional space. In an imaginary experiment he saws the beam into finite pieces, which he then uses to make successive concentric wooden shells that eventually fill the whole of space with wood.

Great clarity characterizes the thoughts and writing of Bernhard BOLZANO, the famous theologician, philosopher and mathematician of Prague (1781–1848). In his definition [1, p. 4] (1847) of a set or "multiplicity" as an "embodiment of the idea or concept which we conceive when we regard the arrangement of its parts as a matter of indifference," we recognize the precursor of our present-day extensional conception, in which a set is completely determined by its elements alone. BOLZANO defends the existence of infinite sets against the critics who deny it. He also shows by means of examples that infinite sets, unlike finite sets, can have the same cardinality as one of their proper subsets [1, p. 28 et seq.]—an insight which Dedekind in 1888 was to make the basis of his definition of finiteness.

Richard DEDEKIND (1831–1916), independently of CANTOR, developed clear ideas on the concept of a set and on its significance for the foundations of mathematics. In 1871 he proposed replacing KUMMER's *ideal numbers*—which in his view were merely figments of the imagination—by the now familiar concept of an *ideal* [3, Vol. III, p. 251] of whose existence there could be no doubt since an ideal is just a certain collection of true numbers. He pursued this idea even more consistently in his book *Stetigkeit und irrationale Zahlen* (whose first edition came out in 1872 but which had been conceived much earlier in 1858) in which the real numbers are, as it were, "created" by Dedekind cuts [3, Vol. III, p. 315 et seq.]. His views are expressed in their purest form in the little tract which appeared in 1888, a book entitled *Was sind und was sollen die Zahlen,* in which the natural numbers as well are defined in terms of sets [3, Vol. III, p. 335 et seq.]. It was through this latter work in particular that DEDEKIND exercised a decisive influence on the development of set theory.

Despite the considerable contributions of others, however, Georg CANTOR who was born in St. Petersburg in 1845 and died in Halle in 1918, must be regarded as the true founder of set theory. His imaginative ideas were responsible for breaking down naive illusions and opened the door to far-reaching developments. By his researches into infinite cardinalities and transfinite ordinals he created, in the words of HILBERT [9, p. 167] "die bewundernswerteste Blüte mathematischen Geistes und überhaupt eine der höchsten Leistungen rein verstandesmäßiger menschlicher Tätigkeit," which might be translated as "some of the most admirable and beautiful creations of the mathematical imagination and, taken as a whole, one of the greatest purely intellectual achievements of the human mind."

CANTOR's set theory is of a vivid, visualizable kind. It is based on conceptual images, which he describes and expresses in various different ways. For him a set is "ein Vieles, welches sich als Eines denken läßt" [a multiplicity which can be thought of as a single entity], an "Inbegriff bestimmter Elemente, welcher durch ein Gesetz zu einem Ganzen verbunden werden kann" (1883, [2, p. 204]), [the essence of certain elements which can be associated by some rule into a single whole], or a "Zusammenfassung von bestimmten wohlunterschiedenen Objekten unserer Anschauung oder unseres Denkens zu einem Ganzen" (1895, [2, p. 282]) [collection into a whole of definite distinct objects of our perception or thought].

In the first decades of this century, in the birth pangs caused by the discovery of the antinomies described in 2.1 below, the intuitive ideas of CANTOR were put into more precise shape and suitable *axiom-systems* were devised for set theory. New and sophisticated techniques, such as the theory of *constructible sets* (GÖDEL 1938) or the *forcing-method* (COHEN 1963), brought about a period of tempestuous development that still continues to this day. For example, by means of these techniques it has proved possible to demonstrate the logical *independence* of the *continuum hypothesis*, first put forward by CANTOR in 1878, to the effect that every uncountable set of real numbers has the same power as the set of all real numbers. In other words, it was shown by COHEN in 1963 that the continuum hypothesis *cannot be proved* from the present-day systems of axioms of set theory, and (by GÖDEL in 1938) that it *cannot be disproved*.

A hundred years or so after CANTOR's pioneering work, set theory has now grown into a full-fledged mathematical discipline of its own; its influence has pervaded the whole of mathematics. Mathematics has more and more taken on a character that bears the imprint of set theory, as the ideas and intentions of DEDEKIND have borne fruit. On the one hand, this has led to sharper and clearer definitions of many mathematical concepts, and on the other hand to a considerable extension of the methods and aids available to the mathematician. HILBERT spoke of the "paradise which CANTOR created for us" [9, p. 170].

In addition, the axiomatization of set theory allows us to close the gaps in the axiomatic construction of mathematical theories, which otherwise had some gaping deficiencies in these respects. Thus, for example, the axiom systems for topological spaces certainly refer to facts from set theory, but do not axiomatize these. And lastly, but by no means least, it is only after the axiom systems of set theory had reached a certain degree of precision that it became possible to prove independence results, such as the independence of the continuum hypothesis.

We now propose in the sections that follow to explain in rather more detail some of the aspects of the relations between mathematics and set theory to which we have alluded. To do this we shall also need to describe an axiomatic structure for set theory, but in doing so we shall have to restrict ourselves to the basic facts, and to leave out many details. We refer

the reader who would like further information to the books [4], [7], [15] and
[19].

## §1. Sets and Mathematical Objects

**1. Individuals and More Complex Objects.** The set theoretical character of present-day mathematics stems mainly from the fact that the objects with which it is concerned can be described as sets. Before tackling
systematically any descriptions of this kind we would like to get a broad
view of the great variety of mathematical objects. To this end we shall begin by first considering a "concrete" theory, say *analysis.* The objects from
which we start are in this case the real numbers. We then go on to $n$-tuples
of real numbers, and to more "complicated" objects such as real functions,
intervals, other sets of real numbers, relations between real numbers, and
so on and so forth.

Real functions possess an inner structure that is of importance for analysis; they represent a special type of relation between real numbers. On the
other hand, the real numbers themselves play the role of "atoms" for an
analyst. Their inner structure is of no interest, and it is only the relations
*between* them, the relations that are formulated in the usual axiom systems for analysis, that are of significance. It is just for this reason that it
is possible to do analysis without knowing what the so-called real numbers
really are. The same applies to the natural numbers in arithmetic, or to
points in Euclidean geometry.

In the theory of sets the objects of a theory which have this "atomic"
character are known as *urelements* (that is, primitive elements) (ZERMELO
1930). The urelements form the lowest level in a hierarchy comprising the
objects of study of a given theory. They are accompanied by so-called *objects of a higher type,* such as properties of urelements, relations between
urelements, sets and functions of urelements or $n$-tuples thereof. Above
these tower ever more complicated objects such as sets of sets of urelements, for example, open coverings in analysis, or rings of residue-classes
in arithmetic. Obviously this process of forming more and more complicated objects can be continued indefinitely, and in this way a hierarchical
structure of mathematical objects can be formed of ever-increasing complexity. To some extent, one can recognize distinct layers (urelements, sets
of urelements, sets of sets of urelements), but one can also define more complicated relations. For example, functions can arise in analysis that map
functions of real numbers onto real numbers, such as those represented by
a definite integral between assigned limits. The technical name for a hierarchy of objects of this kind, which can be built up from a set of individuals,
is a *hierarchy of types.*

In an *abstract* mathematical theory, such as the theory of groups, the
elements of a group play a role comparable to that of the urelement in a
"concrete" theory. However, here the existence of its own urelements is not

required. The situation here is that any mathematical object whatsoever
can be regarded as an element of a group, without one having to postulate
additional requirements or encountering any limitations.

**2. Set Theoretical Definitions of More Complex Objects.** It has
turned out in practice that the properties, relations, and functions that
are mostly used intuitively in mathematics can all be reduced to the set
concept. Consequently, it becomes possible to describe the whole hierarchy
of types derived from a particular field of urelements in terms of set theory.

We shall try to convince ourselves in the following paragraphs of this
possibility, and accordingly we shall make use of a few simple facts from
naive set theory. We begin with *properties.* Let $M$ be a set of urelements
or other objects, say the set of real numbers. Let $P$ be a property over $M$,
that is, a property which can apply to elements of $M$. For mathematical
purposes it now fully suffices to identify $P$ with the set $\{r \in M : P$ applies
to $r\}$ containing those elements of $M$ which have the property $P$. *The*
*properties over $M$ thus correspond to the subsets of $M$.*

This way of looking at things has an interesting consequence. Consider
the property over $\mathbb{R}$ of being the square of a real number: it is the property
of being non-negative, because both properties correspond to the set $\{r \in$
$\mathbb{R} : r \geq 0\}$, since a real number is a square if and only if it is non-negative.
Properties are now defined only by their scope, by the set of individuals
to which they apply, or in a word, by their *extension.* This *extensional*
*conception* is a characteristic feature of the set-theoretic approach because
sets themselves are likewise determined by their elements alone. One is
continually coming up against this in mathematics in all kinds of situations.
For example, one meets it with functions as well; a function defined over
a given domain of definition is completely defined once its value has been
specified for each of its possible arguments. How the value is arrived at
plays no part in the definition of the function.

Another vital idea that is fundamental in making it possible to describe
mathematical objects by the concepts of set theory is the set theoretic definition of *n*-tuples. We begin with the case $n = 2$. Following KURATOWSKI
(1921), one can define an *ordered pair* $(a, b)$ of the two objects $a$, $b$ set
theoretically by

$(*)$                          $(a, b) := \{\, \{a\}, \{a, b\} \,\}.$

It is easily shown that

$(a, b) = (a', b')$   if and only if   $a = a'$  and  $b = b'.$

This equivalence is the only fact about ordered pairs that the mathematician ever really needs; the KURATOWSKI definition therefore does all that
is required of it.

This is perhaps the time to make a remark that basically applies to all set-
theoretic definitions of mathematical objects. *A set-theoretic definition such*

as (∗) *serves no ontological purpose.* It does not establish what an ordered
pair *really* is; it merely provides a *model* for the intuitive idea of an ordered
pair, which suits the requirements of mathematics. This "conventionalistic"
standpoint is also supported by the fact that, as a rule, different definitions
of various kinds are possible, and one would be hard put to give preference
to one definition rather than another on ontological grounds. Thus, for
example, the definition

$$(a, b) := \{ \{ \{a\}, \emptyset \}, \{\{b\}\} \}$$

(WIENER, 1914) serves the same purpose as (∗).

Once ordered pairs have been defined set-theoretically, *triples* can be
defined without further ado by:

$$(a, b, c) := ((a, b), c)$$

and the same idea can be extended to quadruples, quintuples, and so on.

In order to describe the concept of a binary—then similarly an $n$-ary—
relation over a set $M$, we can regard a binary relation between elements of
$M$ as a property of ordered pairs, over $M$. If we define in the usual way
the Cartesian product

$$M \times M := \{(a, b): a, b \in M\},$$

then the binary relations over $M$, in the set-theoretical sense, are simply
the subsets of $M \times M$. For example, the relation $L := \{(r, s): r, s \in \mathbb{R},$
$r < s\}$, the relation which expresses the fact that $r$ is less than $s$, is the
"less than" relation over $\mathbb{R}$, and $2 < 3$ is equivalent to the assertion that
$(2, 3) \in L$.

Similarly, in the well-known way a function $f$ mapping a set $M_1$ into a
set $M_2$ can be defined set-theoretically by its graph

$$f = \{(a, f(a)): a \in M_1\}.$$

In general therefore a function $f$ is a set of ordered pairs such that for
every object $a$, there exists at most one object $b$ with $(a, b) \in f$. The
familiar mathematical notation $f: M_1 \to M_2$ now says that $f \subset M_1 \times M_2$
is a function, so that to every $a \in M_1$ there exists a $b \in M_2$ with $(a, b) \in f$.
For $a \in M_1$, $f(a)$ is the one and only $b$ for which $(a, b) \in f$.

The same procedure can be followed for functions with more variables.

We can see from these examples that the multitude of objects that can
occur in any mathematical theory can be systematically described by using
the concepts of set theory. The starting point is in each case a certain
domain of urelements from which the more complicated objects can be built
up by repeatedly forming new sets. Thus, real functions of one variable are
sets of ordered pairs of real numbers. According to (∗), ordered pairs of
real numbers are sets of sets of real numbers. Therefore, real functions are

sets of sets of sets of real numbers. Ultimately, it is this reduction of the hierarchy of mathematical objects to the notion of set that forms the basis for the set theoretical treatment of mathematics.

Naturally not all the details of a program of this kind, in which mathematical concepts are clarified, sharpened and redefined in terms of the concepts of set theory, are equally essential to the mathematician in his everyday work. The mathematician would hardly need the definition (∗) of an ordered pair and tends to work with the "dynamic" intuitive idea of a function than with the "static" concept defined earlier in terms of sets. The value of formulating mathematical concepts and facts in the language of set theory does not really come about from any *systematic* use of this language. It lies far more in the *possibilities* that are opened up of using the elegant and effective methods of set theory wherever they can be useful. In other words, a set-theoretic formulation should never be a strait jacket but an added weapon in the armory. We shall discuss some further aspects of this question in §2 and in §3.3.

**3. Urelements as Sets.** In the previous section the urelements (numbers, points,...) were still regarded as playing the role of atoms. Their nature remained in the dark, but this need not be a disadvantage from a methodological standpoint. As we have already emphasized, in mathematics the "true nature" of urelements is quite irrelevant. From the working mathematician's point of view, it is even quite natural to retain these as atomic individuals. Moreover, there is no difficulty about incorporating the various facts borrowed from set theory which we have so far used only in a naive way, in a precisely defined *axiomatic set theory with urelements,* — with the same benefits as an axiomatic set theory *without* such individuals was able to give (see the two following paragraphs). On the other hand, it is very tempting to pursue the path described in 2. still further and try to describe the urelements themselves in set-theoretic terms, *in order to make the concept of a set the sole foundation of mathematics.*

Now it is one of the great conceptual achievements of mathematics and set theory that this project has been fulfilled. The pioneering work here was done by DEDEKIND to which we referred in more detail in the introduction. By way of illustration, we mention the *set-theoretic definitions of the natural numbers* by ZERMELO (1908) and by VON NEUMANN (1923). For the moment we shall argue intuitively; later on in 2.3, we give a more precise description in an axiomatic context.

ZERMELO defined the natural numbers by the sequence

$$0 := \emptyset, \quad 1 := \{\emptyset\}, \quad 2 := \{\{\emptyset\}\},\ldots;$$

VON NEUMANN defined them by

$$0 := \emptyset, \quad 1 := \{\emptyset\}, \quad 2 := \{\emptyset, \{\emptyset\}\},\ldots$$

and generally by
$$n + 1 := n \cup \{n\}.$$

His procedure has the technical advantage over ZERMELO's that each number is the set of all the preceding numbers, so that the relation $<$ coincides with the relation $\in$. From the cardinal standpoint, VON NEUMANN's numbers represent a natural measure of the cardinality of finite sets, as the number $n$ is a set with exactly $n$ elements. (This property is also shared by a related definition given by CANTOR (1895; see [2, p. 289 et seq.].).) Lastly, the VON NEUMANN sequence can easily be extended to the transfinite. Thus:
$$0, 1, 2, \ldots, \omega := \{0, 1, 2, \ldots\}, \omega + 1 := \omega \cup \{\omega\},$$

$$\omega + 2 := (\omega + 1) \cup \{\omega + 1\}, \ldots, \omega + \omega := \{1, 2, \ldots, \omega, \omega + 1, \omega + 2, \ldots\}, \ldots.$$

The relations $<$ and $\in$ similarly coincide for the *ordinal numbers* defined in this way.

One now defines, for these set-theoretically defined numbers, a successor function, in the usual way; that is $n \mapsto n \cup \{n\}$ in the VON NEUMANN case (or more precisely in the sense of 1.2 the set $\{(n, n \cup \{n\}): n \in \omega\}$) and $n \mapsto \{n\}$ in the ZERMELO case. It can now easily be shown that the PEANO axioms hold. Since these axioms are sufficient to ensure all the properties of the natural numbers that the mathematician ever needs to use, the set-theoretical definitions of VON NEUMANN and ZERMELO (together with the appropriate successor function) provide adequate models. Of course—repeating once again the point made earlier—a definition of this kind cannot tell us what natural numbers *really* are.

There is now no further difficulty in providing set-theoretical definitions for the arithmetic operations and the other kinds of numbers (integers, rational, real, complex). We have only to follow one of the usual ways of constructing the various number domains.

We have now reached the point where it has become possible to reduce to the single concept of a set the multiplicity of mathematical objects whose very bulk had at first seemed so overwhelming. If we begin, say, from the VON NEUMANN definition of the natural numbers, then we proceed from zero, through the natural numbers, the integers, etc. to any conceivable mathematical object. They are all sets; at the start there is one single atomic individual, the empty set. Rather more concisely, we can say that the whole universe of mathematical objects can be built up "from nothing" by the process of set creation.

## §2. AXIOM SYSTEMS OF SET THEORY

So far we have seen that mathematics can be represented with the help of the single concept of a set, but we have been looking at this only in the context of an *intuitive* set theory. The far-reaching implications of all this invite us to a more thorough-going analysis of the set concept and a more precise statement of our approach. This can best be done by laying down a *system of axioms for set theory*. We propose in this paragraph (§2) to describe a few such systems, and in fact systems for a set theory without urelements. In doing so, we shall also go into some of the difficulties that attended the birth of such systems, and that in a sense strengthened the motivation for their development. By laying down sufficiently powerful axiom systems, a unified axiomatic basis for the objects of mathematics in its entirety has been successfully achieved.

**1. The RUSSELL Antinomy.** Gottlob FREGE (1848–1925) one of the fathers of mathematical logic, gave in the first volume of his *Grundgesetze der Arithmetik* [6] a system of axioms for Cantorian set theory. His goal was to provide a logical/set-theoretical foundation for mathematics. One of his axioms expresses in more precise form the idea of sets as extensions or properties, an idea that appears in CANTOR's visualization of a set as DEDEKIND had often used it. In modern language it states:

**FREGE's axiom of comprehension** (used in the sense of comprising, from the Latin *comprehensio*). *For every property P there exists a set $M_p$ containing all those and only those sets which have the property P.* In the usual notation,

$$M_P := \{x : x \text{ is a set and } x \text{ has the property } P\}.$$

In the summer of 1901 Bertrand RUSSELL (1872–1970) discovered the inconsistency of the comprehension axiom. If one chooses for $P$ the property $R$ of not being an element of itself, then according to the comprehension axiom there exists the set

$$M_R := \{x : x \text{ is a set and } x \notin x\}.$$

For this set, we obviously have

$$M_R \in M_R \Leftrightarrow M_R \text{ is a set and } M_R \notin M_R.$$

Since $M_R$ is a set, we have

$$M_R \in M_R \Leftrightarrow M_R \notin M_R,$$

and thus a contradiction.

A few weeks earlier ZERMELO had told the philosopher HUSSERL of this antinomy. A written note by HUSSERL to this effect was found among the

papers he left on his death. Apparently ZERMELO had not at first attached
any great importance to this discovery, and in any case other antinomies
were known in naive set theory. For example, there was the one published
by BURALI-FORTI in 1897, and expressed in a sharper form by RUSSELL in
1903. The construction of the ordinal numbers by VON NEUMANN's method
described in 1.3, in which the $<$ relation coincides with the $\in$ relation,
leads to the conclusion that the set $\Omega$ of all ordinal numbers—supposing
this set were to exist—would, like $\omega$ or $\omega + \omega$, itself be an ordinal number.
This would imply that $\Omega \in \Omega$ in contradiction to the fact that an ordinal
number cannot be smaller than itself (or also a contradiction to the axiom
of foundation or regularity; see 2).

CANTOR called such "dangerous" sets as $M_R$ or $\Omega$ *absolutely infinite* or
*inconsistent multiplicities* [2, p. 443 et seq.]. For him they were not true sets
in any proper sense; FREGE with his comprehension axiom was therefore
overstepping the boundaries which CANTOR had staked out in the naive
theory dictated by his intuition.

The discovery of RUSSELL's antinomy brought out the opposition of the
reactionary opponents of set theory to the whole program. They saw the
origin of such contradictions as being rooted in the concepts of set theory
and mathematics, which were based on the assumption of the existence of
an actual infinity, and they wished to withdraw to the safety of constructions whose existence could be controlled and verified. One of the precursors
of this attitude (and in this respect an opponent of DEDEKIND—and more
particularly of CANTOR—was Leopold KRONECKER (1823–1891). A quotation from the year 1886 (see [13, p. 336]) illustrates the point "...selbst
der allgemeine Begriff einer unendlichen Reihe ... ist ... nur mit dem Vor-
behalte zulässig, daß in jedem speziellen Falle auf Grund des arithmetischen Bildungsgesetzes der Glieder ... gewisse Voraussetzungen als erfüllt
nachgewiesen werden, welche die Reihen wie endliche Ausdrücke anzuwen-
den gestatten, und welche also das Hinausgehen über den Begriff einer
*endlichen* Reihe eigentlich unnötig machen." [...even the general concept
of an infinite series ... is ... allowable only with the proviso that in each
particular case ... because of the arithmetical laws governing the formation
of the (successive) terms ... certain prerequisite conditions can be shown
to be satisfied which allow the series to be considered as a finite expression and therefore make it strictly speaking unnecessary to go beyond the
concept of a *finite* series.]

The differing epistemological attitudes of CANTOR and KRONECKER not
only led to scientific controversies, but also soured their personal relations
and caused much suffering to CANTOR.

A leading exponent of the critical constructivist tendency in the period
that followed was the Dutch mathematician L.E.J. BROUWER (1881–1966)
who founded the school of thought now known as *intuitionism* [8].

On the other side, numerous mathematicians, including RUSSELL and
ZERMELO, tried to repair the concepts of set theory, which had been ren-

dered untenable by the downfall of FREGE's axioms, and to arrive at a system of axioms free from contradiction that would re-open the possibilities offered by CANTOR. One of the critical intellectual leaders of this movement was David HILBERT (1862–1943); cf. [9].

In the following pages we shall briefly describe the best-known axiom systems. They are regarded nowadays by set theorists as consistent. Until the 1920's, a proof of consistency was thought to be possible. However, by a theorem of mathematical logic, proved in 1931 by GÖDEL, the consistency of these axiom systems cannot be proved without methodological means beyond those they represent (see [5, p. 226 et seq.]). In 3.1 we shall discuss certain mathematical arguments in support of their consistency.

## 2. ZERMELO's and the ZERMELO–FRAENKEL Set Theory.

In 1908 Ernst ZERMELO (1871–1953) proposed a system of axioms which heralded a new approach [22]. With the addition of some later improvements due to FRAENKEL and SKOLEM, it represents the most widely accepted system so far devised. The influence of DEDEKIND's ideas is unmistakeable. ZERMELO described his undertaking in the following words:

Angesichts der RUSSELLschen Antinomie "bleibt ... nichts anderes übrig, als ..., ausgehend von der historisch bestehenden 'Mengenlehre' die Prinzipien aufzusuchen, welche zur Begründung dieser mathematischen Disziplin erforderlich sind ... in der Weise ..., daß man die Prinzipien einmal eng genug einschränkt, um alle Widersprüche auszuschließen, gleichzeitig aber auch weit genug ausdehnt, um alles Wertvolle dieser Lehre beizubehalten." [In the face of RUSSELL's paradox ... "there remains ... nothing else left to us but ... to start out from the historically established 'set theory' and to seek out those principles that are required for the foundation of this mathematical discipline ... in such a way ... that the principles are narrow enough to exclude all contradictions and at the same time wide enough to allow everything of value in this discipline to be kept."]

We shall now present ZERMELO's axioms in a slightly modified form, which is now usual. As to their content, they may be regarded as describing a "universe" of sets; there are no urelements.

*The axiom of existence,* **Ex**. *There exists a set.*

(Instead of this, ZERMELO postulated the existence of the empty set, which on the basis of the other axioms is equivalent to **Ex**.)

*The axiom of extensionality,* **Ext**. *Two sets are equal if and only if they have the same elements.*

**Ext** reflects the extensional conception of a set, in which a set is determined by its members and by nothing else.

*The axiom of separation,* **Sep**. *To every property P of sets and to every set x there corresponds a set y which contains those and only those elements* *of x which have the property P*. The set $y$ is uniquely defined, by virtue of **Ext**. In the usual notation $y = \{z \in x : z$ *has the property P*$\}$.

**Sep** takes over the role played by the axiom of comprehension in FREGE's system, but in ZERMELO's system the comprehension is restricted so that it applies only to already pre-existing sets. By this precautionary measure, ZERMELO achieves his endeavor of not describing a universe that is, so to speak, "finished or completed," but instead conceiving one that can be built-up from below. This same idea underlies most of the remaining axioms: They say how new sets can be formed from those already available. It is easily established, at least on an ad hoc basis, that RUSSELL's contradiction no longer arises.

What properties are allowed in **Sep**? ZERMELO was thinking of particularly "concrete" properties, which he called *definit*, but without giving a satisfactory explanation of precisely what was to be understood by this word. A sharper delimitation was achieved by the Norwegian mathematician and logician Thoralf SKOLEM (1922) who laid down the principle that only those properties should be allowed which could be expressed in the language of first-order predicate logic (see [5]). Here the only non-logical symbol admitted is $\in$, the sign of equality is allowed, and the variables range over sets. A series of examples will be given in 3., but a particularly simple example may already be mentioned here. The property $P$, which applies to a set $z$, when and only when $z \neq z$, satisfies the Skolem requirement. If we choose a set $x_0$, whose existence is guaranteed by the axiom **Ex**, then **Sep** implies the existence of the set $\{z \in x_0 : z \neq z\}$, and thus of the empty set $\emptyset$ (whose uniqueness follows from **Ext**).

*The axiom of pairing,* **Pair.** *If x is a set and y is a set then there exists a set z which contains x and y as elements but no other elements.* (The set $z$ is unique by virtue of **Ext**.)

We write $z = \{x, y\}$ and $\{x\}$ for the set $\{x, x\}$, which has the single element $x$ and is called a *singleton*. By virtue of **Ext** $\{x, y\}$ is always equal to $\{y, x\}$ and therefore $\{x, y\}$ does not have the property of an ordered pair.

*The axiom of union,* **U-Ax.** *For every set* (which in this context is best visualized as a system of sets) *X there exists the set Y comprising all those elements which are elements of at least one of the elements of X* (that is, which belong to at least one of the sets of the given system).

We write $Y = \bigcup X$, or more often in mathematics $Y = \bigcup_{x \in X} x$. Thus, for example, $\bigcup \emptyset = \emptyset$, $\bigcup \{x\} = x$.

*The axiom of power sets,* **Pow.** *For every set x there exists the set y, the so-called power set of x, defined as the set whose elements are the subsets of x.*

We write

$$y = \mathcal{P}(x).$$

The axioms described so far are satisfied by all those (finite) sets which can be derived from the empty set by applying operations of the type $x \mapsto \{x\}$, and/or $x, y \mapsto x \cup y$ a finite number of times. We still want an axiom that ensures the existence of infinite sets in order to allow set theory to extend its sway into the realms of the transfinite. This is provided by:

*The axiom of infinity, **Inf**. There exists an inductive set,* that is, a set containing the empty set $\emptyset$ and the successor of each of its elements. The successor of an element $z$ is $z \cup \{z\}$. (As we have strictly speaking not yet defined the symbol $\cup$, $z \cup \{z\}$ is here simply an abbreviation for the set whose elements are just the elements of $z$ and $z$ itself.)

Intuitively an inductive set must in any case contain the VON NEUMANN natural numbers $\emptyset$, $\{\emptyset\}$, $\{\emptyset, \{\emptyset\}\}$ and so on. As we shall see in §3 **Inf** is definitely needed to prove the existence of the set $\omega$ of these numbers.

We complete the ZERMELO axiom system with

*The axiom of choice, **AC**. To every set corresponds a choice function.* By a choice function corresponding to a set $X$ is meant a function $f$, defined on $X$, such that $f(y) \in y$ for all $y \in X$, $y \neq \emptyset$.

In this formulation we run into a slight initial difficulty. The ZERMELO axiom system is conceived as one in which the only primitive undefined ideas are those of a set and the $\in$-relation between sets. (This same intention also underlies SKOLEM's refinement of the axiom of separation.) In the formulation of **AC** that we have chosen, however, the idea of a function appears (see the equivalent formulation **AC**' in 3.2). However, functions can be defined as sets, as we indicated in an intuitive way in 1.3, so that any reference to the idea of a function can be eliminated from **AC** with the help of the set-theoretic definition. The above formulation of **AC** can then better be regarded simply as a convenient verbal abbreviation for the resulting longer form expressed in terms of sets.

As we shall show by means of a few examples, ZERMELO's axiom system is sufficient to allow us to derive practically all the facts of set theory that the mathematician ever needs. It is only seldom, and in fact in situations where the set-theoretic framework becomes extraordinarily "demanding" (for example, in connection with the definition of CONWAY games and CONWAY numbers in Chapter 13, 2.3 and 7.1 respectively) and in set theory itself that some further axioms are required: in particular, the *axiom of replacement*, **Rep**; and the *axiom of foundation*, **Found** (also known as the axiom of regularity).

**Found** was stated by VON NEUMANN in 1925, but in the formulation given below, is due to ZERMELO (1930). It disallows pathological sets with $x \in x$ or descending chains of sets of the type $\ldots x_2 \in x_1 \in x_0$. It affirms that every non-empty set $x$ has an $\in$-minimal element, that is an element $y$ such that $x \cap y = \emptyset$. The axiom may be stated in words as follows:

**Found.** *To every non-empty set $x$, there is a set $y \in x$ which has no element in common with $x$.*

To see that **Found** always implies $x \notin x$, we form from any given $x$ the set $\{x\}$, by using **Pair**. As $x$ is the sole element of $\{x\}$, $x$ must be the $\in$-minimal element of $\{x\}$, and so in particular $x \notin x$.

Rep (MIRIMANOFF 1917, FRAENKEL 1922, SKOLEM 1923) asserts, expressed informally, that if the elements of a set are replaced "reasonably" by other elements, the result is a set. More formally, we may state the axiom as:

**Rep.** *Let $R$ be a binary relation between sets such that to every set $x$ corresponds at most one set $y$ with $xRy$; then for every set $X$ there exists the set $\{y: \text{there is an } x \in X \text{ with } xRy\}$.*

Similarly, as with the axiom of separation, the allowable relations $R$ in this definition are those which can be expressed in the language of first-order predicate logic. Examples are the relations defined by

$$xRy :\Leftrightarrow x = y; \qquad xRy :\Leftrightarrow y = \{x\}.$$

One can immediately deduce from **Rep** and the latter relation that the singletons of the elements of a set themselves constitute a set. (Instead of using **Rep** one could argue from **Sep** and **Pow**.)

Rep is likewise a special case of FREGE's axiom of comprehension. It is stronger than **Sep**, and indeed **Sep** may be proved from **Rep** and the other ZERMELO axioms.

The so-called ZERMELO–FRAENKEL *system of axioms* ZF comprises the axioms described above with the exception of the axiom of choice. With the inclusion of the latter we have the system ZFC, which is the axiomatic basis most often used for dealing with problems of a set-theoretical nature.

**3. Some Consequences.** We now propose by a few simple examples to show how the ZERMELO axioms already suffice—with a few exceptions—to derive the facts of set theory that a mathematician needs. For the most part, these are the results that we used *intuitively* in 1.2 and 1.3.

(a) The *empty set* $\emptyset$: its existence was proved in 2.

(b) The Boolean *combinations*. Let $x, y$ be given sets.

The *intersection* $x \cap y$, that is, the set $\{z \in x: z \in y\}$, exists by virtue of **Sep**.

The *union* $x \cup y$ consists of the elements of $x$ and the elements of $y$, that is, of the elements of $\bigcup\{x,y\}$. Its existence follows from the fact that $\{x,y\}$ exists by **Pair**, and $\bigcup\{x,y\}$ exists by **U-Ax**.

The *difference* $x \setminus y$, the set of elements of $x$ which are not elements of $y$, exists because $\{z \in x : z \notin y\}$ exists by **Sep**.

(c) The *generalized intersection*. To every non-empty set $X$, which in this context we think of as a family of sets, corresponds the generalized *intersection* $\bigcap X = \bigcap_{y \in X} y$. The existence of this set, written as

$$\bigcap X = \{z \in \bigcup X : z \in y \text{ for all } y \in X\},$$

is seen to follow from **Sep**.

It is not difficult now to derive the well-known laws for $\cap$, $\cup$, $\setminus$, $\bigcap$, $\bigcup$.

(d) *Ordered pairs and Cartesian products*. By applying the axiom of pairing three times, the *ordered pair* $(x,y)$ may be defined as $\{\{x\}, \{x,y\}\}$. The existence of the Cartesian product $x \times y = \{(u,v) : u \in x \text{ and } v \in y\}$ can be derived as follows: if $u \in x$, then $\{u\} \subset x$, and so $\{u\} \subset x \cup y$ whence $\{u\} \in \mathcal{P}(x \cup y)$. If furthermore $v \in y$, then similarly $\{u,v\} \in \mathcal{P}(x \cup y)$.

Since $(u,v) = \{\{u\}, \{u,v\}\}$, it follows that $(u,v) \in \mathcal{P}(\mathcal{P}(x \cup y))$.

Consequently, $x \times y = \{z \in \mathcal{P}(\mathcal{P}(x \cup y)) : \text{there is a } u \in x \text{ and a } v \in y$ *with* $z = (u,v)\}$, so that the existence of the Cartesian product follows from **Sep**. (The condition "$z = (u,v)$" can easily be expressed with the symbol $\in$ alone.)

An extension of these considerations to relations of higher arity presents no difficulty, and the same applies to the derivation of the basic properties of relations and functions defined in terms of set theory (see 1.2).

(e) *Natural numbers*. Intuitively the set $\omega = \{\emptyset, \{\emptyset\}, \{\emptyset, \{\emptyset\}\}, \ldots\}$ of the VON NEUMANN natural numbers is the smallest set which contains $\emptyset$ and with every $z$, the set $z \cup \{z\}$ as well, or in other words, to use the terminology introduced in connection with the axiom of infinity, $\omega$ is the smallest inductive set. If $y_0$ is any inductive set, then we can define $\omega$ — at least intuitively — without the use of dots as:

$$\omega = \bigcap\{y \subset y_0 : y \text{ inductive}\}.$$

In this form the existence of $\omega$ can be *proved* from the ZERMELO axioms: **Inf** affirms the existence of at least one inductive set, say $y_0$. It can easily be shown that inductivity is a property allowed by **Sep**. It follows since $\mathcal{P}(y_0)$ exists, that $\{y \subset y_0 : y \text{ inductive}\}$ which is the same as $\{y \in \mathcal{P}(y_0) : y \text{ inductive}\}$ must exist and hence by (c) so must $\omega$ itself.

It is easy to show that $\omega$ is inductive. Therefore, if $n \in \omega$ then $n \cup \{n\} \in \omega$, and the successor function $\sigma$ can be defined over $\omega$ in much the same way as was used to derive the Cartesian product in (d), since

$$\sigma = \{(n, n \cup \{n\}) : n \in \omega\}$$
$$= \{z \in \mathcal{P}(\mathcal{P}(\omega)) : \text{there is an } n \in \omega \text{ with } z = (n, n \cup \{n\})\}.$$

(The proof of the existence of $\sigma$ is even simpler if we use **Rep.**)

Nothing now stands in the way of demonstrating any desired properties of $\omega$ and $\sigma$. The principle of induction, for example, now becomes trivial: The statement

> *If a property of the natural numbers holds for 0, and if whenever it holds for a natural number n it also holds for the successor $n + 1$, then that property holds for all natural numbers*

becomes, when formulated for $\omega$ and $\sigma$ in terms of set theory

> *If a subset of $\omega$ contains $\emptyset$, and if it contains $z \cup \{z\}$ whenever it contains z, then the subset is $\omega$.* In other words, *every inductive subset of $\omega$ is $\omega$ itself.*

This however follows at once from the definition of $\omega$.

Lastly, the idea of cardinality, or power, can now be defined precisely. Two sets $x, y$ are said to be *equipotent* or to have the same cardinality if there is a bijective function from $x$ to $y$. A set is *finite* if and only if it has the same cardinality as an element of $\omega$; it is *countably infinite* if it has the same cardinality as $\omega$ and *uncountable* if it is neither finite nor countably infinite. It is easily shown that two different elements of $\omega$ are never equipotent. The VON NEUMANN natural numbers therefore represent in a one-to-one fashion the cardinalities of the finite sets and under that aspect they are also called the *finite cardinal numbers.* The set $\omega$ is the smallest infinite cardinal number.

Ordinal numbers and cardinal numbers each can be equipped with an addition, a multiplication, and an exponentiation that generalize the usual arithmetical operations on $\omega$ but differ for infinite arguments. Both the ordinal and the cardinal arithmetic play an important role in the theory of sets and their applications—the first one aiming more at the process of counting and order, and the latter one more at the size of sets. In some sense, the subject would fit well in a book on numbers. It would lead, however, to a lengthy digression. The interested reader is referred to the books on set theory mentioned below.

**4. Set Theory with Classes.** To get around the RUSSELL paradox, ZERMELO weakened FREGE's comprehension axiom to the axiom of separation. His aim was to build up the universe of sets "from below" and to ban any constructions "traversing the whole universe." Nevertheless, it is possible to retain the idea behind the axiom of comprehension, which seems so reasonable and plausible, provided one builds into it a few precautionary features. All that is required is to recognize that the process of comprehension need not lead to the formation of a new set, and to admit that it may lead to objects that are "too large" to be regarded as sets, to collections of elements that in CANTOR's phrase are "absolutely infinite." To distinguish verbally, the objects that arise from FREGE's comprehension axiom are called *classes*. Thus, one talks for example of the class $V$ of all sets, the so-called *universal class*

$$V = \{x : x \text{ is a set}\},$$

or of the *class of all groups*, the class of all sets of the form $(x, f)$ where $x$ is a non-empty set and $f : x \times x \to x$ a function which satisfies the group axioms.

Classes therefore have their origin in the idea of the extensional range of the properties of sets. Thus, the elements of a class are sets, but the classes themselves need not be sets. This terminology does not agree completely with the use of the word "class" in *mathematics*. In mathematics the word *class* is often used for what are quite obviously sets. For example, one talks of *equivalence classes, residue classes* and so forth.

The revised comprehension axiom asserts that to every property $P$ of sets (satisfying certain conditions explained below) corresponds the *class*

$$C_P = \{x : x \text{ is a set and } x \text{ has the property } P\}.$$

The clash with normal mathematical linguistic usage is to a certain extent softened by the fact that, by this axiom, every set is a class, because any set $x$ can always be expressed as $x = \{z : z \text{ is a set and } z \in x\}$. The converse does not always hold good. Thus for the RUSSELL class

$$C_R = \{x : x \text{ is a set and } x \notin x\}$$

one deduces immediately from

$$C_R \in C_R \Leftrightarrow C_R \text{ is a set and } C_R \notin C_R$$

that $C_R$ is not a set, for otherwise we should have $C_R \in C_R$ if and only if $C_R \notin C_R$, which is absurd. Therefore $C_R$ is a *proper class*, that is, a class which is not a set. This shows how the RUSSELL antinomy can be eluded by distinguishing between sets and (proper) classes.

The classes mentioned earlier are also proper classes; if the universal class $V$ were a set then by **Sep** the "RUSSELL class" $C_R = \{x \in V : x \notin x\}$ would be a set and we should again have the RUSSELL antinomy. The proof is even simpler using **Found**: if $V$ were a set we should have $V \in V$ in contradiction to **Found**. In the case of the class of all groups the argument runs: As every non-empty set can be the range of a group it is easily verified (by working back) that $(\bigcup \bigcup G) \cup \{\emptyset\} = V$. Hence, if $G$ were a set, $V$ would be as well. Incidentally, the ordinal numbers constitute a proper class because otherwise we could deduce Burali–Forti's paradox (see 1.3).

In axiomatizing a set theory with classes, one has to lay down not only the rules governing sets as well as the rules governing classes but also the rules of the interplay between them. The revised comprehension axiom here plays an essential role. Just as only certain properties definable in elementary terms are allowed in **Sep** and **Rep**, so only such properties are allowed in the comprehension process. The set axioms (say of **ZFC**) are partially modified. Thus **Ext** is now stated for classes generally. **Sep** says simply that the intersection of a set with a class is again a set. The most important of systems with classes are:

(i) **NBG** *set theory*, based on the work of VON NEUMANN (1925 onwards) and developed essentially by BERNAYS and GÖDEL (1937 onwards);

(ii) KELLEY–MORSE *set theory*, WANG (1949) and MORSE (1939 onwards), which became known through the appendix to KELLEY's textbook on topology [12]. It is distinguished from **NBG** mainly by more liberal conditions on the definability of the properties $P$ in the comprehension axiom.

A set theory incorporating classes proves advantageous in branches of mathematics where proper classes form part of the subject matter under study, as for example in category theory. To a certain extent, however, the advantages are more of a linguistic nature. For a discussion in greater depth we refer the reader to LEVY [16].

## §3. SOME METAMATHEMATICAL ASPECTS

What have we achieved by an axiomatization of set theory? Certainly we have created a more precisely formulated basis for set theoretical investigations and thus raised set theory to the level of an axiomatic theory.

Warned by the inconsistency of FREGE's axiom system, we nevertheless have to ask ourselves whether the systems used today are really consistent. We have already had occasion to mention in 2.1 the so-called *second incompleteness theorem of* GÖDEL, which implies that we can never convince ourselves of this freedom from contradiction by means of a formal proof of consistency. The most we can hope for therefore is to produce arguments that appeal to our intuition, and we shall discuss some of these in **1** below. For the moment, we shall ignore the axiom of choice to which we shall return in **2**. In **3** we shall outline another possibility, which can be useful in mathematics as well. In certain circumstances, by making use of precise systems of axioms for set theory, it is sometimes possible to prove rigorously that certain problems of set theory or mathematics are insoluble.

**1. The VON NEUMANN Hierarchy.** We choose for our remarks the **ZFC** system. The first point to be made in regard to its consistency is that in several decades of intense activity no contradictions have emerged while using it. On the other hand, it has to be admitted that some contradiction might turn up tomorrow.

Next we can say in their favor that the individual axioms of **ZFC** reflect properties that appeal to our intuition as being entirely reasonable. However, the possibility cannot be excluded that the *totality* of these axioms, each reasonable enough *by itself*, may nevertheless be incompatible and thus form an inconsistent whole. Moreover, at first glance, the system appears perhaps to be rather too much determined by isolated individual aspects and accidental features.

Against this last objection, however, there is a convincing argument regarding the *content* of the system, at least for **ZF**. This is that the universe of all sets has what may be called a *cumulative hierarchical* structure. To bring out this feature we consider the VON NEUMANN *universes* $V_\alpha$ for the ordinals $\alpha = 0, 1, 2, \ldots, \omega, \omega + 1, \ldots, \omega + \omega, \ldots$ (see 1.3), which are inductively defined by

$$V_0 := \emptyset,$$
$$V_1 := \mathcal{P}(\emptyset) = \{\emptyset\},$$

and generally

$$V_{\alpha+1} := \mathcal{P}(V_\alpha),$$

and for the so-called *limit ordinals*, such as $\omega$ or $\omega + \omega$, which have no immediate predecessor, $V_\alpha$ is defined as the union of all the preceding $V_\beta$:

$$(*) \qquad V_\alpha = \bigcup_{\beta \in \alpha} V_\beta.$$

The VON NEUMANN universes are thus formed by starting from the null set, iterating the process of power-set formation over all ordinal numbers, coupled with the formation process defined by $(*)$ for the limit ordinals. They constitute a hierarchical structure, the VON NEUMANN *hierarchy*, illustrated diagrammatically below. This hierarchy is *cumulative* in the sense that each $V_\alpha$ is a subset of all $V_\beta$ with $\beta > \alpha$. One can now prove in **ZF** that every set is an element of some $V_\alpha$, or in other words that the VON NEUMANN hierarchy exhausts the universe of sets. It has even been shown by SCOTT (see [4, p. 141 et seq.]) that **ZF** is, in a sense which can be made precise, just strong enough to ensure this cumulative-hierarchical structure of the universe of sets. *Our intuitive conviction that such a structure is an admissible concept thus carries over into a corresponding conviction in regard to* **ZF**.

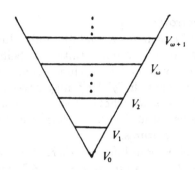

**2. The Axiom of Choice.** The axiom of choice was first mentioned and criticized as a principle of inference at the turn of the last century by the Italians PEANO, BETTAZZI and LEVI. It had already been applied before then by CANTOR and DEDEKIND in the context of set theory. The first explicit formulation by ZERMELO (1904) is then to be found in the form stated in 2.2:

**AC.** *To every set corresponds a choice function.* An equivalent statement is: *if $X$ is a non-empty set of non-empty sets, then the direct product of the elements of $x$ is not void* (because this product consists of the functions $f: X \rightarrow \bigcup X$ with $f(x) \in x$ for $x \in X$, that is to say the choice functions associated with $X$).

For his axiom system of the year 1908, ZERMELO used another form **AC'** whose equivalence to **AC** was proved by RUSSELL in that same year.

**AC'.** *If $X$ is a set of mutually disjoint sets, then there exists a set which has exactly one element in common with each element of $X$. In other words, to every equivalence relation corresponds a system of representatives.*

The equivalence of **AC** to **AC'** is easily proved as follows.

**AC $\Rightarrow$ AC':** If $X$ is a set of mutually disjoint non-empty sets, then the image of a choice function associated with $X$ is a set with the properties required by **AC'**.

**AC' $\Rightarrow$ AC:** Let $X$ be a set, without loss of generality, $\emptyset \notin X$. We can define an equivalence relation $\sim$ between the elements of the set $\{(y, z): y \in X \text{ and } z \in y\}$ by

$$(y, z) \sim (y', z'): \text{ if and only if } y = y'.$$

Let $S$ be a system of representatives corresponding to the equivalence relation $\sim$ which exists by virtue of **AC'**. Then $S$ is the graph of a choice function associated with $X$.                                                     □

A wealth of statements equivalent to **AC**, on the basis of the axioms of **ZF**, are known today. One of these which is of particular significance to mathematics, is a lemma that goes back to HAUSDORFF (1909, 1914), but which became familiar to mathematicians through work of **Zorn** (1935) and is now generally known as

**ZORN's Lemma.** *Any partially ordered set, whose linearly ordered subsets each have an upper bound, contains at least one maximal element.*

The axiom of choice in the formulation **AC** appears intuitively very plausible. Nevertheless, it has given rise to much controversy in set theory and mathematics. The arguments and remarks of ZERMELO in [21] are particularly instructive and refreshing in this connection. We shall mention here a few points which illustrate the special position of the axiom of choice, and try to explain some of the criticisms that have been levelled against it.

(a) "*Lack of constructiveness.*" The **ZF** axioms are all formulated, or can be formulated (as far as **Inf** in particular is concerned) in such a way that the sets whose existence is postulated in the axioms (that is, the pair set in the axiom of pairing, the power set in the axiom of power sets, the set $\omega$ in the reformulated axiom of infinity, and so on) *can all be defined explicitly* starting from the appropriate initial set. With the axiom of choice this is *not* the case: **AC** does not demand the existence of *definable* (in some reasonable sense) choice functions; **AC'** does not demand the existence of *definable* systems of representatives. One can readily appreciate the "non-constructive" nature of the axiom of choice (which can be shown to be unavoidable), if one tries to define a choice function on $\mathcal{P}(\mathbb{R})$, or if one tries to define a system of representatives for the equivalence relation on $\mathbb{R}$, under which two real numbers are equivalent if and only if their difference is rational.

Such systems of representatives provide examples of sets of real numbers that are not Lebesgue-measurable (VITALI, 1905). The use of **AC** here is essential, because it was shown in the other direction by SOLOVAY [17] that a weaker form of the axiom of choice, which suffices for analysis and measure theory, is compatible with the requirement that every subset of $\mathbb{R}$ is Lebesgue-measurable.

The "non-constructiveness" of **AC** has the consequence that proofs which make an essential use of the axiom of choice or ZORN's lemma are themselves in a broad sense, non-constructive. For example, the usual proof that every vector space has a basis gives no inkling of what such a base would look like in an individual case. Thus **ZFC** does not, for example, guarantee the existence of a definable HAMEL basis, that is, a definable basis of $\mathbb{R}$ as a vector space over $\mathbb{Q}$.

(b) *"Paradoxes."* **AC**—in conjunction with the other axioms—has led to some presumably paradoxical consequences. We shall mention the sphere paradox of TARSKI and BANACH (1924). A solid sphere of unit volume can be broken up into a finite number of pieces in such a way that two new unit spheres can be reassembled from the pieces. (Obviously the pieces cannot be measurable, so that the original sphere could not be cut up into pieces with a saw!)

In order to advance the discussion—as with the axiom of parallels in Euclidean geometry—research has been done to find out whether the axiom of choice could be proved or disproved from the other axioms. Neither of these a priori possibilities is true (COHEN 1963, GÖDEL 1938), so that its independence from **ZF** has been demonstrated (of course under the assumption that **ZF** is consistent). In particular, the consistency of **ZF** is not destroyed by assuming **AC** (since otherwise **AC** could be refuted in **ZF**). This justification in terms of proof theory also strengthens mathematics in its judgement: as the numerous and far-reaching applications of ZORN's lemma in the most varied branches of mathematics show, mathematicians seem now to have decided *in favor* of the axiom of choice.

In a few cases, however, the axiom of choice is dispensable. Thus, it can be shown without **AC** that to every *finite* set corresponds a choice function. For countable sets or sets of countable sets this is no longer true in general. On sets of sets of natural numbers a choice function can be defined by choosing the smallest number in each set, and this can be done without the help of the axiom of choice. Furthermore, it can be shown (using GÖDEL's constructible sets) that **AC** is not needed to prove any of the theorems of *number theory*. The reference [11] contains an almost exhaustive account of the axiom of choice.

**3. Independence Proofs.** There is a whole series of problems in mathematics which, despite intensive efforts have so far defied all attempts at finding a solution. Thus, for example, FERMAT's last theorem, the conjecture that for $n \geq 3$, and any positive natural numbers $a$, $b$, $c$

$$a^n + b^n \neq c^n$$

falls into this category. The lack of success in solving such problems may simply be due to trivial reasons: The solutions are there but have not yet been found. On the other hand, there may be more deep-seated causes, for example, reasons connected with the *complexity* of the problem. There may exist solutions, but every solution is so inconceivably lengthy that it could never be found in any foreseeable period of time. Ideas of this kind are suggested by recent results of *complexity theory* (see, for example, [10] or [18]). Finally, there may be reasons of a more fundamental nature: the problem may have no solution in the sense that a solution is impossible in principle.

Such results rest on the assumption that there is a methodological basis for mathematics, to which one can refer. In the axiom systems of set theory, such as **ZFC**, such a foundation is available to us. Meanwhile powerful methods have been developed that allow us to arrive at results on unprovability. Essentially, this requires using the methods of constructible sets and the forcing methods to which we have already repeatedly referred. An account of these methods will be found in [14]. The unprovability results—for example, those in the preceding section—make use of these techniques almost without exception.

One of the earliest single successes was the proof of the independence of the continuum hypothesis. CANTOR had over and over again tried to prove this hypothesis. On more than one occasion he had expressed himself confidently [2, p. 192, 244], particularly as he had been able to achieve some partial results, such as the proof for open and closed sets of numbers. HILBERT placed the continuum problem at the head of a list of twenty-three unsolved problems, which he regarded as promising in a talk given at the International Congress of Mathematics held in Paris in 1900: A proof of the continuum hypothesis would have shown that the continuum possesses the smallest uncountable cardinal number and would thereby have helped to bridge the gulf between the countable and the uncountable. The independence proof shows that CANTOR's efforts were doomed to failure.

Another example of an independent statement is the so-called SOUSLIN *hypothesis*, which postulates that the ordered set of real numbers can be characterized by the properties of being dense, having no first and last element, and being complete and in addition cannot contain any uncountable set of mutually disjoint open intervals.

One cannot so far exclude the possibility that the FERMAT conjecture (FERMAT's last theorem) is independent of **ZFC**. However, unlike the situation in respect of the continuum hypothesis or SOUSLIN's hypothesis a proof of its independence would automatically imply its truth. For if the FERMAT conjecture were false there would have to be a counterexample whose validity could be checked, on the basis of **ZFC**, by simply working out the numbers on either side of the equation. Thus independence could only occur if the conjecture were true. A similar argument applies to all statements about natural numbers consisting of a finite set of universal quantifiers associated with a quantifier-free nucleus, for example, any Diophantine equation or its negation. One can make similar inferences in the same way about statements equivalent to these. Examples are the GOLDBACH conjecture (*that every even number $\geq 4$ is a sum of two primes*) and, despite its "analytic" appearance, the Riemann hypothesis.

EPILOGUE

As we have seen by means of examples the present axiom systems of set theory suffice to provide set-theoretic models for the various objects in mathematics and the techniques for handling them. An understanding of mathematics based on set-theoretical foundations is not only helpful because of the clarification of mathematical concepts, but also because it opens the door to a well-stocked storeroom of methods from set theory, and creates a unified axiomatic basis for mathematics.

How reliable is this basis? We cannot prove that it is consistent; we can only produce intuitive supporting arguments, such as the naturalness of the VON NEUMANN cumulative hierarchy. Suppose it is free from contradiction (as we have hitherto assumed). How far would this take us? We have already seen that there are limits, as the independence of the continuum hypothesis and SOUSLIN's hypothesis have taught us, and many other examples could be adduced.

Furthermore, this *incompleteness* of the axiom systems of set theory is, by a theorem of GÖDEL (see [5, p. 226 et seq.]), inescapable. All we can do is to mitigate some of the consequences each time we try something concrete, say by extending **ZFC** through the addition of axioms which seem reasonable to us. A wealth of proposals in this connection have already been discussed. So far no new principles have emerged that are generally accepted on all sides. Faced with this dilemma, one could perhaps agree to put up with having various different extensions, either because of their reasonableness or plausibility, or because of their methodological interest, perhaps even mutually incompatible extensions. Geometry has shown us how fruitful such a development can be.

A more radical departure might be to cut loose completely from the idea of basing mathematics on set theory in the Cantorian mould, and some interesting experiments in this direction have already been started, for example in category theory or in the so-called *alternative set theory* [19], which is oriented towards the requirements of non-standard analysis. How far these developments will succeed in challenging Cantorian set theory, which is now extending into its second century in full vigor, only the future—perhaps the remote future—can tell.

REFERENCES

[1] BOLZANO, B.: Paradoxien des Unendlichen. Leipzig 1851; English translation: *Paradoxes of the Infinite*. Routledge 1950

[2] CANTOR, G.: Gesammelte Abhandlungen mathematischen und philosophischen Inhalts. Edited by E. Zermelo. Berlin 1933

[3] DEDEKIND, R.: Gesammelte mathematische Werke. Edited by R. Fricke, E. Noether, Ö. Ore. Braunschweig 1932

[4] EBBINGHAUS, H.-D.: Einführung in die Mengenlehre. Darmstadt 1976

[5] EBBINGHAUS, H.-D., FLUM, J., THOMAS, W.: Mathematical logic. New York 1984

[6] FREGE, G.: Grundgesetze der Arithmetik I, II. Jena 1893/1903

[7] HALMOS, P.R.: Naive Set Theory. Springer-Verlag, 1974

[8] HEYTING, A.: Intuitionism, An Introduction. Amsterdam 1956

[9] HILBERT, D.: Über das Unendliche. Math. Ann. 95 (1926)

[10] HOPCROFT, J.E., ULLMAN, J.D.: Introduction to Automata Theory, Languages, and Computation. Reading MA 1979

[11] JECH, J.: The Axiom of Choice. Amsterdam 1973

[12] KELLEY, J.L.: General Topology. Princeton 1955

[13] KRONECKER, L.: Über einige Anwendungen der Modulsysteme und elementare algebraische Fragen. J. Reine Angew. Math. 99 (1886)

[14] KUNEN, K.: Set Theory. An Introduction to Independence Proofs. Amsterdam 1980

[15] LEVY, A.: Basic Set Theory, Heidelberg 1979

[16] LEVY, A.: The Role of Classes in Set Theory. In: Sets and Classes (Edited by G.H. Müller). Amsterdam 1976

[17] SOLOVAY, R.M.: A Model of Set Theory in Which Every Set of Reals is Lebesgue-Measurable. Ann. Math. 92 (1970)

[18] SPECKER, E., V. STRASSEN: Komplexität von Entscheidungsproblemen. Ein Seminar. Heidelberg 1976

[19] VAUGHT R.L.: Set Theory. Boston 1985

[20] VOPENKA, P.: Mathematics in the Alternative Set Theory. Leipzig 1979

[21] ZERMELO, E.: Neuer Beweis für die Möglichkeit einer Wohlordnung. Math. Ann. 65 (1908)

[22] ZERMELO, E.: Untersuchungen über die Grundlagen der Mengenlehre. I. Math. Ann. 65 (1908)

English translations of [9], [21], and [22] can be found in: J. van Heijenoort (ed.): From Frege and Gödel. Cambridge, MA 1967.

# Name Index

Abel, Niels Henrik (1802–1829) 22, 98, 108

Abhyankar, Sheeram 108

Abū Kāmil (ca. 850–ca. 930) 32

Adams, J. Frank 275, 301

Ahmes (ca. 1900 BC) 125

Albert of Saxony (1320–1390) 356

d'Alembert, Jean le Rond (1717–1783) 91, 98, 102, 103, 104, 106, 109

Alexander, James Waddell (1888–1971) 287

Alexandroff, Paul (1896–1982) 222

Apollonius (2nd half, 3rd c. BC) 126

Āraybhata (born 476 AD) 127

Archimedes (287–212 BC) 31, 32, 125, 126, 127

Archytas von Tarent (428–365 BC) 44

Argand, Jean Robert (1768–1822) 3, 60, 67, 98, 103, 108, 111

Aristotle (384-322 BC) 12, 27, 35, 37, 149

Artin, Emil (1898–1962) 227

Atiyah, Michael Francis (1929–) 290

Bachmann, Paul Gustav Heinrich (1837–1920) 34, 44

Bacon, Francis 331

Baltzer, Richard (1818–1887) 129

Banach, Stefan (1892–1945) 376

Beckmann, Petr 125

Bernays, Paul J. (1888–1977) 372

Bernoulli, Jakob (1654–1705) 33, 147

Bernoulli, Johann (1667–1748) 33, 147

Bernoulli, Nikolaus (1687–1759) 100

Berzelius, Jöns Jakob (1779–1848) 60

Bessel, Friedrich Wilhelm (1784–1846) 60, 110

Bettazzi 374

Bézout, Étienne (1739–1783) 65

Boetius (ca. 480–524 AD) 127

Bolyai, Wolfgang (1775–1856) 63

Bolzano, Bernhard (1781–1848) 15, 34, 44, 64, 356

Bombelli, Rafael (1526–1572) 55, 57

Boole, George (1815–1864) 368

Borel, Émile (1871–1956) 49

Bos, W. 272

Bott, Raoul (1923–) 5, 281, 298

Bourbaki, Nicolas 16, 120, 122

Brahmagupta (598–665) 13

Brouncker, Lord W. (1620–1684) 151

Brouwer, Luitzen Erbertus Jan (1881–1966) 35, 364

Bruck 263

Bühler, W.K. (1944–1986) 111

Burali-Forti, Cesare (1861–1931) 364

Cantor, Georg (1845–1918) 3, 5,
  12, 14, 16, 34, 35, 39, 40,
  355, 356, 357, 362, 363,
  365, 374, 377
Cardano, Girolamo (1501–1576) 2,
  13, 57, 98, 101
Cartan, Henri (1904–) 51
Cauchy, Augustin Louis (1789–1857)
  33, 34, 62, 63, 108, 109,
  111, 112
Cayley, Arthur (1821–1895) 89, 181,
  194, 196, 197, 213, 220,
  221, 249
Chrystal, George (1851–1911) 109
Clausen, Thomas (1801–1885) 83
Clifford, William Kingdon (1845–
  1879) 221
Cohen, Paul J. (1934–) 357
Conway, John H. 5, 329, 330, 332,
  337
Copson, Edward Thomas (1901–)
  70
Cotes, Roger (1682–1716) 96
Crowe, Michael J. 193

Daguerre, Louis Jacques Mandé
  (1787–1851) 111
Dedekind, Richard (1831–1916) 2,
  9, 14, 15, 16, 19, 21, 22,
  34, 35, 39, 64, 73, 78, 97,
  120, 181, 183, 187, 332,
  356, 357, 361, 363, 364,
  365, 374
Degen, Carl Ferdinand (1766–1825)
  259
Descartes, René (1596–1650) 13,
  19, 33, 58, 99, 100
Dickson, Leonard Eugene (1874–
  1954) 260
Dieudonné, Jean (1906–) 98, 154
Diophantus of Alexandria (2nd half,
  3rd c. AD) 75
Dugac, P. 34
Dürer, Albrecht (1471–1528) 126

Eckmann, Beno (1917–) 275, 301

Eilenberg, Samuel 205
Engel, Friedrich (1861–1941) 192
Estermann, T. 114
Euclid (ca. 295 BC) 12, 22, 28,
  31, 35
Eudoxos of Knidos (400–347 BC)
  31, 34, 35
Euler, Leonhard (1707–1783) 33,
  55, 59, 67, 69, 91, 93, 94,
  100, 101, 102, 103, 104,
  105, 106, 108, 110, 125,
  147, 151, 209, 219
Eurytos 12
Eutocios (born ca. 480 AD) 126

Fermat, Pierre de (1601–1665)
  209, 267, 376
Ferrari, Ludovico (1522–1565) 13
del Ferro, Scipio (1465–1526) 13
de Foncenex, Daviet (1734–1799)
  107
Fraenkel, Adolf Abraham (1891–
  1965) 5, 365
Frege, Gottlob (1848–1925) 14, 16,
  35, 363, 364, 365
Fresnel, Augustin Jean (1788–
  1827) 64
Frobenius, Georg (1849–1917) 4,
  106, 130, 221, 222, 227,
  228, 229, 230, 266

Galois, Evariste (1811–1832) 22
Gauß, Carl Friedrich (1777–1855)
  13, 60, 62, 63, 67, 98, 102,
  104, 105, 106, 107, 108,
  109, 110, 117, 120, 122,
  181, 192, 210, 221, 267,
  268
Gay-Lussac, Louis Joseph (1778–
  1850) 111
Gelfand, Izrail' Moiseevich
  (1913–) 223, 245
Gelfond, Alexandr Osipovich
  (1906–1968) 153
Gibbs, Josiah Willard (1839–
  1903) 200

Girard, Albert (1595–1632) 99, 100, 101

Gödel, Kurt (1906–1978) 357, 365 372, 378

von Goethe, Johann Wolfgang (1749–1832) 6, 123

Goldbach, Christian (1690–1764) 100, 101, 192, 209

Gordon, I. 287

Grassmann, Hermann Günther (1809–1877) 192, 200, 221

Grauert, Hans (1930–) 172

Graves, John T. 181, 190, 221, 249, 259

Gregory, James (1638–1675) 128

Grimm, Jacob (1785–1863) 111

Grothendieck, Alexander (1928–) 298

Hadamard, Jacques S. (1865–1963) 145

Hamel, Georg (1877–1954) 376

Hamilton, William Rowan (1805–1865) 5, 63, 65, 66, 120, 181, 186, 187, 189, 190, 191, 192, 193, 194, 198, 200, 203, 210, 213, 221, 222, 249, 256, 266

Hankel, Hermann (1839–1873) 22, 98, 106, 109, 119, 120, 181

Happel, Dieter 229

Hardy, Godfrey Harold (1877–1947) 130, 150

Harriot, Thomas (1560[?]–1621) 99

Hasse, Helmut (1898–1979) 165

Hausdorff, Felix (1868–1942) 375

Helgason 272

Hensel, Kurt (1861–1941) 155

Hermite, Charles (1822–1901) 152

Hilbert, David (1862–1943) 22, 35, 47, 152, 266, 356, 357, 365, 377

Hill, Thomas 193

Hippasus (2nd half, 5th c. BC) 28, 29

Hirsch, Morris W. (1933–) 115

Hirzebruch, Friedrich Ernst Peter (1927–) 290

Hölder, Ludwig Otto (1859–1937) 50

Hopf, Heinz (1894–1971) 4, 222, 236, 237, 281, 287, 289, 290, 298

L'Hospital, Guillaume-Francois-Antoine de (1661–1704) 307

Huizinga, J. 353

von Humboldt, Alexander (1769–1859) 110

Hurwitz, Adolf (1859–1919) 4, 78, 152, 182, 265, 266, 268, 274, 275

Husserl, Edmund (1859–1938) 363

Huygens, Christiaan (1629–1695) 58

al-Hwārizmī (ca. 9th c. AD) 127

Iwamoto 150

Jacobi, Carl Gustav Jacob (1804–1851) 108

Jannsen, Uwe (1954–) 177

Juschkewitsch, Adolf Pavlowitsch (1906–) 124

Kant, Immanuel (1724–1804) 63

Kaplansky, Irving (1917–) 271

al Kāšī (died 1429) 127

Kästner, Abraham Gotthelf (1719–1800) 109

Keisler, H. Jerome 327

Kelley, John L. 372, 379

Lord Kelvin (Thomson, William) (1824–1907) 193, 198

Kervaire, Michel A. 5, 281, 290

Klein, Felix (1849–1925) 266

Kleinfeld 263

Kneser, Adolf (1862–1930) 105

Kneser, Hellmuth (1898–1973) 115, 121, 122
Kneser, Martin (1928–) 115
Knopp, Konrad (1882–1957) 109, 143
Knuth, Donald Ervin (1938–) 354
Kolmogorov, Andrei Nikolaevich (1903–) 287
Kronecker, Leopold 9, 19, 22, 63, 129, 266, 364, 379
Kummer, Ernst Eduard (1810–1893) 356
Kuratowski, Kazimierz (1896–1980) 359

Lagrange, Joseph Louis (1736–1813) 103, 104, 105, 110, 122, 209, 267
Lambert, Johann Heinrich (1728–1777) 149, 150, 151
Landau, Edmund (1877–1938) 21, 39, 109, 124, 130
Laplace, Pierre Simon de (1749–1827) 3, 98, 103, 105, 110, 122
Laurent, Pierre Alphonse (1813–1854) 129
Lebesgue, Henri Léon (1875–1941) 376
Lefschetz, Solomon (1884–1972) 288
Legendre, Adrien-Marie (1752–1833) 150, 151, 152, 265, 266
Leibniz, Gottfried Wilhelm (1646–1716) 27, 33, 34, 35, 37, 58, 100, 128, 131, 143, 307
Leonardo of Pisa (1170–1240[?]) 13, 127
Lessing, Gotthold Ephraim (1729–1781) 355
Levi, Beppo (1875–1961) 374
von Lindemann, Carl Louis Ferdinand (1852–1939) 152

Liouville, Joseph (1809–1882) 152
Lipschitz, Rudolf (1832–1903) 35, 40, 78, 109
Liszt, Franz (1811–1886) 111
Littlewood, John Edensor (1885–1977) 114
Liu Hui (ca. 250 AD) 127
Lobachevsky, Nikolai (1793–1856) 110
Ludolph van Ceulen (1540–1610) 127

Machin, John (died 1751) 144
Mal'cev, A.I. (1909–1967) 280
von Mangoldt, Hans Carl Friedrich (1854–1925) 109
Massey, W.S. 280
Mazur, Stanislaw (1905–1981) 4, 223, 238, 243, 244, 245
Mendelssohn-Bartholdy, Felix (1809–1847) 111
Méray, Charles (1835–1911) 34, 39
Milnor, John W. (1931–) 5, 281, 290, 298
Minkowski, Hermann (1864–1909) 130, 266
de Moivre, Abraham (1667–1754) 93, 94, 96
Morse, Anthony P. 372
Morse, Marston (1892–1977) 298

Narmer 9
Napoleon I. (1769–1827) 103
Nelson, Edward 327
Netto 98
von Neumann, John (1903–1957) 16, 361, 362, 364, 368, 372
Newton, Isaac (1642–1727) 58, 91, 94, 121, 131
Niven, Ivan 150, 205
Noether, Emmy (1882–1935) 222, 287

Ögmundsson, Vilhjálmur 272

Ohm, Martin (1792–1872) 22
Opolka, H. 210
Ostrowski, Alexander (1893–1991) 707, 243, 245
Otho, Valentin 127
Oughtred, William (1575–1660) 125

Palais, Richard S. 230
Peano, Guiseppe (1858–1932) 14, 18, 19, 35, 374
Peirce, Benjamin (1809–1880) 193, 221, 222
Peirce, Charles Sanders (1839–1914) 222
Perron, Oskar (1880–1975) 151
Pfaff, Johann Friedrich (1765–1825) 104, 110
Plato (427–348/347 BC) 31, 126
Poincaré, Henri (1854–1912) 286, 287
Poinsot, Louis (1777–1859) 122
Popper, K.R. 126
Ptolemy, Claudius (100–170 AD) 32, 82, 127
Puiseux, Victor (1820–1883) 102

Radon, Johann (1887–1956) 274, 275
Raleigh, Sir Walter (1552–1618) 99
Remmert, Reinhold (1930–) 172
Riemann, Georg Friedrich Bernhard (1826–1866) 64, 124
Riese, Adam (1492–1559) 13
Robinson, Abraham (1918–1974) 5, 306
Rolle, Michael (1652–1719) 96
Roth, Peter (died 1617) 99
Rothe, Hermann (1882–1923) 192
Rüchert, F. 111
Rudio, Ferdinand (1856–1929) 125
Russell, Bertrand (1872–1970) 35, 364, 365, 374

Samelson, Hans 287

Scharlau, W. 210
Schiller, Friedrich (1759–1805) 6
Schneider, Theodor (1911–) 153
von Schlegel, August Wilhelm (1767–1845) 111
Schreier, Otto (1901–1929) 109
Schubert, Hermann Cäsar Hannibal (1848–1911) 266
Scott, Dana S. 373
Shirshov 263
Siegel, Carl Ludwig (1896–1981) 154
Simson, Robert (1687–1768) 85
Slater 263
Skolem, Thoralf (1887–1963) 365, 366
Smale, Stephen (1930–) 115
Solomon 125
Solovay, Robert M. 375
Souslin, M.J. 378
Sperner, A. 109
Springer, Tonny Albert (1926–) 237
Stasheff, James D. 298
Steenrod, Norman E. 205, 297
Steinitz, Ernst (1871–1928) 22
Stevin, Simon (1548–1620) 33, 44
Stibitz, George R. 65
Stiefel, Eduard (1909–1978) 33, 289, 290, 298
Stifel, Michael (1487–1567) 13, 33
Stirling, James (1692–1770) 93
Study, Eduard (1862–1930) 56, 181

Tarski, Alfred (1902–1983) 376
Tartaglia, Niccoló (1499/1500–1557) 57
Tate, John Torrence (1925–) 172
Theodoros von Kyrene (465–399 BC) 31
Tieck, Ludwig (1773–1853) 111
Tropfke, Johannes (1866–1939) 104, 124

Ulug Berg (1394–1449) 127

de Valera, Eamon (1882–1975) 193

Varignon, Pierre (1654–1722) 307
le Vavasseur 98
Vieta, François (1540–1603) 99,
        101, 142
Vitali, Giuseppe (1875–1932) 375
van der Waerden, Bartel Leendert
        (1903–) 98, 266

Wallace, William (1768–1843) 85
Wallis, John (1616–1703) 33, 59,
        125, 143
Wang Fan (228–266 AD) 127
Wang, Hao 372
Weber, Heinrich (1842–1913) 22,
        23
Weber, Wilhelm Eduard (1804–
        1891) 110
Weierstraß, Karl Theodor Wilhelm
        (1815–1897) 3, 33, 34, 39,
        44, 114, 119, 120, 129,
        143, 148, 149, 181, 187,
        221, 266, 305
Wessel, Caspar (1745–1818) 60, 67

Weyl, Hermann (1885–1955) 222
Whitehead, George William
        (1918–) 301
Whitney, Hassler (1907–1988) 287,
        292, 298
Wiener, Norbert (1894–1964) 360
Wingberg, Kay 177
Witt, Ernst (1911–1991) 245
Wright, E.M. 150

Yaglom, Issak Moiseevich
        (1921–) 79

Zassenhaus, H. 98
Zermelo, Ernst (1871–1953) 5, 355,
        358, 361, 362, 364, 365,
        366, 370, 374, 375
Zhang Heng (78–139 AD) 127
Zorn, Max (1906–1993) 4, 226, 249,
        261, 262, 263, 375
Zu Chong-Zhi (430–501 AD) 127,
        151
Zuse, Konrad (1910–) 65

# Subject Index

absolute value, 74
absolutely convergent, 240
absolutely infinite, 364
Adams's theorem, 300
algebra
    alternative, 225, 250
    associative, 184
    Banach, 239
    composition, 267
    division, 182, 186
    normed, 239
    power associative, 184
    quadratic, 227, 250
    real, 182, 183
    vector product, 266, 276
algebraic extension, 68
algebraically closed, 97
alternative algebra, 226
alternative quadratic algebras, 250
alternative set theory, 378
anti-commutative algebra, 184
antipodal map, 283
Archimedean order, 23, 43
Archimedean property, 48
Archimedes, principle of, 38
Argand plane, 60, 69
Argand's inequality, 112, 113
associative law (for algebras), 183
associator, 226
axiom of choice, 367, 374
axiom of comprehension, 363
axiom of existence, 365
axiom of extensionality, 365
axiom of foundation, 367
axiom of infinity, 367
axiom of pairing, 366

axiom of power sets, 367
axiom of replacement, 367
axiom of rotation, 218
axiom of separation, 366
axiom of union, 366

Banach algebra, 239
Banach–Tarski paradox, 376
Bernoulli numbers, 148
Betti numbers, 284, 286
bilinear form (of quadratic algebra), 251
Bott periodicity, 290, 294
Buralli–Forti antinomy, 364

cancellation law, 18
Cantor field, 43
cardinal numbers, 370
Cartesian product, 369
Cauchy criterion, 40, 43, 44
Cauchy–Schwarz inequality, 79, 199
Cauchy sequence, 40, 48, 167
Cauchy's minimum theorem, 112
Cayley algebra, 256, 257
    uniqueness, 262
Cayley–Hamilton theorem, 196
Cayley theorem, 215
center of an algebra, 201
centroid, 80
characteristic classes, 295, 296
choice axiom, 367, 374
choice function, 374
circle group, 75
classes, set theory, 371
classes of games, 338
cohomology, 283, 286

commensurable, 31
commensurable segments, 28
commutative law (for algebras),
        183
complete induction, 15
completeness, 36
completeness of reals, 44
completing the square, 76
        quaternionic, 205
complexity theory, 376
composition algebra, 267
comprehension axiom, 363
conjugation (complex), 71
conjugation (of algebras), 253
conjugation (of octonions), 257
constructible sets, 357
constructivism, 364
continued fraction expansion, 29,
        149
continuum hypothesis, 326, 357
convergence, 40
convex subring, 314
Conway game, 332
Conway numbers, 349, 367
Conway postulates, 346
cosine (complex), 138
countably infinite, 370
covering space, 233
cross product, 198
cross ratio, 80
cubic equations, 57
cuneiform, 10
curl of a vector field, 203
cyclic quadrilateral, 82

Dedekind cut, 36, 43, 330
Dedekind postulates, 331
degree mod 2, 290
de Moivre's formula, 93, 145
density (of $\mathbb{Q}$), 23
Descartes' rule of signs, 97
differential, 321
differential quotient, 321
Diophantine problem, 162
discriminant of polynomial, 103

Disquitiones Arithmeticae, 110
divergence, 203
divisor of zero, 184
DOMINOS, 335
duplication process, 257
duplication theorem, 261

eigenvalues, 117
eight squares theorem, 259
Elements (of Euclid), 12, 28
endomorphism (of an algebra), 202
ENIAC, 65
epimorphism theorem, 90, 134
equal value games, 338
equipotent, 370
Euclidean algorithm, 30
Euclidean distance, 74
Euler characteristic, 288
Euler formula, 59, 90, 138, 146
Euler product formula, 146
Euler series, 147
Eulerian parametrization ($SO(3)$),
        219
Eulerian parametrization ($SU(2)$),
        212
existence axiom, 365
exponential function (complex), 123,
        131
exponential rule, 184
exponential series, 131
extension, 326, 359
extensionality axiom, 365

factorial ring, 118
factorization (of polynomials), 116
factorization lemma, 115
Fermat's last theorem, 377
field extension, 66
filter, 311
finiteness condition, 334
first-order predicate logic, 366
forcing method, 357
foundation axiom, 367
four squares theorem, 209
Frobenius's lemma, 227

fundamental sequence, 34, 39, 40
fundamental theorem (for quaternions), 204, 205
fundamental theorem of algebra, 60, 97, 109

game
    negative of, 339
    positive or negative, 337
games
    isomorphic, 341
    partial ordering, 341
    sum of, 340
game concept, 334
Gelfand–Mazur theorem, 78, 242
generalized intersection, 369
Girard's thesis, 99
Goldbach conjecture, 377
golden section, 30
Grassmann identity, 199
greatest lower bound, 47

Hamilton relations, 194
Hamilton's theorem, 216
Hamiltonian multiplication, 194
Hamiltonian triple, 223, 224
Heine–Borel property, 50
Hensel's lemma, 173
hierarchy of types, 358
hieroglyphs, 9
homology, 283, 286
homotopy groups of spheres, 299
Hopf bundle, 292
Hopf construction, 299
Hopf invariant, 299
Hopf lemma, 232
Hopf's theorem, 236
Hurwitz–Radon theorem, 274
Hurwitz's theorem, 274
hypercomplex numbers, 119

imaginary part (of quaternions), 198
imaginary part, complex numbers, 67

imaginary space, 223
imaginary space (of octonions), 257
imaginary unit, 67
implicit function theorem, 235
incommensurable, 28, 30
incompleteness theorem of Gödel, 372
independence lemma, 224
independence of axioms, 357
independence proofs, 376
induction principle (for games), 335, 344
induction principle for numbers, 347
infinitesimal, 33, 314
infinity axiom, 367
inner automorphism (of an algebra), 201
integers, 20
integral domain, 21
interchange rule, 200
intermediate value theorem, 34, 102
intuitionism, 364
inverse limit, 160
irrational numbers, 32
irrationality of pi, 150
isometric, 85

Jacobi identity, 199

$K$-theory, 281, 294
Kelley–Morse set theory, 372
$KO(X)$, 293

Laplace operator, 203
least upper bound, 47
left class, 330
Leibniz series, 143
Lie algebra, 184, 279
limit ordinals, 373
limit theorem, 320
Lindemann–Weierstrass theorem, 153
linear form, of quadratic algebra, 250

local-global principle, 165
local homeomorphism, 232
Lorentz metric, 207

Malcev algebra, 279
maximal ideal, 155
maximum modulus principle, 110
Möbius strip, 292
modulus, 74
mutation of algebras, 272

*N, 323
nth roots, 77, 89
Nabla operator, 203
natural numbers, 14, 369
negative games, 337
negative number, 42
neighboring elements, 306
nesting of intervals, 32
net, 45
NGB set theory, 372
NIM, 334
non-singular bilinear form, 271
nonstandard analysis, 306
norm, 167
normed algebra, 239
null sequence, 42, 168

O(2), 87
octaves (see octonions), 250
octonions, 250, 257
odd mapping, 282
ontological purpose, none, 360
open mapping theorem, 110, 113
order relation in ℤ, 21
ordered field, 68
ordered pair, 359
ordinal numbers, 362, 370
orthogonal group, 213
orthogonal mapping, 86
orthogonal vectors, 74

p-adic expansion, 156
p-adic integer, 157
p-adic number, 157
p-adic valuation, 166

pairing axiom, 366
parallelizable, 288
partial ordering of games, 341
Peano's axioms, 18
Pentagram, 29
Poincaré duality theorem, 284
polar coordinates (for quaternions),
    204
polar coordinates, complex, 89, 91,
    141
positive games, 337
positive number, 42
power associative algebra, 226
power rule, 133
power set axiom, 367
prime number theorem, 110
prime polynomials, 118
primitive root of unity, 95
projective space, real, 236, 282
properly orthogonal mappings, 214
Ptolemy number, 83
Ptolemy's theorem, 82
Puiseux theorem, 102
purely imaginary, 67
purely imaginary (for quaternions),
    213
purely imaginary elements, of al-
    gebra, 224
Pythagorean quintuplet, 212
Pythagorean triplet, 89
Pythagoreans, 12

quadratic algebra, 227
quadratic equations, 57
quaternionists, 193
quarternions, 63, 194
quaternions lemma, 229
quaternions
    of unit length, 211
    rational, 211

*ℝ, 309
rational net, 45
rational numbers, 22
rational quaternions, 211

rationally convergent sequence, 40
real algebraic curves, 106
recursion theorem, 16
replacement axiom, 367
Riemann hypothesis, 377
right class, 330
root of a polynomial, 98
roots of unity, 94
rotation, 86
Rouché's Theorem, 108
Russell antinomy, 364

scalar product, 71
separation axiom, 366
sexagesimal fractions, 43
sine (complex), 138
$SO(2)$, 87
$SO(4)$, 217
$SO(n)$, 218
Souslin's hypothesis, 377
Span($M$), 288
spectrum, 239, 244
splitting field, 68, 108
squares theorem (general), 265
standard part, 306
Stiefel classes, 289
Stiefel–Whitney classes, 296
Stirling's formula, 94
strategy, 336
$SU(2)$, 212
symmetric matrices, 184
symmetric polynomial, 121

three-party theorem, 80
totally ordered field, 43, 47
transcendence of $e$, 152
transcendence of $\pi$, 151
transfer principle, 319
transference, 318
transfinite ordinals, 355
translation invariant, 81
triangle inequality, 79
triple product identity, 209, 254
two squares theorem, 75

ultrafilter, 311

uncountable, 370
union axiom, 366
unit quaternions, 211
universal class, 371
upper triangular matrices, 185
urelements, 358, 361

valuation, 79
vector bundle, 292
vector field, 288
vector matrices, 263
vector part (of a quaternion), 198
vector product, 184
vector product algebra, 266
Vieta's rule, 77
Vieta's sequence, 144
von Neumann hierarchy, 373, 378
von Neumann sequence, 362

Wallace line, 78, 83
Wallis's formula, 146
Whitney classes, 292
winning strategy, 336

Zermelo axioms, 361
Zermelo–Fraenkel set theory, 365,
    368
zero, 13
zero of polynomial, 98
Zorn's lemma, 375
Zorn's vector matrices, 263

# Portraits of Famous Mathematicians

Leonhard EULER (1707 - 1783)    Carl Friedrich GAUSS (1777 - 1855)

William Rowan HAMILTON
(1805 - 1865)

Richard DEDEKIND (1831 - 1916)

Pen and ink drawings by Martina Koecher

Georg CANTOR (1845 - 1918)

Ferdinand Georg FROBENIUS
(1849 - 1917)

Heinz HOPF (1894 - 1971)

Abraham ROBINSON (1918 - 1972)

# Graduate Texts in Mathematics

1   TAKEUTI/ZARING. Introduction to Axiomatic Set Theory. 2nd ed.
2   OXTOBY. Measure and Category. 2nd ed.
3   SCHAEFER. Topological Vector Spaces.
4   HILTON/STAMMBACH. A Course in Homological Algebra.
5   MAC LANE. Categories for the Working Mathematician.
6   HUGHES/PIPER. Projective Planes.
7   SERRE. A Course in Arithmetic.
8   TAKEUTI/ZARING. Axiomatic Set Theory.
9   HUMPHREYS. Introduction to Lie Algebras and Representation Theory.
10  COHEN. A Course in Simple Homotopy Theory.
11  CONWAY. Functions of One Complex Variable I. 2nd ed.
12  BEALS. Advanced Mathematical Analysis.
13  ANDERSON/FULLER. Rings and Categories of Modules. 2nd ed.
14  GOLUBITSKY/GUILLEMIN. Stable Mappings and Their Singularities.
15  BERBERIAN. Lectures in Functional Analysis and Operator Theory.
16  WINTER. The Structure of Fields.
17  ROSENBLATT. Random Processes. 2nd ed.
18  HALMOS. Measure Theory.
19  HALMOS. A Hilbert Space Problem Book. 2nd ed.
20  HUSEMOLLER. Fibre Bundles. 3rd ed.
21  HUMPHREYS. Linear Algebraic Groups.
22  BARNES/MACK. An Algebraic Introduction to Mathematical Logic.
23  GREUB. Linear Algebra. 4th ed.
24  HOLMES. Geometric Functional Analysis and Its Applications.
25  HEWITT/STROMBERG. Real and Abstract Analysis.
26  MANES. Algebraic Theories.
27  KELLEY. General Topology.
28  ZARISKI/SAMUEL. Commutative Algebra. Vol.I.
29  ZARISKI/SAMUEL. Commutative Algebra. Vol.II.
30  JACOBSON. Lectures in Abstract Algebra I. Basic Concepts.
31  JACOBSON. Lectures in Abstract Algebra II. Linear Algebra.
32  JACOBSON. Lectures in Abstract Algebra III. Theory of Fields and Galois Theory.
33  HIRSCH. Differential Topology.
34  SPITZER. Principles of Random Walk. 2nd ed.
35  WERMER. Banach Algebras and Several Complex Variables. 2nd ed.
36  KELLEY/NAMIOKA et al. Linear Topological Spaces.
37  MONK. Mathematical Logic.
38  GRAUERT/FRITZSCHE. Several Complex Variables.
39  ARVESON. An Invitation to $C^*$-Algebras.
40  KEMENY/SNELL/KNAPP. Denumerable Markov Chains. 2nd ed.
41  APOSTOL. Modular Functions and Dirichlet Series in Number Theory. 2nd ed.
42  SERRE. Linear Representations of Finite Groups.
43  GILLMAN/JERISON. Rings of Continuous Functions.
44  KENDIG. Elementary Algebraic Geometry.
45  LOÈVE. Probability Theory I. 4th ed.
46  LOÈVE. Probability Theory II. 4th ed.
47  MOISE. Geometric Topology in Dimensions 2 and 3.
48  SACHS/WU. General Relativity for Mathematicians.
49  GRUENBERG/WEIR. Linear Geometry. 2nd ed.
50  EDWARDS. Fermat's Last Theorem.
51  KLINGENBERG. A Course in Differential Geometry.
52  HARTSHORNE. Algebraic Geometry.
53  MANIN. A Course in Mathematical Logic.
54  GRAVER/WATKINS. Combinatorics with Emphasis on the Theory of Graphs.
55  BROWN/PEARCY. Introduction to Operator Theory I: Elements of Functional Analysis.
56  MASSEY. Algebraic Topology: An Introduction.
57  CROWELL/FOX. Introduction to Knot Theory.
58  KOBLITZ. $p$-adic Numbers, $p$-adic Analysis, and Zeta-Functions. 2nd ed.
59  LANG. Cyclotomic Fields.
60  ARNOLD. Mathematical Methods in Classical Mechanics. 2nd ed.

61 WHITEHEAD. Elements of Homotopy Theory.

62 KARGAPOLOV/MERLZJAKOV. Fundamentals of the Theory of Groups.

63 BOLLOBAS. Graph Theory.

64 EDWARDS. Fourier Series. Vol. I. 2nd ed.

65 WELLS. Differential Analysis on Complex Manifolds. 2nd ed.

66 WATERHOUSE. Introduction to Affine Group Schemes.

67 SERRE. Local Fields.

68 WEIDMANN. Linear Operators in Hilbert Spaces.

69 LANG. Cyclotomic Fields II.

70 MASSEY. Singular Homology Theory.

71 FARKAS/KRA. Riemann Surfaces. 2nd ed.

72 STILLWELL. Classical Topology and Combinatorial Group Theory. 2nd ed.

73 HUNGERFORD. Algebra.

74 DAVENPORT. Multiplicative Number Theory. 2nd ed.

75 HOCHSCHILD. Basic Theory of Algebraic Groups and Lie Algebras.

76 IITAKA. Algebraic Geometry.

77 HECKE. Lectures on the Theory of Algebraic Numbers.

78 BURRIS/SANKAPPANAVAR. A Course in Universal Algebra.

79 WALTERS. An Introduction to Ergodic Theory.

80 ROBINSON. A Course in the Theory of Groups. 2nd ed.

81 FORSTER. Lectures on Riemann Surfaces.

82 BOTT/TU. Differential Forms in Algebraic Topology.

83 WASHINGTON. Introduction to Cyclotomic Fields.

84 IRELAND/ROSEN. A Classical Introduction to Modern Number Theory. 2nd ed.

85 EDWARDS. Fourier Series. Vol. II. 2nd ed.

86 VAN LINT. Introduction to Coding Theory. 2nd ed.

87 BROWN. Cohomology of Groups.

88 PIERCE. Associative Algebras.

89 LANG. Introduction to Algebraic and Abelian Functions. 2nd ed.

90 BRØNDSTED. An Introduction to Convex Polytopes.

91 BEARDON. On the Geometry of Discrete Groups.

92 DIESTEL. Sequences and Series in Banach Spaces.

93 DUBROVIN/FOMENKO/NOVIKOV. Modern Geometry—Methods and Applications. Part I. 2nd ed.

94 WARNER. Foundations of Differentiable Manifolds and Lie Groups.

95 SHIRYAEV. Probability. 2nd ed.

96 CONWAY. A Course in Functional Analysis. 2nd ed.

97 KOBLITZ. Introduction to Elliptic Curves and Modular Forms. 2nd ed.

98 BRÖCKER/TOM DIECK. Representations of Compact Lie Groups.

99 GROVE/BENSON. Finite Reflection Groups. 2nd ed.

100 BERG/CHRISTENSEN/RESSEL. Harmonic Analysis on Semigroups: Theory of Positive Definite and Related Functions.

101 EDWARDS. Galois Theory.

102 VARADARAJAN. Lie Groups, Lie Algebras and Their Representations.

103 LANG. Complex Analysis. 3rd ed.

104 DUBROVIN/FOMENKO/NOVIKOV. Modern Geometry—Methods and Applications. Part II.

105 LANG. $SL_2(\mathbf{R})$.

106 SILVERMAN. The Arithmetic of Elliptic Curves.

107 OLVER. Applications of Lie Groups to Differential Equations. 2nd ed.

108 RANGE. Holomorphic Functions and Integral Representations in Several Complex Variables.

109 LEHTO. Univalent Functions and Teichmüller Spaces.

110 LANG. Algebraic Number Theory.

111 HUSEMÖLLER. Elliptic Curves.

112 LANG. Elliptic Functions.

113 KARATZAS/SHREVE. Brownian Motion and Stochastic Calculus. 2nd ed.

114 KOBLITZ. A Course in Number Theory and Cryptography. 2nd ed.

115 BERGER/GOSTIAUX. Differential Geometry: Manifolds, Curves, and Surfaces.

116 KELLEY/SRINIVASAN. Measure and Integral. Vol. I.

117 SERRE. Algebraic Groups and Class Fields.

118 PEDERSEN. Analysis Now.

119 ROTMAN. An Introduction to Algebraic Topology.

120 ZIEMER. Weakly Differentiable Functions: Sobolev Spaces and Functions of Bounded Variation.

121 LANG. Cyclotomic Fields I and II. Combined 2nd ed.

122 REMMERT. Theory of Complex Functions. *Readings in Mathematics*

123 EBBINGHAUS/HERMES et al. Numbers. *Readings in Mathematics*

124 DUBROVIN/FOMENKO/NOVIKOV. Modern Geometry—Methods and Applications. Part III.

125 BERENSTEIN/GAY. Complex Variables: An Introduction.

126 BOREL. Linear Algebraic Groups.

127 MASSEY. A Basic Course in Algebraic Topology.

128 RAUCH. Partial Differential Equations.

129 FULTON/HARRIS. Representation Theory: A First Course. *Readings in Mathematics*

130 DODSON/POSTON. Tensor Geometry.

131 LAM. A First Course in Noncommutative Rings.

132 BEARDON. Iteration of Rational Functions.

133 HARRIS. Algebraic Geometry: A First Course.

134 ROMAN. Coding and Information Theory.

135 ROMAN. Advanced Linear Algebra.

136 ADKINS/WEINTRAUB. Algebra: An Approach via Module Theory.

137 AXLER/BOURDON/RAMEY. Harmonic Function Theory.

138 COHEN. A Course in Computational Algebraic Number Theory.

139 BREDON. Topology and Geometry.

140 AUBIN. Optima and Equilibria. An Introduction to Nonlinear Analysis.

141 BECKER/WEISPFENNING/KREDEL. Gröbner Bases. A Computational Approach to Commutative Algebra.

142 LANG. Real and Functional Analysis. 3rd ed.

143 DOOB. Measure Theory.

144 DENNIS/FARB. Noncommutative Algebra.

145 VICK. Homology Theory. An Introduction to Algebraic Topology. 2nd ed.

146 BRIDGES. Computability: A Mathematical Sketchbook.

147 ROSENBERG. Algebraic $K$-Theory and Its Applications.

148 ROTMAN. An Introduction to the Theory of Groups. 4th ed.

149 RATCLIFFE. Foundations of Hyperbolic Manifolds.

150 EISENBUD. Commutative Algebra with a View Toward Algebraic Geometry.

151 SILVERMAN. Advanced Topics in the Arithmetic of Elliptic Curves.

152 ZIEGLER. Lectures on Polytopes.

153 FULTON. Algebraic Topology: A First Course.

154 BROWN/PEARCY. An Introduction to Analysis.

155 KASSEL. Quantum Groups.

156 KECHRIS. Classical Descriptive Set Theory.

157 MALLIAVIN. Integration and Probability.

158 ROMAN. Field Theory.

159 CONWAY. Functions of One Complex Variable II.

160 LANG. Differential and Riemannian Manifolds.

161 BORWEIN/ERDÉLYI. Polynomials and Polynomial Inequalities.

162 ALPERIN/BELL. Groups and Representations.

163 DIXON/MORTIMER. Permutation Groups.